Agricultural Development and Food Security in Developing Nations

Wayne G. Ganpat
The University of the West Indies, Trinidad and Tobago

Ronald Dyer
Grenoble School of Management, France

Wendy-Ann P. Isaac
The University of the West Indies, Trinidad and Tobago

A volume in the Advances in Environmental Engineering and Green Technologies (AEEGT) Book Series

IGI Global
DISSEMINATOR OF KNOWLEDGE
www.igi-global.com

Published in the United States of America by
IGI Global
Information Science Reference (an imprint of IGI Global)
701 E. Chocolate Avenue
Hershey PA, USA 17033
Tel: 717-533-8845
Fax: 717-533-8661
E-mail: cust@igi-global.com
Web site: http://www.igi-global.com

Library of Congress Cataloging-in-Publication Data

Names: Ganpat, Wayne G., editor. | Dyer, Ronald, 1969- editor. | Isaac, Wendy-Ann P., 1971- editor.
Title: Agricultural development and food security in developing nations / Wayne G. Ganpat, Ronald Dyer, and Wendy-Ann P. Isaac, editors.
Description: Hershey : Information Science Reference, 2016. | Series: Advances in environmental engineering and green technologies | Includes bibliographical references and index.
Identifiers: LCCN 2016033133| ISBN 9781522509424 (hardcover) | ISBN 9781522509431 (ebook)
Subjects: LCSH: Agricultural development projects--Developing countries. | Food security--Developing countries. | Food supply--Developing countries. | Sustainable development--Developing countries.
Classification: LCC HD1417 .A4477 2016 | DDC 338.1/8091724--dc23 LC record available at https://lccn.loc.gov/2016033133

This book is published in the IGI Global book series Advances in Environmental Engineering and Green Technologies (AEEGT) (ISSN: 2326-9162; eISSN: 2326-9170)

British Cataloguing in Publication Data
A Cataloguing in Publication record for this book is available from the British Library.

For electronic access to this publication, please contact: eresources@igi-global.com.

Advances in Environmental Engineering and Green Technologies (AEEGT) Book Series

ISSN: 2326-9162
EISSN: 2326-9170

Mission

Growing awareness and an increased focus on environmental issues such as climate change, energy use, and loss of non-renewable resources have brought about a greater need for research that provides potential solutions to these problems. Research in environmental science and engineering continues to play a vital role in uncovering new opportunities for a "green" future.

The **Advances in Environmental Engineering and Green Technologies (AEEGT)** book series is a mouthpiece for research in all aspects of environmental science, earth science, and green initiatives. This series supports the ongoing research in this field through publishing books that discuss topics within environmental engineering or that deal with the interdisciplinary field of green technologies.

Coverage

- Water Supply and Treatment
- Contaminated Site Remediation
- Air Quality
- Industrial Waste Management and Minimization
- Electric Vehicles
- Green Transportation
- Alternative Power Sources
- Green Technology
- Waste Management
- Renewable Energy

IGI Global is currently accepting manuscripts for publication within this series. To submit a proposal for a volume in this series, please contact our Acquisition Editors at Acquisitions@igi-global.com or visit: http://www.igi-global.com/publish/.

The Advances in Environmental Engineering and Green Technologies (AEEGT) Book Series (ISSN 2326-9162) is published by IGI Global, 701 E. Chocolate Avenue, Hershey, PA 17033-1240, USA, www.igi-global.com. This series is composed of titles available for purchase individually; each title is edited to be contextually exclusive from any other title within the series. For pricing and ordering information please visit http://www.igi-global.com/book-series/advances-environmental-engineering-green-technologies/73679. Postmaster: Send all address changes to above address.

Titles in this Series

For a list of additional titles in this series, please visit: www.igi-global.com

Oceanographic and Marine Cross-Domain Data Management for Sustainable Development
Paolo Diviacco (Istituto Nazionale di Oceanografia e di Geofisica Sperimentale (OGS), Italy) Adam Leadbetter (Marine Institute, Ireland) and Helen Glaves (British Geological Survey, UK)
Information Science Reference • copyright 2017 • 425pp • H/C (ISBN: 9781522507000) • US $205.00 (our price)

Examining the Development, Regulation, and Consumption of Functional Foods
Sailas Benjamin (University of Calicut, India)
Medical Information Science Reference • copyright 2017 • 344pp • H/C (ISBN: 9781522506072) • US $185.00 (our price)

Applying Nanotechnology for Environmental Sustainability
Sung Hee Joo (University of Miami, USA)
Information Science Reference • copyright 2017 • 558pp • H/C (ISBN: 9781522505853) • US $225.00 (our price)

Exploring the Nutrition and Health Benefits of Functional Foods
Hossain Uddin Shekhar (University of Dhaka, Bangladesh) Zakir Hossain Howlader (University of Dhaka, Bangladesh) and Yearul Kabir (University of Dhaka, Bangladesh)
Medical Information Science Reference • copyright 2017 • 523pp • H/C (ISBN: 9781522505914) • US $200.00 (our price)

Smart Cities as a Solution for Reducing Urban Waste and Pollution
Goh Bee Hua (National University of Singapore, Singapore)
Information Science Reference • copyright 2016 • 362pp • H/C (ISBN: 9781522503026) • US $190.00 (our price)

Food Science, Production, and Engineering in Contemporary Economies
Andrei Jean-Vasile (Petroleum-Gas University of Ploiesti, Romania)
Information Science Reference • copyright 2016 • 473pp • H/C (ISBN: 9781522503415) • US $240.00 (our price)

Smart Grid as a Solution for Renewable and Efficient Energy
Ayaz Ahmad (COMSATS Institute of Information Technology, Pakistan) and Naveed Ul Hassan (LUMS School of Science & Engineering, Pakistan)
Information Science Reference • copyright 2016 • 415pp • H/C (ISBN: 9781522500728) • US $220.00 (our price)

Biologically-Inspired Energy Harvesting through Wireless Sensor Technologies
Vasaki Ponnusamy (Universiti Tunku Abdul Rahman, Malaysia) Noor Zaman (King Faisal University, Saudi Arabia) Tang Jung Low (Universiti Teknologi Petronas, Malaysia) and Anang Hudaya Muhamad Amin (Multimedia University, Malaysia)
Information Science Reference • copyright 2016 • 318pp • H/C (ISBN: 9781466697928) • US $200.00 (our price)

IGI GLOBAL
DISSEMINATOR OF KNOWLEDGE
www.igi-global.com

701 E. Chocolate Ave., Hershey, PA 17033
Order online at www.igi-global.com or call 717-533-8845 x100
To place a standing order for titles released in this series, contact: cust@igi-global.com
Mon-Fri 8:00 am - 5:00 pm (est) or fax 24 hours a day 717-533-8661

Editorial Advisory Board

Table of Contents

Detailed Table of Contents

This chapter looks specifically at the impact of the ACP agriculture trade agreements on issues of food production, food security and poverty and how they relate in some ways to each other. It is divided into four distinct but inter-related sections. First, poverty in the ACP regions is briefly described. Secondly, the importance of agriculture, particularly the role of small farming systems to economic growth, employment, food production, food security and poverty reduction in the ACP countries is examined. Thirdly, the impact of the acceleration of trade liberalization and agriculture protectionism on ACP countries is reviewed and analyzed. Here, the trade arrangements and networks in the three ACP regions are described and include international, regional and local agricultural trade and market agreements. Further, the paradoxes and dilemmas that emerge are interrogated and their outcomes discussed. Particular attention is paid to small producers and women. The fourth and final section of the chapter concludes that while the impact of agriculture trade liberalization remains a mixed bag of positive and negative possibilities, one thing that is clear is that while trade liberalization is promoted as a cure for food security and world poverty, the jury is still out on whether it has helped or hindered this endeavour. Trade policies and arrangements do substantially change the lives of people in developing countries, but in very complex ways. Recommendations are made to support the South-South initiatives of the ACP and others as well as enhance regional and domestic agriculture production and trade policies. The author, however, warns that agriculture and trade policies must be linked closer to food security and the healthy diets of local populations.

Addressing environmental and social impacts on agricultural development and food security is a global priority since increased food production of 60-70% is estimated to be required by 2050 to feed the growing world population. In developing nations, the situation is more acute since fewer social, technological and financial resources are available to combat climate change, which is expected to have negative effects on

Development has many faces and complete wellbeing of human population is the most important one of them which in more than one ways involves agriculture and the farming population. Providing needed information at the right time to the rural population is the first step in their empowerment and ICTs can play an immensely important role in providing that information by increasing the dialogue between development professionals and rural people at every stage of development process. According to recent statistics released by ITU, over the last 15 years, ICTs have grown in unprecedented ways providing huge opportunities for social and economic development and this growth can be an advantage to rural advisory services. Providing correct and personalized information needs expert opinions and so multi-stakeholder engagement makes the process more efficient and ICTs provide a very unique and important platform for such collaboration, thus bringing together different stakeholders for efficient partnership. The various tools and technologies can also be tailored according to the needs of end users. But inspite of the advantages, ICTs can only be universally accepted and used when the challenges of accessibility, acceptability, funding, and sustainability are overcome. There are no formula for sure success with ICTs and situation is the best determinant of the strategy to be used and so, a balanced and strategic use of ICTs depending on the clients' needs can best utilize its potential for agricultural development and food security in developing nations.

The world's population is increasing rapidly and the food need for the increasing population has been growing also. In the 21st century, efforts for the sustainable use of the natural resources have been raised all around the world. Agriculture has always been on the center of discussions and developments as an activity carried out depending on the nature. Therefore, the sustainable agricultural systems have gained a considerable importance in recent years. This chapter will examine the temporal and spatial development of the organic agriculture in Turkey. The current situation of the agricultural sector in Turkey will be reviewed and the need for a sustainable agricultural system in Turkey be analyzed. The temporal and spatial change of the organic agricultural areas, numbers of producers and products, and organic production amounts in Turkey will be put forward. Also, the organic agricultural potential of Turkey will be evaluated and the reasons for not exploiting this potential will be investigated.

Despite unprecedented technological breakthroughs and rising incomes, the most basic human requirement–food security remains out of reach of many. The countries of Africa are the worst affected, and consistently score lowest on the Global Hunger Index. The downside of technological advance and increasing urbanisation is the decrease in cultivable land and water availability exacerbated by climatic unpredictability. While concerted efforts are on to control the crisis, it is time to take positive action and

Food security for the urban poor has been an important topic for both developed and developing countries over the last 15 years. Although South Bend Indiana is a city in a developed country, declining economic circumstances have caused the city to show significant urban decay somewhat similar to some cities in developing countries. In this chapter, we explore South Bend's history and economic development strategies, and review practices aimed at strengthening food security for the urban poor. The chapter documents how numerous disparate organizations have been trying to help alleviate urban poverty and hunger, and reviews previous strategies used to foster sustainable growth and development. The integration of spatial technologies will become a key factor for promoting community social networks, participatory planning, and collaboration. The case is presented for the use of Geographic Information Systems (GIS) and associated technologies to help organizations, community leaders, local organizations, city planners, higher education institutions and the urban poor, work together to alleviate poverty and malnutrition through networking and sustainable urban agriculture.

Agricultural production in the Caribbean is being threatened by many factors such as decreasing availability of arable land, climate change effects such as increased incidences of flooding and drought, labour shortages, and competition from importers. However, one of the most important threats to agricultural production is the often under-recognised and under-reported area of farm theft (referred to as praedial larceny in the Caribbean). It involves the theft of agricultural produce (crops, livestock and fisheries) and farm equipment. One of the main reasons why this threat is so important is that theft of this type is very hard to prove. If indeed perpetrators are caught, and prosecuted successfully, the penalty is practically negligible. This paper examines the current status of farm theft in the Caribbean region, explores some of the main factors influencing farm theft, reviews some of the strategies attempted in the Caribbean and other places around the world and makes several suggestions to create a more secure food region. While the discussion calls on food producers to take several best practice actions to mitigate losses to praedial larceny, it emphasises that the primary responsibility is with government-led actions in the areas of modernised policies, updated laws and enhanced enforcement efforts.

Preface

Agricultural development has been on the agenda of many countries for quite some time now. While some countries have made progress, other still struggle. The consequences of not making sufficient progress are reflected in the lives of the millions of farmers in developing countries, the lives of farm workers and their families and the nutritional status of all persons particularly the more vulnerable young and the elderly.

Food security has been more recently added to the agenda of countries. Food security and agricultural development are in inextricably linked; there can be no claim to being food secure if your agricultural systems are not experiencing some level of development. In more recent times, the issue of food and nutrition security has risen to the top of developing countries' agendas. It is not sufficient for a nation to be food secure; the nutritional status of its people must also be assured.

Developing nations can be described as those countries that are really seeking to improve the lives of its peoples in all regards. While economic development may be foremost, other development areas are increasingly being added. Many developing nations around the world can be regarded as emerging economies, and they represent at some estimates, approximately 80% of the global population. Developing nations are under some pressure to deliver a lot for its people. However, such development is not without risks in the political and social environments.

The title of this book reflects these three linked concepts. As such, the various chapters were well-chosen and sequenced so as to tell a story of experiences in selected countries worldwide.

Chapter 1 looks at the impacts of trade agreements in the African Caribbean and Pacific (ACP) nations on food production, food security and poverty. Many ACP countries can be described as developing countries and the author firstly reviews the situational content before moving on to present an analysis of the impact of trade liberalization on these economies. As such, poverty is discussed, followed by the importance of agriculture to these countries, then the impact of accelerated trade liberalization and agriculture protectionism. After these experiences are examined, the author concludes that while some benefits of trade liberalization have been achieved, the jury is still out on whether it has helped or hindered agricultural development and by extension the lives of people.

Chapter 2 focuses on the environmental and social impacts of agricultural development. This chapter builds on the first chapter which analyzed selected policy interventions. Sustaining the environment, as well enhancing the social fabric of countries, are being increasing pushed to the top of development agendas. Agriculture based economies no longer want development at any cost; their environment and people's lives are just as important. Authors of this chapter present the world's food need situation in 2050 and argue that while developed nations are making progress, developing nations face many challenges in trying to become food secure. The chapter details the environmental and social issues that affect most stakeholders in the food production sector; farmers, researchers, policy makers, and support institutions among others. At the end, authors address how progress in research in developing countries can be maximized and a cohesive strategy to address challenges is presented. Readers in developing nations struggling with similar issues can find possible entry points to address their own situations.

Chapter 3 puts small holder farmers under the microscope. Over 80% of the farming in developing nations is done by small holders operating on farms of less than 5 acres and in many of the smaller island nations production occurs on farms less than 2 acres. The challenges such small holders face are unique to each country. As such, a global perspective on this issue is not described, but rather authors focus on the small developing Caribbean nation of St. Kitts which has been actively seeking new methods and applying alternative technologies over the last few years in an attempt to hasten agricultural development and attain some measure of food security. Authors discuss various experiences and then offer a range of possible adoptive strategies, policy recommendations and future research areas. Readers have the opportunity to examine all recommendations and make their own judgment on what may be useful to explore in their own country situation.

The book's deliberations then shift slightly from the wider content to examine the key areas of agricultural extension services to support countries in their food security initiatives, and the whole area of Information and Communication Technologies (ICTs) in knowledge exchange.

Chapter 4 focuses on agricultural extension services in African countries; mainly agriculture dependent and more so rain-fed agriculture. Readers are presented with an assessment of extension methodologies used over the years, their successes and short comings. The purpose was to present the case for developing nations not to simply accept extension methodologies developed elsewhere, but to fashion their own approaches to the education of small holders given the particularly acute situations under which such farmers are expected to contribute to the nation's food security goals. Authors, in the final sections of the chapter, call for developing countries to create a sustainable extension system that is socially acceptable, culturally compatible, economically viable and environmentally friendly.

Chapter 5 builds on the previous chapter and focuses on the use of ICTs to provide information in a timely manner to rural populations. ICTs have grown tremendously over the last few years and their use as development tools in agriculture is now widely recognized. Authors show that providing correct and personalized information needs expert opinions and show how engagement with multi stakeholders makes the process more efficient with ICTs by providing a very unique platform for such collaborations. Tools and technologies they suggest can be tailored to meet the needs of end users and several are discussed in the chapter. Readers can then make their own assessment of what is appropriate and useful in other developing countries given their own technological and infrastructural development stage.

The book discussion swings again to address some specific production systems that are top of the agenda for developing economies. Organic agriculture as a potential food security action area, genetically modified crops as an alternative or complementary production system and the use of plant bio technology to speed up efforts using the latest scientific advances.

Chapter 6 deals with the topic of organic food production. The sustainable use of natural resources has come to the fore and sustainable agricultural systems are being pursued in developed countries. Developing countries are now increasing the use of organic production systems. In this chapter, the changes in organic production from both temporal and spatial perspectives are analyzed. Readers can evaluate the experiences of Turkey which has reported both successes and challenges. The challenges present opportunities for growth and authors discuss this issue towards the end of the chapter.

Chapter 7 reviews the status of Genetically Modified (GM) crops in Africa. It discusses the scope for expansion, possible impacts on food security and the validity of concerns about bio-safety. This is an all important topic for food security. GM crops have been promoted for some time now as the answer to hunger and malnutrition; mainly in poorer developing nations. While some have embraced this technology, others are skeptical. Particularly in Africa, where hunger is persistent, the issue of GM crops needs to be fully interrogated. Chapter authors do just that and presents information for readers to make their own decisions. No one can force a county to adopt GM crops. This has to be an informed decision and this chapter offers several discussion points for persons to ponder.

Chapter 8 focuses on the area of plant bio-technology applications. Bio-technology has the potential to deliver solutions to some of the short comings of the green revolution such as conventional breeding which often times result in poor quality and insufficient quantity of planting materials. Readers are exposed to some successful applications of plant bio-technology for sustainable food production. Such applications hold promise to bring about food security in a shorter time and with reduced negative environmental consequences. This chapter presents some to the most up to date techniques being reported and readers have the opportunity to evaluate the applicability of results to their own circumstances given that bio-technology may require support from institutions, particularly financial support for its development if it is to be successful.

The book continues to explore best practices for food security with a special focus on the Caribbean.

In Chapter 9, authors contend that best practices for food security must move beyond the traditional, known and well researched systems of cultivation. They broach the topic of using edible wild plants to complement other strategic crops that are currently being considered to overcome food security challenges. Can insects be used as another source of food? The authors, like others before them, once again raise the issue of local chickens, indigenous livestock and wildlife as sources of food, particularly to meet protein needs. Such systems, they argue have been pushed aside by highly commercialized systems of production and time has shown that developing countries cannot fully adopt many of the practices of developed countries wholesale. As such, they must explore and seek to develop local and indigenous sources of food and associated systems of production.

Chapter 10 presents the specific experiences of the development of the horticulture industry in the Andaman and Nicobar Islands, India. It is useful because its tropical nature is similar to many of the other developing countries in the tropical world. Details are provided for plantation crops, roots and tubers, vegetables, spices and medicinal plants. The development of these crops; successes and challenges, offer the case for active consideration by development planners as they seek to shift consumption patterns towards the use of more locally grown food and the increased utilization of spices and medicinal plants by its people. Farm incomes and the livelihoods of millions of people in emerging economies can be improved.

Following the Asian experiences, Chapter 11 presents an African experience. This chapter explores the factors, based on studies conducted, which impact productivity in the agricultural sector in Uganda. This is a developing country surrounded by many other similarly positioned African nations. As such, the determinants of agricultural productivity can be evaluated by such countries and appropriate actions can be taken. Authors not only describe the poor state of the agricultural sector, but also link it to development efforts which have not been pro-poor and inclusive of the main stakeholders. Several recommendations are made towards the end of the chapter to increase productivity and to make growth process more broad-based, pro-poor, and inclusive.

Chapter 12 is unique. The USA is a developed economy, no doubt about that. Yet, in many states, researchers find areas, cities and towns both urban and rural, with characteristics that bear striking resemblance to those of developing countries. One such place is South Bend, Indiana. It is included in the book because it details the unique experience of using spatial technologies to help develop stronger community networks, participatory planning and greater collaboration among stakeholders. These three issues are major concerns in many developing nations and are often significant limiting factors to the achievement of food security. Readers are presented with the use of Geographic Information Systems (GIS) and associated technologies to help organizations, communities and leaders, local organizations, planners and the urban poor, to work together to alleviate poverty and malnutrition. GIS and associated technologies are becoming more accessible and available for use by developing nations and its applicability to contribute to food security goals cannot be ignored.

The final chapter, Chapter 13, deals with a topic that is not much discussed in the literature, but is one that plagues farmers worldwide and many countries consider as a major threat to the achievement of food security. Farm theft or praedial larceny as it is called in the Caribbean, refers to the theft of agricultural produce and farm equipment. Because of inadequate accounting systems, it is hard to prove and perpetrators have almost a free license in the Caribbean. It is so important simply because the theft of even a small item of farm equipment can mean the end of a small farmer's means of livelihood for a considerable time. Authors not only examine the current status of farm theft in the Caribbean but also explore some of the main drivers of such theft, review some of the current solutions applied and then go on to make several recommendations based on experiences elsewhere in the world to create a more secure food region.

After reading this book, one will appreciate the challenges faced by emerging economies as they seek to become food and nutrition secure. The book is rich in successful experiences that can be considered, innovations that could be embraced, technologies that if applied have the potential to hasten food security goals in developing nations.

Acknowledgement

The editors wish to thank to all the contributors to this book as well as the Editorial Advisory Board for their assistance in providing constructive comments and reviews and for their support in the improvement of the work. We wish to gratefully acknowledge the support of Ms. Sarojini Ragbir for all her insightful comments, ideas and criticism in drafting the book. We wish to thank Dr. Isaac Bekele, Dean of the Faculty of Food and Agriculture for his overall support and also to the entire IGI Global staff involved in making possible this book, particularly Ms. Eleana Wehr, book development editor, IGI Global. Most importantly, we thank God for giving us the strength and fortitude and the support of our family members for their cooperation during the completion of this project.

Wayne G. Ganpat
The University of the West Indies, Trinidad and Tobago

Wendy-Ann P. Isaac
The University of the West Indies, Trinidad and Tobago

Chapter 1
Agriculture, Trade Liberalization and Poverty in the ACP Countries

Theresa Ann Rajack-Talley
University of Louisville, USA

ABSTRACT

This chapter looks specifically at the impact of the ACP agriculture trade agreements on issues of food production, food security and poverty and how they relate in some ways to each other. It is divided into four distinct but inter-related sections. First, poverty in the ACP regions is briefly described. Secondly, the importance of agriculture, particularly the role of small farming systems to economic growth, employment, food production, food security and poverty reduction in the ACP countries is examined. Thirdly, the impact of the acceleration of trade liberalization and agriculture protectionism on ACP countries is reviewed and analyzed. Here, the trade arrangements and networks in the three ACP regions are described and include international, regional and local agricultural trade and market agreements. Further, the paradoxes and dilemmas that emerge are interrogated and their outcomes discussed. Particular attention is paid to small producers and women. The fourth and final section of the chapter concludes that while the impact of agriculture trade liberalization remains a mixed bag of positive and negative possibilities, one thing that is clear is that while trade liberalization is promoted as a cure for food security and world poverty, the jury is still out on whether it has helped or hindered this endeavour. Trade policies and arrangements do substantially change the lives of people in developing countries, but in very complex ways. Recommendations are made to support the South-South initiatives of the ACP and others as well as enhance regional and domestic agriculture production and trade policies. The author, however, warns that agriculture and trade policies must be linked closer to food security and the healthy diets of local populations.

DOI: 10.4018/978-1-5225-0942-4.ch001

Overcoming poverty is not a gesture of charity. It is an act of justice. – Nelson Mandela

INTRODUCTION

International trade and financial transactions are not new phenomena. From time immemorial, people across the globe have exchanged goods and services. Today, societies have become more specialized and no single country produces all that is needed to sustain itself. Thus, the exchange of goods and services is even more necessary than before. However, the formation of cooperative entities and marketing blocs have made trading more political and convoluted. No longer is the simple use of supply and demand economic principles applicable. Currently, the international movements of goods and services have more to do with decisions made by multinationals and groups of developed countries. According to Koroma and Ford (2006), the rapidly changing structure and patterns of trade are heavily influenced by private sector investment and standards. As a result, freedom to trade at all levels, from the farm gate to the stock market, is now much more complicated and is further removed from the power of producers and/or consumers, who are also finding it increasingly difficult to avoid slipping into the clutches of poverty.

The primary purpose of this chapter is to examine the links between agriculture production and trade, food security, small farmer production, and poverty reduction in developing countries. It specifically analyses the impact of agriculture trade policies and outcomes in select African, Caribbean and Pacific countries known as the ACP group. There are 79 African, Caribbean and Pacific states that form the ACP: 48 are in Sub-Saharan Africa, 16 in the Caribbean region, and 15 in the Pacific. The countries are demographically, geo-physically and economically diverse; 40 are considered less developed countries, 37 small island states (or both), and 15 landlocked developing countries (see table 1). The total ACP region population is approximately 727 million and represents 15% of the total population of all developing countries. Ninety-four percent live in Africa, close to 5% in the Caribbean and just over 1% in the Pacific islands. According to ACP Secretary General, Dr. Patrick Ignatius Gomes, in his address to the UN Summit on Sustainable Development at the 70th United Nations General Assembly (2015), "the domains of the ACP group include: rule of law and good governance; global justice and human security; intra-ACP trade, industrialization and regional integration; building sustainable, resilient and creative economies; as well as financing for development." Further South-South and Triangular Cooperation informs the group's approach to all these domains.

In reality marketing and trade arrangements can sometime result in uneven benefits and the extent to which they assist in poverty reduction remains contested. The chapter infers that agricultural trade policies are often bullied by international trade magnates and are not necessarily the answer to economic growth, food security and poverty reduction in most developing countries. Moreover, trade policies, although still important, do have negative impacts on developing countries including marketing preference erosion, decline of tariffs, and the demand for high level of competitiveness which requires a wide range of mitigating factors. Thus, many developing countries are often confronted with substantial disadvantages in implementing current trade policies that necessitate having substantial resources that they simply do not have (UNCTAD, 2013). Since most of these countries including those member states who belong to the ACP Group are agrarian-based, current agriculture trade policies have direct and indirect impacts on poverty, economic growth and the well-being of the people who live in these countries.

Table 1. The ACP states by region and economic classifications

Least Developed Countries (LDC)			Non-LDCs		
Africa (34)	**Caribbean (1)**	**Pacific (5)**	**Africa (15)**	**Caribbean (15)**	**Pacific (9)**
Angola	Haiti	Kiribati	Botswana	Antigua &Barbuda	Fiji
Benin		Samoa	Cameroon	Bahamas	Papua New
Burkina Faso		Solomon	Congo	Barbados	Guinea
Burundi		Islands	Cote d'Ivoire	Belize	Tonga
Cape Verde		Tuvalu	Gabon	Cuba	Marshall Islands
Central African Republic		Vanuatu	Ghana	Dominican Republic	Cook Islands
Chad			Kenya	Grenada	FS of
Comoros			Mauritius	Guyana	Micronesia
Dem. Republic of the Congo			Namibia	Jamaica	Nauru
Djibouti			Nigeria	Saint Kitts & Nevis	Niue
Equatorial Guinea			Seychelles	Saint Lucia	Palau
Eritrea			South Africa	Saint Vincent &	
Ethiopia			Swaziland	the Grenadines	
Gambia			Zimbabwe	Suriname	
Guinea				Trinidad & Tobago	
Guinea-Bissau					
Lesotho					
Liberia					
Madagascar					
Malawi					
Mali					
Mauritania					
Mozambique					
Niger					
Rwanda					
Sao Tome & Principe					
Senegal					
Sierra Leone					
Sudan					
Toga					
Uganda					
United Rep. of Tanzania					
Zambia					

Table taken from Commodities and Trade Technical paper #8, pg. viii, (2006)
LDC= Less Developed Countries

OVERVIEW OF POVERTY AND AGRICULTURE IN THE ACP COUNTRIES

World Bank estimates indicate that on a global level, the first Millennium Development Goal of cutting the 1990 poverty rate by half in 2015 has already been attained. However, despite the progress made there are still large numbers of people whose standard of living fall at or below the poverty line. Close to 17% or 1 billion people in developing countries live on less than US $1.25 a day, that is, about one in seven of the world's population suffer from hunger (World Bank 2015). Similarly, the United Nations Food and Agriculture Organization (FAO) estimates for 2012-14, indicated that approximately 805 million people were chronically undernourished, i.e., one in every nine people in the world had insufficient food for an active, healthy and productive life.

Progress made towards meeting the Millennium Development Goals internationally has also been uneven. The gap between the wealthiest and poorest countries has widened and the level of hunger and poverty varies among developing regions. For example, although developing countries overall experienced a reduction in poverty, a disproportionate amount (just over 80%) of the world's extremely poor live in South Asia (399 million) and in Sub-Saharan Africa (415 million). Fewer than 50 million of extremely poor people live in Latin America and the Caribbean (LAC), the Middle East, North Africa, Eastern Europe and Central Asia, combined. Furthermore, the vast majority (791 million) of people who live in dire poverty reside in developing countries (WorldBank, 2015). Thirty-nine of the 49 developing countries are members of the ACP groupings where approximately 214 million undernourished individuals can be found in Sub-Saharan Africa[1] and 37 million in Latin America and the Caribbean.

There is also the problem where increasingly larger numbers of families in some of the ACP countries are eating too much of the wrong types of food, resulting in overweight and obesity. For example, while hunger is a major problem in sub-Saharan Africa, the excess intake of calories seriously undermines the health of many in the Caribbean and Pacific regions. Chronic non-communicable diseases, several related to poor nutrition, now account for 57% of deaths in the Caribbean. Further, although undernutrition and iron-deficiency anaemia persists in the region, the increasing incidence of obesity is now the major source of nutrition-related problems. In addition, diseases related to obesity such as diabetes, hypertension and heart conditions are chronic and very expensive to manage placing financial burdens on poor families (PAHO 2014).

A similar story can be told for the Pacific region, where half of the adult population is overweight. Pacific Islanders tend to eat a lot of imported canned meats like spam and corned beef, which has a negative effect on their health. In 2007, eight of the 10 heaviest countries were located in the South Pacific. Nauru, the world's smallest republic with just over 9,000 inhabitants, earned the number one spot with over 90% of their adult population considered obese. At the same time, 38% of Papua New Guineans live below the National Basic Needs Poverty Line, which means 2.7 million people are unable to buy sufficient food and meet basic requirements for housing, clothing, transportation and school fees. Even more alarmingly, 61% of the populace does not have access to safe drinking water (Borgen Project Report 2015).

Ironically, the lack of adequate food, proper nutrition and the resulting high levels of diseases and deaths related to poor nutrition in many ACP countries occur in places where agriculture and food production have historically been core to people's livelihood and dietary needs. Currently, ACP countries need to produce increasing quantities of food to satisfy their growing populations. However, food security is not just about growing more food, but also about high quality items that have the nutritional value for healthy diets. Equally important therefore, is that healthy food items should be accessible and affordable to all sections of the population, including those responsible for producing them. It is therefore imperative that domestic agricultural policies and practices be oriented around these understandings and trade negotiations be sensitive to the needs, capabilities, health and well-being of people.

Role of Agriculture in Livelihoods, Employment, and Food Security

Taken as a whole, ACP countries average significantly lower rates of per capita Gross Domestic Product (GDP) growth than most developing countries. However, with the exception of the Caribbean subgroup, agriculture's share in the GDP of ACP countries is normally higher than in developing countries collectively, and in some ACP countries agricultural products form the larger portion of their total trade. In the case of sub-Saharan Africa, the processing of agricultural products also accounts for two-thirds of manufacturing value-added (CTA 2011). Moreover, agriculture still accounts for more than 60%of total employment in ACP countries compared to around 55% in developing countries jointly. Most of the

ACP countries have also sustained increases in per capita food production, but food production has not kept up with population growth nor the dietary needs of any of the three ACP sub-regions (FAO 2006). Specific to Africa, agricultural production has increased steadily, tripling in value and is almost identical to that of South America, but lower than that of Asia. However, the growth of the sector is uneven and the World Bank estimates that there is a wide spread in agriculture's contribution to GDPs across the various African countries. For example, in Libya, Guinea, Gabon, and others, agriculture accounts for over 45% of the GDPs while in Burundi, Republic of Congo, Liberia, and others, agriculture accounts for under 15% of these countries' GDPs. The average for the continent as a whole is approximately 30-40% (World Bank 2012). More importantly, this sector has always absorbed a large percentage of the adult work force. Agriculture creates most of the jobs in Africa, aside from the oil-producing African countries. Half or more of the working population, an estimated 530 million people or 48% of the total African population work in the agriculture sector. A large proportion of all new entrants to Africa's working population (approximately 330 million) have very little or no option but to turn to agriculture (NEPAD 2013).

The main farming system in Africa is quite similar to that of the Caribbean and the Pacific Islands. Africa's agricultural sector is dominated by small family farms that rely mainly on family labor and women's labor. As a whole, Africa has 33 million farms of less than 2 hectares, accounting for 80% of all farms. However, in the last decade large-scale investment contracts have crept, threatening not only the persistence and support to small farmers but also to the continent's natural resources and biodiversity. These large-scale farming systems are estimated to cover close to 20 million hectares, which represents more than the arable area of South Africa and Zimbabwe combined. While the small family farming systems are unable to expand and enhance their production due to the lack of their own investment capacity and secure land tenure, many African countries have bought into the concept of industrial farming as an avenue to meeting the demands of world trade agreements and the food security of their populations. As such, in many African countries it is easy for foreign groups and multinationals to acquire land and set up operations without adequate policies to protect local production by small producers and the countries' natural resources (NEPAD 2013).

In the last 30 years, Africa's population has doubled overall and tripled in urban areas. The most direct consequence of this exponential population growth is that the continent now has more mouths to feed. Yet, cereal production, one of the main staple for Africa, has been unable to keep pace with population growth, since it has only increased by a factor of 1.8. This gap is even wider for processed products and meat, which are increasingly called for by a larger and larger urban population. In fact, Africa has moved from being self-sufficient in cereal production and other traditional crops in the 1960s to becoming a net importer of basic food items that compete with its own production of meat, dairy products, cereals and oils. In 2011, African countries spent close to US $35 billion in food imports (excluding fish), which amounts to 1.7 times the value of its agriculture exports (Africa Progress Report 2014).

A very similar synopsis is found in the Caribbean Region where agriculture is important for the region's GDP, employment, food production and food security, with these varying from country to country. For example, the 2012 data from the World Bank shows that in the Caribbean, the GDP earned from agriculture ranged as high as 21-13%in some countries (e.g. Guyana, Dominica, Belize) to as low as 1-2%in others (e.g. Barbados, Antigua and Barbuda). As in Africa, the sector is an important source of employment where some 3.3 million individuals are employed as agricultural workers. In the past decade this approximated to 32% of the labor force in 15 countries across the region including Haiti (Bourne 2008).[2] In the last decade farmers made up about 8% of the labor force in the least agrarian country such as Trinidad and Tobago, but comprised as much as 20-30% in countries, e.g. Belize, Guyana, Grenada, Dominica and Jamaica, which are more heavily dependent on agricultural production (Potter et al. 2004). Many others are often employed in jobs attached to the forward and backward linkages in the food sector, especially in agro processing and marketing businesses.

Although the agricultural sector in the Caribbean historically focused on wealth generated from the exportation of raw materials (sugar, cocoa, coffee, etc.) to metropolitan states (and some still do), it also benefitted from a fairly decent domestic food sector. According to Rajack-Talley (2015), the production of food items continue to be carried out primarily by small family farms scattered throughout the landscape of the Region. Although there are variations in size of operations, small producers in the Caribbean are mainly family operated enterprises co-existing alongside larger private and state-owned farms. They have limited access to land and their farms are either owned, rented/leased, shared with other family members, or illegally occupied. The size of farms can vary, but small operations range from ½ - 5 acres (.15-2.0 hectares). They practise mixed farming ; a combination of crop production, livestock husbandry, forestry, artisanal fishery and small scale aquaculture, producing food both for the home and the local markets as well as for the export market. Farmers devote a major part of their time to cultivation using mainly their own labor and that of family members, with little paid help from non-family laborers.

Over the years, agricultural development has suffered so much so, that the Region's food importation bill soars higher every year and was estimated at US $3-6 billion in 2012-2013.[3] With the exception of Belize and Guyana, most of the Caribbean countries are net importers of food, averaging 60% in some cases to as high as 80% in others. Such high levels of food importation compete with cost of local food production and food prices. Furthermore, not only have Caribbean people become more dependent on imported foods but the income needed to purchase these items has dwindled. Although the Region has over the years increased minimum wages, food price inflation has outstripped the minimum wage increases in most countries. According to the Director of the Caribbean Food and Nutrition Institute (CFNI), "soaring food prices present a crisis which has devastating consequences on nutrition, particularly in children" (CFNI Annual Report 2008: v).

Statistics on the agricultural earnings or GDP on the third region of the ACP group, the Pacific Island, are not readily available. What is known is that the agriculture and fisheries (forestry in some of the larger countries) sectors remain the economic mainstay of most Pacific ACP countries (PACPs). Commercialized agriculture does provide a substantial income to these countries based on medium, small and micro levels of production and the US dollar value of commercial market products (see table 2).

The agricultural sectors of the Pacific countries are made up primarily of subsistence or semi-subsistence farmers, producing both food crops and cash crop products in the larger countries (palm oil, cocoa, and coffee). Farming is still a family owned and operated system and many families rely on agriculture as

Table 2. Ranking of commercial agricultural production in PACP countries

Value of Top 5 Commercial Market Products	Country
Medium: Above US$100 million	Papua New Guinea; Fiji
Small: US$10 million-US$ 60 million	Solomon Islands, Vanuatu, Timor-Leste, Kiribati, Tonga.
Micro: Under US$7 million	Federates States of Micronesia, Marshall Islands, Cooks Island, Niue, Nauru, Palau, Tuvalu

Source: CTA (2011), Agrictrade, pg.2

their main source of income. Fisheries, especially tuna, remain an important resource in several of these countries but the processing of fish as an industry remains limiting because it is too capital intensive for most of the people in the Pacific Islands. Timber is another natural resource available to some countries but it is unsustainable and under constant threat, as many of the traditional owners of this resource are unable to protect their interests in commercial negotiations and cannot depend on strong support from their governments (CTA 2011; Malua 2003).

Farming remains a very traditional practice tightly linked to family values and culture. Family farms are considered multi-purpose enterprises whereby women and men have historically provided a self-sufficient food system for households and communities (Farming Matters 2013). However, today imported foods constitute a high percentage of household diets, much like the other two ACP Regions. For example, 90% of the diets in the Marshall Islands are dependent on imported foods and 95% of the diary requirements in Fiji are derived from imported foods, particularly powered milk and cheese (CTA 2011). To break this dependency, the PACPs are actively seeking to promote greater national food self-sufficiency. That is, to produce whatever basic foodstuff can be produced locally (e.g. fruits, vegetables, root crops and small livestock) while importing other food stuff. However, because of the small size, remoteness and geographical fragmentation, many Pacific communities are highly vulnerable and suffer disproportionately from external shocks, such as adverse climatic events or market failures.

Natural disasters, reef degradation and deforestation have caused extensive damage to lives and livelihoods. With the exception of Fiji, in the other islands (Papua New Guinea, Samoa, Solomon Islands, Tonga and Vanuatu), agricultural land is very limited and can do little more than support the local population. Traditional ownership patterns govern land use in the Pacific but are becoming increasingly constrained by the shortage of usable land. As a result, there will always be some dependency upon imported food in the Pacific region in spite of its many negative impacts on local economies (UN-Women 2012). A few Pacific island countries have developed tourism and industry but the lack of infrastructure, especially regular transport services and accommodation, prevent these enterprises from reaching full potential (Bulletin on Asia-Pacific Perspectives 2003/04).

Impact on Women Small Farmers

Throughout the discussions so far, the production of food for families and the domestic market has been discussed as an extremely important part of the agriculture sector. Rajack-Talley and Talley (2000) pointed out that any analysis of the domestic food sector in the Caribbean (as well as in Africa and the Pacific) is essentially an analysis of women's role in food crop production. Women farmers tend to concentrate on local food production for their households, communities, society and region primarily as small producers and/or family farms. Globally, women are responsible for 60-90% of total food production in their respective countries with smaller percentages for exports (FAO 2006). In many countries, women are also responsible for fishing in shallow waters and in coastal lagoons, producing secondary crops, gathering food and firewood, processing, storing and preparing family food, and fetching water for the family (UN-Women 2012).

In many African countries women comprise 70% of the agricultural workers and 60-80% of the labor that produces food for household consumption and sale. Women do all the processing for basic food stuffs and provide 80% of the labor for food storage and transport from farm to village, as well as 60% of the harvesting and marketing activities (NEPAD 2013). Although women play a major role in agriculture, their access to land and other resources is governed by customary laws that discriminate against them.

During the last ten years, many African countries have adopted new land laws in order to strengthen women's land ownership rights. This has helped improve the situation of rural women (Mucavele 2015). According to Dr. Lindiwe Majele Sibanda, Chief Executive Officer and Head of Mission Food, Agriculture and Natural Resources Policy Analysis Network (FANRPAN), "women farmers are the pillars of African agriculture. Yet even as the guardians of food security, they are still marginalized in business relations and have minimal control over access to resources such as land, inputs such as improved seeds and fertilizer, credit and technology" (NEPAD 2011:1).

Women in the Caribbean are also highly visible in the production and marketing of food for domestic, household consumption as well as in the regional export food trade, buying directly at farm-gate and exporting to neighboring islands (FAO 2006; Rajack-Talley 2015). The majority of women are smallholders, with the average farm size being significantly smaller in area than that held by men. Rajack-Talley and Talley (2000) argued that production constraints related to land tenure and other resources impact more heavily on women than on men because women are predominantly in the domestic and/or in the regional sphere of agricultural production and marketing, while men typically are more actively engaged in the production and marketing of traditional agricultural commodities for the international markets (see also FAO 2006; Rajack-Talley 2015).

Similarly, women in the Pacific island countries continue to be prominent in traditional agriculture ensuring the absence of absolute poverty and associated hunger in many homes (Bulletin on Asia-Pacific Perspectives 2004/04). They engage in subsistence farming to feed their immediate families and/or grow cash crops to sell in the market. According to a United Nations study on Entity for Gender Equality and the Empowerment of Women in the Pacific Sub-Region (2012:7), "women's agricultural labor is an indispensable part of food production and consumption practices". Although the ways in which the women participate in agriculture vary by island and local cultural norms, they all contribute in planting, tending, and harvesting crops, as well as in marine activities that support the majority of families throughout the region. The important role of women in agriculture on the Pacific islands is similar to Africa and the Caribbean where women are the nurturers of the family and providers of food on land, while men have been providers from the sea. However, they are disadvantaged by their underrepresentation at all levels of society, especially in the decision-making processes. Customary law established centuries ago by colonial administrative systems prevents women from owning and gaining access to land, credit and other resources in many of the Pacific island countries (Balakrishnan and Fairbairn-Dunlap 2005).

In essence, despite the many discriminatory disadvantages, women across Africa, the Caribbean and the Pacific regions play crucial roles throughout the food chain; from food production to distribution and preparation in the homes. ACP countries have historically tailored their agricultural policies to focus on the export of primary 'raw' products, in response to international trade and domination. Current trade liberalization policies that are export-driven will therefore have gendered effects in production and consumption patterns.

AGRICULTURAL TRADE POLICIES AND AGREEMENTS

There is no doubt that the governments of the ACP countries have attempted to resolve the issue of trying to feed their populations, enhance the agricultural sectors and achieve economic growth. They have over the years attempted to take ownership of their economies by adopting the following strategies: (i) developing policies, structures and trading networks among themselves; (ii) reinvigorating the agricultural (and mineral production) systems and reorienting them towards export markets; and (iii) developing strategies and programs aimed at poverty reduction.

In terms of trade, each region has embarked on several unilateral, bi-lateral and regional agreements as well as extra-regional trade and restrictions, all of which affect food production and movement. In Africa, regional trade arrangements include the West African Economic and Monetary Union (WAEMU), the Regional Integration Facilitation Forum (RIFF) formerly the Cross-Border initiative (CBI), the Southern Africa Customs Union (SACU) and the Central Africa Economic and Monetary Community (CAEMC). Regional trade agreements have also been created under the Southern African Development Community (SAADC), the Common Market for eastern and southern Africa (COMESA), and the Commission for East African Cooperation (EAC), the Indian Ocean Commission (IOC) and the Economic Community of West African States (ECOWAS). With the exception of WAEMU, members of the other regional bodies also belong to one or two other trading blocs[4] From these various configurations, three broad patterns of agriculture reform can be discerned: (1) adherence to market reforms (e.g. maize and fertilizer marketing in East and South Africa); (2) countries that resisted reforms and/or re-imposed controls basically opposing trade liberalization (e.g. maize in Zimbabwe after 1998); and (3) states that maintained control while implementing liberalization (e.g. fertilizer market in Zambia and Ethiopia, coffee in Malawi) (FAO 2003).

Similar regional trading networks and blocs can be found in the Caribbean region. For example, the Community Single Market and Economy (CSME) was established by the Caribbean Community (CARICOM) in response to the evolving international environment in areas specifically trade liberalization and globalization. The idea behind the CSME is to create a safe economic space where goods and services along with the factors of production (labor, capital and in the case of land the right of access for establishment purposes) to move freely and be supported by appropriate institutions, and operate in a coordinated fashion. Similar other trade policies emerged in response to the World Trade Organization's (WTO) Agreement on Agriculture such as the Revised Treaty of Chaguaramus (2001), the Regional Transformation Programme for Agriculture (2002), and the Jagdeo Initiative (2004) to name a few. During the same period that some of the agreements were being promoted, the Caribbean reeled from the loss of privileged markets and tariffs which continued in the mid-2000 as the Doha Sanction Waiver[5] expired. In 2009 the Liliendal Declaration on Agriculture, Climate Change and Food Security, as well as the Regional Food and Nutritional Security (RFNSP) were created to focus on sustainable production, processing, preparation, and the commercialization of affordable, nutritious and high quality Caribbean foods, while preserving the region's natural resources. More recently (2011), the Caribbean like its African partners in the ACP group responded to the EU – CAP (Community Agriculture Policy) by establishing its own CAP that encourages the regional building of broad-based and balanced agricultural sectors that could be internationally competitive. The anticipation that these initiatives would address the Region's issues in the agriculture sectors, and by extension Caribbean countries' fiscal stability and sustainability, food security, hunger, poverty and the protection of the region's natural resources, is still doubtful (Rajack-Talley 2014).

Governments of the Pacific Island countries have also established regional trade agreements and processes. The Pacific Island Countries Trade Agreement (PICTA) is based on having free trade among the participating countries, initially for a 10 year period. However, level of trade between the countries has been low due to geographical and logistical constraints. In fact, a 2010 analysis suggested that intra-regional trade agreements could only make limited contributions to economic development (Scollay 2010). In addition to the intra-regional trade agreement, the Pacific group has had bi-lateral Interim

Economic Partnership Agreements (IEPAs) with the EU and a South Pacific Regional Trade and Economic Cooperation Agreement (SPARTECA). Papua New Guinea has had its own free trade agreement with Australia (PATCRA), and Fiji also established an economic framework agreement with Australia (FATERA). In addition, as a group the PACP countries established free trade with Australia and New Zealand – the PACER Agreement. There are several other agreements negotiated as either individual states and/or as a group with the USA and some Asian countries.

Of all the various extra-regional trade arrangements agreed to by the ACP states, one of the largest is that between the African, Caribbean and Pacific countries and those of the European Union (ACP-EU). As far back as 2000 the ACP-EU agreement was signed into what is known as the Cotonou Agreement which entailed a comprehensive framework for ACP-EU relations. The intent was that the ACP-EU arrangement would help with economic development, regional integration, poverty reduction and the smooth and gradual transition of the ACP into the global economy. To achieve these objectives, the Agreement provided an approach for ending the WTO's compatible trading arrangements between the ACP States and the EU. This was to be accomplished through the conclusion of Economic Partnership Agreements (EPAs) that were bargained during 2002 and 2007. The ACP States negotiated the EPAs as regional groups: CEMAC (Economic and Monetary Community of Central Africa; ECOWAS (Economic Community of West Africa States; ESA (Eastern and Southern Africa); SADC (Southern Africa Development Community); CARIFORUM (Caribbean Forum); and the Pacific Forum.

According to Bilal, Dalleau and Lui (2012: ix), the African, Caribbean and Pacific (ACP) Economic Partnership Agreements (EPAs) with the European Union (EU) resulted in various degrees of progress across the three regions. For example, while in the Caribbean comprehensive agreements have been reached, most African and Pacific countries, particularly the least-developed countries (LDCs), are still negotiating comprehensive regional EPAs.[6] Under an EPA, the EU agreed to grant full and immediate market access to ACP/African countries, however, those ACP members who decide to enter an agreement must remove tariffs on "substantially all the trade" between themselves and the EU "within a reasonable length of time", to comply with Article XXIV of the General Agreement on Tariffs and Trade. However, because governments depend on revenues from trade taxes and custom duties, many ACP countries resisted, because the reduced revenues impacted negatively on social expenditures, meeting the Millennium Development Goals (MDGs) and poverty reduction. These concerns obstructed EPA negotiations and regional integration efforts.

Bertow and Schultheis (2007) argued that the opening of the ACP agricultural markets, following the principle of reciprocity, seems to be obligatory for the EU while causing a lot of difficulties for many of the ACP Group of States. For example, while European farmers are able to enter the world market without export subsidies, the farmers in the 'partner' Africa, Caribbean and Pacific countries are unable to do so especially within the timeframe allotted. This is because of underdeveloped infrastructures and little or no subsidies to the farmers in the ACP group. On the other hand, famers in the EU are supported by subsidies, have access to infrastructure, technologies and other resources, and both the negotiations and pace for implementation of trade is on the EU's terms with tenable outcomes for the 'partner' countries. Goodison (2007) explained that the economic partnership agreement (EPA) negotiations are based on a shift from the Lome Convention's non-reciprocity commitment to a basic regime of free trade between the EU and EPA regions. This involves the liberalization of trade in goods as trade-related areas and services. He further pointed out that while Europe has effectively integrated itself, few African (Caribbean and Pacific) countries have strong regional integration.

Impact of Agriculture Trade on Self-Reliance and Self-Sufficiency

It is difficult to decipher at first glance the advantages and disadvantages of extra-regional agriculture trade agreements and market arrangements for the ACP group and developing countries overall. Both governments and academics remain divided on whether or not agriculture trade liberalization contributes to slow economic growth, the demise of small farming households, and restrictions to achieving food security. Some researchers suggest that trade liberalization is good for growth (Dollar & Kraay 2004; Falvey, Foster, & Greenaway 2012; Sachs & Warner 1995; Summers & Heston 1991), and has generally supported agriculture development and food security (See FAO Report 2003 on Trade Reform and Food Security). Others challenge this perspective (Rodriuez & Rodrik 2000), and some believe that trade liberalization centered on comparative advantage is especially challenging for countries producing agricultural commodities, particularly raw materials and where diversification is difficult. For example, Connell and Soutar (2007) pointed out that while the Windward Islands of the Caribbean have reduced banana production and both Barbados and Fiji experienced contractions in the sugar industry, none have found effective substitutes.

Researchers also point out proposed market arrangements have not worked well for small countries. For example, in the Pacific new regional trade agreements have not been unsuccessful and as a whole international agreements have not recognized the disadvantages of small developing countries, "but continue to stress benefits that are nowhere apparent in the small island states" (Connel & Soutar 2007:41). Some see trade liberalization as having both a negative and positive impact on poverty. For instance, research on household income inequality and poverty using detailed microdata across Brazilian states from 1987 to 2005 found that trade liberalization contributes to growth in poverty and inequality in urban areas but reductions in inequality (possibly poverty) in rural areas (Abarche, Dickerson & Green 2004). It was also observed that both state poverty and inequality decrease with rising export exposure and increase with import penetration. Other researchers concluded that there is not enough data on the actual implementation of agriculture trade liberalization, or on reforms that are only partially implemented in some developing countries (e.g. Sub-Sahara), to reach any conclusive assessment (Milner & Morrissey 1999). What is less refuted, is that there are several direct links between trade liberalization and food production and poverty.

Over a decade ago, in an FAO Report (2003) on trade and food security it was suggested that there were two different approaches in addressing issues of agriculture trade and marketing and food importation and food security. First, there is the *self-reliance* approach whereby a country does not need to produce all the food it needs but to export enough goods and services to earn what is required to purchase all its food needs. The other approach, to be *self-sufficient*, is having the ability to produce all the food needs from domestic agriculture production. In this approach, it is necessary to protect and support local farmers. In reality, not all ACP countries can be self-reliant or self-sufficient, particularly if they have limited export opportunities and access to resources as well as high food needs for their populations. This is particularly true for some of the smaller Pacific and Caribbean territories. On the other hand, the World Bank suggested that with certain conditions in place, Africa's farmers can potentially grow enough food to feed the continent, that is, to become self-sufficient. According to the Bank's 2012 Vice President for Africa, Makhtar Diop:

Africa has the ability to grow and deliver good quality food to put on the dinner tables of the continent's families. However, this potential is not being realized because farmers face more trade barriers in getting their food to market than anywhere else in the world. Too often borders get in the way of getting food to homes and communities which are struggling with too little to eat.

As a general rule, ACP member states should first identify the extent to which their agricultural sector has the potential to meet some, most or all of their food needs - and then develop appropriate agriculture reforms.

Interestingly, discussions about self-reliance and self-sufficiency have been held in the Caribbean for decades. Sir Arthur Lewis (1978a, 1978b) suggested that governments needed to support industrialization and revolutionize domestic agriculture, as well as secure better benefits from international trade. He envisioned that export-led growth could be either detrimental or supportive to the economic development of Caribbean societies, depending on the relationship between the export and domestic sectors. Lewis warned that international trade arrangements should be linked to the specific domestic sectors of a country, including agriculture. Other Caribbean scholars further explained that self-reliance is based on domestic food production as well as local industrial development, to meet the consumption needs of the masses (Beckford 1972; Girvan 1973; Rodney 1974; Thomas 1982). This was offered as an alternative to export-led growth and import substitution. Here, the idea is not to isolate Caribbean economies from the international market but to reduce the external diversion of resources and instead invest in domestic food production and link-industries. For example, Thomas (1982) proposed the re-orientation of agriculture from producing raw materials for the world market, to the production of foodstuffs and needed local industrial inputs. Industry, in its turn, should be required to produce basic consumption goods, in particular those needed to raise agricultural efficiency (see also Amin 1977; Mansour 1979).

Self-sufficiency in food involves the development of the domestic food sector and the ability of the region to produce an adequate amount of the region's most basic food items, such as rice, sugar, citrus, fruit and vegetables, milk, eggs, poultry, fish and meats from small animals. This requires that people have access to suitable agricultural lands and other resources to more than just subsist. Given the current high food importation bills and poor diets of the vast majority of people in the ACP Nation States it is imperative that countries grow more of what they can and promote a change in food culture. This will result in decreased dependency on high levels of imported processed foods and an addiction to fast foods, especially those with high fat, sugar and salt contents. However, a move in this direction necessitates increases in government involvement, greater participation by the people, and the buy-in from the private sector. The success of self-sufficiency is heavily reliant on building the capacity of the people so that they can take advantage of the knowledge, technology and the decision-making processes needed throughout the chain of events (Rajack-Talley 2014).

Both approaches, self-reliance and self-sufficient, are linked to policies that support domestic agriculture with the possibility of one of three outcomes. First, countries can become self-reliant as net food exporters where food security is not an issue. In this case, however, access to foreign market opportunities can be daunting because of the challenges of meeting new and emerging market requirements, especially the more stringent food safety and quality requirements for export products. Additionally, many of the new trade agreements have mandated the removal of guaranteed markets between some ACP countries and the EU. As a result, in reference to the sugar industry (as well as banana and others) in the Caribbean (and other ACP countries), the Chairman of the Sugar Industry Board in Jamaica commented that:

We have moved from a situation of a guaranteed market, guaranteed price, and guaranteed access to one in which we still have guaranteed access but nothing else and we now have to be competitive because that guaranteed access has been expanded to include a number of other countries (Heaven 2010).

Second, countries can become self-sufficient in food meeting local food needs and even engaging in export of some food items. However, to do this requires minimal biases against domestic agriculture and small farm production as well as strong agricultural sector reforms. It is dependent upon the extent to which small producers and the network of family farms in ACP countries can easily access new technologies and other resources that are necessary to increase yields and profitability from food exports, and at the same time lower domestic food prices. If this were to take place, then both rural and urban poverty would be reduced. In the real world however, most foods produced from industrialized countries are sold at subsidized and therefore cheaper prices on the open markets. This common occurrence presents a disadvantage to ACP countries that are aiming for self-sufficiency but who are discouraged by the World Bank and the IMF from subsidizing their own farmers. This practice, along with the mandate of food importation that is tariff-free, creates unfair competition against the small (and some large) domestic producers in the ACP countries. The distortion of world food prices for these reasons is detrimental to the small producers and family farms of developing countries. Consequently, the less efficient farmers who do not have access to technology and other resources are eventually forced out of the domestic and international markets.

Third, some ACP countries may decide that because they are unable to be self-sufficient in local food production aim to generate enough earnings from other agricultural exports to meet food import bills. In this case, it is important to not be dependent on the export of primary commodities with low income elasticity and high price volatility. As Ambassador Derrick Heaven, the former executive chairman of the Sugar Industry Authority (SIA), pointed out in the Jamaica *Gleaner* newspaper, the 2010 divestment of the local sugar factories has not produced the desired result because the new players in the sector are now struggling to keep pace with what is unfolding on the wider global stage. According to Heaven:

We have been weaned on the situation, where for the past several hundred years, we have been making one product - raw sugar, particularly for the mother country (England) and Europe...the situation has changed, but I don't think we have changed as quickly as we should, even though we were aware of the implications.

SUMMARY AND CONCLUSION

There is no doubt that the agriculture sector is important to the 3 regions which comprise the ACP for the following reasons: economic development and employment; food production for domestic use; and foreign exchange earnings through agriculture export production. Nonetheless, most of the ACP countries are experiencing a decline in their agriculture sectors and worse yet are increasingly unable to feed their growing populations. As a result, food import bills in these countries continue to climb. Equally dangerous, is that high levels of food importation are usually accompanied by a fast food culture and the use of unhealthy, highly processed foods – both of which contribute to increases in obesity, hypertension, and other health-related conditions. Consequently, ACP governments are grappling with exorbitant food import bills, loss of employment in the agriculture sector, and a decline in the availability and accessibility of nutritious food that can support healthy and productive populations. Many ACP countries are neither self-sufficient in food production nor self-reliant, that is, having the ability to grow and/or purchase what is required to feed their people. Faced with this situation, international trade policies and market arrangements are often touted as the life line for agriculture growth, food security and poverty reduction for developing countries.

The benefits of agriculture trade liberalization, however, are not straight forward. The impact of these trade policies on food production, food security and food dependency is different for ACP and most developing countries compared to the more developed, industrialized nations. Moreover, there is no 'typical' or 'model' ACP country within and across the three regions. The information and discussions presented in this chapter suggest that individual countries should exercise caution when dealing with trade liberalization policies as no 'one size fits all'. Nevertheless, there are some general concerns about the negative effects of agriculture trade liberalization on the well- being of low-resource households, rural families, small farmers and women farmers in all of the ACP member States.

One glaring feature is that the removal of guaranteed protected markets and subsidies for farmers, as well as the stringent market product mandates (sanitary and phytosanitary measures, standardization and labelling, etc.) has made it difficult for the ACP countries to compete for the same export products on the 'open' and 'free' market. As a result, the agriculture sectors in these countries that are dependent on international trade of 'cash' products find themselves on a downward tailspin as many small producers, farming, rural and working families can no longer generate adequate incomes to support their families. This includes the ability to adequately educate their children, purchase food and other imported items (regardless of cost), afford proper health care and improve their living conditions overall. In other words, the move towards a self-reliant way of life for certain groups in many ACP countries is arduous.

The situation is further compounded because the removal of farm subsidies combined with an increase dependency on imported agriculture inputs, makes it difficult for ACP farmers to grow food for themselves and the domestic markets, at affordable prices. The end result, is that, rather than becoming self-sufficient in food production many of the ACP countries are importing their food, including traditionally grown poultry and other meats, milk and dairy products, some fruit and vegetables, as well as traditionally grown staples (starches, grains, cereals, etc.). So not only do we see local farmers struggling with sales for export crops but also in the production of food for home, community and the domestic market. Currently, many small producers are forced to seek other forms of employment or migrate in search of non-agricultural work. The inability of small producers to maintain their farms also translates to loss of freedom and independence. As farm and rural families migrate, the physical, economic, social and cultural life of entire villages also diminishes.

Equally important, not only are more families unable to grow or purchase adequate food to feed themselves but they are also unable to maintain nutritious diets. Cheaper food does not necessarily mean healthier choices and diets. The quality of food items imported, in particular those that are offered at lower prices tend to have high fats, salt, sugars and carbohydrates. As a result, many ACP countries continue to become dependent on food imports "that contravene their health, nutritional, environmental and cultural norms, including Genetically Modified crops, but without the option of refusing it" (Hurungo 2006:21). Citizens have no control over the price of consumer items, what items should be subsidized by their governments and what foods are imported. Food cultures are changing, health concerns are increasing and the human capabilities of both children and adults are diminishing.

Current trade liberalization strategies are not altogether dissimilar from the old Plantation Model in that they are export production centered. Similarly, they pose serious competition for resources for domestic food production and by extension small producers and women farmers in these countries. As many of the ACP countries adopt policies that support trade liberalization, small producers, especially women farmers are dislocated as land and the support services that should be directed towards local food production are diverted to larger farms producing for the export market. Since men tend to concentrate on cash crop production they are more likely to benefit. Even in cases, Uganda for instance, where the use of women in the production of agriculture goods for export did increase, their access to and control of agricultural resources did not. Neither did the increase in female labor in the export market sector resulted in increased incomes for farming families. Moreover, where some traditional export products (e.g. bananas, sugar, etc.) are 'in trouble' because of the removal of protected markets, there is a tendency for men to shift to growing the same crops as women and compete for limited resources.

There are not many studies on the impact of current agriculture trade liberalization on small farmers and women, but the little evidence that does exist show that there tends to be more harm than benefits. As market operations become more globalized and concentrated at the same time, small producers and women farmers overall are increasingly excluded from decision-making and the advantages of markets (ADB 2013; FAO 2006). International trade agreements, the EU-ACP arrangements in particular, need to be sensitive to the constraints faced by small producers and women farmers in Africa, the Caribbean and the Pacific. Longstanding limitations such as low levels of technical capacity, limited land allocations, credit, extension services, infrastructure and market support continue to inhibit the capacity of small producers and/or women farmers to engage with trade opportunities and switch to higher-return crop or animal production. Worse yet, as more productive resources are reallocated for 'cash crops', the production of food for the domestic market is stymied even though food security and poverty reduction remain top priorities for ACP countries.

On a global level, the ACP group, and developing countries as a whole, are more or less weak partners who are unable to take full advantage of the 'intended' benefits from agriculture trade liberalization and 'open' market arrangements. Truth be told, trade policies including the use of tariffs and protectionism have not been very rewarding for African, Caribbean and Pacific countries. While these work for the benefit of developed countries, they do not have the same impact for developing countries and poor people across the globe. To simply follow or respond to the terms of agreement led by groups of select developed countries, multinationals and financial institutions can result in the ACP countries remaining food insecure and income generation from agriculture is stunted. Eventually, some ACP countries may shift from being food producers to becoming reliant on food aid from outside.

Because ACP preferences will continue to be eroded by the WTO, governments should be prepared to adjust to the changes as they expand their own trade. Stronger negotiations are critical so that the agriculture policies and strategies of ACP countries are better integrated into the globalized agriculture and economic trade agreements. According to Koroma and Deep Ford (2006), adjustment assistance should be an accepted dimension of ongoing negotiations by the ACP group on specific products such as rice, sugar, fish, and other items produced largely in these countries. Similarly, it would be advantageous if regional and sub-regional cooperation are strengthened in areas of management, control and surveillance, research and in overall development capabilities. Stepped-up regional economic integration and trade policies, revisiting those that already exist, and addressing issues of natural barriers and transactional costs are all important recommendations. In addition, similar benefits/privileges offered to farmers in developed countries should be given to those in developing countries.

Interestingly, although it is implied that agriculture trade liberalization will have a positive effect on economic growth and poverty, few studies actually examined the distributional effect on both the small farmers and the poor in developing countries, including those of the ACP group. FAO's latest projections indicate that global food production will increase by 60% between 2005/07 and 2050 under its baseline scenario. This represents a downward revision, based on updated data and information, from the 70% increase projected for the same period in 2009 (ESA E Working Paper 2012, No. 12-03). A rise in food production is possible in developing countries, albeit not in all. For example, in Brazil, food production increased from 12.5% in 2003 to 48% in 2008. Likewise, malnutrition declined in Rwanda from 24.5% in 2000 to 15.8% in 2005 (CTA 2011). But this requires that the agriculture in ACP countries to not be driven by international trade liberalization policies, but of meeting the Sustainable Development Goals (SDGs),[7] specifically those related to poverty reduction, ensuring gender equality, economic growth and improving the health and well-being of people. Such gains are more likely to occur from regional cooperation and domestic reforms, rather than from international trade and market arrangements (IFAD 2007).

Nevertheless, it is not recommended or is it practical to simply reverse international trade policies and agreements. However, until the proverbial playing field is levelled, agriculture trade liberalization is not 'fair' nor 'free' and 'open'. The ACP Group of States have proposed South-South Programmes to support food security and nutrition through sustainable food systems and social protection. Governments from these States, while focusing on their local and regional efforts at self-sufficiency and/or self-reliance, must step up their support for this South-South initiative. This is key to the impact of current trade liberalization policies as history has shown us that both wealth and poverty can be derived from agriculture's relationship between the debtor (developing nations) and the creditor (groups of industrialized nations, multinationals and financial institutions).

On the domestic front more attention is required to produce food of high nutritional quality. This requires a range of interventions, including policies that empowers women's decision-making in the sector and providing them with greater access to land, credit, extension services, technology and other resources. Women farmers and small producers are key to both self-reliant and self-sufficient approaches to food production ad food security. They should be encouraged to produce a diversity of grains, roots and tubers, vegetables and fruits, fish and livestock. In doing so, they can provide themselves, and supply local markets, with a diverse diet containing the minerals, vitamins, proteins and calories essential for a healthy life. Raising smallholder productivity is one of the measures being advocated by the Pan-African Nutrition Initiative and the Africa Regional Nutrition Strategy developed by the New Partnership for African Development (NEPAD) and the African Union. Effective implementation of these initiatives in Africa as well as in the Caribbean and the Pacific should improve the nutritional security of the most vulnerable populations in many of the ACP countries. But this calls for a different vision, one that sees the urgent need to create stronger links between agriculture production, food security and health and nutrition. To build stronger partnerships with school feeding projects, local food establishments and cafeterias, with hospitals and other social programs. The end goal is to utilize the agriculture sectors to reduce poverty, hunger, and poor nutrition and to build the capacity, wealth and health of all groups in the ACP countries.

REFERENCES

Abarche, J. B., Dickerson, A., & Green, F. (2004). Trade liberalization and wages in developing countries. *The Economic Journal, 114*(493), 73–96. doi:10.1111/j.0013-0133.2004.00188.x

Amin, S. (1977). Self-reliance and the new international economic order. *Monthly Review (New York, N.Y.)*, *3*(29), 1–21. doi:10.14452/MR-029-03-1977-07_1

Asian Development Bank. (2013). *Gender equality and food security: Women's empowerment as a tool against hunger.* Author.

Africa Progress Report. (2014). *Grain, fish, money, financing Africa's green and blue revolutions*. Author.

Agrictrade. (2011). *Informed analysis, expert opinions- executive brief: Pacific, agric- trade policy debates and developments.* Author.

Asian Development Bank (ADB). (2013). Gender equality and food security: Women's Empowerment as a Tool against Hunger. Manila, Philippines: Author.

Balakrishnan, R., & Fairbairn-Dunlap, P. (2005). Rural women and food security in Asia and the Pacific: Prospects and paradoxes. UN-FAO. RAP, 2005/30.

Beckford, G. (1972). *Persistent poverty: underdevelopment in plantation economies of the Third World.* Oxford, UK: Oxford University Press.

Bertow, K., & Schultheis, A. (2007). *Impact of EU's agricultural trade policy on smallholders in Africa.* Germanwatch.

Bilal, S., Dalleau, M., & Lui, D. (2012). *Trade liberalization and fiscal adjustments: The case of EPAs in Africa.* European Center for Development Policy Management.

Borgen Project Report. (2015). Retrieved from http://borgenproject.org/9-facts-poverty-south-pacific/

Bulletin on Asia-Pacific Perspectives. (2003/04). *Beyond "subsistence affluence"*. Poverty in Pacific Island Countries.

Commodities and Trade Technical Paper #8. (2006). *The agricultural dimension of the ACP-EU Economic Partnership Agreements*. United Nations, Food and Agriculture Organization.

Connell, J., & Soutar, L. (2007). Free trade or free fall? Trade liberalization and development in the Pacific and Caribbean. *Social and Economic Studies, 56*(1/2), 41-66.

Caribbean Food and Nutrition Institute. (2008). *Annual Report.* Retrieved from ris.paho.org/xmlui/bitstream/handle/.../AnnualReport08%20CFNI.pdf

Dollar, D., & Kraay, A. (2004). Trade growth and poverty. *The Economic Journal, 114*(493), F22 F49.

Farming Matters. (2013). *Family farming – a way of life*. ILEIA, Center for Learning and Sustainable Agriculture. Retrieved from www.farmingmatters.org

Falvery, R., Foster, N., & Greenaway, D. (2010). *Trade liberalization, economic crises, and growth.* Bond University.

Food and Agriculture Organization. (2003). Trade reforms and food security: Conceptualizing the linkages. United Nations, FAO.

Food and Agriculture Organization. (2006). *Agriculture, trade negotiations and gender.* Rome: United Nations FAO Publications.

Girvan, N. (1973). The development of dependency economics in the Caribbean and Latin America: Review and comparison. *Social and Economic Studies*, *1*(22), 1–33.

Goodison, P. (2007). EU trade policy and the future of Africa's trade relationship with the EU. *Review of African Political Economy*, *34*(112), 247–266. doi:10.1080/03056240701449646

Heaven, D. (2010). *Local sugar facing an uncertain future-divestment not as sweet as was projected.* Retrieved from http://jamaica-gleaner.com/article/news/20151002/local-sugar-facing-uncertain-future-divestment-not-sweet-was-projected

Hurungo, J. (2006). An outline and analysis of EU export subsidies on production of export interest to ESA countries. Academic Press.

International Food Policy Research Institute (IFAD). (2007). Impact of trade liberalization on agriculture in the Near East and North Africa. Rome, Italy: Author.

Karingi, S. (2008). *Webpate.* Retrieved from http://www.oecd.org/tad/agricultural-policies/39596692.pdf.karingi

Koroma, S., & Deep Ford, J. R. (2006). *The Agricultural dimension of the ACP-EU Economic Partnership Agreements.* Rome, Italy: United Nations, Food and Agriculture Organization.

Lewis, A. (1978a). Growth and Fluctuations 1870-1913. London: Allen & Unwin.

Lewis, A. (1978b). The Evolution of the International Economic Order. Princeton University Press.

Lindiwe, M. S. (2011). *FANRPANO 2011.* Academic Press.

Malua, M. B. (2003). Case Study: The Pacific Island in *Turning Losses into Gains.* SIDS and Multilateral Trade Liberalization in Agriculture. *UN Conference on Trade Development*, (pp. 171-204).

Milner, C., & Morrissey, O. (1999). Measuring trade liberalization in Africa. doi:10.1007/978-1-349-14307-8_3

Morrissey. (Ed.). (n.d.). *Evaluating economic liberalization.* London: MacMillan.

Mucavele, S. (2015). The role of rural women in Agriculture. In *MuGeDe - Women, Gender and Development, Republic of Mozambique - Southern Africa.* Retrieved from http://www.wfo-oma.com/women-in-agriculture/articles/the-role-of-rural-women-in-agriculture.html

NEPAD. (2013). *African agriculture: Transformation and outlook.* Johannesburg, South Africa: NEPAD.

Pan-American Health Organization (PAHO). (2014). *Countries pledge action to reduce child obesity in the Americas.* Author.

Rajack-Talley, T. A. (2015). The role of the small farmer, family farms and women in food production, food security, and sustainable development in the Caribbean. In Sustainable Food Production Practices in the Caribbean (vol. 2). Ian Randle Publishers.

Rajack-Talley, T. A. (2014). *Poverty is an economic crime not a moral sin: A history of Globalized food production, foreign dominated economics, trade liberalization and dependency in the Caribbean*. Paper presented at SALISES 15th Annual Conference, St. Augustine, Trinidad.

Rajack-Talley, T.A., & Talley, C. (2000). The role of farming women in Caribbean agriculture: A historical and contemporary perspective. *Caribbean Journal for Agriculture and Natural Resources, 2*(2), 4-13.

Rodriguez, F., & Rodrik, D. (2000). *Trade policy and economic growth: A skeptic's guide to the cross-national evidence*. Academic Press.

Rodney, W. (1974). *How Europe underdeveloped Africa*. Washington, DC: Howard University Press.

Sachs, J. D., & Warner, A. (1995). Economic reform and the process of integration. Brookings Papers on Economic Activity.

Scollay, R. (2010). *South-South and North-South trade agreements: The Pacific Islands Case*. United Nations University.

Summers, R., & Heston, A. (1991). *The Penn World Table (Mark5): An Expanded Set of International Comparisons, 1950-1987*. NBER Working Paper No. R1562.

Technical Center for Agriculture and Rural Development (CTA). (2011). *Policy brief 3, improving nutrition in AC countries and the role of agriculture*. Retrieved from http://www.cta.int/images/docman-files/policypapers/3_Improving_nutrition_in_ACP_countries.pdf

Thomas, C. Y. (1982). From colony to state capitalism: Alternative paths of development in the Caribbean. *Transition, 5*, 1–20.

UNCTAD. (2013). *Global supply chains: trade and economic policies for developing countries*. Geneva: United Nations.

World Bank. (2012). *Press Release*. Retrieved from http://www.worldbank.org/en/news/press-release/2012/10/24/africa-can-feed-itself-earn-billions-avoid-food-crises-unblocking-regional-food-trade

World Bank. (2012). *Little Data book*. Author.

World Bank. (2015). Retrieved from http://www.worldbank.org/en/topic/poverty/overview

Falvey, R., Foster, N., & Greenaway, D. (2012). Trade Liberalization, Economic Crises, and Growth. *World Development, 40*(11), 2177–2193. doi:10.1016/j.worlddev.2012.03.020

World Agriculture Towards 2030/2050. (n.d.). *The2012revision ESA E Working Paper No. 12-03*. Retrieved from http://www.fao.org/economic/esa/esag/en/

ENDNOTES

1 One in every four undernourished people lives on the African continent as a whole.

2 In the Caribbean there is a total of 1.1 million agricultural workers or 17% of the employed labor force, if Haiti is excluded.

3 CARICOM Workshop on Regional Policy for Food and Nutrition Security, Georgetown, Guyana, September 20, 2010.

4 Not all of the African countries however, participate in the regional trade protocols and agreements.

5 The Doha Round is the latest round of trade negotiations among the WTO membership. Its aim is to achieve major reform of the international trading system through the introduction of lower trade barriers and revised trade rules. The work programme covers about 20 areas of trade aimed to improve the trading prospects of developing countries. The Round was officially launched at the WTO's Fourth Ministerial Conference in Doha, Qatar, in November 2001. The *Doha Ministerial Declaration* provided the mandate for the negotiations, including on agriculture, services and an intellectual property topic, which began earlier. In Doha, ministers also approved a decision on how to address the problems developing countries face in implementing the current WTO agreements. The Doha Sanction Waiver expired in 2008 resulting in the loss of tariffs and privilege markets.

6 EPAs represent a fundamental shift in the trading relations between the two parties, from a non-reciprocal preferential trading regime – under which ACP countries could export almost freely to the EU while maintaining their own restrictions on EU imports – to one requiring reciprocity in liberalization, albeit with a certain degree in asymmetry in commitments, in line with rules of the World Trade Organizations (Bilal, Dalleau and Lui, 2012).

7 The Sustainable Development Goals, otherwise known as the Global Goals, build on the Millennium Development Goals (MDGs), eight anti-poverty targets that the world committed to achieving by 2015. The SDGs include a set of 17 Sustainable Development Goals (SDGs) to end poverty, fight inequality and injustice, and tackle climate change by 2030.

Chapter 2
Social and Environmental Impacts on Agricultural Development

Frances Bekele
The University of the West Indies – St. Augustine, Trinidad and Tobago

Isaac Bekele
The University of the West Indies – St. Augustine, Trinidad and Tobago

ABSTRACT

Addressing environmental and social impacts on agricultural development and food security is a global priority since increased food production of 60-70% is estimated to be required by 2050 to feed the growing world population. In developing nations, the situation is more acute since fewer social, technological and financial resources are available to combat climate change, which is expected to have negative effects on agricultural production, and there are other constraints to achieving food security. This chapter explores the social and environmental issues affecting agricultural production facing farmers and other agricultural practitioners, policy makers, institutions and stakeholders in the developing world. It will also address how progress in research in emerging economies can be put to maximum benefit in the face of these existing social and environmental challenges. A cohesive strategy to address these challenges is presented.

INTRODUCTION

The world's population is expected to increase to 9 billion by 2050 (Hubert, Rosegrant, Van Boekel & Ortiz, 2010), and it is estimated that a 60-70% global increase in food production is needed to feed this growing population (Consultative Group for International Agricultural Research (CGIAR), 2015; FAO, 2007, 2014; Mba, Guimaraes, & Ghosh, 2012). Currently, approximately 2 billion persons are food insecure since they do not meet one or several of the Food and Agriculture Organisation's (FAO's) dimensions of food security (access to adequate food, availability and utilization of nutritious food and stable supply (Hubert et al., 2010; Wheeler & von Braun, 2013)). Furthermore, 805 million persons did not have access to sufficient food during 2012-2014 (Maggio, Van Criekinge, & Malingreau, 2015). Food

DOI: 10.4018/978-1-5225-0942-4.ch002

availability must increase to meet demand by 100% in developing countries (CGIAR, 2015). This can be achieved through increasing production as well as reducing losses (food waste). Improving agriculture[1] to meet the Millennium Development Goals[2] of the United Nations, which include halving extreme poverty and hunger by 2015 and eliminating it by 2030 (under the sustainable development agenda), requires optimisation of agricultural practices and systems, and dealing effectively with technological, social, environmental as well as economic issues that influence the sustainability of agricultural production. The United Nations Framework Convention on Climate change (UNFCCC) (2007) described the developing nations as most vulnerable to climate change impacts that are expected to have negative effects on agricultural production and food security, as described by the International Food Policy Research Institute (IFPRI, 2009). This is attributed to the limited social, technological and financial resources available to address climate change in these countries.

Currently, sustainable food security and agricultural development are constrained globally by degradation and loss of agricultural land, the loss of biodiversity, depletion of water and other resources and pollution (Chaudhury, Vervoort, Kristjanson, Ericksen, & Ainslie, 2103; Sonnino, Moragues Faus, & Maggio, 2014). These factors are exacerbated by the negative impacts of climate change and may also contribute to climate change.

This chapter explores the social and environmental issues facing agricultural practitioners (producers and entrepreneurs alike) in the developing world, who are striving to contribute to food security in their communities and, by extension, the world. It will also address how the progress in research in emerging economies can be put to maximum benefit in the face of existing social and environmental challenges in developing nations, as described by Ejeta (2009).

Heat stress could affect developing countries by 2030 (Inter-Governmental Panel on Climate Change (IPCC), 2007, 2013). Expenditure of USD 200 and 250 billion a year may be required to address negative impacts of climate change in developing countries (Peterson, 2011). These impacts can significantly reduce sustainability of livelihoods and the well-being of citizens in developing nations (Noble et al., 2014).

A multi-sector approach is advocated to deal with the challenges and potential impact of social and environmental factors such as climate change on food security. The proposed model encompasses the role of farmers, entrepreneurs, social policy makers, governments, the private sector, scientists and organisations in a cohesive strategy. The role of education and extension in preparing for the impacts of climate change, as described by Bekele and Ganpat (2014) in the context of small island developing states (SIDS), is also discussed.

In order to illustrate the current situation, constraints and putative solutions to social and environmental impacts on agricultural development and food security in developing countries, several staple crops in Africa, Latin America and the Caribbean (LAC) and the commodity, cocoa, are highlighted in this chapter.

CURRENT STATUS OF AGRICULTURAL DEVELOPMENT AND FOOD SECURITY IN DEVELOPING NATIONS

1. Africa

In Africa, the agriculture sector employs 60% of the total population, contributes 20% to the gross domestic product (GDP) and earns more than 10% of export revenues (Sasson, 2102). It is expected that access to adequate and secure food would be a priority in Africa. However, in 2010, the number of food insecure persons in Sub-Saharan Africa was estimated at 239 million, and in North Africa it was 37 million (Sasson, 2012). The main cause of food insecurity in Africa was identified as insufficient food production (Beddington et al., 2012; Sasson, 2012). Soaring world food prices and rural migration have exacerbated the situation, and the impacts of climate change could be detrimental. High food prices have resulted in a significant percentage of the African population spending more than half their income on food. There have been recurrent droughts in the Horn of Africa since 2009 as well as a history of famines due to drought. Mba et al. (2012) described Africa as one of the most vulnerable continents to climate change and variability. The expected decrease in the length of the growing seasons of African crops (Thornton & Herrero, 2015) will result in the further loss of production in marginal agriculture.

Crop yields in Africa can decrease by as much as 50% by 2020 due to climate change, and this could result in lower calorific intake by humans (by as much as 500 calories per capita in 2050, a 21% decline) (Toulmin, 2009). Jarvis, Ramirez-Villegas, Campo, and Navarro-Racines (2012) found that while cassava is likely be affected positively by climate change in many areas of Africa (−3.7% to +17.5% changes in climate suitability across the continent), other major food staples in Africa are expected to face negative impacts, with the most significant predicted for beans (−16% ± 8.8), potato (−14.7 ± 8.2), banana (−2.5% ± 4.9), and sorghum (−2.66% ± 6.45). For resource-poor farmers in developing countries such as Africa, the use of fertilizers that could improve plant vigour, root growth and access to soil moisture, and irrigation to combat drought may not be feasible options to address climate change (Beebe et al., 2011).

Mixed farming (crop-livestock) systems form the mainstay of African agriculture and have certain benefits (Thornton & Herrero, 2015). The adaptation potential of these systems must be examined in the face of climate change since they are complex. Furthermore, there is very little data currently available on putative climate change impacts on these systems.

2. Latin America and the Caribbean

In Latin America and the Caribbean (LAC), approximately 15.9% of the labour force is directly involved in agriculture (Andersen et al., 2014). LAC's agricultural market share increased from 9.5% in 1980 to 18.1% in 2010 (Flachsbarth et al., 2015). The largest agricultural sectors in LAC are based on livestock, feedstuff and biofuel crops (Flachsbarth et al., 2015).

The contribution of agriculture to overall GDP ranges from 3.1% in Mexico to more than 5.5% in Brazil and to 7.8% in Peru (Andersen et al., 2014). From 1961 to 2007, LAC had the highest growth rate in agricultural productivity among developing regions (1.9%) (Ludena, 2010). The LAC countries with the highest land availability, such as Brazil, Argentina, Chile, Colombia, Mexico and Venezuela, have the most significant agricultural production (Ludena, 2010). Caribbean countries experienced limited growth by comparison (0.5%) (Ludena, 2010). Most of that growth occurred in the livestock sector, especially for non-ruminants (pigs and poultry). The most valuable agricultural export products from LAC currently are soybeans and cane sugar (Flachsbarth et al., 2015).

Wickham (1992) underscored the need for further development in vegetable production in the Caribbean to foster self-sufficiency in this area. Priority areas identified were germplasm enrichment and enhancement, selection of superior cultivars, development of more efficient production systems (including disease and pest management) and application of post-harvest technologies to promote the shelf-life of vegetables. Several other issues related to agricultural development and food security have been identified as requiring attention in LAC. These include improved partnerships among public sector, agencies, research institutions and the private sector to deliver food production programmes and transfer post-harvest and other technologies to farmers to assure food quality and availability (Roberts, Ganpat, Narine, Heinert, & Rodriguez, 2015). Farmer training through extension services is essential.

Climate change can also adversely affect crop production in LAC. Andersen et al. (2014) modelled the impact of climate change in the major agricultural producing countries in LAC, *viz*., Brazil, Peru and Mexico. They projected that in Brazil, the effect on soybean production could be of the magnitude of –0.5 and +0.3% per annum. In Mexico, maize is the main agricultural product and accounts for 14 to 22% of the regional agricultural GDP. A decrease of 0.3% per annum was predicted for this crop. In Peru, the production of potatoes was forecasted to change by –0.6 to +0.2%.

Jarvis et al. (2008) predicted reductions in rice yields in Latin America by the 2020s, but increases in soybean production due to possible increases in carbon emissions. They also forecasted:

- Reduced acreage of land available for coffee production in Brazil,
- Lower coffee production in Mexico,
- Salinization and desertification of agricultural lands in drier areas of Latin America, and
- Probable increased incidence of the coffee leaf miner (*Perileucoptera coffeella*) and the nematode *Meloidogyne incognita* in Brazilian coffee-growing areas.

In addition, the risk of *Fusarium* head blight in wheat may increase under climate change in southern Brazil and Uruguay (Fernandes et al., 2004).

Overall, the combined impacts of climate change are estimated to amount to less than 1% of household incomes during the next 40 years in Brazil, Peru and Mexico (Andersen et al., 2014). The negative impacts on agricultural production in these countries were forecasted to be somewhat balanced by increased commodity prices. This was based on the premise that any large increases in the global prices of agricultural products, caused by climate change, will tend to benefit these agricultural producers of the respective commodities. However, for the land-constrained Caribbean countries, any further reduction in agricultural productivity will restrict the likelihood of achieving food security and poverty reduction. These countries are currently mainly net food importers. The Windward Islands have suffered significantly with the removal of the European Union preferential trade tariffs for banana imports. According to Ganpat and Isaac (2014) and Hutchinson et al. (2013), these SIDS are very vulnerable to the impact of climate change.

ENVIRONMENTAL ISSUES CRITICAL TO AGRICULTURAL DEVELOPMENT AND FOOD SECURITY

The following are some of the pertinent environmental issues to be considered with regard to agricultural development and food security in developing nations:

- Environmental impacts on production, productivity and product quality;
    - Impacts of climate change on agriculture and post-harvest production;
- Ensuring a reliable supply of produce or commodity under climate change;
- Use of superior (in terms of yield/productivity, vigour and pest and disease resistance) and climate-smart crop varieties;

- Improving energy efficiency of crop and livestock production (reduction in carbon emissions and prudent use of energy):
  - Application of eco-friendly sustainable practices such as organic fertilisation and pruning of crops;
  - Choosing the most efficient crop production system for specific crops and environments (such as terrains on steep slopes and multi-cropping on impoverished soils (Bekele & Ganpat, 2014)); and
- Monitoring land conversion/biodiversity losses.

Climate-Change Impacts on Agricultural Development and Food Security

According to FAO (2007), climate change impacts can be classified into two groups, *viz.*, biophysical impacts:

- Physiological effects on crops, forests and livestock (quantity, quality) and availability of fodder and pastures;
- Changes in land, soil and water resources (quantity, quality);
- Increased weed and pest challenges;
- Shifts in spatial and temporal distribution of impacts;
- Sea level rise, changes to ocean salinity and
- Sea temperature rise causing fish to inhabit different ranges.

and socio-economic impacts:

- Decline in yields and production;
- Reduced GDP from agriculture;
- Fluctuations in world market prices;
- Changes in geographical distribution of trade regimes;
- Increased number of people at risk of hunger and food insecurity and
- Migration and civil unrest (increased crime).

Climate Change Impacts on Crop Production

Global

Global warming and the El Niño effect, with the attendant drought, floods, increased temperature of the atmosphere and oceans and salinity and shifts in weather patterns, have the capacity to significantly reduce yields of food crops (refer to Figure 1) including staples such as rice, wheat, corn and barley (Beddington et al., 2012; Rosegrant, 2011). The potential impacts in small island developing states (SIDS) have been discussed by Roberts and Rodriguez (2014). The situation is compounded by the need to produce bio-fuels to reduce the carbon footprint of nations and meet the increasing demand for energy/fuel. This implies competition for food crops with biomass attributes favoured for biofuel production. Crop production will also be constrained by the spread of invasive species such as pests and fungi that may proliferate under altered climatic conditions (Leishman & Gallagher, 2015). It has been reported

Figure 1. The effects of climate change on agricultural production
(Source: http://croplife.org/news/infographic-how-does-climate-change-impact-agriculture/

that even a 2-degree change in temperature over time can affect agricultural productivity adversely in the absence of adaptation (Ramirez-Villegas & Thornton, 2015; Wheeler & von Braun, 2013).

In Africa

In a Working Paper of the CGIAR Research Program on *Climate Change, Agriculture and Food Security* (CCAFS), Ramirez-Villegas and Thornton (2015) have provided a detailed overview of the potential (based on published studies) and projected (based on crop model simulations) effects of climate change (extreme and changing weather conditions) on the production of maize, common bean, sorghum, cassava, yam, pearl millet, groundnut, banana and coffee in rainfed Africa. Projections were analysed and it was concluded that climate change affects crops and regions differently in Africa. For instance, in the Sahel, it was predicted that some crops may no longer be suitable for cultivation and a shift to drought and heat-resistant varieties of crops such as cassava (Jarvis et al., 2012), yams and sorghum may be a necessary adaptation measure to consider. By comparison, the production of maize in South Africa may be reduced by 30% (Sasson, 2012), but northern South Africa, Botswana, Namibia, Zimbabwe and Lesotho are not expected to experience any significant change in maize production (Ramirez-Villegas & Thornton, 2015).

The yield of common bean is very responsive to climate and the areas suitable for its cultivation in Africa can be reduced by unfavourable climatic conditions (Ramirez-Villegas & Thornton, 2015). The same applies for maize, banana and finger millet as well as coffee in different regions of Africa. For coffee, a reduction in suitable growing area for *Coffea arabica* in Ethiopia is projected with increased temperatures. Cultivation of this premium variety may need to be shifted towards higher elevations and more heat-tolerant *Coffea robusta* may be required at altitudes of less than 1500 m above sea level. Maize productivity (yield) could be reduced by 5-10% in Africa for every degree of warming (Knox, Hess, Daccache, & Wheeler, 2012). This is alarming since maize contributes the greatest portion of calories in the African diet (a mean of 16% and range of 0 – 60% across the continent). The current production is roughly 42 million tons per year (Ramirez-Villegas & Thornton, 2015). The most at risk area in terms of maize production under climate change is the Sahel.

STRATEGY TO MAINTAIN SUSTAINABLE CROP PRODUCTION INTENSIFICATION (SCPI) IN DEVELOPING COUNTRIES

Increased agricultural production requires higher yields and improved management strategies to reduce abiotic and biotic stresses. However, in order to avoid the depletion of land, energy and water resources, sustainable intensification has become crucial (Flachsbarth et al., 2015). Achieving SCPI may be realised through the following practices:

- Maintaining healthy soil to enhance soil-related ecosystem services and crop nutrition;
- Cultivation of a wider range of species and varieties in rotations, sequences and combinations to minimize overall risk of crop failure;
- Using quality seeds and planting material of well-adapted, high-yielding varieties;
- Adopting integrated management of nutrients, pests, diseases and weeds; and
- Efficient water use (FAO, 2013).

SCPI requires effective use of technological and scientific innovations so that food supply may be increased without converting larger acreages to agricultural production or unleashing negative impacts on the environment and human health (Dempewolf et al., 2014; FAO, 2011; Sonnino et al., 2014).

ADAPTATION AND MITIGATION STRATEGIES TO OFF-SET THE IMPACT OF CLIMATE CHANGE ON AGRICULTURAL DEVELOPMENT AND FOOD SECURITY

Several adaptation strategies have been proposed to combat the potential negative impacts of climate change such as rising temperatures, intense and more frequent storms, droughts *etc.* These include diversification of livelihoods through agroforestry, integration approaches such as crop-livestock systems and rice-fish systems (Palombi & Sessa, 2013) and cultivation of heat, drought and other abiotic stress tolerant crops. The latter, described as climate-smart varieties, can be developed through crop varietal selection and breeding and should be cultivated under improved crop and eco-system management systems through precision agriculture, tailored to optimise production of particular crops under specific growing environments (FAO, 2009a; Palombi & Sessa, 2013).

Improved crop production management techniques can contribute to mitigating climate change by reducing greenhouse gas (GHG) emission. This may be achieved by limiting the use of inorganic fertilisers, avoiding soil compaction or flooding to reduce methane emissions, as in paddy rice systems, and sequestering carbon by planting perennial crops, forest trees (Dawson et al., 2011) and grass species (Palombi & Sessa, 2013). The latter practices fall under the umbrella of conservation agriculture.

Agricultural Approaches and Practices that Contribute to Climate Change Adaptation

Certain approaches and practices for sustainable crop production can contribute to climate change adaptation. They provide options for location-specific scenarios and should be adapted with local farmers/farming communities. Examples, provided by Palombi and Sessa (2013, p. 203), include SCPI listed above (FAO, 2013) and others such as:

- Mulch and cover cropping;
- Landscape-level pollination management; and
- Land fragmentation (riparian[3] areas, forest land within the agricultural landscape) (FAO, 2008, 2009b, 2012; FAO-PAR, 2011; Lin, 2011).

Tol (2002) presented comparative values of adopting versus not adopting measures to off-set climate change impacts on agriculture for a 2.5°C increase in global mean temperature. In LAC, the projected change in GDP from agriculture without climate change adaptation was –0.8 (0.6) whereas it was 0.6 (0.7) with adaptation measures. In Africa, the projected GDP adjustment was –0.2 (0.2) without adaptation measures compared to 0.5 (0.3) with adaptation measures. This research effectively underscores the value of adaptation measures.

Agricultural Approaches and Practices that Contribute to Climate Change Mitigation

As with climate change adaptation, approaches and practices for climate change mitigation can provide options for specific locations and should also be adapted in conjunction with local farmers/farming communities. Examples, provided by Palombi & Sessa (2013, 204), include in addition to some of the adaptation strategies listed above:

- Soil compaction management;
- Promotion of legumes in crop rotations;
- Restoration of cultivated peaty soils and degraded lands;
- Soil management practices that reduce fertilizer use (*e.g.* deep placement of urea);
- Growing nutrient-use efficient crop varieties;
- Integrated crop and livestock systems (also recommended by Boland et al., 2013; FAO, 2010a);
- Dedicated production of energy crops to replace fossil fuel use;

- Emission control and reduction (combustion engines, animal waste);
- Improved rice cultivation techniques; and
- Agroforestry (FAO, 2004, 2008, 2009b, 2012).

FAO (2007) has also recommended harnessing the knowledge and experience of farmers to select types of animal breeds or varieties that can best withstand changing climatic conditions. In addition, FAO (2007) also advocated mitigating the negative impact of ruminants on GHG emissions via animal husbandry practices such as adjusting ruminant diets and stocking ratios. Monitoring and rapid appraisal of the impacts of disasters on local fishing communities and aquatic ecosystems so that immediate and longer-term remedial action could be implemented in a timely manner were also recommended. FAO has adopted an Ecosystem Approach to management of fishery resources and the ecosystems on which they depend, in the face of climate change.

Climate-adapted companion cropping is another sustainable agricultural approach that can contribute to climate change mitigation. In sub-Saharan Africa, a push-pull companion cropping practice has been adopted to address constraints of the parasitic weed, *Striga hermonthica*, stem borer (lepidopterous insect) attack and poor soil fertility (Midega et al., 2015). The climate-adapted, drought-tolerant *Brachiaria* cv mulato (bred by the International Center for Tropical Agriculture (CIAT)) is used as a border crop with cereals such as maize, and greenleaf desmodium is used as an inter-crop. Desmodium fixes nitrogen, and thus provides some fertilizer to the crop. Highly significant reductions in *Striga* and stem borer damage to maize along with improved plant height, grain yield and soil fertility were achieved via this climate-adapted, push-pull system studied (Midega et al., 2015). *Brachiaria* cv mulato also serves as feed for livestock and, as stated previously, is drought-tolerant unlike traditional fodder, Napier grass (*Pennisetum purpureum*). Similarly, in St. Vincent and the Grenadines, the research of Isaac, Brathwaite, Cohen, and Bekele (2007) demonstrated that *Desmodium heterocarpon* controls weeds such as *Commelina diffusa* and nematodes and improves soil fertility when intercropped with banana (*Musa* spp.).

This push-pull approach provides a renewable, organic, affordable source of herbicide, insecticide, fertilizer and fodder for rural farmers such as those in sub-Saharan Africa and LAC. Thus it can improve food security, allow diversification of the farming system and contribute to poverty alleviation in the face of climate change.

Implementation of Early Warning and Risk Management Systems

Early warning and risk management systems can facilitate the timely implementation of adaptation measures to address climate variability and change. These are listed by FAO (2007) as:

- A historical climate data archive; an archive on climate impacts on agriculture;
- Monitoring tools using systematic meteorological observations;
- Climate data analysis (to determine the patterns of inter-annual and intra-seasonal variability and extremes);
- Information on the characteristics of system vulnerability and adaptation;
- Monitoring evidence of resilience under climate change as well as critical meteorological and environmental thresholds for manifestation of negative impacts of climate change; and
- Crop weather insurance indices to reduce the risk of climate impacts for lower-income farmers.

Adaptation Strategies for Crop Production in Africa

The crucial role of adaptation in countering the putative negative impacts of climate change in Africa, particularly that of global warming, was emphasised by Ramirez-Villegas and Thornton (2015). In general, these authors advocated complex systemic and transformational changes in farming systems in Africa to cater for increased temperatures. Shifts in cropping patterns to allow for cultivation of new crops that are more drought-tolerant are recommended along with a concomitant adjustment in diets and trade policies.

Yield losses for maize can be minimised by shortened cropping seasons and heat stress during the crop's reproductive phase (Ramirez-Villegas & Thornton, 2015). For the common bean (highly sensitive to climate change), Ramirez-Villegas and Thornton (2015) projected decreases in yield across Africa. However, a heat-tolerant accession of a wild relative of the common bean (*Phaseolus vulgaris*), tepary bean (*Phaseolus acutifolius*) was identified at CIAT for use in breeding heat-tolerant common bean lines (Beebe et al., 2011). Thus far, breeding lines in the greenhouse have been found to maintain productivity at +3 °C over the current temperature limit of common bean. The potential for such heat tolerant lines will be achieved with further investment in research and incentives for farmers to test and adopt promising new varieties in developing countries.

THE ADOPTION OF CLIMATE-SMART AGRICULTURE

Climate smart agriculture and landscapes are widely regarded as the panacea for addressing social and environmental impacts that threaten agricultural development and food security in the face of climate change. The widespread cultivation of climate-smart crops that are well-adapted to climatic change and changing growing environments will assure food security in a sustainable manner (Harvey et al., 2014; Neufeldt et al., 2013). Smart crops are better able to withstand drought than wheat or rice and do not require as much water to thrive (Swaminathan, 2012). They can be grown by rural families without irrigation. They are also nutritionally rich and have much higher protein and thus nutritional value. Consequently, these crops, many of them currently orphan crops (Cheng, Mayes, Dalle, Demissew &, Massawe, 2015), have ecological, social and nutritional benefits. Swaminathan thus recommended diversification of cropping patterns and the cultivation of more climate-smart crops to address the potential impact of climate change and year-to-year variability in weather patterns.

Features of Climate-Smart Agricultural Landscapes

Climate-smart agriculture will mitigate against the impact of climate change by harnessing natural biological processes to improve production in a more environmentally-friendly way and avoiding degradation of natural resources (Branca, McCarthy, Lipper and Jolejole, 2011; Palombi & Sessa, 2013). This sustainable system will be less vulnerable to shocks and stresses. It will harness the best of crop varieties and livestock breeds under optimum agro-ecological and agronomic management (Beddington et al., 2012).

Mba et al. (2012) recommended that in order to achieve food security under worsening climatic conditions and with constrained natural resources, higher yields must be achieved per unit of input; yield efficiency must be optimised. In Brazil, Martinelli, Naylor, Vitousek, and Moutinho (2010) emphasised the importance of this sustainable approach of increased productivity on existing agricultural land without further environmental degradation and deforestation. The cultivation of diverse, climate-smart crop varieties that can produce more with less is the objective of climate-smart agriculture. High yielding, well-adapted, resilient crop varieties that use inputs such as water and nutrients efficiently are required. They must combine superior genetic constitutions and be managed optimally. The case of the New Rice for Africa (NERICA) (West African Rice Development Association, 2001) was cited as an example of climate smart agriculture, which has already accounted for significant increases in rice production in sub-Saharan Africa.

Strategy for Achieving Climate-Smart Agricultural Landscapes

CGIAR's new strategy and research programmes: Answering to poverty, health and climate change (CGIAR, 2015) is an excellent example of a strategic and integrated approach towards achieving food security in developing countries (refer to Figure 2).

THE APPLICATION OF SCIENCE AND TECHNOLOGY AND KNOWLEDGE TRANSFER IN SECURING FOOD PRODUCTION

Pollinator-Friendly Management Practices

Pollinator-friendly management practices are also essential for climate-smart landscapes since they will enhance yields, quality, diversity and resilience of crops and cropping systems. Such practices include:

Figure 2. CGIAR's new strategy and research programmes: Answering to poverty, health and climate change
(CGIAR, 2015)

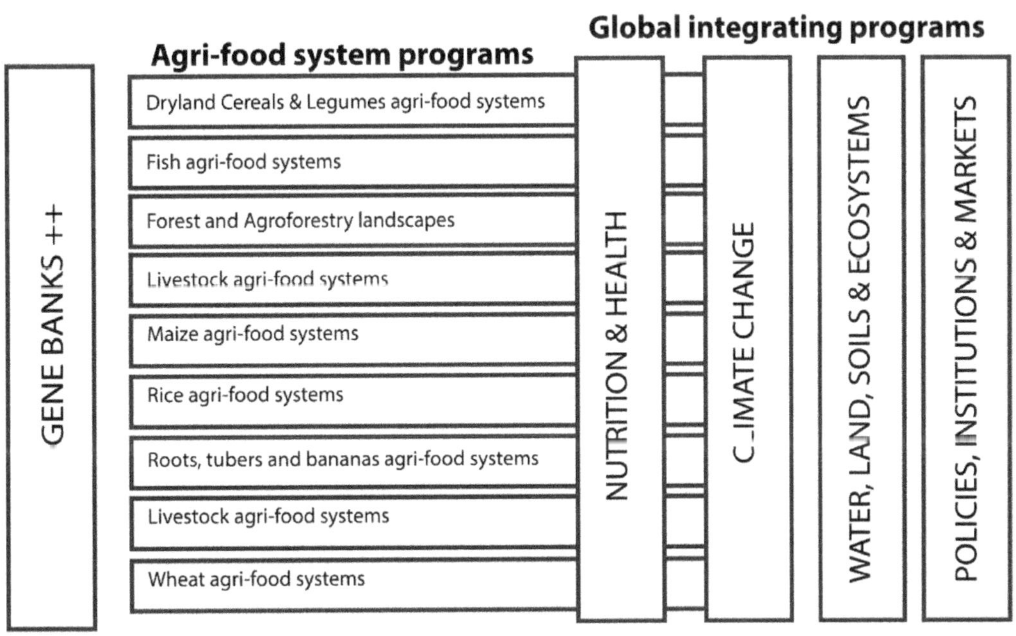

- Conservation of wild habitats;
- Managing of cropping systems and environs to protect habitats and forage;
- Cultivation of shade trees;
- Protection of bee and other pollinator nest sites;
- Reduced applications of pesticides and other hazards in production systems and
- Establishing landscapes with natural habitats and resources that favour pollinator population survival and services (FAO, 2009a).

In Costa Rica, it was found that coffee farms located in close proximity to forested areas benefit from greater diversity and number of pollinators visiting the coffee plants. This translates into increased coffee yields and improved coffee quality (FAO, 2009a). Furthermore, the maintenance of forests provides other ecosystem services such as carbon sequestration.

Maintaining pollination biodiversity in agricultural landscapes serves the dual role of ensuring pollination of crop species while serving as insurance against pest and disease threats among managed pollinator populations (FAO, 2009a). However, the system requires that farmers be trained about the value of adopting this strategy and the resources that pollinators require so as to encourage their proliferation in niches in the landscape.

Management of Soils: Nutrient Status, Porosity and Structure

Climate-smart landscapes must be characterised by healthy, fertile soils that support intensive production systems. Optimal crop production cannot be achieved on degraded soils with nutrient deficiencies and soil-borne pests and diseases. The cultivation of pulses is recommended in climate-smart landscapes since legumes fix nitrogen and thus reduce the need for inputs such as nitrogen fertilizers.

A case for conservation agriculture that includes zero soil tillage has been advanced by FAO (2007, 2012) and Mba et al. (2012). Reichert et al. (2015) recommended the latter for common beans based on their observations under highly variable weather conditions on sandy soils in a sub-tropical environment (Brazil). Conservation agriculture involves the maintenance of soils in an undisturbed state with a sufficient supply of organic matter that provides a good habitat for soil fauna. The avoidance of mechanical tillage increases the populations of earthworms, millipedes and other micro fauna that take over the role of tillage and build soil porosity and, with organic matter, improve soil structure. This makes land less susceptible to flooding and erosion (Palombi & Sessa, 2013). Knowledge transfer to farmers of these sustainable soil management practices is thus crucial.

Developing Climate Smart Crop Varieties Through Research in Breeding and Technological Advances

The threat of crop failures due to biotic and abiotic stress under climate change may be addressed through breeding climate-smart varieties (refer to Table 1). In 1982, Swaminathan advocated that Third World countries should master all recent developments in genetics. This should be widened now to embrace all areas of advanced crop breeding and biotechnology including molecular biology, next-generation sequencing (Bansal, Lenka and Mondal, 2014; Edwards & Batley, 2010); genomics, genomic selection and marker-assisted selection and breeding (Tester & Langridge, 2010), gene pyramiding (Gennaro et

Table 1. Adaptive traits to introduce through crop improvement/breeding in select climate-smart crop varieties

Crop	Adaptive traits
	Drought
Rice	Early flowering, deeper and thicker roots, root pulling resistance, root penetration, osmotic adjustment (OA), rapid stomatal closure, water-use efficiency (WUE), membrane stability, leaf rolling score, and leaf relative water content
Wheat	Early flowering, stem water-soluble carbohydrate (SWSC), green flag-leaf persistence, WUE, transpiration efficiency (TE)
Maize,	Early flowering, anthesis silking interval (ASI), ears per plant, stay-green chlorophyll content, osmotic adjustment (OA), root traits, leaf abscisic acid (L-ABA)
Barley	Early flowering, plant stature, ear type, WUE, OA, high biomass combined with SWSC, water extraction, and transpiration efficiency (TE)
Sorghum	Early flowering, stay-green, SWSC, TE, rooting depth and patterns, epicuticular wax
Pearl millet	Early flowering, few tillers, low biomass and high harvest index including panicle harvest index, grain yield, individual grain mass Profuse rooting in the deeper layers of soil
Chickpea	Early flowering, deep rooting and higher root length density
Groundnut	Early flowering, WUE, transpiration, TE, specific leaf area (SLA), and chlorophyll meter reading
	Heat
Rice	Time of day when flowering (TDF) commences (early TDF protects fertility from high temperature), high temperature tolerance at grain filling (pollen shedding, pollen germination and pollen tube extension) Higher leaf transpiration rate to maintain lower leaf temperature combined with opening of flowers in morning (*O. glaberrima*)
	Salinity
Rice	Early vegetative stage: salt exclusion or low uptake, compartmenting of toxic ions in structural and older tissues, higher tissue tolerance by compartmenting salt into vacuoles, stomata that close faster upon exposure to salt stress Reproductive development stage: salt exclusion from flag leaves and developing panicles
Wheat and barley	Na+ exclusion, K+/Na+ discrimination, sheath retention of ions, tissue tolerance, ion partitioning into different-aged leaves, OA, enhanced vigour, WUE, and early flowering
Maize	OA and abscisic acid (ABA)
Sorghum	Whole plant tolerance resulted from reduced shoot Na+ concentration, a major mechanism involved in salt tolerance
Pearl millet	Whole plant tolerance associated with reduced shoot N content, increased K+ and Na+ contents; with shoot Na+ concentration a potential non-destructive selection criterion at vegetative-stage screening
Chickpea	Ability to maintain large number of filled pods; shoot Na+ or K+ not related to salinity
Pigeon pea	Reduced shoot Na+ concentration
Groundnut	Pod numbers per plant
	Flood Tolerance/Submergence
Rice	Vigorous seedling growth, elongation ability, submergence tolerance, resistance to lodging or recovery from lodging after water level reduction

(Adapted from Jarvis et al., 2008)

al., 2007) and tissue culture. Landscape genomics can be employed to identify the environmental factors that affect adaptive genetic variation and gene variants that facilitate local adaptation (Anami et al., 2015; Rellstab, Gugerli, Eckert, Hancock, & Holderegger, 2015). Mba et al. (2012) concluded that a re-invigorated plant breeding approach is necessary to translate climate-smart heritable variations into crop varieties to achieve increased yields and enhanced adaptation to abiotic and biotic stresses due to climate change.

Fereres, Orgaz, and Gonzalez-Dugo (2011) identified the recommended strategy towards developing smart crop varieties as combining research on plant breeding, crop physiology, and agronomy, and exploiting their interactions. Furthermore, they advocated more balanced funding of agricultural research disciplines to achieve success.

Conservation of Biological Diversity: *In Situ* and *Ex Situ*, Protection of Landraces and Farmers' Selections

One of the strategies towards achieving climate-smart crop varieties is the widening of the genetic base or sources of heritable variations in crops (FAO, 2009c; Mba et al. 2012). This is in keeping with the Global Crop Diversity Trust's efforts to promote the collection, conservation and use of wild relatives of crops in pre-breeding and breeding (Dempewolf et al., 2014; Guarino & Lobell, 2011). The value of national and national genebanks for *ex situ* conservation and conservation plots in farmers' fields as well as of wild types in their natural habitats (*in situ* conservation) cannot be over-emphasised (FAO, 2010b).

The advantage of using the tepary bean (a wild relative of the common bean) in developing heat-tolerant common bean lines at CIAT has already been discussed (Beebe et al., 2011). Gur and Zamir (2004) reported on the success achieved using a wild relative of tomato, *Solanum pennelli*, to introduce genes for drought tolerance into the commercial *Solanum lycopersicum* tomato variety with an attendant yield increase of 50%. At IITA, Nigeria, several genes of wild relatives of cassava were used to introduce genes to enhance cassava mosaic disease resistance, as well as to improve nutritional quality and extend shelf life of a commercial variety (Mba et al., 2012).

An Example of an *Ex-Situ* Collection Used to Cater for the Impact of Climate Change: The *Seeds for Needs* Project in Ethiopia

The Seeds for Needs project in Ethiopia uses *ex-situ*.collections to help farmers, mainly women, cope with impact of climate change on crop production (Bioversity International, 2013). Accessions, maintained at the Ethiopian National Genebank's Institute of Biodiversity and Conservation, were catalogued in terms of climate profiles. Accessions most likely to be adapted to future growing conditions in areas with similar environmental profiles were identified using the collection, characterisation and evaluation data from germplasm collections and Geographic Information System (GIS) technology. Varieties were then tested by selected farmers under their respective growing conditions, and based on their indigenous knowledge and adaptation strategies. After the testing process, the best performing and most adaptable varieties were distributed to farming communities for multiplication and observation.

Utilisation of Crop Genetic Diversity in Selecting Climate Adaptation Traits

Conservation, characterisation, evaluation and utilisation of diverse crop genetic resources should facilitate the identification (selection) and breeding of climate-smart crop varieties (Heisey & Rubenstein, 2015; McCouch et al., 2013). Plants under abiotic stress are affected physiologically and biochemically and undergo changes in gene expression and regulation, in the short and long term, in response to unfavourable environmental conditions (Saibo, Lourenço, & Oliveira, 2009). Adaptation to water deficits (drought stress) can thus be mediated through morphological, physiological and biochemical mechanisms in crops (Beebe et al., 2011) and requires a diverse genetic base for the identification of favourable genotypes with adaptive traits for drought tolerance. The latter includes deeper root systems, stomatal control and improved translocation of nutrients within the plant. Beebe et al., (2011) advocated research to *fit the right root system to specific common bean production environments* under drought conditions, and reliable screening methods to evaluate waterlogging tolerance of genotypes under flooded conditions.

With regard to heat stress, acclimation to high temperatures in certain genotypes may favour selection of heat tolerant varieties. A method of screening pollen to determine heat tolerance in soybean was developed by Salem, Kakani, Koti, and Reddy (2007) and may be useful for screening of the common bean (Beebe et al., 2011).

Detailed data on the responses to climate change (changing temperature and rainfall patterns) of the pests and diseases affecting crops in developing countries are necessary. An understanding of these phenomena and the identification of adapted or tolerant genotypes will facilitate the implementation of adaptation strategies to deal with the biotic stress due to pests and diseases.

Marker-Assisted Selection and Genomic Breeding

Molecular markers are currently the most powerful tools for elucidating the inheritance of target regions of plant and animal genomes in breeding material, and environment-neutral markers allow marker-assisted selection (MAS) and genomic breeding (Edwards & Batley, 2010). Marker assisted selection could be a useful tool to identify (abiotic) stress tolerant traits such as specific rooting depth in common bean, once quantitative trait loci (QTL) for this quantitative trait have been located. It leverages the burgeoning available molecular data on completely or nearly completely sequenced plant genomes to select for candidate genes, protein sequences, and favourable genotypes, while taking into account genotype by environment interaction and targeting adaptive/morpho-physiological traits (Jha, Bohra, & Singh, 2014; Lasky et al., 2015). The advances in bioinformatics and computational molecular biology facilitate valid inferences to be made from the large and complex volumes of data involved in these processes. The usefulness of MAS is particularly outstanding in breeding programmes that involves polygenic traits. MAS allows quick progress in accumulating two or more genes (pyramiding) for desirable traits in segregating breeding material compared to conventional breeding methods. The latter is constrained by the length of the breeding cycle and the dependence on screening, based on phenotypes, which often has to be conducted on mature plants when reproductive traits are involved. The time required to breed a new crop variety may be decreased by half when MAS is employed (Mba et al., 2012). The negative side is the cost involved (Ribaut, Vicente, & Delannay, 2010). However, high-throughput molecular biology assays are becoming increasingly available and costs are decreasing. Furthermore, human capacity building in using these techniques and putting the results at the disposal of farmers in developing countries is in progress.

High-Throughput Phenotypic Evaluations

High throughput imaging of living plant organs can also be valuable for unravelling complex traits affected by climate change such as drought and salinity tolerances (Ghanem, Marrou, & Sinclair, 2015). Access to the technology required for thermal infra-red and near-infrared fluorescence spectrometry and magnetic resonance imaging allow non-destructive physiological, morphological and biochemical assays that can be used to investigate these complex traits (Fahlgren et al., 2015). However, the high establishment costs and technical expertise required to conduct high-throughput phenotyping may also restrict access to it, in the short-term, by developing countries.

CASE STUDY

Harnessing Research and Technology to Combat Climate Change in Cacao (Theobroma cacao L.)

Research and application of research findings and technology to combat the threat of climate change in cacao[4] have been ongoing (Stigter, 2008). The impact of soil, water and climate (drought, flooding, high temperature and irradiance) pose a threat to sustainable cocoa production (Oyekale, Bolaji, & Olowa, 2009). There is a need to increase efficiency to manipulate water use, soil and microclimate and to use improved (climate-smart/well adapted) stress tolerant planting material. Technological approaches such as genome-wide association studies and breeding with genomics and other technological advances may be applied in developing superior cacao varieties to combat the impact of climate change. The completed mapping of the cacao genome will facilitate the identification of marker-trait associations in cacao germplasm that can be used to breed varieties with biotic and abiotic stress tolerance.

The selection of well-adapted varieties with drought, heat (high irradiance) stress and other forms of abiotic stress tolerance is a major goal of cocoa breeding in West Africa (Läderach, Martinez-Valle, Schroth, & Castro, 2013). Studies on the effects of soil water deficit (Antwi, 1994) and irradiance (Galyuon, McDavid, Lopez & Spence, 1996a, b) have been conducted in Trinidad & Tobago, and an investigation on climate change adaptation to drought and heat in cacao is currently in progress.

In Brazil, dos Santos et al., (2014) found that the best traits to screen for drought tolerance were leaf and total dry biomass, relative growth rate and magnesium content of leaves. These authors also found increased expression of drought tolerance candidate genes (as those associated with ABA biosynthesis and biosynthesis of proteins of PSII of the photosynthetic pathway) in genotypes classified as non-tolerant after exposure to soil water deficit for a period of time. dos Santos et al. (2014) also observed that in tolerant genotypes, there were significant increases in guaiacol peroxidase activity, reflecting a more efficient antioxidant metabolism. This increased activity of oxidative stress enzymes is a response to soil water deficit.

In Ghana, where drought is the major abiotic stress affecting cacao (Anim-Kwapong & Frimpong, 2005), Ofori, Padi, Acheampong, and Lowor (2015) examined genetic variation and the relationship of traits associated with drought tolerance in cacao under shade versus non-shade conditions. These authors reported that stem growth, percentage survival, leaf chlorophyll content and plant vigour were important traits to screen for drought tolerance under these conditions.

In order to mitigate the potential adverse effects of climate change in cacao, Stigter (2008) stressed the role of good managerial practices (proper shade management, application of organic fertiliser and bio-control of pests), inter-planting with agro-forestry species and extension to educate farmers on good agricultural practices (GAPs), in Climate Field School classes (CFSs), to cope with the impact of climate change. In addition, biodiversity friendly cacao cultivation can be practised such that a favourable balance between yield, farmer income and biodiversity conservation is achieved, as prescribed by Waldron, Justicia, Smith, and Sanchez (2015). Farmer income can be supplemented by sustainability certification (eco-labelling) and incentives awarded to small farmers for managing eco-friendly cacao agroforestry systems based on the United Nations REDD (reducing emissions from deforestation and forest degradation) initiative (Millard, 2011).

SOCIAL REFORMS FOR OPTIMISED AND SUSTAINABLE AGRICULTURAL PRODUCTION AND DEVELOPMENT

Scherr, Shames, and Friedman (2012) identified several key social factors required in strategies for securing food production and achieve food security. These include:

- Strengthened technical capacities;
- Institutional and political support for multi-stakeholder planning;
- Good governance;
- Spatial targeting of investments and
- Multi-objective impact monitoring.

Sasson (2012) identified several social constraints to be addressed in order to increase agricultural production and achieve food security. These include lack of easy access to land, lucrative employment opportunities in the rural environment, lack of proper access to local markets (with proper road infrastructure and loan arrangements) and inadequate access to training and technologies that will increase production, productivity, food safety, improve food handling and storage, post-harvest activities (including packaging), distribution and marketing and promote sustainable value-addition initiatives. Access to training in best practices and technology (including agricultural diagnostic applications on smart phones) is particularly critical to implement climate change adaptation techniques to address drought, extreme temperatures and other abiotic stresses. Research and development along with dissemination of the resulting information to agriculturists are thus crucial (Ejeta, 2009). Gender inequality was also regarded as a constraint to achieving food security (Flachsbarth et al., 2015; Perez et al., 2015).

Thus a summary of social reforms that can impact positively on agricultural development and food security includes:

- Poverty alleviation;
- Creating a shorter, more efficient supply and value chain for farmers
- Improving farmer livelihoods;
- Diversification of incomes;
- Adequate access to credit;
- Strengthening of women's position;
- Greater access to social amenities;
- Greater access to an adequate and reliable labour force;
- Improvement in production standards with regards to child labour;

- Attracting youth to agriculture;
- Stability of land ownership rights and land tenure;
- Improving infrastructure (particularly of access roads and bridges);
- Access to equipment and technology for farm/production efficiency;
- Improvement of access to schooling and education;
    - Access to extension services and training in GAPs and good manufacturing practices (GMPs) – through participatory learning;
    - Improving the business acumen of farmers – through farmer field schools (FFS);
- Adoption of food safety and quality assurance practices;
- Avoiding trade-offs between quality of produce and short term revenues;
- Improving access to market information and support services;
- Access to efficient transport/wholesaling markets, refrigeration and post-harvest practices to avoid food spoilage and wastage;
- Creation and strengthening farmer's organisations such as co-operatives to facilitate optimisation of all activities along the food value chain and facilitate sharing of costs among food producers and promoting value addition; and
- Addressing the disconnect between food producers and consumers.

Promoting food security is deemed feasible through the spread of modern farming, crop research and food processing in developing nations (Sasson, 2012). Farming communities must be educated in modern farming through knowledge and technology transfer and equipped to practise such methods.

Investment and resources must be supplied to rural farming communities, particularly of small farmers, and policies implemented to bring about the requisite social and environmental changes and human capacity building to combat climate change. Policy, legal and market environments that encourage innovation and capital investment must be fostered. The role of the International Convention for the Protection of New Varieties (UPOV) is also critical so that there is an equitable balance between Intellectual Property Rights (IPR) and contributions to public good (Mba et al., 2012).

THE ROLE OF EDUCATION AND INSTITUTIONS

Training and education are key catalysts to bring about environmental and social reform to promote agricultural development and food security, particularly in SIDS (Bekele & Ganpat, 2014). The areas of particular concern are:

- Participatory approaches to social learning to persuade farmers to adopt new and improved agricultural approaches such as the climate-adapted push-pull system (Isaac et al., 2007; Midega et al., 2015) and integrated pest management;
- Technology transfer to food producers along the value chain;
- Sensitization of rural and urban communities to encourage proper nutrition, avoidance of food wastage (according to the FAO, one third of the food produced globally is discarded or spoilt) and adoption of healthier and more energy efficient eating habits;

    - including the acceptance of a diet with protein from vegan and non-red meat sources to reduce the consumption of red meat that contributes to carbon emissions (14% of the global total) and utilisation of large volumes of water; and
- Adoption of modified and healthy diets based on climate-smart agriculture.

Education to change public awareness and practices that affect agricultural production and food security is crucial. According to Bailey, Froggatt and Wellesley (2014), governments and environmental groups are reluctant to pursue policies or programmes to shift consumer behaviour.

Much research has and is being conducted in developing nations to facilitate more sustainable food consumption patterns. Boland et al. (2013) and Lukefahr (2005) identified rabbits as a suitable alternative for animal protein in developing countries, but the populace has to be convinced to utilise this and other alternative animal protein sources such as the collared peccary/quenk and agouti (Mollineau, Garcia, Samayah, Kissoonsingh, & Procope-Garcia, 2000) in future animal protein production and consumption systems. Alternatives for imported staples such as wheat are available in the developing world and consumer appreciation for tropical food crops such as breadfruit (*Artocarpus altilis*, Moraceae) and chataigne (*A. camansi*) is already fairly high (in excess of 61% in Trinidad and Tobago) (Roberts-Nkrumah & Legall, 2013). In other cases, consumer acceptance will have to be encouraged, taboos and fear of the new (neophobia, as explained by Cox and Evans, 2008) will have to be displaced, and ethical issues resolved. A concerted education drive is thus required for potential producers and consumers. Furthermore, accountability and responsibility are necessary to preclude public distrust (Boland et al., 2013).

Social systems could be valuable in addressing food waste. A well organised system for collection of food waste with acceptable nutritional value that is safe for animal consumption can serve the dual role of reducing food waste while providing a source of animal feed. There should be a review of legislation that currently forbids the use of human food waste in animal production (Boland et al., 2013). Education of personnel to sort and manage food waste and to monitor its safety for animal consumption is thus necessary. There will also be safety issues for humans operating the system and for consumers of the resulting animal products that will have to be addressed by research and knowledge transfer.

A key role of education and institutional support is to link science and policy on food security under climate change. This requires multi-stakeholder, participatory scenarios to bring the diverse actors and organizations together to create the necessary policy-making tools, as described by Chaudhury et al., (2013). The knowledge made available by scientists can thus be transferred to policy makers to facilitate implementation of necessary changes and programmes to address food insecurity and the negative impacts of climate change. Institutions such as regional universities and international organisations based in the developing world can undertake the credible, salient and legitimate boundary work required in this process. They can provide fora for assembling farmer representatives, scientists, policymakers, civil society, government and private sector so that multiple perspectives may be aired, scenarios created and policies drafted to implement approved practices and technology. In East Africa, the CGIAR programme on Climate Change, Agriculture and Food Security (CCAFS) has adopted this approach to promote climate-smart agriculture and food security, bridging science and non-science boundaries, and have forged the link between knowledge and action (Chaudhury et al., 2013). At multi-stakeholder workshops, they were able to determine common outcomes of interest for food security, *viz.*, food affordability, regional production, food distribution and nutritional value for food security, water and soil quality, forest cover, biodiversity status, water sufficiency for environment, and financial wealth, social capital, health, and knowledge and skills for livelihood.

CASE STUDY: SOCIAL ISSUES AFFECTING COCOA PRODUCTION

The cocoa industry, which supports the projected (by 2016) 98.3 billion USD chocolate and cocoa-based industry (World Cocoa Foundation, 2015), has been beleaguered by social as well as environmental challenges. Despite this, the International Cocoa Organisation (ICCO) recognised the importance of the cocoa sector for economic development and poverty eradication, particularly in West Africa, which produces 70% of the world's cocoa. The ICCO affirmed that to secure this important role, stakeholders in the cocoa chain are jointly facing the challenge to promote sustainable development in cocoa production, commercialization, processing, manufacturing and consumption (Ebai, 2004). To rise to these challenges, ICCO launched a Roundtable on a Sustainable Cocoa Economy in 2007 in Accra, Ghana, and a Second Roundtable was convened in Trinidad in 2009. The social issues of serious concern in the cocoa sector pertain to labour, farmer well-being and livelihoods and training in best practices. There is justification in concluding that the monetary benefits of the thriving global cocoa and chocolate industry have not trickled down to the farming communities adequately.

Child Labour: A Serious Social Issue

Child labour in cocoa production became the focus of international attention and disdain in 2002 (Mustapha, 2009). The International Labour Office (ILO) and the International Program on the Elimination of Child Labour (IPEC) presented statistics that indicated that 284,000 children were working under hazardous conditions in West Africa. Since then, the industry has made concerted efforts to address the situation and stop all forms of exploitative or the worst forms of child labour in cocoa production (Grossman-Greene & Bayer, 2009; Schrage & Ewing, 2005). Several initiatives have resulted and groups formed to achieve the goals of the Harkin-Engel law adopted in 2001 to address the worst forms of child labour (Berlan, 2009) as well as assure sustainability of cocoa production. These include the Sustainable Tree Crop programme (STCP) in West Africa and South-east Asia (Abbott, 2003), CocoaAction and Transforming Education in Cocoa Communities (TRECC), which was launched in May, 2015.

CocoaAction was launched in May 2014 to deliver a comprehensive package of services in both productivity and community development to cocoa farmers, their families and their communities. TRECC will reach nearly 200,000 individuals through a series of interlinked interventions in research, capacity building, policy formulation and influence, fundraising, and supporting matching grants to complement CocoaAction. It is an initiative of the Swiss Charitable Foundation, The Jacobs Foundation that will be executed in Côte d'Ivoire. A key element will be education to restrict child labour, which is currently still a serious problem in Côte d'Ivoire.

Farmer Education to Promote Sustainability in Cocoa Production

Farmer education is included in the aforementioned *CocoaAction*, *TRECC* and *STCP* programmes. In addition, the *Learn To Grow* programme was launched by Hershey and will include more than 60,000 West African cocoa farmers. It is intended to create both better farming practices and better cocoa communities. Hershey has a goal to purchase 100% certified cocoa by 2020. This is envisaged to provide premiums to assist farmers and farmer organizations in adopting modern farming practices, including organic farming, and to develop their communities.

In South-east Asia, (Sulawesi, Indonesia and Papua New Guinea), the Australian Centre for International Agricultural Research (ACIAR) conducted a project, SMAR/2005/074, entitled *Improving cocoa production through farmer involvement in demonstration trials of potentially superior pest/disease resistant genotypes and IPM*). The objectives were to select and test local cacao clones at different locations for disease and pest resistance (Vascular Streak dieback and Cocoa Pod Borer), evaluate IPM options and assess technology uptake by farmers (Neilson et al., 2014). The importance of addressing social and economic constraints in cocoa production, and in educating farmers on how to utilise the available information, based on empirical research as well as technology and improved methods, such as IPM, was emphasised.

Fairtrade in Cocoa

Approximately eighty-five percent of the world's chocolate comes from farms of 5 hectares or less, mainly in the developing countries of West Africa, South-east Asia and Latin America. Cocoa farming families earned as little as US$30 - $110 per family member per year, according to Knapp (2005). Fairtrade chocolate, made from Fair Trade certified cocoa, offers a solution for small farmers and their dependents to break the cycle of poverty. By cutting out middlemen and brokers, Fair Trade allows cocoa farmers to receive a fair share of the final market value of their products (Liu, Byers, & Giovannucci, 2008; Pay, 2009).

Fairtrade certification creates direct trade links between farmer-owned cooperatives and buyers, and it provides access to affordable credit. Fairtrade gives farmers the tools to access the market and to farm in a sustainable manner. In addition, under the terms of Fairtrade, there are strict labour standards that foster healthy working conditions and allow children under the age of 15 to work on their family's farm only if their education is not jeopardized. Children under the age of 18 are not allowed to work with machetes (or other dangerous tools) or to apply pesticides.

A Multi-Stakeholder, Cohesive Approach to Ensure Sustainable Cocoa Production

A structured, multi-sector, international approach to conserve cacao germplasm as a safeguard against the impact of climate change and other threats and to meet emerging needs of cocoa farmers has already been adopted. The United Nations Common Fund for Commodities/International Cocoa Organisation/ Bioversity International (CFC/ICCO/BI) executed a project entitled *Cocoa Germplasm Utilization and Conservation: A Global Approach* (1998 – 2004) (Eskes, Engels, & Lass, 1998; Eskes & Efron, 2006). The main objective for this project at the Cocoa Research Unit (now Cocoa Research Centre, CRC), a partner in the project, was germplasm enhancement for Black Pod (BP) and Witches' Broom (WB) disease resistance in cacao. Several primary germplasm varieties were identified, and enhanced genotypes (through pre-breeding (Guarino & Lobell, 2011)) produced at CRC with combined resistance to both WB and BP diseases. The yield potential, bean traits, precocity, vigour and flavour profiles of these individuals were or are still being assessed at CRC (Iwaro et al., 2010). The most promising enhanced cacao varieties were transferred to the International Cocoa Quarantine Centre, Reading University, U.K. to facilitate distribution to the other project partners in cocoa producing countries. Selected progeny are also being used in the Ministry of Agriculture, Trinidad and Tobago's successful Trinidad Selected Hybrid cacao breeding programme (Maharaj et al., 2011) to accumulate resistance genes for BP. This will facilitate the development of further improved, superior planting material to distribute to cocoa farmers in Trinidad and Tobago.

Launching of *Farmer Field Schools* (FFS) and *Farmer Participatory approaches* to improving cocoa production have also received considerable attention within the last decade. The CFC/ICCO/BI project entitled *Cocoa productivity and quality improvement: a participatory approach* was conducted from 2004 to 2009. This project was successful in educating farmers and assisting them in recognising and selecting the best trees within their fields for multiplication and conservation. Techniques to screen for disease and pest resistance and to assess traits of economic interest as well as quality and flavour of cocoa were also put at the disposal of rural cocoa farming communities (Eskes, 2011).

These two, aforementioned CFC/ICCO/BI projects involved 14 cocoa-producing countries, including Trinidad and Tobago, as well as manufacturers (consumers), government bodies and institutions. They are models for a successful, cohesive approach towards achieving agricultural development and food security.

In West Africa, South-east Asia, Latin America and the Caribbean, FFS are now routinely hosted. In Trinidad & Tobago, a project, funded by the Centre for the Development of Enterprise and implemented by the Cocoa Research Centre (CRC), resulted in the training of farmers and entrepreneurs in GAPs and GMPs practices (including for assuring food safety) in six Caribbean countries.

A COHESIVE APPROACH TO ACHIEVE AGRICULTURAL DEVELOPMENT AND FOOD SECURITY

Any strategy to achieve social reform and to foster agricultural development and food security in developing nations requires:

1. A cohesive approach to achieve agricultural development and food security in developing nations;
2. Proper fiscal/economic planning;
3. Social programmes and networking – encompassing collaborative, iterative social learning;
4. A proper environmental framework and
5. Public, private and community based approach.

A comprehensive, multi-sector approach towards addressing the challenges and potential impact of climate change on food security is advocated (Chaudhury et al., 2013; Maggio et al., 2015; Mba et al., 2012). Sonnino et al., (2014) urged that a sustainable food security framework should address the complex relationships between the various stages and actors in the food supply chain (refer to Figure 2) in a holistic manner. The conventional focus on individual components (supply and demand) will not ensure sustainable food security. Participatory, multi-disciplinary and demand-driven breeding programmes, which are nurtured by appropriate policy and legal environments and utilise scientific and technological tools to exploit crop and animal genetic resources, are recommended (Mba et al., 2012). Governance and policy structures must be put in place to support food security and sustainability efforts (Sonnino et al., 2014). Without the social support of extensionists, financiers, administrators and policy makers, the positive outcomes of participatory food production initiatives will not become transformative across the agricultural sector, but be confined to isolated sections of the rural and research communities.

Participatory breeding considers the views of farmers, consumers, extensionists, vendors, rural co-operatives and industry in crop improvement. The needs of the growers and stakeholders thus inform the design of the breeding activities, rendering them relevant and efficient. Furthermore, the role and achievements of farmers in selecting and conserving crop landraces can be harnessed in the process of selecting favourable candidates to use as parents in breeding. In addition, the adoption of improved varieties by farmers is facilitated and accelerated when farmers are part of the process.

The benefits of such an integrated approach have been supported by research findings based on analyses of various scenarios (Vervoort et al., 2014). The latter authors advocated more effective and efficient transformation of scientific results into a policy agenda for adaptive action on food security under climate change. Such an approach in SIDS has also been recommended and outlined by Maximay (2014). Integrated adaptation and mitigation strategies, under the umbrella of integrated landscape level planning, were recommended by Harvey et al., (2014) in order to avoid trade-offs that limit success.

There is thus compelling evidence that to achieve food security under climate change in developing countries, an organised programme to breed climate-smart varieties and practise climate-smart agriculture can only be achieved through collaboration. This should also involve international organisations such as the Global Crop Diversity trust, FAO, CGIAR and the Integrated Breeding Platform, the World Bank and the World Trade Organisation. Sonnino et al., (2014) presented a detailed graphic that depicts the key issues to be addressed for a future research agenda that can assure sustainable food security (refer to Figure 3). The broad scope of this agenda demands a multi-sector approach. McCouch et al. (2013) have costed the systematic, concerted, collaborative global effort to feed the future. They estimated it as US $200 million annually.

CONCLUSION

It has been projected by Flachsbarth et al. (2015) that the progress of achieving food security will vary from one developing region to the next. Their outlook for LAC and Southeast Asia and the Pacific is better than for Central-West Africa and other regions of Sub-Saharan Africa. Lobell et al. (2008) also predicted that South Asia and Southern Africa will most likely face negative climatic impacts by 2030 on several important crops if adequate adaptation measures are not adopted. This scenario may be attributed to the fast population increase relative to food production growth, particularly in Africa, as well as continued lack of access to safe water along with limited improvement in female secondary education. Policies are required to remove gender bias that *limits resilience of female farmers* (Perez et al., 2015).

The environmental and social priorities for increasing agricultural development and achieving food security in developing countries, which were identified in this chapter, are included in the CGIAR's strategy (Rijsberman, 2015) (refer to Table 2).

Thus, there are numerous factors that need to be urgently addressed in order to assure agricultural development and food security in developing nations. These include:

Social Factors

- Reform of social conditions for rural communities – improvement of basic amenities – health, education, housing and infrastructure;
- Technological knowledge transfer and farmer training and education in GAPs for primary production and GMPs for value chain initiatives along with training to ensure better health and fiscal management as well as socialisation;

Figure 3. Sustainable food security pillars- key issues for future research
(adapted from Sonnino et al., 2014)

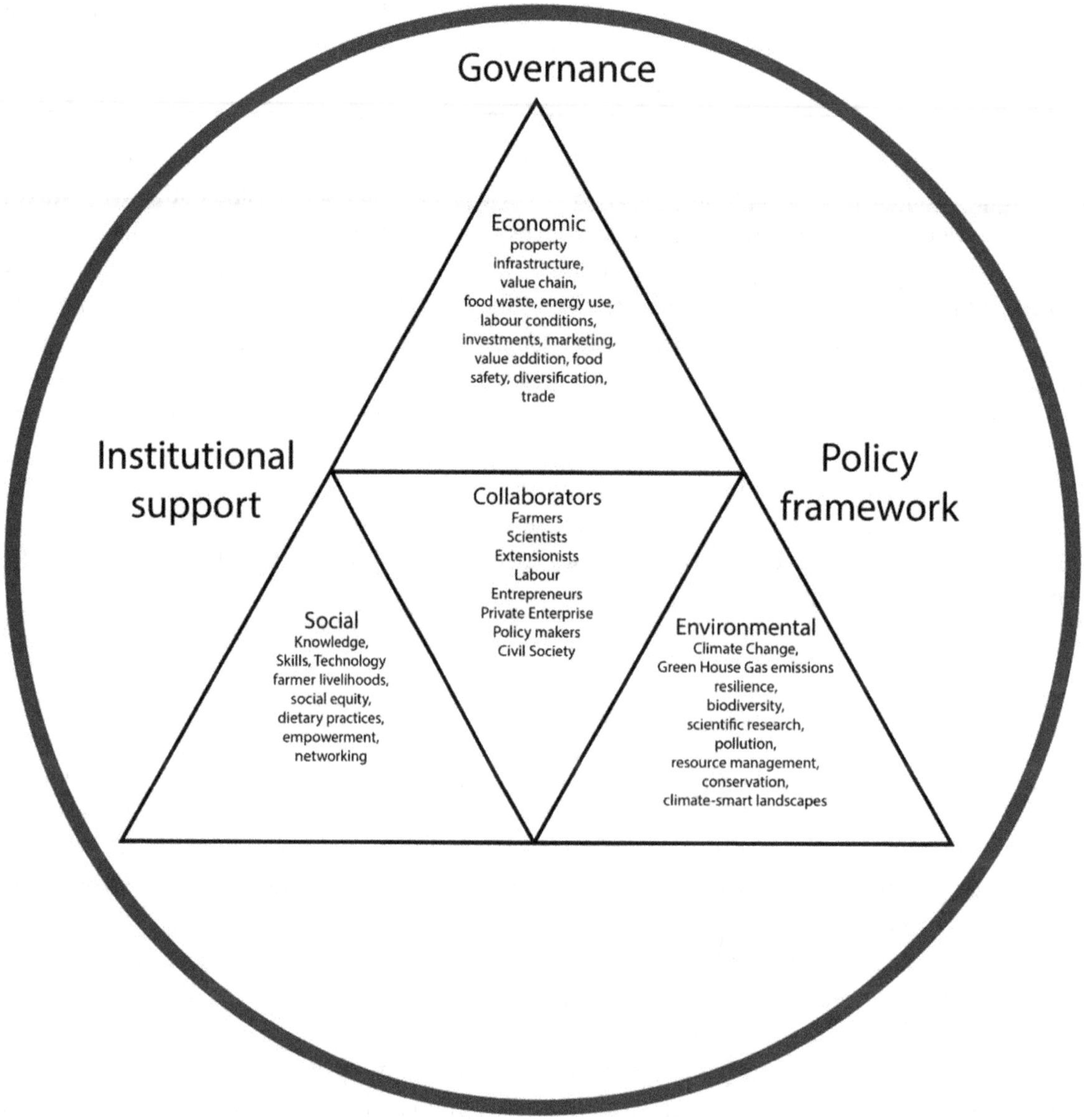

- Support of rural communities in terms of financial aid and farmer livelihood enhancement programmes to optimise income generation, eradicate poverty and improve subsistence farming (as for the establishment of co-operatives and farming enterprises that shorten the value chain);
- Provision of technological tools to modernise food production;
- Public co-ordinated programmes to address inadequate access to credit and capital and insecurity of land tenure and
- Application of science and technology in education to re-engineer the farming culture and inculcate innovative approaches to food production and value-addition enterprises that will attract the youth.

Table 2. Consultative Group for International Agricultural Research (CGIAR's) strategy for 2016-2030

Genetic improvement of crops, livestock, fish and trees, to increase productivity, resilience to stress, nutritional value and efficiency of resource use.
Agricultural systems, adopt a systems approach to optimize economic, social and environmental co-benefits in areas with high concentrations of poor people.
Gender and inclusive growth, creating opportunities for women, young people and marginalized groups.
Enabling policies and institutions, to improve the performance of markets, enhance delivery of critical public goods and services, and increase the agency and resilience of poor people.
Natural resources and ecosystem services, focusing on productive ecosystems and landscapes that offer significant opportunities to reverse environmental degradation and enhance productivity.
Nutrition and health, emphasizing dietary diversity, nutritional content and safety of foods, and development of value chains of particular importance for the nutrition of poor consumers.
Climate-smart agriculture, focusing on urgently needed adaptation and mitigation options for farmers and other resource users.
Nurturing diversity, ensuring that CGIAR in-trust plant genetic resources collections are safely maintained, genetically and phenotypically characterized to maximize the exploitation of these critical resources for food security, productivity, nutrient rich crops and resilient farming systems.

(CGIAR, 2015)

Environmental Factors

- Climate-smart agriculture and landscapes – identifying climate adapted or resilient planting material and adopting energy efficient, sustainable agricultural production practices;
- The application of technology in climate adaptation strategies (next-generation sequencing, genomic selection, marker-assisted selection and breeding with genomics to develop climate-smart varieties); and
- Harnessing conservation and utilisation of crop genetic resources in the face of climate change and other emerging threats such as new pests and diseases.

Meeting the challenges to secure agricultural development and food production in developing nations is indeed daunting. The complex social, environmental as well as economic issues that contribute to the dilemma cannot be addressed without a collaborative approach. Policies must be developed to engender trust and co-operation among the key stakeholders. Furthermore, an agriculture-energy-water nexus approach (Rasul and Sharma, 2015) must be adopted since the impacts of climate change dictate that agriculture can only develop and be sustainable when plant and animal genetic resources and materials, water, energy and land resources are not limiting (Hubert et al., 2010).

Fiscal and rural planning is a key parameter for success in achieving agricultural development and food security in developing nations. The strategies to be adopted require adequate financing along with research and development to benefit concerted genetic conservation efforts, human capacity building, technological applications in plant breeding and selection of adapted and superior plant varieties and animals, ongoing applied research such as genomics and a sophisticated network of food quality, safety, handling, processing, storage, marketing and transportation facilities.

The successful implementation of the aforementioned strategies should guarantee a stable supply of nutritious food, readily available to growing populations in developing countries into the future. Achieving food security cannot be accomplished without ending poverty in developing nations. Farming communities must be educated, well-equipped, able to access and utilise technology and superior planting material, livestock or fisheries, and practise sustainably intensive (based on natural agricultural processes and biodiversity), modern agriculture on a lucrative scale with the benefit of secure land tenure, co-operative structures, extension services and appropriate financial support.

REFERENCES

Abbott, P. (2003). *Towards more socially responsible cocoa trade.* International Agricultural Trade Research Consortium, working paper 03-3. Retrieved June 20, 2015, from http://nbcgib.uesc.br/cicacau/arquivos/producao_tecnico_cientifica/wp03-03.pdf

Anami, S. E., Zhang, L. M., Xia, Y., Zhang, Y. M., Liu, Z. Q., & Jing, H. C. (2015). Sweet sorghum ideotypes: Genetic improvement of stress tolerance. *Food and Energy Security, 4*(1), 3–24. doi:10.1002/fes3.54

Andersen, L., Breisinger, C., Mason-D'Croz, D., Jemio, L. C., Ringler, C., Robertson, R. E., . . . Wiebelt, M. (2014). *Agriculture, incomes, and gender in Latin America by 2050: An assessment of climate change impacts and household resilience for Brazil, Mexico, and Peru.* IFPRI Discussion Paper 1390. Retrieved June 15, 2015, from http://ssrn.com/abstract=2539588

Anim-Kwapong, G. J., & Frimpong, E. B. (2005). *Vulnerability of agriculture to climate change-impact of climate change on cocoa production.* Final Report submitted to the Netherlands climate change studies assistance Programme: NCAP, 2. Retrieved June 12, 2015, from http://www.nlcap.net/fileadmin/NCAP/Countries/Ghana/COCOA_DRAFT_FINAL_REPORT.pdf

Antwi, A. (1994). *The effects of water deficit on growth and development in young cocoa plants (Theobroma cacao L).* (Unpublished MPhil thesis). University of the West Indies, St. Augustine, Trinidad and Tobago.

Bailey, R., Froggatt, A., & Wellesley, L. (2014). Livestock – Climate change's forgotten sector. *Global Public Opinion on Meat and Dairy Consumption.* Retrieved June 11, 2015, from http://www.gastronomiaycia.com/wp-content/uploads/2014/12/estudio_consumo_carne.pdf

Bansal, K. C., Lenka, S. K., & Mondal, T. K. (2014). Genomic resources for breeding crops with enhanced abiotic stress tolerance. *Plant Breeding, 133*(1), 1–11. doi:10.1111/pbr.12117

Beddington, J., Asaduzzaman, M., Fernandez, A., Clark, M., Guillou, M., Jahn, M., . . . Wakhungu, J. (2012). *Achieving food security in the face of climate change: Final report from the Commission on Sustainable Agriculture and Climate Change.* Copenhagen, Denmark: CGIAR Research Program on Climate Change, Agriculture and Food Security (CCAFS). Retrieved June 2, 2015, from http://www.ccafs.cgiar/commission

Beebe, S., Ramirez, J., Jarvis, A., Rao, I. M., Mosquera, G., Bueno, J. M., & Blair, M. (2011). Genetic improvement of common beans and the challenges of climate change. In S. S. Yadav, R. J. Redden, J. L. Hatfield, H. Lotze-Campen, & A. E. Hall (Eds.), *Crop adaptation to climate change* (pp. 356–369). Richmond, Australia: John Wiley & Sons, Ltd., Published by Blackwell Publishing Ltd. doi:10.1002/9780470960929.ch25

Bekele, I., & Ganpat, W. (2014). Education, extension, and training for climate change. In W. Ganpat & W.P. Isaac (Eds.), Impacts of Climate Change on Food Security in Small Island Developing States (pp. 361-388). Hershey, PA: IGI Global.

Berlan, A. (2009). Child labour and cocoa: Whose voices prevail? *The International Journal of Sociology and Social Policy*, *29*(3/4), 141–151. doi:10.1108/01443330910947516

Bioversity International. (2013). *Seeds for needs project.* Retrieved June 7, 2015, from http://www.bioversityinternational.org/research/sustainable_agriculture/seeds_for_needs.html)

Boland, M. J., Rae, A. N., Vereijken, J. M., Meuwissen, M. P., Fischer, A. R., van Boekel, M. A., & Hendriks, W. H. et al. (2013). The future supply of animal-derived protein for human consumption. *Trends in Food Science & Technology*, *29*(1), 62–73. doi:10.1016/j.tifs.2012.07.002

Branca, G., McCarthy, N., Lipper, L., & Jolejole, M. C. (2011). *Climate-smart agriculture: a synthesis of empirical evidence of food security and mitigation benefits from improved cropland management.* Retrieved June 13, 2015, from http://www.fao.org/climatechange/29764-0aa5796a4fb093b6cfdf05558c6dd20bb.pdf

CGIAR (Consultative Group for International Agricultural Research). (2015). *CGIAR's strategy and results framework 2016-2030.* Retrieved June 21, 2015, from http://www.cgiar.org/consortium-news/cgiars-new-strategy-and-research-programs-answering-to-poverty-health-and-climate-change/#

Chaudhury, M., Vervoort, J., Kristjanson, P., Ericksen, P., & Ainslie, A. (2013). Participatory scenarios as a tool to link science and policy on food security under climate change in East Africa. *Regional Environmental Change*, *13*(2), 389–398. doi:10.1007/s10113-012-0350-1

Cheng, A., Mayes, S., Dalle, G., Demissew, S., & Massawe, F. (2015). Diversifying crops for food and nutrition security – a case of teff. *Biological Reviews of the Cambridge Philosophical Society*, n/a. doi:10.1111/brv.12225 PMID:26456883

Cox, D. N., & Evans, G. (2008). Construction and validation of a psychometric scale to measure consumers' fears of novel food technologies: The food technology neophobia scale. *Food Quality and Preference*, *19*(8), 704–710. doi:10.1016/j.foodqual.2008.04.005

Dawson, I. K., Vinceti, B., Weber, J. C., Neufeldt, H., Russell, J., Lengkeek, A. G., & Jamnadass, R. et al. (2011). Climate change and tree genetic resource management: Maintaining and enhancing the productivity and value of smallholder tropical agroforestry landscapes. A review. *Agroforestry Systems*, *81*(1), 67–78. doi:10.1007/s10457-010-9302-2

Dempewolf, H., Eastwood, R. J., Guarino, L., Khoury, C. K., Müller, J. V., & Toll, J. (2014). Adapting agriculture to climate change: A global initiative to collect, conserve, and use crop wild relatives. *Agroecology and Sustainable Food Systems*, *38*(4), 369–377. doi:10.1080/21683565.2013.870629

dos Santos, I. C., de Almeida, A. A. F., Anhert, D., da Conceição, A. S., Pirovani, C. P., Pires, J. L., & Baligar, V. C. et al. (2014). Molecular, physiological and biochemical responses of *Theobroma cacao* L. genotypes to soil water deficit. *PLoS ONE*, *9*(12), e115746. doi:10.1371/journal.pone.0115746 PMID:25541723

Ebai, S. (2004). Global coordination towards a sustainable cocoa economy: an update. In *Proceedings of the APASTT Seminar – Exhibition entitled Re-vitalisation of the Trinidad & Tobago Cocoa Industry*, (pp. 17-21). St. Augustine, Trinidad: APASTT.

Edwards, D., & Batley, J. (2010). Plant genome sequencing: Applications for crop improvement. *Plant Biotechnology Journal*, *8*(1), 2–9. doi:10.1111/j.1467-7652.2009.00459.x PMID:19906089

Ejeta, G. (2009). Revitalizing agricultural research for global food security. *Food Security*, *1*(4), 391–401. doi:10.1007/s12571-009-0045-8

Eskes, A. B. (2011). *Collaborative and participatory approaches to cocoa variety improvement. Final report of the CFC/ICCO/Bioversity project on cocoa productivity and quality improvement: a participatory approach (2004–2010)*. London, UK: Bioversity International.

Eskes, A. B., Engels, J. M. M., & Lass, R. A. (Eds.). (1998). Working procedures for cocoa germplasm evaluation and selection. In *Proceedings of the CFC/ICCO/IPGRI Project workshop* (*vol. 1*, no. 6). Rome, Italy: IPGRI.

Eskes, B., & Efron, Y. (Eds.). (2006). *Global Approaches to Cocoa Germplasm Utilization and Conservation. Final Report of CFC/ICCO/IPGRI Project on Cocoa Germplasm Utilization and Conservation (1998-2004)*. Rome, Italy: Bioversity International.

Fahlgren, N., Feldman, M., Gehan, M. A., Wilson, M. S., Shyu, C., Bryant, D. W., . . . Baxter, I. (2015). A versatile phenotyping system and analytics platform reveals diverse temporal responses to water availability in Setaria. *Molecular plant*. Retrieved June 2, 2015, from http://www.sciencedirect.com/science/article/pii/S1674205215002683

FAO. (2007). Adaptation to climate change in agriculture, forestry and fisheries: Perspective, framework and priorities. Rome: Interdepartmental Working Group on Climate Change, Food and Agriculture Organization of the United Nations (FAO).

FAO. (2008). *Climate change adaptation and mitigation in the food and agriculture sector. High-level conference on food security – the challenges of climate change and bioenergy*. Retrieved June 17, 2015, from ftp://ftp.fao.org/docrep/fao/meeting/013/ai782e.pdf

FAO. (2009a). *Increasing crop production sustainability, the perspective of biological processes*. Retrieved June 18, 2015, from http://www.fao.org/docrep/012/i1235e/i1235e00.htm

FAO. (2009b). *How to feed the world in 2050*. Rome. Retrieved June 18, 2015, from www.fao.org/fileadmin/templates/wsfs/docs/expert_paper/How_to_Feed_the_World_in_2050.pdf

FAO. (2009c). *Climate change and biodiversity for food and agriculture: taking systemic and second order effects into account*. Rome: Commission on Genetic Resources for Food and Agriculture. Retrieved from June 18, 2015, from ftp://ftp.fao.org/docrep/fao/meeting/017/ak223e.pdf

FAO. (2010a). An international consultation on integrated crop-livestock systems for development, the way forward for sustainable production intensification. *Integrated Crop Management*, *13*, 208.

FAO. (2010b). *Gardens of biodiversity – conservation of genetic resources and their use in traditional food production systems by small farmers of the Southern Caucasus*. Rome: Food and Agriculture Organization of the United Nations.

FAO. (2011). *Save and grow: a policymaker's guide to the sustainable intensification of smallholder crop production*. Retrieved June 17, 2015, from http://www.fao.org/ag/save-and-grow/

FAO. (2012). *What is conservation agriculture?* Retrieved June 18, 2015, from http://www.fao.org/ag/ca/1a.html

FAO. (2013). *Climate-smart agriculture sourcebook*. Rome: Food and Agriculture Organization of the United Nations.

FAO. (2014). *The state of food insecurity in the world*. Rome. Retrieved June 17, 2015, from http://www.fao.org/3/a-i4030e.pdf

FAO (Food and Agriculture Organization). (2004). *Carbon sequestration in dryland soils*. Retrieved August 30, 2015 from http://www.fao.org/fileadmin/templates/nr/images/resources/pdf_documents/wsrr102.pdf

FAO-PAR (Platform for Agrobiodiversity). (2011). *Biodiversity for food and agriculture: contributing to food security and sustainability in a changing world*. Retrieved June 16, 2015, from http://agrobiodiversityplatform.org/files/2011/04/PAR-FAO-book_lr.pdf)

Fereres, E., Orgaz, F., & Gonzalez-Dugo, V. (2011). Reflections on food security under water scarcity. *Journal of Experimental Botany*, *62*(12), 4079–4086. doi:10.1093/jxb/err165 PMID:21624976

Fernandes, J. M., Cunha, G. R., Del Ponte, E., Pavan, W., Pires, J. L., Baethgen, W., ... Travasso, M. I. (2004). Modelling Fusarium head blight in wheat under climate change using linked process-based models. In *Proceedings of the 2nd International Symposium on Fusarium Head Blight* (pp. 441-444). East Lansing, MI: Michigan State University.

Flachsbarth, I., Willaarts, B., Xie, H., Pitois, G., Mueller, N. D., Ringler, C., & Garrido, A. (2015). The role of Latin America's land and water resources for global food security: Environmental Trade-Offs of Future Food Production Pathways. *PLoS ONE*, *10*(1), e0116733. doi:10.1371/journal.pone.0116733 PMID:25617621

Galyuon, I. K. A., McDavid, C. R., Lopez, F. B., & Spence, J. A. (1996a). The effect of irradiance level on cocoa (Theobroma cacao L.): I. Growth and leaf adaptations. *Tropical Agriculture*, *73*(1), 23-28.

Galyuon, I. K. A., McDavid, C. R., Lopez, F. B., & Spence, J. A. (1996b). The effect of irradiance level on cocoa (Theobroma cacao L.): II. Gas exchange and chlorophyll fluorescence. *Tropical Agriculture*, *73*(1), 29-33.

Ganpat, W. G., & Isaac, W. P. (Eds.). (2014). *Impacts of climate change on food security in Small Island Developing States*. Hershey, PA: IGI Global.

Gennaro, A., Forte, P., Carozza, R., Savo Sardaro, M. L., Ferri, D., Bitti, A., & Ceoloni, C. et al. (2007). Pyramiding different alien chromosome segments in durum wheat: Feasibility and breeding potential. *Israel Journal of Plant Sciences*, *55*(3-4), 267–276. doi:10.1560/IJPS.55.3-4.267

Ghanem, M. E., Marrou, H., & Sinclair, T. R. (2015). Physiological phenotyping of plants for crop improvement. *Trends in Plant Science*, *20*(3), 139–144. doi:10.1016/j.tplants.2014.11.006 PMID:25524213

Grossman-Greene, S., & Bayer, C. (2009). *A History of child labor, child rights, & the Harkin-Engel Protocol*. New Orleans, LA: Tulane University.

Guarino, L., & Lobell, D. B. (2011). A walk on the wild side. *Nature Climate Change*, *1*(8), 374–375. doi:10.1038/nclimate1272

Gur, A., & Zamir, D. (2004). Unused natural variation can lift yield barriers in plant breeding. *PLoS Biology*, *2*(10), 1610–1615. doi:10.1371/journal.pbio.0020245 PMID:15328532

Harvey, C. A., Chacon, M., Donatti, C. I., Garen, E., Hannah, L., Andrade, A., & Wollenberg, E. et al. (2014). Climate-Smart Landscapes: Opportunities and Challenges for Integrating Adaptation and Mitigation in Tropical Agriculture. *Conservation Letters*, *7*(2), 77–90. doi:10.1111/conl.12066

Heisey, P. W., & Rubenstein, K. D. (2015). *Using Crop Genetic Resources To Help Agriculture Adapt to Climate Change: Economics and Policy (No. 202351)*. Washington, DC: United States Department of Agriculture, Economic Research Service.

Hubert, B., Rosegrant, M., Van Boekel, M. A., & Ortiz, R. (2010). *The future of food: scenarios for 2050*. Retrieved June 11, 2015, from https://dl.sciencesocieties.org/publications/cs/abstracts/50/Supplement_2/S-33

Hutchinson, S., Gomes, C., Alleyne, D., & Phillips, W. (2013). *An assessment of the economic and social impacts of climate change on the agriculture sector in the Caribbean*. Port of Spain: UNECLAC. Retrieved September 27, 2015, from http://200.9.3.103/portofspain/noticias/documentosdetrabajo/5/49705/Agriculture.pdf

IFPRI (International Food Policy Research Institute). (2009). *Climate change: impact on agriculture and costs of adaptation*. International Food Policy Research Institute. Retrieved July 1, 2015, from http://www.ifpri.org/publication/climate-change-impact-agriculture-and-costsadaptation

IPCC. (2013). *Intergovernmental Panel on Climate Change*. Retrieved June 2, 2015, from http://www.ipcc.ch/

IPCC (Intergovernmental Panel on Climate Change). (2007). Technical Summary. In *Climate change (2007): mitigation. Contribution to Working Group III to the Fourth Assessment Report of IPCC*. Cambridge, UK: Cambridge University Press.

Isaac, W. A. P., Brathwaite, R. A., Cohen, J. E., & Bekele, I. (2007). Effects of alternative weed management strategies on *Commelina diffusa* Burm. infestations in Fairtrade banana (*Musa* spp.) in St. Vincent and the Grenadines. *Crop Protection (Guildford, Surrey)*, *26*(8), 1219–1225. doi:10.1016/j.cropro.2006.10.019

Iwaro, A. D., Bekele, F. L., Butler, D. R., Singh, V., Holder-John, A., Bharath, S., & Bidaisee, G. G. et al. (2010). Recent progress in breeding for specific traits in cocoa to meet challenges to production. In *Proceedings of the International Congress on Tropical Agriculture: Overcoming Challenges to Developing Sustainable Agri-Food Systems in the Tropics* (pp. 43-52). Port of Spain, Trinidad: The University of the West Indies.

Jarvis, A., Ramirez-Villegas, J., Campo, B. V. H., & Navarro-Racines, C. (2012). Is cassava the answer to African climate change adaptation? *Tropical Plant Biology*, *5*(1), 9–29. doi:10.1007/s12042-012-9096-7

Jarvis, A., Upadhyaya, H. D., Gowda, C. L. L., Agrawal, P. K., Fujisaka, S., & Anderson, B. (2008). *Climate change and its effect on conservation and use of plant genetic resources for food and agriculture and associated biodiversity for food security*. Retrieved June 4, 2015, from http://oar.icrisat.org/5810/

Jha, U. C., Bohra, A., & Singh, N. P. (2014). Heat stress in crop plants: Its nature, impacts and integrated breeding strategies to improve heat tolerance. *Plant Breeding*, *133*(6), 679–701. doi:10.1111/pbr.12217

Knapp, S. (2005). *Rich chocolate and poor farmers*. Wellington Aotearoa, New Zealand: Dev-Zone.

Knox, J., Hess, T., Daccache, A., & Wheeler, T. (2012). Climate change impacts on crop productivity in Africa and South Asia. *Environmental Research Letters*, *7*(3), 034032. doi:10.1088/1748-9326/7/3/034032

Läderach, P., Martinez-Valle, A., Schroth, G., & Castro, N. (2013). Predicting the future climatic suitability for cocoa farming of the world's leading producer countries, Ghana and Côte d'Ivoire. *Climatic Change*, *119*(3-4), 841–854. doi:10.1007/s10584-013-0774-8

Lasky, J. R., Upadhyaya, H. D., Ramu, P., Deshpande, S., Hash, C. T., Bonnette, J., . . . Morris, G. P. (2015). Genome-environment associations in sorghum landraces predict adaptive traits. *Science Advances*, *1*(6), e1400218. Retrieved August 5, 2015, from http://advances.sciencemag.org/content/1/6/e1400218.full-text.pdf+html

Leishman, M. R., & Gallagher, R. V. (2015). Will there be a shift to alien-dominated vegetation assemblages under climate change? *Diversity and Distributions*. Retrieved August 12, 2015, from http://onlinelibrary.wiley.com/doi/10.1111/ddi.12338/full

Lin, B. B. (2011). Resilience in agriculture through crop diversification: Adaptive management for environmental change. *Bioscience*, *61*(3), 183–193. doi:10.1525/bio.2011.61.3.4

Liu, P., Byers, A., & Giovannucci, D. (2008). *Value-adding standards in the North American food market-trade opportunities in certified products for developing countries*. Rome, Italy: FAO Trade and Markets Division.

Lobell, D. B., Burke, M. B., Tebaldi, C., Mastrandrea, M. D., Falcon, W. P., & Naylor, R. L. (2008). Prioritizing climate change adaptation needs for food security in 2030. *Science*, *319*(5863), 607–610. doi:10.1126/science.1152339 PMID:18239122

Ludena, C. E. (2010). *Agricultural productivity growth, efficiency change and technical progress in Latin America and the Caribbean. Inter-American Development Bank*. IDB Working paper 61, Series No. IDB-WP-186. Washington, DC: Inter-American Development Bank.

Lukefahr, S. D. (2005). Sustainable and alternative systems of rabbit production. In *Proceedings of the 8th World Rabbit Congress* (pp. 1452-1464). World Rabbit Science Association (WRSA).

Maggio, A., Van Criekinge, T., & Malingreau, J. P. (2015). *Global Food Security 2030. Assessing trends with a view to guiding future EU policies*. Foresight Series. European Commission DG Joint Research Centre. Retrieved on August 18, 2015, from http://publications.jrc.ec.europa.eu/repository/bitstream/JRC94867/lbna27252enn.pdf

Maharaj, K., Maharaj, P., Bekele, F. L., Ramnath, D., Bidaisee, G. G., Bekele, I., & Sankar, R. et al. (2011). Trinidad selected hybrids: An investigation of the phenotypic and agro-economic traits of 20 selected cacao cultivars. *Trop. Agric*, *88*(4), 175–185.

Martinelli, L. A., Naylor, R., Vitousek, P. M., & Moutinho, P. (2010). Agriculture in Brazil: Impacts, costs, and opportunities for a sustainable future. *Current Opinion in Environmental Sustainability*, *2*(5), 431–438. doi:10.1016/j.cosust.2010.09.008

Maximay, S. (2014). The Caribbean's Response to Climate Change Impacts. In W. Ganpat & W. P. Isaac (Eds.), *Impacts of Climate Change on Food Security in Small Island Developing States*. Hershey, PA: IGI Global.

Mba, C., Guimaraes, E. P., & Ghosh, K. (2012). Re-orienting crop improvement for the changing climatic conditions of the 21st century. *Agriculture and Food Security, 1*(7).

McCouch, S., Baute, G. J., Bradeen, J., Bramel, P., Bretting, P. K., Buckler, E., & Zamir, D. et al. (2013). Agriculture: Feeding the future. *Nature*, *499*(7456), 23–24. doi:10.1038/499023a PMID:23823779

Midega, C. A., Bruce, T. J., Pickett, J. A., Pittchar, J. O., Murage, A., & Khan, Z. R. (2015). Climate-adapted companion cropping increases agricultural productivity in East Africa. *Field Crops Research*, *180*, 118–125. doi:10.1016/j.fcr.2015.05.022

Millard, E. (2011). Incorporating agroforestry approaches into commodity value chains. *Environmental Management*, *48*(2), 365–377. doi:10.1007/s00267-011-9685-5 PMID:21556936

Mollineau, W. M., Garcia, G. W., Samayah, D., Kissoonsingh, W., & Procope-Garcia, A. (2000). The wildlife industry in Trinidad: a case study towards developing a sustainable industry model for a small Twin Island State (Trinidad and Tobago). In *Proceedings of the Conference on Managing Space for Sustainable Living in Small Island Development States*. St. Augustine: The University of the West Indies.

Mustapha, K. (2009). Taste of child labor not so sweet: A Critique of Regulatory Approaches to Combating Child Labor Abuses by the US Chocolate Industry. *Wash. UL Rev.*, *87*, 1163.

Neilson, J., Susilo, A., Mulia, S., Guest, D., Lambert, S., Munawar, N. L., . . . Muhajir, A. S. (2014). *Improving cocoa production through farmer involvement in demonstration trials of potentially superior and pest/disease resistant genotypes and integrated management practices*. Retrieved June 29, 2015, from http://aciar.gov.au/files/smar-2005-074_final_report.pdf

Neufeldt, H., Jahn, M., Campbell, B. M., Beddington, J. R., DeClerck, F., De Pinto, A., & Zougmoré, R. et al. (2013). Beyond climate-smart agriculture: Toward safe operating spaces for global food systems. *Agric Food Secur*, *2*(12), 1180–1186.

Noble, I. R., Huq, S., Anokhin, Y. A., Carmin, J., Goudou, D., Lansigan, F. P., & Villamizar, A. et al. (2014). Adaptation needs and options. In C. B. Field, V. R. Barros, D. J. Dokken, K. J. Mach, M. D. Mastrandrea, T. E. Bilir, & L. L. White et al. (Eds.), *Climate Change 2014: Impacts, Adaptation, and Vulnerability. Part A: Global and Sectoral Aspects* (pp. 833–868). Cambridge, UK: Cambridge University Press.

Ofori, A., Padi, F. K., Acheampong, K., & Lowor, S. (2015). Genetic variation and relationship of traits related to drought tolerance in cocoa (Theobroma cacao L.) under shade and no-shade conditions in Ghana. *Euphytica*, *201*(3), 411–421. doi:10.1007/s10681-014-1228-8

Oyekale, A. S., Bolaji, M. B., & Olowa, O. W. (2009). The effects of climate change on cocoa production and vulnerability assessment in Nigeria. *Agricultural Journal*, *4*(2), 77–85.

Palombi, L., & Sessa, R. (2013). Climate-smart crop production system Module 7. In *Climate-smart agriculture: Sourcebook*. Retrieved June 23, 2015, from http://www.fao.org/3/a-i3325e/i3325e07.pdf

Pay, E. (2009). *The market for organic and fair-trade cocoa*. Rome, Italy: FAO. Retrieved August 5, 2015, from: http://www.fao.org/fileadmin/templates/organicexports/docs/Market_Organic_FT_Cocoa.pdf

Perez, C., Jones, E. M., Kristjanson, P., Cramer, L., Thornton, P. K., Förch, W., & Barahona, C. (2015). How resilient are farming households and communities to a changing climate in Africa? A gender-based perspective. *Global Environmental Change*, *34*, 95–107. doi:10.1016/j.gloenvcha.2015.06.003

Peterson, S. (2011). International technology transfer-not quite the silver bullet in international climate policy. *Climate Change: Current Issues*, 20-21. Retrieved August 13, 2015, from https://www.ifw-kiel.de/pub/e-books/climate_change.pdf

Ramirez-Villegas, J., & Thornton, P. K. (2015). *Climate change impacts on African crop production*. CCAFS Working Paper no. 119. Copenhagen, Denmark: CGIAR Research Program on Climate Change, Agriculture and Food Security (CCAFS). Retrieved June 22, 2015, from https://cgspace.cgiar.org/bitstream/handle/10568/66560/WP119_FINAL.pdf?sequence=1

Rasul, G., & Sharma, B. (2015). The nexus approach to water–energy–food security: an option for adaptation to climate change. *Climate Policy*, 1-21. Retrieved August 11, 2015, from http://www.tandfonline.com/loi/tcpo20

Reichert, J. M., Rodrigues, M. F., Awe, G. O., Riquelme, U. F. B., Kaiser, D. R., & Reinert, D. J. (2015). Common bean in highly variable weather conditions, on sandy soils, and food security in a subtropical environment. In *Food and Energy Security*. John Wiley & Sons Ltd. and the Association of Applied Biologists. Retrieved August 7, 2015, from http://onlinelibrary.wiley.com/doi/10.1002/fes3.65/epdf

Rellstab, C., Gugerli, F., Eckert, A. J., Hancock, A. M., & Holderegger, R. (2015). A practical guide to environmental association analysis in landscape genomics. *Molecular ecology*. Retrieved August 9, 2015, from http://onlinelibrary.wiley.com/doi/10.1111/mec.13322/epdf

Ribaut, J. M., de Vicente, M. C., & Delannay, X. (2010). Molecular breeding in developing countries: Challenges and perspectives. *Current Opinion in Plant Biology*, *13*(2), 213–218. doi:10.1016/j.pbi.2009.12.011 PMID:20106715

Rijsberman, F. (2015). *CGIAR's new strategy and research programs: Answering to poverty, health and climate change.* News from the CGIAR Consortium. Retrieved August 3, 2015, from http://www.cgiar.org/consortium-news/cgiars-new-strategy-and-research-programs-answering-to-poverty-health-and-climate-change/

Roberts, T. G., Ganpat, W. G., Narine, L., Heinert, S. B., & Rodriguez, M. T. (2015). Trinidad Food Producers' Training Needs for Food Security and Implications for Extension. *Journal of International Agricultural and Extension Education, 22*(1), 7–20. doi:10.5191/jiaee.2015.22101

Roberts, T. G., & Rodriguez, M. T. (2014). An Overview of climate change and impacts on food security in small island developing states. In W. Ganpat & W. P. Isaac (Eds.), *Impacts of Climate Change on Food Security in Small Island Developing States* (pp. 1–30). Hershey, PA: IGI Global.

Roberts-Nkrumah, L. B., & Legall, G. (2013). Breadfruit (*Artocarpus altilis*, Moraceae) and Chataigne (*A. camansi*) for Food Security and Income Generation: The Case of Trinidad and Tobago. *Economic Botany, 67*(4), 324–334. doi:10.1007/s12231-013-9250-7

Rosegrant, M. W. (2011). Impacts of climate change on food security and livelihoods. In *Food security and climate change in dry areas* (pp. 24-26). Aleppo: International Center for Agricultural Research in the Dry Areas (ICARDA).

Saibo, N. J., Lourenço, T., & Oliveira, M. M. (2009). Transcription factors and regulation of photosynthetic and related metabolism under environmental stresses. *Annals of Botany, 103*(4), 609–623. doi:10.1093/aob/mcn227 PMID:19010801

Salem, M. A., Kakani, V. G., Koti, S., & Reddy, K. R. (2007). Pollen-based screening of soybean genotypes for high temperatures. *Crop Science, 47*(1), 219–231. doi:10.2135/cropsci2006.07.0443

Sasson, A. (2012). Food security for Africa: an urgent global challenge. *Agriculture & Food Security, 1*(2), 1-16. Retrieved June 12, 2015, from http://www.biomedcentral.com/content/pdf/2048-7010-1-2.pdf

Scherr, S. J., Shames, S., & Friedman, R. (2012). From climate-smart agriculture to climate-smart landscapes. *Agriculture and Food Security, 1*, 12. Retrieved June 17, 2015, from http://www.biomedcentral.com/content/pdf/2048-7010-1-12.pdf

Schrage, E. J., & Ewing, A. P. (2005). The cocoa industry and child labour. *Journal of Corporate Citizenship, 2005*(18), 99–112. doi:10.9774/GLEAF.4700.2005.su.00013

Sonnino, R., Moragues Faus, A., & Maggio, A. (2014). Sustainable food security: An emerging research and policy agenda. *International Journal of Sociology of Agriculture and Food, 21*(1), 173–188.

Stigter, K. (2008). Cocoa and climate change. *Can the lame help the blind.* Retrieved June 19, 2015, from http://sitp.rpn.co.id/uploads/riset/kakao/Prosiding%20Kakao%202008%20Cocoa%20And%20Climate%20Change.pdf

Swaminathan, M. (1982). Biotechnology research and third world agriculture. *Science, 218*(4576), 967–972. doi:10.1126/science.218.4576.967 PMID:17790573

Swaminathan, M. S. (2012). Role of genetic modification in developing climate smart agriculture to ensure sustained food security. *Agricultural Research, 1*(4), 295–298. doi:10.1007/s40003-012-0041-7

Tester, M., & Langridge, P. (2010). Breeding technologies to increase crop production in a changing world. *Science*, *327*(5967), 818–822. doi:10.1126/science.1183700 PMID:20150489

Thornton, P. K., & Herrero, M. (2015). Adapting to climate change in the mixed crop and livestock farming systems in sub-Saharan Africa. *Nature Climate Change*, *5*(9), 830–836. doi:10.1038/nclimate2754

Tol, R. S. J. (2002). Estimates of the damage costs of climate change, Part I: Benchmark Estimates. *Environmental and Resource Economics*, *21*(1), 47–73. doi:10.1023/A:1014500930521

Toulmin, C. (2009). *Climate change in Africa*. London: Zed Books.

UNFCCC (United Nations Framework Convention on Climate Change). (2007). *Climate change: impacts, vulnerabilities and adaptation in developing countries*. Bonn, Germany: UNFCCC Secretariat. Retrieved June 4, 2015, from http://unfcc.int/resource/docs/publications/impacts.pdf

Vervoort, J. M., Thornton, P. K., Kristjanson, P., Förch, W., Ericksen, P. J., Kok, K., & Jost, C. et al. (2014). Challenges to scenario-guided adaptive action on food security under climate change. *Global Environmental Change*, *28*, 383–394. doi:10.1016/j.gloenvcha.2014.03.001

Waldron, A., Justicia, R., Smith, L., & Sanchez, M. (2012). Conservation through chocolate: A win-win for biodiversity and farmers in Ecuador's lowland tropics. *Conservation Letters*, *5*(3), 213–221. doi:10.1111/j.1755-263X.2012.00230.x

West African Rice Development Association. (2001). NERICA: Rice for life. *WARDA/ADRAO, Bouaké*. Retrieved June 18, 2015, from http://www.warda.org/publications/NERICA.pdf

Wheeler, T., & von Braun, J. (2013). Climate change impacts on global food security. *Science*, *341*(6145), 508–513. doi:10.1126/science.1239402 PMID:23908229

Wickham, L. D. (1992). The potential for increasing the productivity and quality of vegetables produced in the Caribbean region: UWI research and perspectives. In G. Villanueva, J. Corven & A. Campos (Eds.), Proceedings of Taller regional centroamericano y consulta sobre planificacion de investigacion horticola (pp. 245-251). San José, Costa Rica: IICA.

World Cocoa Foundation (WCF). (2015). *President's message*. Retrieved June 27, 2015, from: http://worldcocoafoundation.org/wcf-may-june-2015-2/#2&utm_source=WCF+May+%26+June+2015+Newsletter&utm_campaign=WCF+Newsletter&utm_medium=email

KEY TERMS AND DEFINITIONS

Climate Change Adaptation: Human actions or responses in nature taken to adjust to risks posed by *climatic* changes or to benefit from any advantages.

Climate Change Mitigation: Action taken to permanently eliminate or reduce the long-term risk and hazards of climate change to the environment, human life and property.

Climate-Smart Agriculture: Agricultural practices that sustainably increase agricultural productivity and incomes, facilitate adaptation and build resilience to climate change and reduce or remove emission of greenhouse gases.

Conservation Agriculture: Encompasses cost-effective soil management practices that minimize the disruption of the soil's structure, composition and natural biodiversity. It takes advantage of soil fauna and properties that improve fertility that can improve crop yields while protecting the environment.

Ex Situ **Conservation:** Protecting species of a plant or animal outside its natural habitat (*in situ*).

Functional Genomics: Study of genes, their resulting proteins, and the role played by the latter in the biochemical processes of plants and animals.

Gene Pyramiding: A process of combining or assembling multiple desirable *genes* from multiple parents into a single genotype.

Genomewide Association Analysis: Involves scanning markers across the complete sets of DNA or genomes for a particular species such as *Oryza sativa* subspecies (e.g. *indica*), examining many common genetic variants in different individuals to see if any variant is associated with a trait of interest.

Genomic Selection: A process used to identify potentially favourable organisms (animals or plants) based on specific genetic information they harbour.

Genotype: Genetic makeup of a particular organism or group of organisms.

Marker Assisted Selection: A process where a trait of interest is *selected*, not based on the physical trait itself (such as plant height), but on a molecular or genetic *marker* linked to it.

Phenotype: Observable characteristics of an individual (such as seed size) resulting from the interaction of its genotype with the environment.

ENDNOTES

1. Agriculture includes food crops, pastures and livestock, industrial crops and biofuels, forestry (commercial forests), aquaculture.
2. The Millennium Development Goals (MDGs) are the world's time-bound and quantified targets for addressing extreme poverty, hunger, disease, lack of adequate shelter, and exclusion while promoting gender equality, education, and environmental sustainability.
3. Riparian vegetation is composed of plant habitats and communities along river/stream margins and banks, and a riparian area is the interface between water and land.
4. 'Cacao' is the term used to describe the plant of *Theobroma cacao* L. 'Cocoa' describes the commercial product - the nibs/seeds.

Chapter 3
Strengthening Food Security with Sustainable Practices by Smallholder Farmers in Lesser Developed Economies

Leighton Naraine
Clarence Fitzroy Bryant College, St. Kitts-Nevis

Kevin Meehan
University of Central Florida, USA

ABSTRACT

Smallholder farming has played a key role in reducing world hunger and the focus now turns to consolidating these gains. What practices and policies will enable smallholders to sustain their livelihoods and strengthen food security? Following a brief discussion of agricultural models, and the global shift to sustainable development goals, this chapter examines practices of smallholder farmers in St. Kitts with examples from elsewhere. Assessing what has worked and what barriers continue to limit smallholder success, the chapter offers a list of adaptive strategies, policy recommendations, and areas for future research that can maximize smallholder farming impact on food security. Suggested priorities include: focusing research and policies on individual farm enterprises; moving toward an integrated farm model; adopting best practices in marketing, distribution, and accounting; improving government support; and adapting education programs to include modernized curricula, non-formal education, and the use of ICTs for training and extension at all levels.

DOI: 10.4018/978-1-5225-0942-4.ch003

INTRODUCTION

Improved agricultural productivity, especially by small and family farmers, leads to important gains in hunger and poverty reduction. (Food and Agriculture Organization, 2015)

Beginning in 1980s, lesser-developed countries (LDCs) witnessed a trend of declining production of agricultural food products, both crops and livestock, while at the same time more developed countries (MDCs) experienced increased production. As such, LDCs became more dependent on imported food from MDCs. Because of this, and despite their former status as agrarian societies, emerging economies have had to divert scarce economic resources away from other imports, investments, and foreign debt payments to pay for imported food. Even worse, when a dependency syndrome combines with government instability and social conflict, the result has often been hunger, malnutrition, and even famine in extreme cases.

With these dynamics in mind, governmental and non-governmental entities have worked to combat world hunger within a comprehensive United Nations framework known as Millennium Development Goals (MDG). In 2015, the UN Food and Agriculture Organization reported that 79 out of 129 countries had met the MDG benchmark of reducing hunger by one half. Worldwide, hunger has fallen by almost 20% in the past decade with most of those gains happening in the developing world mainly through the increased productivity of smallholder farms, operated mostly as family enterprises. In order to understand this success, and to envision ways of enhancing it to address the problems of the 796 million people who remain in hunger worldwide, this chapter approaches the topic of food security in emerging economies from several angles. After briefly touching on how MDCs have transformed their agricultural sector to become competitive in world food production, the chapter focuses primarily on the concerns of food production and distribution facing the Caribbean and other emerging economies with the input factors of cultural, socio-politico-economic, technological, and environmental conditions, including climate change. The chapter proposes how such economies may be transformed so that food security may be achieved. It presents examples of best practices and scenarios using a simple model of decision-making based on this range of factors.

In exploring food production and distribution in emerging economies, a central claim is that agricultural diversification is essential to achieve local food security and that diversification rests on four pillars: productivity, flexibility, competitiveness, and sustainability. Therefore, this chapter will examine concerns about food security from the perspectives of these four pillars. While some areas may realize high productivity, there is scarcity in other areas, which prompts a concern with food distribution networks in the context of trade and marketing.

All of the concerns can be addressed systematically by the application of a model of the farming system adopted from Waugh (2009). The unit of analysis is the individual enterprise or individual farmer level. The proposition is that if individual enterprises are successful, then national and regional agricultural economies will be successful. The model will address the prevailing conditions in terms of input factors and what decisions can be made in those respects. It also recommends the evaluation of profit and/or loss and a feedback mechanism to the decision-making process.

Main Characteristics of Agriculture in MDCs

Virtually every industrialized nation, for example, Australia, France, Japan, and the United States, has experienced an agricultural revolution (Nebel and Wright, 1996). The agricultural revolution in such countries had one thing in common, that is, significantly higher utilization of farm machinery not only for land preparation but also for planting, application of fertilizers and pesticides, and harvesting. They brought additional land into production, increased the use of chemical fertilizers and pesticides, increased use of irrigation, and substituted new varieties of crops or hybrids. According to Nebel and Wright (1996), all these developments carried an environmental cost, and expanding production by those methods has reached or exceeded sustainable limits. With competitiveness resulting from such significant inputs by MDCs, there has been a worldwide trend since the 1970s in which food production in MDCs has been rapidly increasing while there has been a rapid decrease in LDCs, resulting in greater export of food from MDCs and higher import of food by LDCs.

From findings based on an analysis of 15 select countries using 5 variables (GDP per capita; Percentage of population engaged in agriculture; Percentage of GDP derived from Agriculture; Amount (kg) of fertilizer used per hectare of agricultural land; and Number of tractors per 100 ha of land) Waugh (2009) postulates that:

- The less developed a country, the greater the percentage of its population involved in agriculture.
- The less developed a country, the greater the percentage of its GDP/GNP is made up from agriculture.
- The less developed a country, the less fertilizer it will use.
- The less developed a country, the less mechanized will be its farming.

The lesson here is that MDCs have transformed their agricultural sector with the incorporation of high capital input for the use of improved technology and by bringing larger expanses of land under cultivation, which is in stark contrast to the input made by LDCs. It also indicates that MDCs generally thrust forward with economic development as a priority over the environment, resulting in larger impact of climate change events and environmental degradation. Once the industry is developed, such countries then place increasing emphasis on environmental integrity.

It is not proposed here that LDCs require the same transformations that MDCs have gone through. Nonetheless, with changing conditions and under different circumstances, it is proposed that LDCs build adaptive capacities, while introducing locally innovative, sustainable, and appropriate technologies for mechanization, irrigation, plant science, and non-traditional growing systems to increase their food security as a high priority. Within this scenario, there are opportunities to satisfy local market demands while building niche markets that enable producers to enter the global market.

Main Characteristics of Agriculture in LDCs

Numerous scholars, economists, and leaders of LDCs have lamented the various ways in which the plantation system, imposed under colonialism, has hindered the development of local food production. As such, some main characteristics of agriculture in LDCs are as follows:

- Small-sized farms (less than 5 acres, with the majority being 2 acres or less).
- Parcels are fragmented.
- Land is marginal, on the periphery of former plantations.
- Population is mainly rural-based peasantry.
- Land tenure is such that the majority of farmers are "landless."

While acknowledging the lasting impact of plantation systems, this section focuses less on that legacy and more on the current availability of land during the past four to five decades or so since many countries have become independent of colonial rule. In general, LDCs have not had transformations in their agricultural sector as did MDCs. This is perhaps in part due to the lack of comparative advantage, vis-à-vis their "later arrival on the scene" and having to compete with an already advanced agricultural economy in MDCs. Moreover, such transformation requires high capital input that is essentially scarce in LDCs.

With rising food prices and increasing impacts of climate change on food production, LDCs are increasingly challenged to maintain basic nutrition from food, alleviate poverty in rural communities, and reduce their relatively high food import bill. Food security has emerged in recent years as a high priority in LDCs and is often tied to building resilience to the effects of climate change. A hypothesis proposed is that the better the livelihoods of people in communities, the more resilient they would be to the vagaries of climate change and environmental degradation.

According to the CARIBSAVE Partnership, based on participation in CARIBSAVE National Consultation Workshops in Caribbean countries, LDCs' needs are ranked as follows, with food security as the highest priority (CARIBSAVE, 2011):

1. Food security (agriculture with improved methods; less dependent on weather – floods and drought, higher productivity on less land; more reliability of production, less impact on environment).
2. Water supply (rationing already exists in some places).
3. Medical supplies and facilities.
4. Education and awareness (at all levels).
5. Food storage (for extended periods of time).
6. Water harvesting and storage and replenishment of aquifers (cisterns, reforestation, conservation).
7. Shelters and Institutional support (some organizations, public services, and disaster management plans already in place).
8. Political will (not only from leaders but also from citizenry – building consensus - can draw political salience and influence public agenda).
9. Legislation enforcement (appears that legislation is already in place).
10. Reduce sedimentation by reforestation and by limiting construction sites near the sea.
11. Avoid building in high risk areas.
12. Fuel storage and alternative sources of energy (solar, wind, geothermal).
13. Fish sanctuaries and fish farms.

Small Island Developing States (SIDS) are perhaps the most vulnerable to the effects of climate change and sea-level rise, given that their populations are largely located on low-lying, coastal communities, and their relatively large percentage of rural populations are living in poverty and engaged in subsistence agriculture. The observations and predictions are pointing to the Caribbean as an area among the most vulnerable to the effects of climate change. Other SIDS in the South Pacific and Indian Ocean are equally vulnerable with respect to geographical characteristics. Food security goals are seriously at risk!

The climate in SIDS and LDCs is mainly tropical, with some variations from equatorial to tropical maritime, and sometimes with a Mediterranean-type effect. More importantly, it is characterized by high rainfall and hot conditions. Heavy rainfall often results in flooding, but also renders our soils vulnerable to leaching and erosion. Higher rainfall and higher temperatures result in higher rates of bio-decay. It is a misconception to think that tropical soils are the richest; this is perhaps due to the year-round and lush vegetation, and also to its history of agriculture as the main source of GDP. Compared to chernozem and peat soils in temperate regions, tropical soils are relatively impoverished. While there are greater opportunities for reforestation in temperate regions, a critical need exists for soil conservation or soil production, for example, by composting or integrated crop-livestock farming, in the tropics in order to maintain optimum nutrients in soils for cultivation. Volcanic ash contributes to high soil fertility, but this component also makes these soils easily drained and thus prone to excessive drying in drought conditions.

In light of these characteristics, many of which the Caribbean shares with other developing societies, we can also outline a set of specific vulnerabilities and opportunities for enhancing food production and food security in emerging economies. These areas are examined in more detail in the next section in the context of translating them into "Areas for Strategic Improvement". Additionally, concrete recommendations are given for smallholder farming. Two mini-case studies that highlight what may be termed a dialectic of vulnerability and opportunity are presented. As these two examples make clear, vulnerabilities often create related opportunities that stakeholders of many interests must strive to understand and are challenged to use for individual and collective advantage.

Example

Nigeria: Government Regulation as Barrier and Incentive

In Nigeria between 1970 and 1985, the government implemented a system of complex bureaucratic requirements that made it difficult for businesses to expand (Balogun, 2003). Furthermore, in the early 2000s, regional Nigerian governors established arbitrary areas where business activity was restricted (Ewah & Ekeng, 2009).

In spite of all the difficulties of business management in developing areas, these regions offer several opportunities and benefits to new firms, including undiscovered markets, government subsidies, a burgeoning middle class, and cheap production inputs (Ewah & Ekeng, 2009).

- Population growth. The size and growth of populations in developing countries presents a huge potential market for local businesses, and a ripe opportunity for marketing to help overcome the preference for foreign goods.
- Absence of competition. Due to their underdevelopment and large populations, developing countries have substantial undiscovered or partially served markets. Often these markets are so large that concerns about competition are nearly nonexistent because there is room for so many entrants.
- Government incentives. Developing countries often provide welcoming business environments through lenient tax codes and subsidies for local enterprise.

- Growing middle class. Incomes are rising in numerous developing countries, bringing with them increased consumption and marketing opportunities.
- Cheap production inputs. Labor and raw materials are usually cheaper in developing countries, making it easier to market and operate local businesses (Ewah and Ekeng, 2009).

Example

Latin America: Supermarkets as Threats to Smallholders and Catalysts for Increased Local Demand

The rise of Latin American supermarkets provides an example of the two-sided nature of marketing in developing countries. Grocery stores and markets emerged as dominant players in Latin America over just ten years, increasing from a population-weighted average of about 10 percent in 1990 to over 50 percent in 2000. This can be compared to the transformation in the U.S. that took over 50 years. Large scale supermarkets and manufacturers' entrance into agrifood markets excluded small Latin American farms because bulk purchase methods lowered the farmers' margins. As a result, only farms with sufficient production to reach substantial economies of scale remained profitable. However, the rise of supermarkets provided opportunities as well, in terms of increased local demand and changing consumer patterns (Reardon & Berdegué, 2002).

SUSTAINABLE FOOD PRODUCTION

Sustainability

Sustainable development is development that meets the needs of the present without compromising the ability of future generations to meet their own needs (World Commission on Environment and Development, 1987)

This quote from the 1987 Bruntland Report, while greeted with controversy by some, is still the most popular general definition of sustainable development. How, though, does the key concept of *sustainability* apply more specifically to food security, smallholder farming, and adaptations in the Caribbean and other developing regions? In assessing the progress made toward meeting the Millennium Development Goals, the UN Food and Agriculture Organization noted several trends that affect developing societies broadly and the Caribbean specifically. For instance, much of the success in reducing world hunger during the MDG period (2000-2015) was attributed to gains in productivity among smaller farms similar to those that define Caribbean agriculture today. The FAO report also noted that growth in hunger eradication was steady but slower in the Caribbean region. This report correlates with the emphasis on smaller compared to large-scale farming that generates growth through a multiplier effect. The report notes as well, a lengthy transition process from former plantation economies to newer systems where stakeholders search for the right balance in allocating resources among local food production, tourism and other services.

As the millennium development goals are superceded by a new set of 17 interlocking benchmarks known as the Sustainable Development Goals (SDG), food security remains a top priority, with "zero hunger" listed as SDG2. The FAO calls for continued emphasis on smallholder and family farms, social protection for small-scale farming such as government purchase of food supplies, easier access to credit and land, and inclusive growth. Effectively meeting these goals depends on gathering more concrete information about what smallholders are actually doing, and identifying recent innovations that have been more (as well as less) successful. Information should also ascertain the next-step scenarios that seem promising based on a review of actually existing practices, and what policy decisions will best protect, support, and spur the sustainable growth of small-scale Caribbean farming.

The next section will highlight current practices with examples drawn mainly from the Eastern Caribbean. This is, followed by discussion of areas for strategic improvement that can immediately begin to enhance smallholder production. A series of policy recommendations and suggestions for future research needs is presented after the discussion.

Smallholder Practices

According to the FAO, "the term 'smallholder' refers to their limited resource endowments relative to other farmers in the sector" (Dixon et al., 2004). Globally, there are about 2.5 billion people who rely on these farms for their livelihoods. Moreover, smallholders produce over 80 percent of the food in parts of the developing world, specifically Southern Asia and sub-Saharan Africa (IFAD, 2013).

New pathways to sustainable food production demand a holistic approach to understanding sustainability at the local and livelihood level. However, any appreciation of the importance of transforming current agricultural livelihoods into sustainable food production systems must include a deeper understanding and appreciation of the role of major assets in transforming the lives of farmers and their households. Financial capital, land, and knowledge of modern farming techniques are key variables and assets for successful application of small-scale back yard gardening. According to the FAO (A Tool for Gender-Sensitive Agriculture and Rural Development, 2013), inequalities in distribution is linked to women's underperformance in agriculture, but increased investment in smallholder farmers (women in particular) is a pathway out of poverty.

Home Food Production: St. Kitts-Nevis Case Study

A 2010 survey of former sugar workers in St. Kitts-Nevis (five years after sugarcane production ceased) revealed that the best route to alleviating poverty amongst the vulnerable grouping was to reconsider land distribution options and to promote backyard farming as a viable option for food security and livelihood resilience (Clarke, 2013; Naraine, 2005). Former sugarcane workers presented a unique opportunity for understanding how deeply engrained cultures of resource distribution can affect use of land and potential vulnerability.

The growing concerns for land are not only shared by those with livelihoods directly dependent on the asset, but also by the population at large. The former female sugarcane workers were also assigned to farms. The success of this program was short lived with many women distrusting the process of large scale group farming. It is clear that campaigns at the end of sugarcane production which sought to distribute land to every Kittitian did not do much to solve the problem of vulnerability amongst former workers. Larger plots of land distributed to women were burdensome and demanded help, money and access to social capital that was weak or non-existent in their households and extended community networks. Land as a route to diversification and resilience was therefore seen as a burden, not an asset. Increased access to land did not equate with a reduction in vulnerability or an increase in the sustainability profile of livelihoods or production systems, since land was not put to productive use. The lesson learnt was that land resources, a crucial asset in sustainable production and livelihoods, must be distributed according to capital access that makes for optimum use. Based on this, recommendations were made for backyard farming as a viable option.

A sustainable approach to backyard farming starts with a food security oriented approach to land distribution. The FAO (2013) outlined clearly that women face a range of constraints in their access to productive resources such as land, inputs, training and financial services. Giving them access will very well translate into lifting millions out of food insecurity! Land distribution policy must, therefore, include distribution of larger sized residential plots at reasonable rates to provide ample room for backyard gardening.

Backyard gardening must be carefully and delicately guided through instruction in using small spaces for growing vegetables and other food crops. However, backyard gardening can harness all the techniques and skills of the local population without heavy reliance on external inputs. This may encourage more organic approaches to food crop management, as well as non-traditional agricultural systems (for example, hydroponics, organoponics, aquaponics, and hybridponics) which are adaptable to rural, sub-urban and even urban areas.

Protected Agriculture: Shadehouse Model

Protected agriculture has been reintroduced recently into the Caribbean region with the construction and commissioning of tropical shadehouses and greenhouses to increase year-round local production and in an effort to reduce imports.

At the Clarence Fitzroy Bryant College (CFBC) in St. Kitts, researchers, students, government agencies, NGOs, business and community-based stakeholders have partnered to develop a new Climate-Smart System that blends shadehouse protected agriculture with a combination of hydroponics, organoponics, and hybridponics growing systems. The term hybridponics was coined to describe a system in which a seedling is removed from a hydroponic system and transplanted in an organoponics system to complete the growth process. Proven benefits from the Climate-Smart System include the following: year round rather than seasonal production; no leaching of nutrients to the environment; no use of pesticides; no weeds; high nutritional values; maximum production in confined areas; efficient use of water; less labor intensive; can be used in any local location; sustainable; and promotes gender equity. Subsequent to developing a best-practices routine, and commercialization model for a 24' x 20' enclosure, the CFBC system was funded by the Organization of American States for implementation in Guyana, Haiti, Barbados, Trinidad and Nevis.

Pig Farming

Over the years Caribbean pig farmers (for example, Bounty Farms, Guyana; St. Catherine, Jamaica; Plum Tree Farms, St. Kitts) have witnessed a slow but needed transition from poorly kept operations in backyards where pigs were left outdoors in muddy pens to formal operations with enclosed housing complete with concrete floors (see Figures 1-3).

This transition has seen the improvement in techniques for waste management including, use of sewer systems, and heavy reliance on water pumped to the farm site for washing pens.

Figure 1. Outdoor pig pen

Figure 2. Sewage and drainage system

Figure 3. Old style of pig housing

Local breeds of pigs have been known to mature with a high fat content. The introduction of Yorkshire White and the Duroc (and cross-breeding of the two for meat quality) has drastically increased stock and stock quality. Along with the improvement on stock is the improvement in the variety of pig rations to include starter and grower ration. These upgrades are supplemented with regular visits by Veterinarians to the farmers, especially for disease control, breeding and farrowing. Agricultural staff undergo continued training in disease monitoring and controls. These key inputs have been integral in the cycle of sustainable livestock production.

As these practices make evident, even accounting for the systemic challenges associated with emerging economies, important opportunities exist for smallholder farmers to position themselves more favorably, particularly with strategic assistance from government agencies. In this dynamic, based on sustainable agricultural practices, more consumers of food become producers, and food security is increasingly based on a "many-to-many relationship," with the many feeding the many.

AREAS FOR STRATEGIC IMPROVEMENT

Vulnerabilities and Opportunities for Food Security in Emerging Economies

Sound fundamental business practices are key to building successful agricultural operations. In preceding sections, this chapter explored sustainability, integration of different farm components, and efficiency as they relate to agricultural development. However, creating and managing profitable farms requires additional considerations outside the scope of these internal elements. Here, the conversation turns to the role of basic economic and financial concerns, with an emphasis on marketing and distribution.

Before delving into areas for strategic improvement of smallholder business practices, it is important to better understand the context of operating in a developing economy. Developing economies exhibit three principal characteristics: low incomes, high variance in consumer preferences and infrastructure quality, and relatively cheap labor costs, leading to substitution of labor for capital (Dawar & Chattopadhyay, 2002). Managing a farm in these countries raises unique challenges that revolve around the local social, political, economic, and cultural factors. Specifically, these operational obstacles include low business education, preferences for imported goods, deficient infrastructure, strict government regulations and interference, political instability, and social conflict (Ewah & Ekeng, 2009).

Although there are inevitably several challenges to doing business in developing areas, these regions also offer several opportunities and benefits to small farmers. Consider, for example, that 70 percent of the global poor live in rural areas and rely on agriculture for income and employment (World Bank, 2015). In spite of this gloomy statistic, local farmers are in a unique position to be successful due to their self-employment (free labor), agricultural expertise, and familiarity with the local environment (FAO, Trends and Impacts, 2012). Several of these vulnerabilities and opportunities are now discussed with a focus on identifying concrete recommendations for strengthening food security in emerging economies.

Marketing and Distribution

The act of marketing a good or service is prevalent in nearly all aspects of life. The American Marketing Association (2005) defines marketing as "the activity, set of institutions, and processes for creating, communicating, delivering, and exchanging offerings that have value for customers, clients, partners, and society at large." Marketing helps farmers to find the customers who need their product most (efficient resource allocation), and enables two-way signaling between farmers and customers so that both sides can communicate needs and offerings (Timmer et al. 1983).

Although marketing is a crucial step in the communication between potential producers and consumers, sales cannot occur without the physical distribution of produce. To understand how agricultural firms transport their goods from the farm to shelves or tables, it is important to appreciate the role of distribution in the larger context of food security.

Food security literature focuses on resource access, distribution, and closer connections between producers and consumers as principal issues of concern (Vervisch et al. 2013; Murphy, 2001; Edelman, 2014; Kelly & Schulschenk, 2011; McKenzie & Woodruff, 2015). Several factors help determine the efficacy of distribution, including markets, transportation infrastructure, patterns in consumer demand, and the characteristics of supply. Governments in developing countries have traditionally intervened in food distribution networks by purchasing produce from growers and selling it to consumers nationwide, often at subsidized prices.

Government involvement in production and distribution had both positive and negative effects on farmers. For instance, African governments had interfered with farmers for years by purchasing agricultural produce at less than the cost of production in order to provide cheap food to nearby urban centers. When reforms disbanded this handout program, farmers were initially grateful. However, growers soon realized the restructuring of government activity left them in a void with no alternative to the state-run marketing and distribution arrangement. As a result, farmers either were unable to take their products to market or were forced to rely on private companies to fulfill all the functions previously managed by the government (Murphy, 2001).

Example: The Internet

The Internet not only simplifies distribution for both producers and consumers, but also other market stakeholders including food processors, restaurants, and retailers. Online connectivity helps the entire agribusiness market chain to stay abreast of price changes and production estimates. Furthermore, the Internet facilitates cooperation among farmers, helping them to protect their interests even when separated geographically. As a result, developing countries have seen the emergence of farmer-collaboratives that depend on web marketing to connect directly with consumers (Murphy 2001).

Farmers in South Africa's Stellenbosch region used virtual networks and communication to develop a stronger local food structure by helping local farmers produce more of what was actually needed, and closing the gap between consumers and producers. Moreover, food networks that operate at a community level, including community supported agriculture initiatives (CSAs), farmers' markets, consumer cooperatives, and local distribution centers, can shorten the value chain and help farmers retain more of the retail price (Kelly & Schulschenk, 2011).

Local communities are more receptive to direct marketing, but many poor consumers are not aware of the nutritional content of their food and produce. They also lack knowledge on how to prepare and incorporate new ingredients into meals. Thus, education is an important part of marketing in developing countries. Offering classes on cooking with healthy ingredients, providing free samples, and suggesting convenient recipe ideas are creative ways to persuade local customers to incorporate less familiar fruits and vegetables as ingredients in their diets (Hu et al. 2013).

Smallholder distributors might also consider the existing social network as a favorable marketing prospect. These avenues allow farmers to interact with larger groups and build a sense of pride around locally grown food. For example, small farmers working to attract potential clients can use community centered events as great venues to showcase their produce in educational and interactive ways. Farmers' markets also provide excellent social opportunities for farmers to engage customers and discuss their products, growing methods, and food preparation ideas (Hu et al. 2013).

On a smaller-scale, farmers can take advantage of the fact that community leaders and older citizens are often influential voices for younger generations. Utilizing these groups as marketing allies, lends credibility to the case for healthy eating. Furthermore, these members are in a unique position to model healthy lifestyles for younger generations that would otherwise be unresponsive to these lifestyles. Building support among younger generations adds a new level of consumers that will continue to purchase goods into the future. Moreover, they broaden the word-of-mouth marketing team and help to access a new segment of local consumers (Hu et al. 2013).

Convenient distribution plays a significant role in augmenting innovative marketing efforts. Anticipation for products builds in direct correlation with a farmer's ability to demonstrate affordable pricing, appropriate location, and high frequency. In their surveys of poor areas, Hu et al. (2013) found many participants to be interested in a wide variety of mobile distribution mechanisms, for example, food carts, farm stands. Local customers also identified consistency, price-competitiveness, and frequent advertisements as contributing to their reasons for patronizing particular sellers. According to buyers, there was a huge benefit to knowing in advance what produce would be offered by which farmers (Hu et al. 2013).

Case studies focusing on horticultural marketing abound, and readers interested in the more detailed analysis based on a range of cases can access the references cited in this chapter. For summary purposes, several common threads unify the discussion about marketing for small farmers:

1. Perform basic market research and incorporate the results into decisions made about production and distribution. Dixie (2005) and Mafaru et al. (2007) both offered extensive commentaries on how farmers and farmer co-operatives in Bangladesh, India, Mozambique, South Africa, Tanzania, Yemen, and elsewhere effectively analyzed market conditions across a range of products and scales.
2. Cooperate in order to pool resources and experience, achieve economies of scale, ensure smooth supply, and increase market negotiating strength. Efforts of women farmers and craft workers in Bangladesh are detailed by Dixie (2005).
3. Innovation and forward thinking. Mafaru et al. (2007) provided details on the ability of Tanzanian farmers to effectively control and optimize sales of tomatoes and other crops to local and regional hotels. This was achieved by assigning leadership roles and building an organizational structure. A case study from Uganda provides a discussion of effective tribal leadership resulting in improved onion production in Yemen (Alimuga et al. 2007; Dixie 2005).
4. Cut out middlemen and engage in direct sales whenever possible. Dixie (2005), in his discussion of onion farming in Yemen, provided an insightful discussion of how to increase and benefit from direct sales.

Agro-Processing

After all costs have been paid for land preparation, irrigation, planting, maintenance, and harvesting, and such like, an enterprise is expected to achieve profits from the sale of produce. Post-harvest loss continues to pose high risk to farmers when produce is not sold due to lower market demand. One way to reduce post-harvest loss is by processing of produce, while adding value to the items. It appears that locally packaged produce are not competitive and the perception of farmers or entrepreneurs is that they have to lower their prices to compete with imported packaged foods. However, indications by agro-processors in some Caribbean island nations are that there is a demand for locally packaged foods with growing preference by tourists and local residents alike. Moreover, they are likely to be more "fresh" than the imported packaged foods which have generally longer shelf-life and higher quantities of preservatives.

Furthermore, when foods that can otherwise be produced locally are imported, the local economy loses out on the multiplier effect of farming activities. Therefore, it is prudent to build an economy around the primary sector, with import substitutes where possible, while reducing dependency on imported foods. Although many emerging economies have had their primary products as raw materials for trade, agro-processing adds value to primary products either for local consumption or for exporting.

Food processing is an extension of the postmodern agricultural industry that involves different sectors such as; research and development, equipment investment, production and marketing management. The processing industry needs to integrate different technologies and production equipment to make mass production possible. Data from the Agro-processing Unit, St. Kitts, show that juices and snacks have a high market demand. The juice revenue ratio is 37.2%, while that of snack foods is 29.2%. The St. Kitts and Nevis food service sector country report of 2006 stated that snack food is the third largest imported food in terms of volume and juices are the fifth (the sequence is; meat, seafood, snack food, processed fruits and vegetables, and juices). Based on this information, there is an opportunity to create economies of scale in juices and snacks in St. Kitts and Nevis.

The St. Kitts-Nevis experience in agro-processing, with infrastructural and technical support from the Republic of China (Taiwan), has been growing steadily over the past few years and still has significant room to grow the meet local demand. There is potential to grow multiple times even with local demand.

ADDITIONAL CHALLENGES AND ADAPTIVE STRATEGIES: TOWARDS A MODEL FOR SUCCESSFUL SMALLHOLDER FARMING

In a 2005 study in the Caribbean small island nation of St. Kitts-Nevis it was found that the characteristics of farmers are representative of SIDS (Naraine, 2005). These findings are detailed below under three general headings: A model for smallholder farming that is 1) diversified and scalable; 2) operational adaptations; and 3) educational and organizational adaptations. Each individual finding will be followed by an adaptive strategy or strategies taken from case study findings of best practices.

Diversified and Scalable

Finding: *Of all commercial farmers, 60% operate on a part-time or seasonal basis.*
Finding: *Farming households are generally larger (about 65% between 3-5 persons) as compared to the national average of 3 per household.*
Finding: *About 60% of farmers operate individually, not as part of a company.*

The first data point indicates that a majority of farmers are not fully-involved in farming and therefore are not likely to depend on farming alone as a source of income. Related to that finding, 30% of all farmers have other occupations. Instead, there is a growing trend in "backyard" and urban gardening on a small scale requiring less time for operation. Production can be in containers where land is not available. It will also maximize productivity with minimal time and with optimum soil nutrients, moisture, and solar radiation (lighting and heat intensity) in the containers. Such a growing system is referred to as organoponics (Naraine, 2005). It can further save time through less weeding and tillage and can adapt to varying weather conditions. It can also expand to varying scales with some not requiring shade, or shade can be introduced by trees. Organoponics can progress to full time farming with the greater attributes and benefits than on a small scale, such as the opportunity to diversify crop types. As it increases in scale, the advantage will grow exponentially, as in the case study of ALT enterprise in St. Kitts, an integrated system with livestock to produce manure and recycling of nutrients among crops and livestock. Scaling up still higher towards commercial and more diversified cropping, a mix of farming types and systems can be introduced, for example, hydroponics, hybridponics, open-air organoponics, and low tillage and point tillage. This facilitates increased sustainability, flexibility, efficiency, productivity and risk reduction. Sustainability and flexibility are key features of adaptation to effects of climate change. A more diversified system is also suitable for natural-biological pest control mechanisms by promoting a farm ecosystem with natural predator-prey relationships and symbiotic feeding relationships. It also promotes the recycling/reusing of water for various farming activities. Moreover, the housing for livestock can be utilized for water harvesting. Many organic materials, whether from crops or livestock, can be composted or processed into ground cover to enhance soil water retention capacity. Compost has the unique ability to improve the chemical (nutritional), physical (structural) and biological characteristics of the soil or growing media which is beneficial to plant growth.

With higher productivity and less labor input required in non-traditional agricultural system, this will enable the formation of true family farms with small households. Higher productivity can also attract other household labor or business partners for shared benefits. The data shows a 60% preference for individual enterprise, which may in fact be due to inadequate training in business techniques. Notwithstanding this, the ALT enterprise in St. Kitts teams up with young entrepreneurs, for example, Standard Hydro, to introduce a hydroponics facility as an addition to its already integrated farming concept. Existing employees partner with others in the various activities or aspects of the total enterprise. In such a partnership, all parties benefit from training in business techniques, including marketing and entrepreneurship.

The Diversity of Jobs and the Multiplier Effect

Finding: *Among the almost 150 farms surveyed, only about 50 persons are employed on about 30 of those farms. There is an almost even distribution of male and female employees working on an occasional basis among those 30 farms.*

Farms should not be viewed only for employment directly on farms. The economy around farming has a multiplier effect in which employment and commerce is created for related businesses, whether directly or indirectly, for example, insurance, retail and/or wholesale shops, traders, and transportation. Oftentimes, food supply is not always the problem for food security and availability, but the problem is marketing and distribution of produce. There is a great potential, based on the results of a needs' assessment conducted in St. Kitts-Nevis, Dominica, and St. Lucia, for labour market information (LMI) to inform the introduction of a new agri-business curriculum at local colleges. The LMI indicated a need for entrepreneurs devoted to marketing and distribution, while creating more employment related to the agricultural sector.

Income and Livelihoods

Finding: *Over 80% of all farmers reported annual household income of less than EC$35,000 (USD$12,900) consistent with the national median income of about $20,000 (USD$7,400) per person with two persons in the household employed.*

Finding: *About 70% of all farmers indicated that less than 25% of their household income is attributed to farming activities.*

Backyard gardening and non-traditional growing systems on a small scale are excellent for increasing household income, given the part-time nature of farming activities on a small scale. An important consideration in the income equation is that household food bills are reduced when they can supplement their diet with food produced by themselves or bartered with other growers. Moreover, it is a way of building resilience to climate change through better livelihoods.

Operational Adaptations

Land Tenure and Use

Finding: *Very significant is that about 75% of land occupied by farmers is government owned (they are illegal occupiers).*

This is perhaps the most challenging problem for farming to achieve food security in the Caribbean region. Simply put, land is tied to politics. With all the adaptive strategies that farmers use to reduce dependence on land by increasing productivity using non-traditional growing systems, that alone is not sufficient to achieve food security and make agriculture a viable economic activity. Ultimately, agriculture requires more land under production and the development of irrigation. Farmers are less likely to invest in infrastructure and technology when they do not own the land or have secure tenancy agreements. It is also the case that some farmers have the mindset that if they acquire land either by renting or illegal occupation that they would eventually own the land or have infinite access when it may become regularized. There is widespread evidence that some farmers hold onto relatively large parcels of farmland but with little farming activity. In this scenario government, on the one hand, has limitations to its land management policy and, on the other hand, with some farmers not utilizing farmland to the optimum when they have access. In some cases, farmlands remain totally unproductive for years and even decades. Moreover, prime agricultural land cannot compete with other land uses such as for tourism, housing, and manufacturing. Land-use planning would show that some of the best agricultural soils are not always used for farming. There are adequate policy tools and options available to government, but it is unequivocally a matter of choices that rests solely on the political will and ethical principles with regards to land management practices. Where such practices fail, the option for farmers is only with non-traditional growing systems, such as hydroponics, organoponics, and aquaponics, as adaptive strategies to increase productivity. Non-traditional systems can operate in diverse environments such as marginal land, low freshwater availability, backyards, and urban centers. This option when practiced on a small individual scale but on a more widespread basis can have a cumulative and positive impact on food security.

Housing and Losses from Theft, Nuisance, and Predators

Finding: *Many farmers do not live in the same parish as the farm location which poses a problem for surveillance, theft, security.*

Finding: *Monkeys, theft/praedial larceny, and stray roaming livestock, in addition to diseases and pests are significant problems for crop production, but dogs and theft are problems for livestock.*

Losses from theft, nuisance, and predators are perennial problems with smallholder agriculture, especially when the farmer's plot is located at a distance from home. Where government policies prevent housing of people on farms, this presents a barrier to preventing such losses. It is also not feasible for persons to build homes on land on which they are renting or illegally occupying. Backyard farming is definitely one way to combat the issue of theft and praedial larceny, but more extensive farming also has a role to play in food security. On larger smallholder farms, one way of increasing the human presence is to build temporary cabins where workers can dwell. This solution can help to lower living expenses of farm workers, while having mutual benefits to workers and owners if occupation is rent-free, in exchange for surveillance. Cabins have added benefits to their uses by providing a mechanism for water harvesting from rooftops. Additionally, living in cabins on the farm site gives the option of flexibility of work hours so that work can be performed in the cool hours of the morning and late afternoons, as well as on weekends when some routines have to be performed.

While small ruminants are vulnerable to dog attacks, poultry and pigs are protected in pens or cages. It is not feasible to raise ruminants on small plots of land, and this type of farming is more suitable for large operations or designated areas in small countries where they are protected and monitored.

Irrigation

Finding: *About 50% of all commercial farms are below 1 acre in size, with a combined total of about 70% being less than 3 acres, or 75% being less than 5 acres and are predominantly rain-fed or lack consistent water supply.*

Finding: *About 60% of all farms use irrigation and about 70% of these use overhead sprinklers. However, about 60% of those irrigating depend on rain-fed supply.*

In a traditional setting, those farm sizes are too small. It is also not feasible to make the input of costly infrastructure and irrigations systems when farms are small and fragmented. Followed by land availability, water is of utmost importance to increasing productivity and food security, and therefore, government support is essential on both fronts.

The introduction of non-traditional growing reduces the amount of land needed to be productive enough for both subsistence and production at higher scales. However, communal infrastructure may be beneficial to serve several farmers, for example, the system of empoldering found in Guyana and Suriname. Empoldering is essentially a technique of Dutch origin, involving irrigating large expanses of land (thousands of acres) with irrigation and drainage canals running in parallel but with varying levels. The water is supplied from lakes or conservancies (where lakes do not exist, dams can be introduced to form a conservancy) into the irrigation canals and utilized on farms. The spent water is discharged in the drainage canals that are at a lower level than the irrigation canals. The water is regulated by sluices at various intervals along the canals. An overseer for the commune or empoldered area monitors fairness in distribution and usage, with the added benefit of surveillance on farms. In environmental conditions where there is sloping land or porous soils that do not retain surface water, irrigation pipes with open-shut valves can be used to regulate water from artificial reservoirs in higher elevations where rainfall is available. Furthermore, in environments where freshwater is scarce or prone to drought, the problem is typically not a lack of water entering the drainage basin but insufficient water retention or harvesting. In SIDS, the water drains efficiently and rapidly into the sea. Artificial dams and reservoirs are required to increase water retention and availability to replace rain-fed agricultural practices that are prevalent in SIDS.

Where freshwater supply is scarce, drip-irrigation is most effective to conserve water. Drip irrigation is focused on the root systems and minimizes evaporation that is higher when water sprinklers are used. Mulching is also recommended to minimize evapotranspiration, with the caution that plastic mulch also reduces utilization from direct rainfall from which some crops can benefit, although plastic mulch prevents leaching and erosion. Mulch from biomass has beneficial effects in water retention, and also in the control of certain pests. Using mulch is also a way of recycling nutrients from the farm, and it can also be added to compost and then utilized as organic fertilizer instead of mulch. A technique that is used on a large scale in arid and semi-arid conditions is the development of swales simply by creating a depression on the land and piling organic matter from plant cuttings. It is allowed to mature by decomposition and then used for planting of crops. Swales thus enhance soil water retention and increase available organic nutrients.

Livestock, Soil Maintenance, and Feed Supply

Finding: *Livestock are traditionally of five main types: cattle, goats, pigs, poultry, and sheep. Poultry is used mainly for eggs with an insignificant production of broilers resulting in a heavy dependence on imported chicken.*

Finding: *There is a prevailing high demand for food crops and livestock products, but typically an inadequate domestic supply. As such, there is hardly any difficulty in finding a market or needing long-term storage.*

Mixed farming with integration of crops and livestock are found to be more successful (Naraine, 2005). The key to their success is the mutually beneficial effects with recycling of nutrients by which crops benefit from organic fertilizers and livestock benefit from plant matter or unsold produce. As noted previously, in item #1, such farms have achieved optimum diversification with as many as 5 types of crops and livestock combined. Diversification reduces risk.

There are opportunities on mixed farms to produce soil and reduce the need to buy fertilizers. Additionally, aquaponics or fish ponds on farms are excellent for irrigation of plants instead of using chemical fertilizers, while productivity is increased with fish supply, for example, tilapia as a source of protein in our diet. It is not unusual to see soil imported from MDCs for use in LDCs, a practice that is appalling in view of the high feasibility of producing soils locally. Mixed farming is one way to produce soils on a small scale, but there is also the potential in LDCs to produce soils on a large scale, considering that biomass and topsoil from clearing land for construction of residential and commercial properties end up in landfills and dumpsites.

The major issue with production of pork and poultry is the cost of feed. In most cases, there is no local production of feed for pigs and poultry, and when it is produced there is a higher demand than supply. The cost associated with importation of feed, storage, and the short shelf-life before degradation, is essentially too high a risk to render it feasible for farmers. Some pig farmers utilize the spent grain from local breweries as a supplement to imported feed. Poultry for egg production is generally more feasible compared to broilers for meat that require high amounts of protein in a relatively shorter time span than for layers that continue to produce eggs well beyond their rapid growth stage. The production of feed remains more feasible in large countries, for example, Brazil, Guyana, and Suriname with large production of rice, corn, and soybeans. A better system of marketing and distribution is key to supplying SIDS with meat or raw materials from which countries can manufacture feed locally instead of importing feed for which shelf-life and transportation cost present challenges. The local production of feed, even if it were from imported ingredients, can be an opportunity to ensure reliability of fresh supply and higher-quality meat than those imported and frozen for extensive periods.

Capital Investment and Land

Finding: *With regards to farm assets, real property averages only about EC$12,000 (USD$4,444).*

This level of investment is very low in view of the importance of the role of local food production to achieve food security. Essentially, farmers are not willing to invest, maintain, or improve land that they do not own. However, this low level of investment is generally adequate for backyard gardening and non-traditional growing systems at small scales, particularly in small spaces that they own or lease long-term.

Capital Investment and Machinery

Finding: *Heavy equipment averages about $25,000 (USD$9,300) while light equipment averages about $3,000 (USD$1,100), with almost all farms having light equipment.*

Heavy machinery is typically owned and operated by government agriculture departments to give support to farmers. However, these operations are typically inefficient and their machines are often inoperable or in need of maintenance. Therefore, farms, both small and large scale, are not highly mechanized. In isolated instances, there are private owner-operators who contract with various farmers, but there is still a general scarcity of heavy machinery when needed. Where agriculture associations or co-operatives exist, there can be opportunities for community owned machinery with a fee structure that is proportionate to hours of use, but with the fees as primarily a means of covering operation, maintenance, and replacement costs.

Access to Credit

Finding: *Less than 15% of all farmers receive farm input credit, with the main source of credit being private lending institutions, with as much as 13% interest rate.*

Vulnerability in this aspect stems from lack of business skills. Education in business practices and entrepreneurship, addressed below, will lead to better outcomes with banks or other lenders or mechanisms.

The Bangladesh model of micro-financing with the now controversial figure of Muhammad Yunnus has had extensive impact in bringing millions out of poverty. Each nation or community has its own circumstances and a solution in one area may not apply in other areas. The lesson to be learnt is that micro-financing is essential for development of communities practicing small-scale farming. Mechanisms can range from formal programs by banking institutions or development banks, to informal small circles of people in communities who pool monies and lend to members on a rotational basis.

Educational and Organizational Adaptations

Agricultural Education and Youth Engagement

Finding: *About 70% of all farmers are between the ages of 35-54 years old, contrary to the literature that suggests most farmers in St. Kitts are older or retired and younger people are not interested in farming.*

However, youth involvement is not fully utilized and has a great potential with non-traditional growing systems such as hydroponics. This was noted in higher enrolments in college programs, for example, at the Clarence Fitzroy Bryant College, in St. Kitts and Nevis with emphasis on this increasingly popular technology. Youth engagement can be achieved by educating more persons to get better results and also by using social media to network farmers to share experiences and knowledge of best practices. According to Simoes et al. (2010), education is a key to adaptive capacity and building resilience through greater understanding of the natural environment and knowing how to develop sustainable systems to increase productivity. Furthermore, with formal and non-formal education, all age-groups, socio-economic groups, and women can be targeted so that a greater percentage of any human population can enter into economically productive activities. The OAS-funded project under the theme, "Transfer of knowledge in science and engineering, and the promotion of innovations and entrepreneurship" engages all age groups, male and female, from formal and non-formal educational backgrounds in non-traditional agricultural systems and is networked through social media. The non-traditional growing systems are suitable for greater inclusion of women, youths, and those physically challenged or with health issues. This is possible because of the lower physical stress required to operate such systems.

Popular Education and Vocational Training

Finding: *Less than 15% of all commercial farmers have attained technical and/or vocational training.*

The OAS-funded project, referred to previously, includes an educational response through a toolkit, a formal curriculum, and a non-formal educational documentary on non-traditional agricultural systems that can be delivered as a short-course with a competency-based approach.

Entrepreneurial Training

Finding: Economic: *About 80% of all farmers do not keep business records and over 90% do not have a farm business plan.*

Related to improved credit access, and arguably a more important adaptive strategy, is business education to do better planning and minimize risks. Formal and non-formal competency-based training with incubation support is prudent, irrespective of whether or not it is from micro-financing sources or self-funded enterprises.

ICTs and Extension Services

Finding: *The majority of farmers by far, receive farming information from radio and other farmers; not from the Agriculture Department.*

With recent developments in smart phones and widespread availability and connectivity on the Internet, there is greater opportunity for information sharing and networking around farming interests. Agriculture Departments can utilize ICTs to disseminate quality information to farmers. One effective arrangement is to have a space in each local community or farming area designated for Extension Officers to set up on-going sessions for demonstrations, experimentation, information dissemination and training on current issues relevant to that area. Too often, Extension Officers spend more time at their offices at a central location, where they also conduct sessions.

Professional Organizations, Political Impact, and Extension Services

Finding: *About 90% of all farmers do not have any membership or affiliation with agricultural based organizations, and there is scarcely any organized farming group.*

With more education, people are more likely to understand the importance of organizing around agriculture. Formal and continuing non-formal education is perhaps the most effective strategy for engaging a generation of more knowledgeable farmers. Although there may be several registered organizations in some communities, membership is low and meetings are infrequent and not effectively organized around issues that are significant to the development of the sector. Encouraging more of such organizations is now likely to yield greater success if guided to overcome these constraints. For this reason, there is a greater need for government agriculture departments to play a more active supporting role and be held more accountable to farmers.

POLICY RECOMMENDATIONS

We yearn for economic systems within our control, within our comprehension and that once again provide space for human interaction – and yet we are constantly overwhelmed by finding ourselves trapped into vast global economic systems that are corrupting and corrupt. (Bunting)

Recommendation for Policy Design and Implementation Strategy

At the apex of the hierarchy of needs—and the most important ethical responsibility faced by policy makers in emerging economies—is food security. Efforts to achieve this goal must begin at the household or individual farm enterprise level to meet basic nutritional needs and to combat hunger and malnutrition. The issues, case studies and examples, and adaptive strategies detailed in the previous sections reveal that land reform is essential to transform agricultural economies.

Land reform in LDCs has not been getting traction either because there is no perceived prospect of competing with large agricultural economies in MDCs, or the power is in the hands of resistant private or public elites, or both. Numerous policies at regional and national levels as well as those outlining sectoral strategies have been formulated with attempts for improvement of this situation. However, they have generally not been fulfilled due to failures at implementation, primarily because of insufficient budgetary support for implementation at the individual farm level, compounded by poor management and governance at all levels.

There has been growing success of smallholder farmers and backyard gardening throughout emerging economies of the world. This phenomenon has had a significant positive impact on world hunger despite myriad challenges. Based on this, policies should be focused in support of adaptive strategies, at least in the short- to medium-term, so that the trend can be continued and multiplied.

The approach should be at the enterprise level where individual farms can be assessed in terms of sustainability along input factors of land tenure, culture, technology, economy and trade/markets, and environment. Consideration should also be given to their output in terms of household benefits such as nutritional requirements, livelihoods, and building resilience to climate-related vulnerabilities (Waugh, 2009). Incubation centers can be introduced as a mechanism to minimize risks of farms that are not successful as well as to support new (and hence vulnerable) farming enterprises. Extension efforts should be restructured in terms of location and delivery formats to aid implementation in this framework or model. Extension should also expose smallholders to the benefits of integrating additional components and supporting efforts to create more integrated enterprises through effective communication. There must be policies to incentivize cooperatives and other strategies for expanding the enterprise to the next level in scale of operations.

In the long-term, in the face of long-range forecasts about climate change impacts, farmers and communities need to continue their fight. They should make representation to their elected officials on the basis of ethical responsibilities for decisions to meet the basic needs of food and nutrition of the general citizenry. They require factors such as; land reform, freshwater catchments and irrigation, machinery, fertilizers, pesticides, transportation, and improved varieties of crops and breeds of livestock and improved storage facilities.

Future Research

In emerging economies, food security may well depend on the success of smaller individual farming enterprises. As such, all interested parties need to develop an awareness of different enterprise scales as they learn to understand and adapt larger-scale forms of agriculture to local conditions. For scholars, policy makers, smallholder farmers and other stakeholders, consolidating recent gains in hunger eradication will depend on future research that focuses on six related areas:

- Building a model for larger-scale integrated farm enterprises where individual farms can achieve sustainability, higher productivity, and flexibility.
- The effective and innovative use of information and communication technologies (ICTs) in extension work.

- The development of government policies that promote social protection for small-scale agriculture through regulation of land tenure, credit, state purchasing, and other factors that influence the success of smallholder farming.
- Quantitative assessments that measure the impact of projects in categories such as productivity, profitability, nutrition, and inclusive growth.
- Qualitative assessments that measure factors such as stakeholder uptake of new approaches and techniques, local knowledge, and feedback about extension service and policies.
- Redefining key concepts such as sustainability, subsistence, adaptive capacity and resilience for emerging economies.

CONCLUSION

This chapter has outlined the challenges and opportunities facing smaller, emerging economies as they attempt to improve food security in an ever-changing global economy. While dealing with a range of agricultural sectors, including crop farming, livestock, fisheries, and integrated enterprises that combine elements of some or all of those sectors, several common threads unify the analysis.

First, emerging economies typically depend on smaller individually-owned enterprises and will continue to do so into the future. Research and policy-making must recognize and respond to this reality by seeking to better understand and support the success of smallholder farming in all its permutations.

Second, local, national, and regional governments need to do a better job of supporting individual agriculture enterprises, and educators. Additionally, extension agents need to effectively and creatively target such enterprises with a range of formal and non-formal training and information services. Farmers themselves need to treat their farming—in whatever sector—as a business. This means adopting and maintaining sound techniques in marketing, accounting, communications, and political advocacy. While it goes beyond the scope of this chapter, developing political will on the part of smallholder farmers and the public who rely on them is an essential driver for shaping policies that will benefit smallholder success and ensure food security in developing societies.

Third, there is an assumption in this chapter that smallholder farming successes have a multiplier effect. This idea is validated by the recent FAO report that largely attributes a 20% drop in world hunger to advances in smallholder production. This dynamic is even more important in Small Island Developing States, where individual successes can add up more quickly to positive changes. These can manifest in less vulnerability to food imports and climate variability, more positive indicators in livelihoods and nutrition, and overall strengthening of societal resilience.

As the next generation of agricultural producers and policy-makers takes shape in the Caribbean and across the globe, hopefully this chapter will play a part in highlighting the role of small-holder farmers in sustaining nutritional health and overall food security for emerging economies.

REFERENCES

Alimuga, L., Magala, D., & Lwasa, S. (2007). *Regoverning Markets: Small-scale producers in modern agrifood markets, Uganda--The case of the Nyabyumba United Farmers Group in Kabale district.* Uganda Agricultural Economics Association.

American Marketing Association. (2005). *Definition of Marketing*. American Marketing Association. Retrieved from https://www.ama.org/AboutAMA/Pages/Definition-of-Marketing.aspx

Balogun, E. D. (2003). Foreign Private Investment in Nigeria. *Central Bank of Nigeria Economic and Financial Review*, *37*(2), 31-81. Retrieved from http://www.cenbank.org/OUT/PUBLICATIONS/EFR/RD/1999/EFRVOL37-2-3.PDF

Bunting, M. (2011). Small Is Beautiful—An Economic Idea that Has Sadly Been Forgotten. *Guardian UK*. Retrieved from http://www.theguardian.com/commentisfree/2011/nov/10/small-is-beautiful-economic-idea

CARIBSAVE. (2011). Caribbean Climate Change, Tourism & Livelihoods: A sectoral approach to vulnerability and resilience. The CARIBSAVE Climate Change Risk Atlas – St. Kitts Final Draft Country Risk Profile. The CARIBSAVE Partnership.

Clarke, J. (2013). *Sugar, Land and Sustainable Livelihoods: Global Change and Local Vulnerability in St. Kitts*. (Ph.D. Thesis). University of the West Indies, Mona Campus.

Dawar, N. & Chattopadhyay, A. (2002). Rethinking Marketing Programs for Emerging Markets. *Long Range Planning*, *35*(5), 457–474. Retrieved from http://www.sciencedirect.com/science/journal/00246301

Dixie, G. (2005). Horticultural Marketing. Rome: Food and Agriculture Organization of the United Nations. Retrieved from http://www.fao.org/docrep/008/a0185e/a0185e05.htm#bm05

Dixon, J., Taniguchi, K., Wattenbach, H., & Tanyeri-Arbur, A. (2004). Smallholders, Globalization and Policy Analysis. Rome: Food and Agriculture Organization of the United Nations. Retrieved from http://www.fao.org/docrep/007/y5784e/y5784e02.htm

Edelman, M. (2014). Food Sovereignty: Forgotten Genealogies and Future Regulatory Challenges. *The Journal of Peasant Studies*, *41*(6), 959–978. doi:10.1080/03066150.2013.876998

Ewah, S. O. E., & Ekeng, A. B. (2009). Problems and Prospects of Marketing in Developing Economies: The Nigerian Experience. *International Journal of Business and Management*, *4*(9). doi:10.5539/ijbm.v4n9p187

FAO. (2012). *Trends and Impacts of Foreign Investment in Developing Country Agriculture*. Rome: FAO. Retrieved from http://www.fao.org/fileadmin/templates/est/INTERNATIONAL-TRADE/FDIs/Trends_publication_12_November_2012.pdf

FAO. (2013). *FAO Policy on Gender Equality: Attaining Food Security Goals in Agriculture and Rural Development*. Rome: FAO.

FAO. (2013). *A Tool for Gender-sensitive Agriculture and Rural Development Policy and Programme Formulation: Guidelines for Ministries of Agriculture and FAO*. Rome: FAO.

FAO. (2015). *Saint Kitts and Nevis and FAO*. Food and Agriculture Organization of the United Nations. Retrieved from http://www.fao.org/3/a-av273e.pdf

Hu, A., Acosta, A., McDaniel, A., & Gittelsohn, J. (2013, January). Community Perspectives on Barriers and Strategies for Promoting Locally Grown Produce from an Urban Agriculture Farm. *Health Promotion Practice*, *14*(1), 69–74. doi:10.1177/1524839911405849 PMID:22102602

IFAD. (2013). *Smallholder farmers key to lifting over one billion people out of poverty.* International Fund for Agricultural Development. Retrieved from http://www.ifad.org/media/press/2013/27.htm

Kelly, C., & Schulschenk, J. (2011). Assessing the vulnerability of Stellenbosch's food system and possibilities for a local food economy. *Development Southern Africa*, *28*(4), 563–578. doi:10.1080/0376835X.2011.605575

Mafuru, J. M., Babu, A. K., & Matutu, T. F. (2007). Tanzania: The impact of market links on horticultural production in the Mara region. *Regoverning Markets.* Lake Zone Agricultural Research and Development Institute. Retrieved from <http://pubs.iied.org/pdfs/G03265.pdf>

McKenzie & Woodruff. (2015). Business Practices in Small Firms in Developing Countries. *International Labour Organization*. Retrieved from http://www.ilo.org/wcmsp5/groups/public/---ed_emp/documents/publication/wcms_360368.pdf

Murphy, S. (2001). The Global Food Basket. *Forum for Applied Research and Public Policy*, *16*(2), 36–42.

Naraine, L. (2005). *Agricultural Diversification in St. Kitts: Policy and Implementation. St. Kitts-Nevis*, Department of Agriculture.

Nebel, B., & Wright, R. (1996). *Environmental Science: The Way the World Works*. Upper Saddle River, NJ: Prentice Hall.

Reardon, T., & Berdeagué, J. A. (2002). The Rapid Rise of Supermarkets in Latin America: Challenges and Opportunities for Development. *Overseas Development Institute: Development Policy Review*, *20*(4), 371–388.

Simoes, A. (2010). Enhancing adaptive capacity to climate change: The case of smallholder farmers in the Brazilian semi-arid region. *Environmental Science and Policy,* (13), 801-808. Retrieved from www.sciencedirect.com

Timmer, C. P., Falcon, W. P., & Pearson, S. R. (1983). *Food Policy Analysis*. The John's Hopkins University Press. Retrieved from http://web.stanford.edu/group/FRI/indonesia/documents/foodpolicy/chapt4.fm.html

Vervisch, T. G. A., Vlassenroot, K., & Braeckman, J. (2013). Livelihoods, power, and food insecurity: Adaptation of social capital portfolios in protracted crises—case study Burundi. *Disasters*, *37*(2), 267–292. doi:10.1111/j.1467-7717.2012.01301.x PMID:23278356

Waugh, D. (2009). *Geography: An Integrated Approach.* Cheltenham, UK: Nelson Thornes.

World Bank. (2010). *Enterprise Survey: Saint Kitts and Nevis.* The World Bank. Retrieved from http://www.caribank.org/uploads/2014/11/2014-St.-Kitts-and-Nevis-PSAR.pdf

World Bank. (2015). *Data: Agriculture and Rural Development.* The World Bank. Retrieved from http://data.worldbank.org/topic/agriculture-and-rural-development

World Commission on Environment and Development. (1987). *The Bruntland Report: Our Common Future*. Oxford, UK: Oxford University Press.

KEY TERMS AND DEFINITIONS

Emerging Economy: Economies that have passed their early growth stage, i.e. from their primary-based economy, and rapidly expanding to an industrialized base to become more developed. Some economies that have not become industrialized may rapidly expand into the service-based economy through tourism and/or offshore banking.

Food Security: The ability of a country to meet its basic food requirements from local agricultural production and without heavy dependence on imported foods.

Resilience: The ability of a community to recover from natural and technological disasters, and external shocks relating to commerce and trade. The better the livelihood of people in communities, the more resilient they are to disasters and external shocks.

Small Island Developing States (SIDS): A category of lesser-developed countries of the world that is mostly located in the tropics, such as the Caribbean Sea, Indian Ocean, and the Pacific Ocean. Their communities are primarily located in low-lying coastal areas and are susceptible to natural disasters, including volcanoes, as well as tropical storms and sea-level rise that have intensified due to climate change.

Smallholder: A farmer who operates a farm of no more than 5 acres on either a subsistence level or semi-commercial level. At such level, the farm is usually not highly mechanized.

Sustainability: In the context of the environment, it is the ability to utilize natural resources without exceeding the capacity to regenerate themselves to meet current needs of a community and without compromising the needs of future generations.

Chapter 4
Appropriate Extension Methodologies for Agricultural Development in Emerging Economies

B. S. Famuyiwa
Cocoa Research Institute of Nigeria (CRIN), Nigeria

O. A. Olaniyi
Ladoke Akintola University of Technology, Nigeria

S. A. Adesoji
Obafemi Awolowo University, Nigeria

ABSTRACT

Over two-thirds of the world's poorest people are located in rural areas and engaged primarily in agriculture and agricultural-related activities for their livelihood. Therefore, the future of most underdeveloped and developing nations depends on agriculture. Most African Countries have agricultural dependent economies that are hinged to rain-fed agriculture and based on smallholdings. These countries are referred to as countries with emerging economies and characterized as having low to middle per capita income and represent 20% of the world's economies. This chapter discusses; concepts of agricultural extension methodologies assessed from past to present, appropriating extension methodologies to encourage agricultural development, identifying roles of agricultural extension activities in agricultural development and factors influencing the choice of appropriate extension methodologies in emerging economies. It concludes with the constraints to sustainable agricultural development and extension methodologies which if removed will have the potential for progression towards economy development. Developing countries should fashion a sustainable extension system that will be socially acceptable and culturally compatible, economically viable and environmentally friendly.

DOI: 10.4018/978-1-5225-0942-4.ch004

INTRODUCTION

More than two-thirds of the poorest of the poor populations of the world are located in rural areas and derive their livelihood from the rural environment. More than 70% of such people in developing nations are engaged in agriculture and agricultural-related activities. Therefore, the future of most underdeveloped and developing nations depends on agriculture. Most African nations have agricultural dependent economies that are hinged to rain-fed agriculture and based on smallholdings. The smallholders depend mostly on traditional tools which, though easily maintained, are energy sapping and make agricultural production laborious. Many of the innovations that are developed by researchers, especially in the universities and research institutes, are wasting away on their shelves. Agricultural Extension agencies that are tasked with disseminating the developed innovations, are either moribund, the innovations/technologies have not been transferred to them or innovations are inappropriate. In some instances, scientists are not ready to release their innovations because they think that the extension agencies that disseminate the innovation to farmers will take their glory. All ofthese contribute to continued poverty among farmers.

Extension is a service or system which assists farm people, through educational procedures, in improving farming methods and techniques, increasing production efficiency and income, bettering their levels of living and lifting the social and educational standards of rural life (Maunder,1973). Agricultural extension can be defined simply as the transfer of useful information from a source to the farmers or rural dwellers. Swanson et al. (1997) defined agricultural extension as the extending of relevant agricultural information to people. Piercell and Anderson (1997) defined agricultural extension as a process of assisting farmers to be aware and adopt improved technology from reliable sources to enhance production efficiency, income, and welfare of farmers. The main goal of agricultural extension is to improve the socio – economic condition of farmers and rural dwellers. As a discipline, it finds itself articulating appropriate knowledge and decision-making capacity to determine how factors of production can be best utilized to improve the standard of living of rural dwellers through appropriate extension methodologies (Famuyiwa, 2013).

The general objectives of extension are to:

1. Assist farmers to discover and analyze their problems and identify their felt needs.
2. Develop leadership among farmers and help them in organizing groups to solve their problems.
3. Disseminate research information of economic and practical importance in a way farmers would be able to understand and use.
4. Assist farmers in mobilizing and utilizing the resources which they have and to identify what they need from outside.
5. Collect and transmit feedback information for solving management problems.

Agricultural extension, therefore, provides the indispensable elements that farmers need to improve their agricultural activities. This has made agricultural extension to be a crucial input in agricultural development, especially in Nigeria.

Most developing nations of the world especially in Africa have implemented various agricultural extension methodologies. Most of the methodologies are transferred from a country where it has been successfully implemented, to other countries regardless of the farming system and culture of the country. In addition, some funding agencies also sponsor and fund agricultural extension methodologies in developing countries. As soon as the funding ceases, the project also ceases. A good example is Agricultural

Development Project (ADP) which was jointly funded by the World Bank, Federal and State governments in Nigeria. The ADP operated the Training and Visit (T&V) agricultural extension approach in Nigeria. This project was implemented between 1985 and 1995. But, after the cessation of the World Bank counterpart funding in 1995, technology delivery by the ADP progressively declined to the extent that State's ADPs found it difficult to pay salaries and to provide funding for other operating transaction costs (Agwu &Chukwuone 2005). This decline in the World Bank funding according to Havnevik et al. (2007) was not limited to Nigeria, it was part of the Structural Adjustment Policy of the Bank that adversely affected Africa. Though successes were recorded during the implementation of the project, it was not economically viable. If developing nations could create sustainable technologies, these would support their systems over a long period.

In the light of this, developing countries should fashion a sustainable extension system that will be socially acceptable and culturally compatible, economically viable and environmentally friendly. It has been discovered that most agricultural extension approaches and methodologies that were borrowed or sponsored by external agencies in developing countries did not involve farmers in a participatory process. Any sustainable agricultural extension approach or methodology to be adopted in developing countries, must guard against the above mentioned issues if rapid growth in agriculture is to be delivered and to enhance food security.

It might be useful at this point to differentiate between the terms 'approaches' and 'methods' in agricultural extension as used in this chapter.

Approaches: An approach in agricultural extension connotes the way or avenue to get things done. It can also be seen as the manner in which a problem is solved. Anandajayasekeramet al. (2002) referred to the concept "approach" as the "essence of agricultural extension system". They added further that "approach is the style of action within a system and embodies the philosophy of the system". In an agricultural extension system, an approach is the framework and structure of the system which serves as the pathway that guides in a program of action.

Method: This refers to the process and techniques used by extension agencies to get a task done. Examples include visits by extension agents to a group or a family, demonstration of skills to a group by extension agent and so on.

This chapter reviews the past and present agricultural extension approaches and methods. The understanding of these along with an in-depth view of their success and failures will assist in building appropriate and sustainable agricultural extension approaches and methods for the future. These will bring rapid development to the wellbeing of farm families, agriculture and the environment.

FACTORS INFLUENCING CHOICE OF APPROPRIATE EXTENSION METHODOLOGIES IN EMERGING ECONOMIES

Emerging economies are defined and characterized as economies having low to middle per capita income and represent 20% of the world economies. However, they can further be described as countries having low levels of human capital, high levels of poverty and under-nutrition, and high population growth rates. There are also the predominance of agriculture, low levels of industrialization, and urbanization but rapid rural-to-urban migration, dominance of informal sector, underdeveloped labor, financial, and other markets (Kumar, 2015 and Wikiversity, 2015).

For easy dissemination and diffusion of innovation, some cogent factors have to be appropriately considered in choosing the method to be used among the array of methods available for extension activities. The majority of agricultural projects and programs that are extension based, fail or are moribund because some of these factors are not carefully considered.

Technological Applicability

Technology improvement is an indication of development. However, challenges are faced in the areas of applicability, adaptability, and acceptability into the system. Before now, dissemination of innovation was undertaken using a top-down approach which makes diffusion difficult and adoption almost impossible. The alternative Farmers' Participatory Approach (FPA), has so far yielded better results. This approach promotes the involvement of the target group, the farmers or the end-users, right from the conception of the idea through to the completion. This approach is being employed in National Fadama programs implemented in Nigeria. This approach empowers the target group to input their ideas for sustainable adoption of the innovation. The general aim is to see if the innovation fits into their felt needs.

Economic Feasibility

Economic feasibility of a method is very important for agricultural development. The final goal of development has to do with the cost effectiveness of the method and its approach. A sustainable method is the one that is market and farmer driven; meeting the needs of consumers and profitable to the farmers. In the case of technology or innovation to be adopted by farmers, it will only be appropriate if it adds to the income of farmers, either by cost reduction or by yield increase.

Another area of economic feasibility advanced by Campbell and Barker (1998) asks the question: Do farmers have financial resources to implement the innovation, in order to derive the benefit that will accrue? In the case of drying cocoa beans, this has been a very serious problem in developing countries that produce 75% of the world cocoa. Does the individual farmer have the financial capability to buy a Simona dryer, which was developed as an alternative and useful method for drying cocoa beans in humid areas?

Environmental Safety

There are inter dependencies in the systems that make up an environment; the social system and the ecosystem (Marten, 2009) as shown in Figure 1. These two systems interact with each other and they both transmit energy, materials and information to each other. The social system is represented as the human activities, while the ecosystem is made of biotic and abiotic organisms. These interactions need to be guided through appropriate methods to enhance environmental safety.

Serious environmental issues have developed in emerging economies because of farmers' low income, high population, and high illiteracy level. Indiscriminate use of pesticides and other chemicals on the farm have been identified by different studies (Sosan and Akingbohungbe, 2009; Ogunjimi and Farinde, 2012 and Famuyiwa, Torimiro, and Adesoji, 2013). The diagram below as described shows the interaction between environmental elements; social system and ecosystem.

Technology in the social system brought about the use of pesticides for the control of pests on cocoa farms. Farmers, because of lack of knowledge on the use of pesticides or their anxiety to increase yields,

Figure 1. Interaction of the human social system with the ecosystem (Marten, 2009)

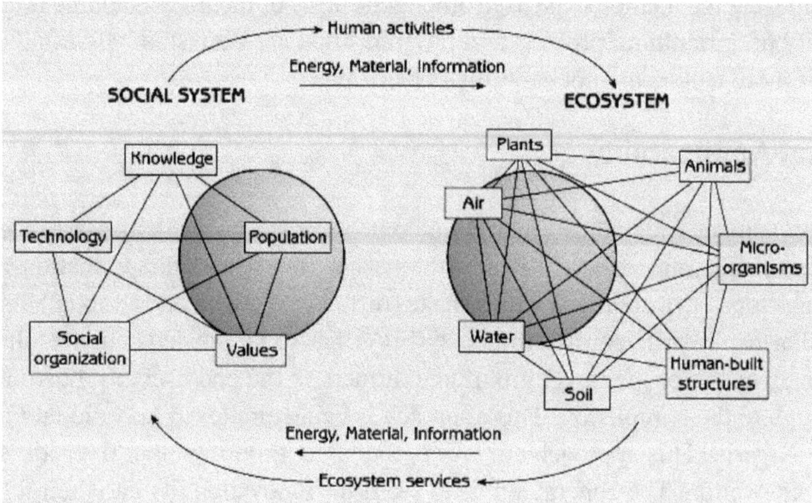

tended towards the indiscriminate use of pesticides. The indiscriminate use of chemical/pesticides which is a human activity from the social system consequently pollute the ecosystem and brings about chemical residue in farm products like cocoa beans that are made into chocolate and consumed by human beings. These chemical residues have been identified as causative agents of diseases such as cancer.

The goal should, therefore, be to carefully design dissemination methods of non-pesticide procedures or organic farming approaches to increase farmers' yield and the marketability of their farm produce.

Socially Acceptable

A social system is an interrelationship between the linkages that exist among all the elements of a society namely; individuals, groups, and institutions that make these elements form a whole. Since these elements are knitted together, moving the society from the status of developing to developed, will require methods that will be acceptable to them. However, introducing a paradigm shift needs a well-designed methodology to achieve acceptability.

Culturally Acceptable

The cultural system is located within the larger society or the social system. It is defined as interrelationship of six elements namely; traditions and norms; language; arts and literature; religion and belief systems; forms of government; and economic systems. Agricultural productivity that leads to development depends largely on how knowledge is adequately managed, shared and applied. Early theorists emphasized that behavioral change among farmers is a prerequisite of the agricultural modernization process. Hence, in order to achieve development in emerging economies, the method of dissemination and appropriate diffusion of innovation need to be carefully chosen. No matter how beautiful an innovation is, if it is not culturally acceptable the required adoption level will not be attained. This factor is highly considered in designing extension programs.

Sustainably Suitable

Sustainability is the ability to undertake a project or process using the present resources, without compromising the future resources. ICCO (2008), posited that practices of cocoa farming should be considered as sustainable in terms of economic, environmental and social. However, the United Nations (2015) in its meeting (25th – 27th September 2015) deliberated on 17 points Sustainable Development Goals (SDGs) by member countries, the majority of which are emerging economies. The goals SDGs discussed include the issues of;

1. No poverty
2. Zero hunger
3. Good health and well-being
4. Quality education
5. Gender equality
6. Clean water and sanitation
7. Affordable and economic growth
8. Decent work and economic growth
9. Industry, innovation, and infrastructure
10. Reduced inequalities
11. Sustainable cities and communities
12. Responsible consumption and production
13. Climate action
14. Life below water
15. Life on land
16. Peace, justice and strong institutions
17. Partnerships for the goals

Audience Configuration

A proper understanding of the audience is another key factor in program development. Messages need to be specific as audiences are also specific. There is divergence in culture, gender, education and access to resources. The International Food Policy Research Institute (IFPRI, 2011), identified that gender relations are culture and context specific. It was specifically stated that addressing gender issues is one of the most effective ways of achieving development and reducing poverty in the society.

Content Relevance

The content flow between research and farmers is termed "information"; information about methods, innovation or advice. Information to meet the competitiveness of the stakeholders in agricultural development has to be content-specific to achieve desired sustainable economic growth. Studies according to Adedeji et al. (2013), and Mwangi and Kariuki, (2015), showed that the adoption rate of most technological innovations disseminated in emerging economies is low. The majority of these failures (low adoption rates) are attributed to inappropriate methods of extension delivery of the technology; innovations are not economically feasible, socially acceptable, technologically adaptive, culturally compatible, environmentally compliable and hence the innovations are not sustainably practiced. Technology in isolation, without regular guidelines by the extension professionals and the use of appropriate methodologies, cannot help in enhancing the productivity of any economy. Hence, it has to be content relevant to the specific environment where it is to be promoted.

Role of the Agricultural Extension in Agricultural Development

The role of agricultural extension in the past was to transfer new technologies from researchers to farmers. This role has not been able to meet the developmental challenges in developing countries.This is because most of the technologies disseminated to farmers have not been able to solve farmers' production problems. Agricultural extension's role in the recent past in most developing countries was seen to help farmers make their own decisions. Extension only increases the number of options from where the farmers can choose and also highlights the consequences of each option. At this point, it is the role of extension to answer questions that may arise when the knowledge of the farmers is limited. In a way, extension still guides in decision making.

Agricultural extension promotes and enhances the learning process among farmers. Farmers have many avenues by which they can learn. For example, people say, 'experience is the best teacher'. Farmers learn from their past experiences, from neighbours, from extension demonstrations as well as discussions with other experienced farmers. Agricultural extension must assist farmers in this learning process. Farmers are known to be knowledgeable in their area of production; extension should not look down on the farmers, but should be able to also learn from the farmers.

For extension to achieve its own set objectives, it is important to do what other agencies have left undone. For example, extension should include sources for agricultural inputs as well as marketing information in their programs or activities. It will be a waste of time and effort for an extension agent to introduce the use of fertilizer to a group of farmers who do not know the source of the input, fertilizer. Also, it will not be helpful for both the extension agent and the farmer to have a bountiful harvest without a market for the harvest. Providing guidance to the farmers in this regard would afford credibility to the extension agent and hasten the economic development of the farmers. Extension may also assist in supervising the appropriate use of loan/ credit and ensuring regulation adherence.The approach of the Cocoa Research Institute of Nigeria (CRIN) in solving coffee farmers' problems is designed as market and farmer driven approach to change farmers' attitude towards coffee cultivation. When farmers' problems are addressed with effective solutions, it is always easy to adopt innovations.

It has been observed that valuable local knowledge and resources are underutilized. Extension should harness the knowledge of 'indigenous' and rural people that are often termed as 'primitive' or 'unscientific', then pass them to researchers to transform and improve. This is important because what the farmers know and believe can develop them.

Appropriateness and Selection of Extension Methodologies for Agricultural Development

Extension is an ongoing process of getting useful information to people and then assisting those people to acquire the necessary knowledge, attitudes and skills (ASK) to utilize this information or technology for the purpose of improving their quality of life. The term, 'agricultural extension', narrows the focus and defines the area to which the extension process is applied. Without extension, the farmers would lack access to the support and services required to improve their livelihoods and other productive activities.

Extension is not concerned directly with generating knowledge. This is done in specialized institutions such as agricultural research centres, agricultural colleges, and Universities of agriculturc.

Researchers neither have the time nor are they equipped and trained for the job of ascertaining from the farmers their field problems and of persuading farmers to adopt technologies. Similarly, it is not always practical for the farmers to visit the research stations and obtain first-hand information. Thus, there is a need for an agency that could interpret the findings of research to the farmers and to carry the problems of the farmers to researchers for solutions. This process is guided and supported by Extension Agents of the extension agency who work at the field levels and have direct contact with farmers and their families. The task of extension personnel is bringing scientific knowledge to farm families in the farms and homes. The objective of the task is to improve the efficiency of agriculture.

AGRICULTURAL EXTENSION SYSTEMS/ APPROACHES IN NIGERIA

A number of systems have been developed and applied in the process of agricultural development through various programs and projects with extension outfits in Nigeria. The extension system/approaches are categorized into two:

1. General Clientele Approach: These include a variety of approaches such as
   a. Ministry-based general extension
   b. Training and visit extension (T&V)
   c. Integrated (Project) approach
   d. University-based extension
2. Selected Clientele Approach: These include:
   a. Commodity based extension
   b. Commercial extension service
   c. Client-based and client-controlled extension

In an attempt to realize the expectations of increase in food production for the teeming population, successive governments in Nigeria have, at one time or the other, put in place many agricultural policies and programs as well as projects. Such programs that were implemented include, as shown in Box 1.

All these programs and projects were targeted at enhancing food security and improving the standard of living of farmers through the use of various agricultural extension approaches and methodologies.

Extension Methods Incorporated in Various Agricultural Development Programs in Nigeria

Successes recorded in Nigeria in relation to agricultural development, could be traced to the implementation of agricultural programs with extension components and coupled with the selection of appropriate extension methodologies. Agricultural Extension has two dimensions (a) Educational and (b) Communicational. Both of these dimensions must be incorporated into agricultural development programs to achieve successful results.

Box 1.

The National Accelerated Food Production Programme (NAFPP, 1972); First generation Integrated Agricultural Development Projects (1st generation ADPs, 1975); Operation Feed the Nation (OFN, 1976); River Basin Development Authorities (RBDAs, 1976); National Committee on Green Revolution (NCGR, 1979); the World Bank Assisted Agricultural Development Projects (2nd generation ADPs, 1986); Better Life Programme (BLP, 1989); National Agricultural Land Development Authority (NALDA, 1991); Family Support Programme (FSP, 1992); Family Economic Advancement Programme (FEAP, 1997); Fadama Development Programme (FDP, 1982); National Special Programme on Food Security (NSPFS, 1994) and the Poverty Eradication Programme (PEP) since 1999. Other efforts of note are the involvement of young school leavers and graduates in agricultural production through programs such as Farm Settlement Scheme (FSS, 1957); National Directorate of Employment (NDE, 1986); School to Land and Back to Land programme (SLBLP, 1985), National Economic Empowerment and Development Strategy (NEEDS, 2004) and recently the Farmer fields School (FFS) and Growth Enhancement Scheme (GES) under the Agricultural Transformation Agenda (ATA) (Olaniyi, 2010).

The *educational dimension* involves the change agent (extension worker) attempting to build human capacity through the use of effective teaching methods to improve knowledge and develop the skills of the clientele (farmer) along with acquiring knowledge on improved practices to enhance production and marketing activities (Williams et al.1981). The purpose of education is for immediate action or to solve a problem, hence it is different from the formal education.

The *communicational dimension*: This involves dissemination of information to the clientele on new technologies or innovations. It also involves exchange of ideas that could provide the basis for research or feedback on the use of previous innovations. Therefore, extension is involved in dissemination of useful information using the appropriate methods.

With various extension approaches, the choice of the communication channel or method generally depends on the number and location of the target audience and the time available for communication. The extension methods are categorized into: individual methods, group methods, and mass methods. Each of these methods has both advantages and limitations. The extension agent has to choose a particular appropriate method or combination of methods according to the needs of the situation. Let us now look at different extension methods in detail and their importance for small farmers in developing countries.

Individual Methods

Individual methods, also known as face-to-face methods, are the most universally used extension methods in both developed and developing countries especially in Nigeria. The extension agent meets the farmer at home or on the farm and discusses issues of mutual interest, giving the farmer both information and advice. The atmosphere of the meeting is usually informal and relaxed, and the farmer is able to benefit from the agent's individual attention. Individual meetings are probably the most important aspect of all extension work and are invaluable for building confidence between the agent and the farmer.

Learning is very much an individual process and, although group methods enable the agent to reach a greater number of farmers, personal contact and the individual attention of the extension agent are important supports for a farmer. The personal influence of the extension worker can be a critical factor in helping a farmer through difficult decisions, and can also be instrumental in getting the farmer to participate in extension activities. A farmer is often likely to listen to the advice given by the extension agent and will be grateful for this individual attention.

The individual contact method between the extension agent and the farmer may take a number of forms. These include:

Farm and Home Visits

Farm and home visits are the most common form of personal contact between the agent and the farmer and often constitute over 50% of the agent's extension activities. Because they take up so much of the agent's time, it is important to be clear about the purpose of such visits and to plan them carefully. Farm and home visits can: familiarize the extension agent with the farmer and family; enable provision of specific advice or information to the farmer; build up the agent's knowledge of the area, and of the type of problems which farmers face; permit explanation of a new recommended practice or follow up and observe results to date; arouse general interest among the farmers and stimulate their involvement in extension activities.

The following are involved in Farm and home visits.

1. **Planning for the Visit:** This include: extension agent makes an appointment with farmers if possible, decide the purpose of the visit; review previous records and information; prepare specialized subject matter that might be required; and schedule the visit into the overall work plan.
2. **Making the Visit:**
    - Be punctual for the visit.
    - Greet the farmer and family.
    - Praise the farmer's work.
    - Encourage the farmer to explain and discuss any problems.
    - Provide any technical or other information required.
    - Record the details of the visit.
    - Plan with the farmer the time and purpose of the next meeting.
3. **Recording and Follow-Up:** The purpose of farm and home visit will be lost if its content and conclusion are not recorded and no follow-up action taken. Therefore, record the purpose of the visits and decision resulting from visit; arrange for any follow-up information or advice to be sent and schedule the next visit to the farmer.

Advantages and Limitations of Farm and Visits

Farm and home visits are perhaps the single most important aspect of the agent's work in terms of establishing rapport with the farmers in their area and of building the trust and confidence that are vital to the success of their work. However, farm visits take up a lot of time and only a few farmers can be reached. Farm visits are, therefore, a costly extension method and for this reason they must be carefully thought out and planned. The visits must make an impact and must lead to positive agricultural development if they are to justify their cost. Finally, the agent should beware of visiting some farmers repeatedly. This would not only severely limit the range of activities, but could also arouse the resentment of other farmers who might feel excluded.

When is the methodology for personal contact used? It is worthy to note that individual methods are useful when there is sufficient time or in the slow periods of farming. Also, when the farm family is operating as a production unit.

Office Calls

Along with the extension agent visiting the farmer, it is expected that from time to time the farmer will visit the agent in the office. Such a visit is often a reflection of the interest which the agent may have aroused among the local farmers. The more confidence local farmers have in an extension agent, the more likely they are to visit the extension office. Such office visits are less time consuming for the extension worker, and offer some of the advantages of a farm visit. While no extension agent would wish to be overwhelmed by such visits every day, farmers should be encouraged to drop in, if it is convenient, for them to make the exchange of visits two-way. As with farm visits, office visits similarly have to be prepared. Although the agent may not know when a farmer is likely to drop in, the extension office can be arranged in such a way that the visitor feels at ease and can understand the activities of the office.

Telephone Calls

The deployment of information and communication technology in extension delivery service makes the use of mobile telephones relevant. Telephone calls and office visits serve a similar purpose. It is improbable that the extension worker will deal with many of the farmers in the area (if at all) by telephone. If the telephone is used, however, it should not be for long discussions but for passing on specific advice or information. Whatever the reason, it is important for the agent to speak clearly, to note the main points discussed and to enter them on the farmer's record.

Group Methods of Extension

The extension agent should consider the use of the group approach in work with farmers. The use of groups in extension has become more common over the past decade, and indeed, a number of new ideas have emerged about how groups may be used most effectively. For example, the widespread Small Farmer Development Programme (SFDP) in Southeast Asia was based upon group methods and has produced two manuals which detail the approach of group extension work. Furthermore, in Latin America, work with extension groups in Brazil and Colombia has shown the usefulness to extension of the formation of extension groups, and how these groups can support extension activity. Empirical evidence shows that a group of 5–7 members usually works best for achieving optimum productivity and participation. However, due to cost constraints, normally a group is formed comprising around 20–25 members. Some situations, such as microcredit, input acquisition, and marketing, may demand a larger group for initiatives to be cost effective. Various projects indicate that optimum membership ranges between 20 and 50 people, but groups of less than 20 farmers can also function well as reported by Sanginga et al. (2001), and Pretty, (2003). It is worthy to note that as with any social unit, conflicts within groups may arise. What matters most is how conflict is managed (Anandajayasekeram et al. 2008). Conflict, depending on the types, may be harmful or beneficial to the group. Generally, it is poor management of conflicts that weakens groups as it may lead to dissatisfaction of members, withdrawal or total collapse of the group. This is also applicable to extension work in many countries of the world.

Advantages of Group Methods

It has been seen that individual extension methods can be costly both terms of time and scarce extension resources, and that they reach only a limited number of people. There is also the danger that too much emphasis upon individuals can lead to undue concentration on progressive farmers to the detriment of the poorer farmers. The group method has advantages in the following aspects.

Coverage

The group method offers the possibility of greater extension coverage, and is therefore more cost-effective. Using the group method, the extension worker can reach more farmers and in this way make contact with many more farmers who have had no previous contact with extension activities.

Learning Environment

Extension groups offer a more reflective learning environment in which the farmers can listen, discuss and decide upon their involvement in the extension activity. The support of the group helps an individual farmer to make decisions and determine a course of action. The group creates a supportive atmosphere, and individual farmers can gain greater self-confidence by joining others to discuss new ideas and try out new practices.

Action

The group method brings together farmers with similar problems. Often, these problems demand concerted action (tackling the erosion of a hillside, for example), and such action can be taken more effectively by a group rather than by an individual, who may be overwhelmed by the enormity of certain problems.

Types of Group Extension Methods

Group Meetings

Calling the members of a group or the inhabitants of a local community together for a meeting is the most common group extension method. Although there may be an air of informality about such meetings, they will nevertheless need to be carefully thought out and planned. The group or community meeting is a useful educational forum where the agent and farmers can come together, and ideas can be openly discussed and analyzed. The agent may have information about a new government policy, or agricultural idea or practice. The agent will want to introduce this new information, to seek the opinions of community members and gain their support for extension activities. Indeed, there are a whole range of objectives for such community or group meetings:

Planning the Meeting

There are two important decisions to make regarding the time and location. A date and time for the meeting must be decided and announced. The time should be convenient to all concerned and should

Table 1. Forms of group meetings in agricultural extension systems in developing economies

Types/ Form of Group Meetings	Description	Purpose
Information meetings	The agent calls the group or community together to communicate specific new information which is thought will benefit them and upon which their advice is sought.	Review a particular problem, suggest a number of solutions and decide upon a course of action.
Special interest meetings	Topics of specific interest to a particular group of people (e.g., horticulture, bee keeping, or dairy farming) are presented and discussed in detail at a level relevant to those participating.	Discuss specific issues that are related to particular interest group based on their needs.
General community meetings.	Men, women and young people of a community are invited to attend and discuss issues of general community interest. It is important to hold such general meetings occasionally so as to avoid any community group feeling that it is excluded from extension activities. Whatever the case, however, the agent should only call a meeting if it can be useful. If farmers feel that their time has been wasted in coming to a meeting, they may refuse to come to subsequent meetings and thus frustrate the agent's work. The agent should make careful preparations and check a number of important arrangements which will be necessary to ensure a successful meeting.	Mainly for the purpose of general community interests especially as the situation demands..
Smaller meeting	Smaller meetings are more likely to meet the specific needs of those who attend. When plans are to be made or decisions taken, a small number of representatives will usually achieve more than a large gathering of all community members. On other occasions, it will be important for the meeting to be open to as many people as possible.	Ad-hoc committee to proffer solutions to a particular problem for a group of people.
Lecture and Discussion	A lecture or talk allows the agent (or other speaker) to convey a detailed, well-prepared message to the audience on a specific issue; for example, a new piece of technology, can be presented in this way, and illustrated by visual aids. It should be remembered, however, that the lecture is a particularly tedious approach to meetings and care must be taken to ensure that the audience will not get bored. Alternatively, in a discussion, many people are able to express points of view and ask questions. Discussions may be completely open and unstructured, or based on a prepared agenda of discussion points.	Presentation and deliberation of specific issues.
Formal Meeting	A formal meeting, with a chairperson, agenda and written record of proceedings. The chairperson keeps the meeting to the central issues, and the decisions of the meeting are recorded accurately so that they cannot be disputed later.	Appropriate when specific business/ issues has to be dealt with or decisions are to be reached for a particular group of people.
Informal Meeting	In an informal meeting, people feel free to express their own point of view and less dominated by the structure and formality of the proceedings. However, an open, unstructured discussion, although it allows all to participate, may result in a few people dominating the proceedings.	The same as for formal meeting

Source: Oakley & Garforth (1997)

avoid clashes with other events or activities. The meeting-place should be well-known, easy to access and appropriate for the form of meeting. The meeting-place should also be comfortable and have the necessary facilities. An extension agent should never hold a meeting at midday, on a very hot day, in the open sun. Such a meeting could be disastrous, as well as cause considerable discomfort.

After the above two issues have been considered, it may be useful for the agent to draw up a list of other arrangements to be made in preparation for the meeting. Such a check-list could include:

- Publicity for the meeting
- Seating arrangements
- Audio-visual equipment and material, or other educational aids
- Agenda, and order of events
- Guest speakers or other specialists who will contribute to the meeting
- Chairperson to take charge of the meeting; should be elected by the community
- Refreshments for speakers and, where necessary, other participants.

Conducting the Meeting

Even the most carefully prepared meeting can fail if it is not conducted in the right manner. While the above arrangements are important, the way the actual meeting proceeds will determine whether it will be a success or not. The agent must be conscious that interaction is with adults who do not want to sit for hours listening to a speaker talk endlessly. The agent should try to vary the agenda of the meeting: for example, a short talk, accompanied by visual aids, followed by comments and questions.

Variety of content, as well as a chance for the farmers to participate, will be important. In addition, the meeting must not extend too long. One-and-a-half hours are probably sufficient for a group or community meeting. It is better to have a highly productive, short meeting than one which rambles and loses effect.

The agent's role in the meeting should also respond to the circumstances. The agent should encourage the community to appoint a chairperson and should allow the chairperson to conduct business. The agent's role should be to inform and support, but not to dominate the meeting. Furthermore, the meeting should not resemble a class-room with the agent as teacher and the farmers as pupils. The agent should make every effort to ensure that during the proceedings the community members feel that it is their meeting and that they have a role.

As a guide to the proceedings of the meeting, the agent should keep the following points in mind. Start the meeting on time, then welcome community members and special guests, explain the purpose of the meeting and the programme to be followed, and begin the programme. Later, the agent should encourage questions and discussion, and be prepared to summarize the main points and note important decisions. The meeting should be closed with thanks to all concerned.

As the most commonly used form of group extension method, the group or community meeting will be most effective if carefully thought out and planned. After each meeting, the agent should make a brief record of the proceedings and the principal decisions taken and should also take any prompt follow-up action that has been decided.

Demonstrations

Farmers like to see how a new idea works, and also what effect it can have on increasing their production. Both purposes can be achieved by means of a farm demonstration. A good, practical demonstration is an invaluable method in extension work. The demonstration is a particularly powerful method to use with farmers who do not read easily. A demonstration will give such farmers the opportunity to observe, at first hand, the differences between a recommended new crop practice and traditional practices. The strength of the demonstration should lie in its simplicity and its ability to present the farmers with concrete results.

There are two principal types of demonstration used by extension agents - method demonstration and result demonstration.

Method Demonstration

Method demonstrations show and explain to farmers detailed procedures for implementing a technology. In the method demonstration, the farmer is shown step by step how, for example, to plant seeds in line, to use a mechanical duster to control insects, or to 'top' tobacco. The agent will probably be dealing with farmers who have already accepted the particular practice being demonstrated, but who now want to know how to do it themselves.

The main advantage of the method demonstration is that the extension agent can explain simple farming skills to a large number of people, thus increasing the impact of extension work. Moreover, as farmers are able to participate, there is a greater chance that they will benefit from the demonstration than if they were passively hearing it in a lecture.

The main limitation of a method demonstration is that, if there are too many farmers present, only a few get a chance to see, hear and do. The agent must be conscious that the demonstration is a learning experience and prepare the event accordingly. It is also vital that the demonstration be well thought out and competently conducted.

Result Demonstration

The main purpose of a result demonstration is to show local farmers that a particular new recommendation is applicable and beneficial under local conditions. Comparison is the important element in a result demonstration, for example, comparison: between using compost and no compost, between poor seed and selected seed, or between use of fertilizer and no fertilizer. "Seeing is believing" is an age-old expression, but one appropriate to a result demonstration. Until a farmer has actually seen the results of, for example, the application of a fertilizer, he will not be convinced by the agent's recommendation. By showing tangible results of a new practice recommended by the extension service, the agent can help to create confidence among the farmers and can greatly encourage them to try the practice themselves.

A result demonstration is an ideal way to present to farmers a comparison between traditional and new practices. It can also help to establish confidence in more scientific farming methods and increase the farmers' confidence in ideas originating from research stations. It shows proof of the value of a new practice. A result demonstration is also a useful tool that an agent can use to establish confidence among farmers in a new area.

Its major limitation is that it takes a long time to come to fruition and is thus a costly use of extension resources. If, in the end, for whatever reason, the new practice should fail, it could have disastrous consequences. Often such failures (for example, because of lack of rain) are outside the control of the agent.

Both method and result demonstrations are extension activities that require a lot of thought, careful planning and efficient execution. Although the two demonstrations differ somewhat in their purposes, they share a lot of common points and, in terms of their preparation and execution, they can be considered together.

Basic Principles for Demonstrations

Before the agent begins to plan and prepare for a demonstration, clarity about a number of key points will guide preparation and handling of the demonstration.

Participation. Where possible, demonstrations should be carried out on local farms with farmers' participation rather than on an extension plot or research station. Farmers will have more confidence if a demonstration is held on a neighbour's land, or if a new practice is shown by a fellow farmer, than if it is carried out by agents on extension's land. The more the local farmers can be involved in the whole process of a demonstration, the greater will be their self-confidence and readiness to learn.

Simplicity. Simple, clear-cut demonstrations of a single practice or new idea will be far more effective than ambitious and over-complex demonstrations that demand too much of the farmer. It is better to proceed step by step with a number of demonstrations than to try do to everything at once.

*Learning.*The demonstration is a learning environment and should be conducted in such a way that the farmers do in fact learn. A demonstration is a type of class-room, and the agent must be conscious of classroom requirements in terms of space, time, equipment and the teaching method to use.

Preparation. An extension agent should never contemplate holding a demonstration without careful planning and preparation. A demonstration hastily given could have disastrous consequences.

Planning the Demonstration

When the agent decides that a demonstration would be useful, time should be dedicated to planning and preparing for it. In this respect, a number of questions should be considered.

- What is the objective of the demonstration?
- Why is the demonstration the most suitable extension method?
- What would be the usefulness of the new idea to be demonstrated?
- When should the demonstration be held? When is the most convenient date and time both for the farmers and in terms of the application of the new idea?
- Where is the demonstration to be held? Which suitable location is the most convenient for the farmers?

The agent should work out in some detail answers to the above questions before proceeding any further. It is very important that the reasons for the demonstration be appropriate and clearly understood and that there is a realistic expectation that the demonstration will be of benefit to the farmers involved.

Preparing the Demonstration

The more carefully the agent can prepare all the details of the demonstration, the greater the likelihood that it will run smoothly. The following are the key areas of preparation.

1. Consult the local people and seek their help and advice in the preparation of the demonstration.
2. Prepare a detailed plan of the demonstration, the main issues to be covered, the sequence of events, the resources needed and the contributions required from other people.
3. Collect information and material available on the new idea or practice to be demonstrated, and make sure that the topic is familiar and that questions can be answered.
4. Check that all the support material is ready (for example, audio-visual aids, implements).
5. Select those farmers who will take part in the demonstration and brief them on the outline of events.

6. Ensure that the demonstration has been publicized and that the farmers know exactly when and where it is to take place.
7. Visit the demonstration site beforehand to make sure that all is in order and that the site is appropriate.

Supervising the Demonstration

During the demonstration, the agent's role should be to supervise but not to dominate. Actively support the farmer who may be assisting in the demonstration, and encourage the others to participate as much as they can. The agent should be keen to ensure that all those present benefit from the demonstration. During the demonstration, therefore, the agent should:

Welcome the participants, make them feel at ease and ensure that they have all they require to benefit fully from the demonstration.

Explain the purpose of the demonstration, what is hoped to be achieved, and outline the various stages to be followed. Distribute any literature or other material which was prepared as a guide for the participants.

Conduct the demonstration in person or be ready to help the demonstrator farmer. Proceed at a pace the farmers can follow, and be prepared to repeat explanations or answer questions from participants. Emphasize key points and explain the practice step by step in simple words. In a method demonstration, ensure that all those who wish to do so, have a chance to practice the technique themselves.

Summarize the main issues or points which have arisen, encourage questions from the farmers and make sure that the participants have had every opportunity to try out or examine the practice being demonstrated.

Conclude the demonstration with a vote of thanks to all concerned, and with a few comments about any follow-up activities planned.

Follow-Up

It is important that any interest generated, or decisions taken at the demonstration be followed through. Farmers will feel let down if the agent does not do so. This follow-up will be useful for the agent as well. Demonstrations can often result in good contacts with local farmers, and the agent may be able to enlist their support for future activities. It is also important that the agent reflect upon the demonstration and evaluate its effectiveness. The agent should, therefore, write a report and prepare a record of the demonstration, noting the names of the participants, the effect achieved and personal impressions of the usefulness of the demonstration.

Field Days

Field days are usually opportunities to hold method or result demonstrations on a slightly larger scale, and are usually run in a more informal and less highly structured manner. The purpose is often to introduce a new idea and a new crop, and to stimulate the interest of as many farmers as possible. Experimental stations or other government centers may be used for field days, but it is more usual and profitable for them to be held on the land of a local farmer. There is a greater chance of making an impact if the field day is held on a farmer's land, and if the farmer plays a part in running it and explaining the purpose. Field days can range in size from a small group to annual events attracting hundreds of farmers. Since the aim is a general introduction to some new idea, there is less need to be concerned about limiting the numbers. The extension agent's role on the field day is to support the farmer on whose land it is being held, to offer general guidance to ensure that things run smoothly and to be available to answer questions and queries. It is probably better not to over-organize the field day but to try to create an atmosphere in which visiting farmers can inspect, inquire, question and generally get to know what is available. Although the agent will try to encourage an open and informal atmosphere for the field day, there is still a considerable amount of preparation needed to ensure that it runs well. The issues which the agent must consider are very similar to those noted under demonstrations and will not be repeated here. It may be useful, however, to bear a few additional points in mind.

- ***Limit*** the numbers to the capacity of the field, to avoid overcrowding.
- ***Ensure*** a good layout of field-day activities, with easy access and facility of movement around the field.
- ***Encourage*** the demonstrator farmer to take most of the initiative; give support but do not take over the field day.
- ***Provide*** suitably large visual material and also, if necessary, a loudspeaker, to ensure that all can hear. Check that extension literature and other material are available for information.
- ***Conclude*** the field day by bringing all the participants together, reviewing the day's proceedings and the main items seen and discussed, and explain any future relevant extension activities. A field day is a day out for farmers and is often a welcome relief from their daily hard work. The agent should, therefore, provide an interesting and well-presented exhibition, suitable refreshments and points of rest, and generally create an atmosphere in which the farmers will feel at ease and will be eager to know what is going on.

Tours

Farmers like to visit farms in other districts to see how they operate, what they grow and what kinds of problems the farmers there are facing. A tour is a series of field demonstrations on different farms, or at different centres, and can often attract a lot of interest from local farmers. The tour should give local farmers a chance to see how other farmers cultivate their land, and to exchange ideas and experiences with them. It is important, therefore, that the area to be visited be in some way similar agriculturally to that of the visiting farmers.

As with all other forms of extension, tours have to be well thought out, planned, prepared and conducted. The five stages of; determining the objective, planning the content, preparing the arrangements, conducting the tour and arranging for appropriate follow-up, will be a guide to the extension agent. However, it may be useful to add these points:

- ***Visit*** the area first to become familiar with local conditions, the farms to be visited, the route and road conditions
- ***Limit*** the tour to what is possible. It is better to do a short tour in which visitors can have a good look at local farms than to arrange an ambitious tour and be pressed for time. Don't tire the visitors out.
- ***Encourage*** the host farmers to do all the explaining and to take charge of the tour.

- ***Arrange*** for food and drink during the tour
- ***Conclude*** the tour with a short summary of the main events and note any comments or conclusions.

A field tour is an ideal method of involving farmers and of stimulating genuine interest in extension activities. It is also very useful in bringing farmers together to discuss common problems, and to gain useful experience of other areas.

Advantages of Group Methods (Based on the Experience of the Authors)

1. Enables Extension Agent to have face-to-face contact with a larger number of people at the same time.
2. Enables the Extension Agent to reach a selected part of the target group.
3. Facilitates sharing of knowledge and experience and thereby strengthen learning of the group members.
4. Reaches relatively large number of people, also offers more opportunities for interaction and feedback.
5. Satisfies the basic urge of people for social contacts.
6. Motivates people to accept change due to group influence. More effective than mass method in stimulating action. Less expensive than individual method owing to wider number of people reached at a time (Refer to Table 2).

Limitations of Group Methods

1. Wide diversity in the interest of group members may create a difficult learning situation.
2. Holding the meeting may be regarded as an objective itself.

Table 2. Summary of extension methods used in developing countries: Characteristics and functions

Characteristics	Individual Methods (Farm and Home visits, telephone calls, office calls)	Demonstrations (Method and Result)	Mass Media (radio, Television, Newspapers, Magazine, Bulletins)	Group Methods (group meetings, Lectures)
Creating Awareness of Innovation	0	0	XXX	0
Creating Awareness of own problems	XXX	XX	0	XXX
Knowledge transfer	XXX	XX	XXX	XXX
Behavioral change	0	0	0	XXX
Activating Learning processes	XXX	XX	0	XXX
Cost per farmer reached	AAA	AA	A	AA

(Van den Ban and Hawkins 2002)

0 --- Indicates Unsuitable X----The number of crosses indicates suitability

A ---- The number of As indicates level of cost incurred.

N.B. The points outlined in the Table 2 are based on research results (Olaniyi et al. 2006; Chaudhry et al. 2006; Cho and Boland, 2004) rather than the ordinary impression and opinions of individuals.

3. Vested interests, power groups and village factions may hinder free interaction and decision making by group members.

Mass Methods

When Are Mass Methods Used in Extension?

Mass methods are used to

- Provide farmers with new awareness and draw their attention to certain new technology.
- Sharing experiences among successful farmers to other farmers in the community.

In this method, the Extension Agent communicates with a vast and heterogeneous range of people, without taking into consideration their individual and group identity. The method is used when a large and widely dispersed audience is to be communicated within a short time. There may be a few communicators and Subject Matter Specialists involved. The size of the audience may vary depending on the technique used.

Advantages

1. Suitable for creating general awareness among people.
2. Helps in transferring knowledge and, forming and changing opinions.
3. Large number of people may be communicated within a short time.
4. Facilitates quick communication in times of emergency.
5. Reinforces previous learning.
6. Cost effective due to large coverage.

Limitations

1. Less intensive method.
2. Little scope for personal contact with the audience.
3. Little opportunity for interaction with and amongst the audience.
4. Generalized recommendations hinder their application by individuals.
5. Little control over the responses of audience.
6. Difficulty in getting feedback information and evaluation of results.

CONSTRAINTS TO SUSTAINABLE AGRICULTURAL DEVELOPMENT AND EXTENSION METHODOLOGIES

Emerging countries are usually faced with constraints which are consequences of the factors that identified them as developing countries as earlier highlighted. More importantly, the 17 Sustainable Development Goals of the United Nations can only be achieved if the majority of the constraints are concertedly attended to by member countries. Conclusively, the way forward to solving the challenges facing the emerging countries will only be attained through the Sustainable Development Goals which are embedded in the issues stated below. These are the major constraints jointly summarized by the authors based on their experiences and studies; affecting extension activities in Nigeria.

1. **Lack of Strategic Planning that Indicates Future Directions:** Madukwe (2008) in his Professorial inaugural lecture describes agricultural extension in Nigeria as 'practice without policy'. This situation makes it difficult for extensionists to operate effectively since the majority of their activities are not legislated.
2. **Inadequate Monitoring and Evaluation Systems:** Effective monitoring and evaluation bring about performance and early indication of deviation from program objectives. The majority of federal and state agricultural programs such as ADP, OFN, Green Revolution, NAFPP and RBDAs became moribund due to lack of effective monitoring and evaluation.
3. **Top-Heavy, Bureaucratic Procedures:** A lot of bureaucratic bottle necks and red tape entangle activities by delaying, while waiting on instructions that may not come at the right time
4. **Insufficient Resources for Effective Implementation of Priority Research:** Resources in the form of capital, equipment and humans are either, insufficient, inadequate or are not released to meet timeliness of priority research. Agricultural projects are time and season specific, so that efforts should be made by agricultural policy makers to be conscious of the importance of time and season
5. **Lack of Effective External Linkages:** A perfect linkage system involving the research-extension-farmers, results in solving pressing issues in the agricultural sector. The Nigeria story has a linkage system called Weak research-extension-farmers-inputs-linkage system which needs effective performance of all the stakeholders in the linkage system for better coordination.
6. **Inward Looking Attitude:** For progressive performance of agriculture in developing countries, all stakeholders along the agricultural value chain must demonstrate effective attitudes in every stage of agricultural production in which they operate.
7. **Lack of a Culture of Evaluation of Performance:**The majority of projects die within a short period, because of lack of evaluation to improve performance. The problem of not having performance indicators to evaluate, before commencement of a project leads to project failure.
8. **Decreasing Investors' Confidence:** Decreasing attitude of the investors sets in when researchers work more on basic research rather than applied research. On this note the targets of researchers and investors are at variance. While some researchers concentrate on basic research for self-aggrandizement and publication recognition, investors are interested in problem solving research that can address their needs; which is applied research. This brings about low private sector participation in research activities
9. **Policy:** Without a well-defined Agricultural policy by the Federal Government, Agricultural practices in the country looks like a journey without a destination
10. **Infrastructural**
    a. Equipment. There is inadequate modern equipment to work with.
    b. A good Research environments needs stable electricity, regular water supply and uninterrupted internet services

11. **Information Dissemination**: Extension plays important roles in creating innovation platform conducive for establishing an interactive linkage between research and End-Users (Roling, 1997). However, this unit is faced with many challenges in performing this role
    a. Lack of modern equipment to disseminate information
    b. Insufficient manpower to meet the increasing End-Users requests
12. **End-Users – Challenges faced by Research Institute are;**
    a. Poor participation and cooperation of End-Users in research activities
    b. Lack of adequate stakeholders participation and responsiveness
13. **Inadequate Training for Extension Agents:** There is need for on-the-job training of extensionists to bring about better job performance. The majority of the extensionists never attended on-the-job training. Extensionists should be exposed to the use of Geographical Information System (GIS), web Social Media for development, photographic and video equipment and open broadcast van, to mention few. All these will put value to extension work and make dissemination easy. More importantly, there is need to note that at every point of diffusion level, there are specific skills and approach in the use of equipment to facilitate adoption.

CONCLUSION

This chapter discussed several concepts of agricultural extension methodologies assessed from past to present and actions taken to make extension methodologies more appropriate to encourage agricultural development. It also identified roles of agricultural extension activities in agricultural development and the factors influencing the choice of appropriate extension methodologies in emerging economies. It concluded with the advice that developing countries should fashion a sustainable extension system that will be socially acceptable and culturally compatible, economically viable and environmentally friendly. Such an extension system will help hasten agricultural development in emerging economies.

REFERENCES

Adedeji, T.O., Nosiru, M.O., Akinsulu, A.A., Ewebiyi, I.O., Abiona, B.G. & Jimoh, T.S. (2013). Adoption of new rice for Africa (NERICA) technology in Ogun State, Nigeria. *Journal of Development and Agricultural Economics, 5*(9), 365-371.

Agwu, A. E., & Chukwuone, N. A. (2005). Funding of Agricultural Extension in a Democratic and Deregulated Economy: The Cost Sharing Approach. *Journal of Agricultural Extension, 8*, 90–98.

Anandajayasekeram, P., Puskur, R., Workneh, S., & Hoekstra, D. (2008). *Concept sand practices in agricultural extension in developing countries: A source book.* Nairobi, Kenya: International Livestock Research Institute.

Anandajayasekeram, P., Torkelsson, A., & Dixon, J. (2002). *Emerging participatory approaches to technology development and transfer. What is new? Occasional paper.* Harare, Zimbabwe: FARMESA.

Campbell, D. A., & Barker, S. C. (1998). Selecting appropriate content method in program delivery. In Improving agricultural extension: A reference manual. FAO.

Chaudhry, K. M., Muhammad, S., & Ashraf, I. (2006). Alternative Extension Approaches to Technology Dissemination for Sustainable Agriculture in The Punjab, Pakistan. *International Journal of Agriculture and Biology*, *8*(6), 832–835.

Cho, K. M., & Boland, H. (2004). *Education and Extension for Multi-Functional Agriculture: Extension Concepts for Sustainable Agricultural Development in Myanmar*. Dublin, Ireland: AIAEE.

Famuyiwa, B. S. (2013). *Assessment of Environmental hazards associated with cocoa farming in Nigeria.* (Unpublished Ph.D. thesis). Department of Agric., Extension and Rural Development, O.A.U. Ife.

Famuyiwa, B.S., Torimiro, D.O., & Adesoji, S.A. (2013). Determination of Cocoa Farmers' Knowledge on Environmental Hazards Associated with Cocoa Farming Operations in Nigeria. *International Journal of Knowledge, Innovation and Entrepreneurship*, *1*(1-2), 58-70.

Havnevik, K., Bryceson, D., & Birgegard, L.-E. (2007). *African agriculture and the World Bank: development or improvement?* Retrieved from www.eldis.org/go/topics&id=34737&ttpe=Document#.Vws_DjY1gwg

ICCO. (2008). *Manual of best known practices in cocoa production. Version 1.* Consultative Board on the world cocoa economy sixteenth meeting Berlin, Germany.

IFPRI. (2011). *Priority setting Research & extension adoption evaluation; Engendering agricultural research development, and extension.* 10.2499/9780896291904as

Kumar, A. (2015). *Lecture 2 Major characteristics of developing countries.* Retrieved from http://web.uvic.ca/~kumara/econ420/characteristics-dev.pdf

Madukwe, M.C. (2008). *Practice without policy: The Nigeria agricultural extension services.* 30th Inaugural Lecture of the University of Nigeria.

Marten, G. G. (2009). *Human ecology.* Earthscan Publications.

Maunder, A. H. (Ed.). (1973). *Agricultural extension: A reference manual.* Rome: Food and Agriculture Organisation of the United Nations.

Mwangi, M., & Kariuki, S. (2015). Factors determining adoption of new agricultural technology by smallholder farmers in developing countries. Journal of Economics and Sustainable Development, 6(5), 208-217.

Oakley, P., & Garforth, C. (1997). Extension methods. In *Guide to Extension Training*. FAO. Retrieved from www.fao.org/docrep/t0060e/T0060E07.htm#5

Ogunjimi, S. I., & Farinde, A. J. (2012). Farmers' Knowledge Level of Precautionary Measures in Agro-Chemicals Usage on Cocoa Production in Osun and Edo States, Nigeria. International. *Journal of Agriculture and Forestry.*, *2*(4), 186–194. doi:10.5923/j.ijaf.20120204.10

Olaniyi, O. A. (2010). *Assessment of Utilization of Agricultural Information on Selected Arable crops among Rural youth in Oyo and Osun states, Nigeria.* (Unpublished Ph.D Thesis). Department of Agricultural Extension and Rural Development, Ladoke Akintola University of Technology, Ogbomoso, Nigeria.

Olaniyi, O. A., Ogunleye, Y. K., & Eniola, P. O. (2006). Extension Delivery Methods Preferred by Women Farmers in Oyo State, Nigeria. *Journal of Agricultural Extension Management*, *7*(2), 1–6.

Pretty, J. (2003). *Social capital and connectedness: Issues and implications for agriculture, rural development and natural resource management in ACP countries*. Review paper for CTA. CTA Working Document Number 8032.

Purcell, D. L., & Anderson, J. R. (1997). *Agricultural Research and extension: Achievementand problems in national systems. World Bank Operations Evaluation Study*. Washington, DC: World Bank. doi:10.1596/0-8213-3878-1

Roling, N. (1997). *Extension Role in Sustainable Agricultural Development in Improving Agricultural Extension a Reference Manual*. FAO Rome.

Sanginga, P.C., Lilja, N. &Tumwine, J. (2001). *Assessing the quality of participation in farmers' research groups in the Highlands of Kabale, Uganda*. PRGA; CGIAR, Future Harvest. Working document/PRGA 19.

Sosan, M. B., & Akingbohungbe, A. E. (2009). Occupational Insecticide Exposure and Perception of Safety Measures among Cacao Farmers in Southwestern Nigeria. *Archives of Environmental & Occupational Health*, *64*(3), 185–193. doi:10.1080/19338240903241077 PMID:19864221

Swanson, B., Bentz, R., & Sofranko, A. (1997). *Improving agricultural extension: A reference manual*. Rome, Italy: Food and Agriculture Organization of the United Nations.

United Nation. (2015). *Sustainable Development Goal*. Retrieved from http://www.un.org/sustainabledevelopmentas

Van den Ban, A. W., & Hawkins, H. S. (2002). Agricultural Extension (2nded). Blackwell Science publishers.

Wikiversity. (2015). *Characteristics of developing countries*. Retrieved from https://en.wikiversity.org/wiki/Characteristics_of_developing_economies

Williams, S.K., Fenley, T.M., & Williams, C.E. (1981). *A Manual for Agricultural Extension Workers in Nigeria*. Academic Press.

Chapter 5
ICTs for Agricultural Development and Food Security in Developing Nations

Bhattacharjee Suchiradipta
Independent Researcher, India

Raj Saravanan
National Institute of Agricultural Extension Management (MANAGE), India

ABSTRACT

Development has many faces and complete wellbeing of human population is the most important one of them which in more than one ways involves agriculture and the farming population. Providing needed information at the right time to the rural population is the first step in their empowerment and ICTs can play an immensely important role in providing that information by increasing the dialogue between development professionals and rural people at every stage of development process. According to recent statistics released by ITU, over the last 15 years, ICTs have grown in unprecedented ways providing huge opportunities for social and economic development and this growth can be an advantage to rural advisory services. Providing correct and personalized information needs expert opinions and so multi-stakeholder engagement makes the process more efficient and ICTs provide a very unique and important platform for such collaboration, thus bringing together different stakeholders for efficient partnership. The various tools and technologies can also be tailored according to the needs of end users. But inspite of the advantages, ICTs can only be universally accepted and used when the challenges of accessibility, acceptability, funding, and sustainability are overcome. There are no formula for sure success with ICTs and situation is the best determinant of the strategy to be used and so, a balanced and strategic use of ICTs depending on the clients' needs can best utilize its potential for agricultural development and food security in developing nations.

DOI: 10.4018/978-1-5225-0942-4.ch005

INTRODUCTION

Global development has many faces and all of them concern human development – taking people out of poverty, assuring food security for everyone, and ensuring a fairly comfortable life. This is a difficult task considering one in nine people in the world suffer from hunger. The problem becomes complicated with the uneven distribution of the famished population with higher concentration in countries of Sub-Saharan Africa where one in four people suffer from chronic hunger, as well as Southeast Asia which has the largest number of undernourished persons (FAO, 2015). One of the largest missions taken worldwide in recent times to ensure human development was the Millennium Development Goals (MDGs) adopted at the Millennium Summit of 2000 of the United Nations, where the member countries promised global partnership and set time bound quantified targets with a deadline of 2015. These MDGs were formulated with the goal of all round human development by eradicating hunger and poverty, providing universal primary education, promoting gender equality and women empowerment, reducing child mortality, improving maternal health, combating major life threatening diseases across the world, ensuring environmental sustainability, and developing a global partnership for development (Millennium Project, 2006). The MDGs targeted world poverty in its many dimensions and by 2015, though the target of halving extreme poverty has been met, it was far from being well distributed across the continents. While a few Asian countries accounted for most of the decline, the absolute number of poor has risen in Sub Saharan Africa.

Poverty and food insecurity in the world right now is predominantly rural with 78 per cent of the world's extreme poor living in rural areas with the majority dependent on agriculture. In addition, almost 60 per cent of child labour worldwide is found in agriculture. The world has the capacity to produce enough food to feed everyone equally, even with the estimated 60 per cent increase in production to meet the projected demands of 2050, yet still, the rural poor - mostly consisted of subsistence producers, family farmers, and landless labourers – live on less than $1.25 USD per day. After 2015, the development agenda shifted from goals and targets to measurements and means. As a result, discussions on building on innovative ways to address the world's most pressing challenges began and Sustainable Development Goals (SDGs) were implemented. One of the five agents of change identified for transforming development economies is family farmers and small scale producers. Agricultural growth is up to five times more effective than any other sectoral growth in resource poor low income countries in reducing poverty (FAO, 2015) but the problem is more than sustainable agricultural production; it is of connecting the farmers and making them informed, and helping them to organize themselves to get the most out of their investment in their agricultural fields. In so doing, Information and Communication Technologies (ICTs) can be a very important aid to rural advisory services.

INFORMATION EMPOWERMENT

Information is a critical input in agriculture and given at the right time, it can prove to be the most important one (Saravanan and Suchiradipta, 2015b). Providing needed information at the right time to the rural population is the first step in their empowerment and ICTs can play an immensely important role in providing that information by increasing the dialogue between development professionals and rural people at every stage of the development process. Information empowerment of rural people helps them to be active partners in development efforts and not mere beneficiaries. ICTs make the job easier through TV, radio, computer, mobile phones, and recently social media. But in developing countries, information poverty is one of the more significant and insidious obstacles to effective exploitation of information processing and other types of technology (Sadowsky, 1996).

According to recent statistics released by ITU, over the last 15 years, ICTs have grown in unprecedented ways providing huge opportunities for social and economic development. Since 2000, internet penetration has increased 7 fold from 400 million users in 2000 to 2.2 billion users in 2015. Mobile subscriptions are up from 738 million to 7 billion and developing countries provide a wide opportunity for further development with new users and more population to reach (ITU, 2015). Social media, the recent addition to the world of ICTs, is gaining popularity very fast with a global penetration of 29 per cent, a 12 per cent increase from 2014 with a major boost from increased mobile subscriptions. Facebook is the most popular social media platform with about 1.5 billion users worldwide (Kemp, 2015) and the number increasing every day. But conditions in least developed countries (LCDs) are still challenging. Internet penetration is 9.5 per cent for population and 7 per cent for households, mobile broadband subscription is only 12 per cent of the population and fixed broadband subscription negligible at 0.5 per cent. Also, the fact that monthly fixed broadband prices are three times higher and mobile broadband prices are two times higher in developing countries than in developed countries (ITU, 2015) is only adding to the cause with poor households finding it difficult to invest in internet services.

ICTs IN Rural Advisory Services (RAS)

With increasing diversities in agriculture, more stakeholders and increasing need to produce more to reach farmers at the right time with correct and needed information is not limited only to the extensionists. Moreover, information being one of the most critical inputs in agriculture, farmers are willing to pay for it, thus making extension a profitable business. But providing correct and personalized information needs and expert opinions on a timely basis is a complex process and so multi-stakeholder engagement makes it more efficient. ICTs provide a very unique and important platform for such collaboration, thus bringing together different stakeholders for efficient partnership.

Philosophy and Principles of ICTs for Agricultural Development (Adopted from Saravanan et al., 2015a)

ICTs are better enablers for information and knowledge access and sharing among agricultural innovation system actors compared to conventional methods. They also complement the conventional extension advisory methods. The guiding principles (World Bank, 2011; Saravanan 2011b; Saravanan, 2013) of ICTs for a more efficient extension and advisory services (EAS) are:

- **Relevant Content:** Contextualized or farmer specific, need based, timely and quality content is the major aim of ICT based EAS.
- **Appropriate ICT:** Among a variety of ICTs, choose the formats, channels, tools, devices and applications that best match the purpose, content and clientele.
- **Integration of Methods, Actors and Services:** Integrating ICTs with other conventional extension methods (like Farmer Field Schools (FFS), Community Knowledge Workers (CKWs), etc), pluralistic actors (public, private, Farmer Based Organisations (FBOs), etc) with their services along the value chain will create synergy in extension and advisory services.

- **Information PLUS:** To convince the clientele, (i.e. show and tell). ICT based information alone is not enough and needs to be combined with field demonstrations, exposure visits, group discussions etc. Not just advisory information, but a complete resource package across the agricultural value chain[1] needs to be provided.
- **Human Element:** Commitment of the extension stakeholders to use ICTs, development of ICT champions with the legacy of promoting continuous leaders and followers are important.
- **ICTs Complementary to EAS:** ICTs can play only a complementary role in extension. If they are used appropriately, they create synergy and better impact with conventional extension efforts.
- **Institutionalizing ICTs:** Institutional policy and guidelines to use, developing the ICT literacy and competency among the human resource and infrastructure development should be an integral part of the institutional set-up.
- **Reasonably Long Term and Continuous Engagement with ICTs**: To see the better outcome, ICTs need to be integrated with the conventional extension approaches for a reasonably (at least for 5 years) long period.

Broad Areas of ICT Implementation for Agricultural Development (Adopted from Saravanan et al., 2015a)

ICTs based extension advisory methods are implemented along the agriculture value chain activities such as pre-production, production, post-harvest and marketing, financial services (credit, payment, savings, insurance, etc.), climate and other data. Five broad areas[1] of ICTs implementation are:

1. **Offering Localized and Customized Information, Advisory and Other Services:** Farmers call centres (FCCs), mobile apps, radio, TV etc.
2. **Helping to Create, Document, Store, Retrieve, Share and Manage the Information:** Web portals, crop specific portals, knowledge banks, expert systems and agricultural information management systems etc.
3. **Enabling Collaboration, Sharing and Partnerships for Innovation Among Extension Actors:** Social media, Dgroups etc
4. **Enabling Farmers and Others to "Gain a Voice":** Community radio, tele-centres, videos, communities of practice (COPs) etc.
5. **Facilitating Capacity Development of Farmers, Extension Professionals and Other AIS Actors:** E-learning mechanisms (Open Distance Learning (ODL), Learning Object Repositories (LOR), Massive Open Online Courses (MOOCs) and other e-learning mechanisms etc.), training by using ICTs, survey and monitoring tools and applications.

ICTs have been used in various forms depending on the need of the clientele, location specificity, availability and accessibility of technology and so on. Major forms of ICT tools and applications in Rural Advisory Services (RAS) are discussed below:

- **Radio and Television:** Radio and TV are the earliest forms of ICTs used in agricultural extension and advisory services and are the most popular with the rural communities. They have been very helpful in dissemination of information and technologies among the farmers in rural areas but their very nature inhibits interactivity or location specificity which restricts their scope in information dissemination in an era of technology. In recent times though, community radio stations have tackled the problems of location specific information and have made the rural people major stakeholders in information dissemination and discussions about community development.

- **Interactive Multimedia Compact Discs (IMCDs), Decision Support Systems (DSS), and Expert Systems:** Multimedia is a combination of text, graphics, art, sound, animation, and video elements where the viewers can control what elements are delivered and when. This is known as interactive multimedia. IMCDs offer learners complete and individual control over their learning and encourages better understanding between individual learners and the subject matter. Decision support system (DSS) is a computer program application that analyzes data and presents it so that users can make decisions more easily. DSS uses Artificial Intelligence (AI) and expert systems to present data in a format suitable to the users. Expert systems are tools for information generation from given information by simulating human reasoning about a problem domain. The knowledge acquisition process for building an expert system facilitates the integration of knowledge and experiences of different specialties which help generate advice based on its knowledge base and reasoning mechanism from all developed extension documents.
- **Web Portals:** Web portals are digital platforms that provide organized gateways to information or act as aggregators of knowledge from various stakeholders. They are specifically designed single access points to information collected from diverse sources. In agricultural extension and advisory services context, web portals can be of two types – those providing technical and market knowledge to end users at the grassroots level, and those helping with capacity development of extension personnel. Financially and socially, portals can have far reaching impact on users. Illiteracy – both educational and technological – can be a barrier in accessing web portals by grassroots level users but with the help of extensionists, they can be of much help to the rural communities (Saravanan et al., 2015b). A few examples of some successful initiatives to use web portals for agricultural communication are discussed below.

University Web Portal

TNAU Agrictech Portal has been catering to the needs of farmers, extensionists and other stakeholders in agriculture and allied sectors since 2009. It offers a diverse range of information from crop related to weather information, daily market prices, schemes and programs for farmers, daily news, events and publications supported by multimedia, expert systems and many others. The portal can also be accessed in Tamil and English languages and offers a keyword search facility. Farm technology portal (http://agritech.tnau.ac.in) was designed by integrating the allied sectors of agriculture, horticulture, agricultural marketing and agriculture business, agricultural engineering, sericulture, seed sector, forestry, fisheries and animal husbandry. This dynamic portal page holds more than half a million pages in Tamil and English with multiple media content. According to Google Analytics, from April 1st 2015 the daily visitors ranged between 26000-28000, new visitors numbered 1,44,100 at the rate of 57%, the new sessions were 2,33,168 with an average session duration of 4:09 minutes, number of page views were 7,39,706 with a bounce rate of 62%. The average page/session was 3:17 minutes.

The online input information service ensures timely sourcing of critical inputs to farmers in the state. By using the last mile connectivity, Short Message Messages (SMS), more than 10,000 million SMS have been sent to farm beneficiaries in English and Tamil through m- kisan mobile linked interface.

The agritech portal is linked to social media networks like blogs, twitter and YouTube, and Facebook messages are updated on agriculture related news on a daily basis both in English and Tamil.

Crop Specific Portal

"Rice Knowledge Management Portal" (www.rkmp.co.in), as the name exemplifies, facilitates efficient management of the enormous knowledge on rice crop on a single click. With more than 16,000 pages of validated and localised content in seven vernacular languages and 18 stakeholder platforms, this is considered as the largest repository of knowledge on any single crop across the globe. Today, a rice farmer from any corner of the country is able to get all the rice related reliable information specific to their region in the language of his choice. The dream of any extension professional to provide the right information at the right time and context in the local languages to the farmers is realized through Rice Knowledge Management Portal - the one stop shop for rice related information.

Web Portal for Agriculture Videos

Access Agriculture (www.accessagriculture.org) is a video based web portal started in 2012 with training videos in agriculture and allied sectors, business skill development and market development. It has something to offer to all agricultural stakeholders. The online videos can also be downloaded and language of the audio can also be requested as per preference of the user. The web portal also has social media presence in Facebook and LinkedIn and has a video social media site called AgTube.

Web Portal by the Extension Organisation

eXtension (www.eXtension.org) is an interactive learning environment delivering the best, and most researched knowledge from the land-grant university minds across America. eXtension connects knowledge consumers with knowledge providers - experts who know their subject matter inside out. eXtension is unlike any other search engine or information-based website. It's a space where university content providers can gather and produce new educational and information resources on wide-ranging topics. Because it is available to students, researchers, clinicians, professors, farmers as well as the general public, at any time from any Internet connection, eXtension helps solve real-life problems in real time.

Web Portal for Integrated Rural Advisory Services

As part of the India Development Gateway initiative of Government of India, Vikaspedia portal (www.vikaspedia.in) is aimed at creating a versatile collective knowledge repository and demand driven information in the development oriented sectors including Agriculture. This multilingual portal is serving as a collaborative content creation, knowledge sharing and utilization platform for the stakeholders in agriculture and allied sectors[14]. Currently, Vikaspedia is one of the largest knowledge portal hosting information/knowledge in 10 Indian languages, offering information on success stories, best practices, government schemes, technologies, market information and related value added services in Agriculture.

- **Online Learning Platforms:** Learning platforms are interactive online services that provide learners with resources to support and enhance educational delivery and management. They are virtual learning environments that make the process interactive with integrated discussion boards, chat rooms and other such facilities where contents are both expert suggested and user generated. These kinds of learning platforms also serve as online repositories and can be very helpful for all agricultural stakeholders to increase their expertise on practical topics while sharing experiences with their colleagues across the globe and leaning from others' experiences. A good example of online learning platforms is Massive Open Online Courses (MOOCs) that have recently become very popular among all agricultural stakeholders.

- **Mobile Phones:** Mobile phones are devices that can create, store, access and share information anytime, anywhere and teamed up with extension and advisory services can help improve the livelihood of rural people by taking timely information to their fingertips at potentially low cost. They also provide the unique opportunity of instant feedback and one on one interaction with experts. mExtension can facilitate the creation of scalable, replicable and commercially sustainable advisory services for rural clients and women can especially benefit from the services if access to mobile phones can be ensured (Saravanan and Suchiradipta, 2015c).
- **mExtension:** Agricultural extension and advisory services are provided through a number of modes – push and pull SMS, interactive voice response (IVR), mobile apps (mApps), and so on – sometimes individually and sometimes integrated. While SMS and IVR services are accessible from both feature and smart phones, mApps require smart phones. Services can be free or subscription-based. Cost does not seem to affect popularity as shown by services such as IKSL in India, iCow in Kenya, Kilimo Salama in Kenya and Rwanda, and e-Krishok in Bangladesh. Mobile-based advisory services are mostly targeted at farmers and the rural population but collaboration among stakeholders in agricultural innovation systems (AIS) for providing content is not unknown. The advisory services also vary from providing solely agricultural information (e.g. Gobi Sahana Sarana in Sri Lanka) to providing micro insurance to rural people (Kilimo Salama in Kenya and Rwanda), real time market information (e-soko active in 10 African countries), farmer-specific fertiliser recommendations (NMRiceMobile in Bangladesh, China, India, Indonesia, Phillipines and West Africa) or integrating agricultural and weather information along with entertainment to attract large numbers of rural people (Nokia Life Tools) (Saravanan and Suchiradipta, 2015c).
- **Social Media:** Social media are web based tools of electronic communication that allow users to personally and informally interact, create, share, retrieve, and exchange information and ideas in any form (text, pictures, video, etc.) that can be discussed, archived and used by anyone in virtual communities and networks (Saravanan and Suchiradipta, 2015d). While social media has been playing a significant role in personal communication and development communication across the globe, in the agriculture sector its popularity has also been increasing. Increased use of social media in agvocacy (agvocacy is the combination of two words agriculture and advocacy, and means talking for and about agriculture) can increase the dialogue among agricultural stakeholders and make information access much easier for rural smallholders. Also, with proper planning, a social media communication strategy can be scalable across geography (local, regional, national, global), topics of interest (e.g. business, career, agronomic practices, crops, etc.), and type of clients (women, young people, smallholders, etc.) (Saravanan et al., 2015c).

Extension professionals are using social media to form networks and farmers are taking to social media to talk to peers and consumers. All big things in agriculture – new technology or innovations, seminars and meetings, workshops and trainings, reports, publications – get tweeted or hashtagged. Facebook, Twitter, YouTube, and blogs are the major platforms for agricultural information dissemination. The uses of socially integrated messaging apps are also increasing in the rural areas. But still, there is a difference in the intensity of use in developed and developing countrics – basically due to economic conditions and infrastructure availability. In spite of the differences, use of social media is picking up in rural areas of developing and least developed countries as well (Suchiradipta and Saravanan, 2016)

GFRAS Global survey on use of social media in agricultural extension and rural advisory services conducted online during 2015 across 62 countries and 229 respondents provided interesting responses. Facebook was found to be the most popular social media platform used by extension professionals. The major activity on social media was searching for news events and sharing information. A major impeding factor for social media use was the lack of authenticity of information shared online. Social construction of information (development and publication of information socially by the users) was considered the most important feature of social media (95.1%). Ninety five percent of the respondents believed social media can play an important role in bridging the gap between stakeholders in Agricultural Information services (AIS). Reaching clients (77.4%) was a major use of social media in AEAS. Training in social media use was uncommon, and 71% of the respondents said they need training on social media use. If and when there was training conducted by the respondents' organizations, it mainly focused on the specifics of different social media platforms and awareness creation on the use of social media in agricultural extension. But on an organizational level, social media is still not given much importance by higher authority and social media policy restricts rather than encourages its use. Also, weak or non-existent connectivity in rural areas, high cost of data charges, illiteracy of the clients and low participation and lack of interest of clients are reported to be major problems. Overall, the survey found that social media is a very useful tool in agricultural extension and rural advisory services. To quote one respondent, "social media is not only a tool for reaching large audiences; it is also an opportunity to develop relationships." (Suchiradipta and Saravanan, 2016)

The use of social media is becoming more and more pronounced in agriculture and a few of the popular social media accounts and handles across various platforms, as in Table 1, will give a better idea into their works.

ICTs have provided many opportunities for collection, processing, storage, retrieval, managing and sharing of information in multiple format for providing a wide range of services (information, awareness, promotional, advisory, knowledge, technology transfer, training, education, and much more) to farmers and other agricultural innovation system (AIS) actors in a timely, comprehensive, cost-effective, and interactive manner (Saravanan et al., 2015a). But one important factor that impacts the success of ICT based projects is the suitability of the technology used. In areas where the majority of the end users are illiterate, TV, radio, voice based mobile advisory services and video- based web portals are more appropriate; community radio and mobile phones are more suitable media to engage women farmers. Young farmers can be most engaged on social media platforms like Facebook, Twitter and Whatsapp, and depending on the electronic device availability with the client, the development organizations need to provide basic infrastructure like internet kiosks to access the information provided. To make ICTs a real success and not a piecemeal approach in RAS, more attention needs to be paid to the 'I' and 'C' of ICTs rather than the 'T' only (Saravanan and Suchiradipta, 2015a). This requires shifting the focus to specific target groups and engaging them fruitfully to ensure successful implementations of ICTs. Youth everywhere can be the most important stakeholder in this process.

Table 1. Examples of use of social media in agriculture

Name of Group/ Community/ Pages	Description	Target Users	Region
FACEBOOK			
By farmers			
Livestock Information and Marketing Centre (https://www.facebook.com/groups/Livestock.TN/)	Members (farmers, extension personnel, scientists, market functionaries, consumers, local leaders, *etc.*) of this group share information related to livestock production, management, marketing, *etc.* A separate page is also on Facebook related only to marketing of livestock. (https://www.facebook.com/Livestock.Market)	Agricultural stakeholders related to livestock	India
Mkulima Young (Young Farmer) (https://www.facebook.com/mkulima.young)	This page is an information sharing platform for young farmers started by Joseph Macharia, a young farmer himself. Mostly agro-advisory and market information are shared.	Young farmers	Kenya
Turmeric Farmers' Association of India (https://www.facebook.com/turmeric.farmers)	This page was created by turmeric farmers to stabilize the price of turmeric in the market. To date, the farmers connect through the page and share information to keep turmeric price stable and increase marketing opportunities of turmeric.	Turmeric farmers	India
National Ecological Producers Association (APNE) (https://www.facebook.com/anpe.peru)	Information related to ecological farming is shared through the page.	Farmers	Peru
By extension centres			
Krishi Vigyan Kendra, Namakkal (https://www.facebook.com/krishi.namakkal)	Krishi Vigyan Kendra, Namakkal communicates information related to farmers' training programmes, availability of inputs *etc.* through this account	Subject Matter Specialists of KVK, farmers, agricultural stakeholders	India
By extension professional networks			
Agricultural Extension in South Asia (AESA) (https://www.facebook.com/groups/428431183848161/)	Members post links to relevant publications on extension and advisory services, announcements of workshops and conferences, major policy decisions on extension, reports of meetings/ workshops and blogs relevant to the broader theme of extension	Agricultural Extension stakeholders	South Asia
Global Forum for Rural Advisory Services (GFRAS) (https://www.facebook.com/groups/gfras/)	This page provides information related to advocacy and leadership on pluralistic, demand-driven rural advisory services.	AEAS Professionals and others	Global
By extension personnel			
Vivasayam Karkkalam (Let us Learn Agriculture) (https://www.facebook.com/groups/madhualan)	Mr. Madhu Balan, a public extension officer started a Facebook group to cater to the information needs of famers in 2012. This group exchanges information on improved farm technologies, initiates discussion with other farmers and extension personnel, share information and photos on best practices by other farmers, government schemes, *etc.* Question and answers, information on Terrace garden and hydroponics are the most discussed topics in this group.	Farmers and others who are interested in agriculture	India

continued on following page

Table 1. Continued

Name of Group/ Community/ Pages	Description	Target Users	Region
TWITTER			
Farmers			
AgChat (https://twitter.com/agchat)	The AgChat (Twitter online discussion group by the AgChat Foundation) started in 2009 by a group of American farmers is widely used in USA, UK, Australia, New Zealand and Ireland for facilitating discussions of industry issues between farmers and agribusinesses	Farmers, entrepreneurs, farm product consumers	USA, UK, Australia, New Zealand, Ireland
Agriculture Proud (https://twitter.com/AgProud)	Twitter handle of Ryan Goodman, a young farmer and rancher from Montana, US. Through his Twitter account he shares his experiences of farm life and answers questions from fellow farmers, agriculture enthusiasts, and consumers.	Agriculture enthusiasts, consumers, and fellow farmers	USA
Young Farmers (https://twitter.com/F4YFKenya)	Information shared through this Twitter handle of the Foundation for Young Farmers shares information for better agriculture with the objective to attract more youth to farming.	Young farmers	Kenya
Extension centres			
USDA (https://twitter.com/USDA)	The Twitter handle of U.S. Department of Agriculture shares latest news, events, and information in agriculture	Farmers, extensionists, development practitioners	USA
INGENAES (https://twitter.com/INGENAES)	This Twitter handle of Feed the Future initiative Integrating Gender and Nutrition within Agricultural Extension Services shares information and gender-appropriate, nutrition-enhancing technologies to improve life and livelihood of women farmers	Researchers, extensionists, farmers	Global
eXtension4U (https://twitter.com/eXtension4U)	Twitter handle of eXtension.org, a research based learning network of cooperative extension of USA. Sound research based information is shared through the handle.	Farmers, researchers, policy makers of USA related to Agric. Research and Dev	USA
Professional networks			
MEAS (https://twitter.com/MEAS_extension)	Twitter handle of the project Modernizing Extension and Advisory Services shares good practice strategies and related information to ultimately raise farm income and enhance livelihood of rural poor of 12 selected countries of Asia and Africa.	Development practitioners	Global
GFRAS (https://twitter.com/infogfras)	This page provides information related to advocacy and leadership on pluralistic, demand-driven rural advisory services.	Extensionists, development practitioners, researchers, policy makers	Global
e-Agriculture (https://twitter.com/e_agriculture)	Twitter handle of e-Agriculture, a global initiative to enhance sustainable agricultural development and food security by improving the use of ICTs. Information shared is related to recent developments, efforts, publications and stories of ICT use in agriculture.	Farmers, researchers, development practitioners	Global
BLOGS			
Individual blog			
Gate to Plate Blog (Michele Payn-Knoper) (http://www.causematters.com/blog/)	Through this blog, agriculturist, entrepreneur and founder of Cause Matters Corp. shares her views about food and agriculture.	Farm product consumers, agriculture enthusiasts, farmers	USA

continued on following page

Table 1. Continued

Name of Group/ Community/ Pages	Description	Target Users	Region
Ecoagriculturist (Oluwabunmi Ajilore) (https://ecoagriculturist.wordpress.com/)	The blogposts are related to sustainable agriculture, environment, youth involvement in agriculture, ICT4Ag, and other related topics. The blog was also a winner of the YoBloCO Awards of CTA in 2014.	Farmers	Nigeria
The Unconventional Farmer (Gil Carandang and Patrick Gentry) (http://theunconventionalfarmer.com/flog/)	Featured in top 50 farm blogs by www.seametrics.com, this blog covers natural farming techniques from Japanese and Korean natural farmers. Topics range from farming techniques to animal care for urban and rural farmers.	Farmers, agriculture enthusiasts	Global
Institutional blog			
AGRF Blog (African Green Revolution Forum) (http://www.agrforum.com/blog/)	The AGRF was established in 2010 to initiate discussions and develop concrete plans for achieving Green Revolution in Africa. The blog is a part of the initiative where issues related to African agriculture and ways to develop are discussed by various authors working in different capacities in the agriculture sector. The posts are by invitation only to maintain professionalism.	Policy makers, private actors, civil society actors, researchers, farmers, agribusinesses	Africa
Agricultural entrepreneurship (Penn State Extension) (http://farmbusiness.blogspot.in/)	This blog is especially helpful for agripreneurs for getting information on marketing, economics, and other recent news in agricultural industry. Since 2008, 348 blog posts have been made by the eight contributors.	Agripreneurs	USA
TNAU Agritech Portal blog (Tamil Nadu Agricultural University) (http://tnauagritechportal.blogspot.in/)	The blogs of TNAU Agritech Portal deals with everything agriculture – from sowing to harvesting, crop protection to crop management, weather, recent happenings in the agriculture industry, schemes and programs for farmers, ICTs, and many more. A total of 940 blog posts have been made since 2012, 541 of which are made in 2015 itself, by 43 members consisting of extensionists, researchers, academicians, and farmers.	Farmers, agripreneurs, extensionists	India
YOUTUBE			
Farming First (https://www.youtube.com/user/FarmingFirst/)	This channel highlights the mission of the agricultural development coalition of the same name, founded in 2009. Made up of 131 organizations worldwide, Farming First prioritizes the protection of natural resources, knowledge sharing, local infrastructure, harvests, market access, and innovative research (www.foodtank.com).	Policy makers, researchers, agricultural enthusiasts and practitioners	Global
CGIAR Research Program on Climate Change, Agriculture and Food Security (CCAFS) (https://www.youtube.com/user/CCAFS)	The videos shared by the channel features stories of smallholder farmers, interviews with leading agricultural experts across the globe, and innovative information on climate-smart agriculture (www.foodtank.com).	Researchers, farmers, policy makers	Global
IFADTV (https://www.youtube.com/user/IFADTV/)	It is a well produced and highly engaging channel from the International Fund for Agricultural Development (IFAD). Videos feature news stories about smallholder farmers in addition to interviews with agriculture experts (www.foodtank.com).	Policy makers, farmers, extensionists	Global
Farmers Weekly Video (https://www.youtube.com/user/FarmersWeeklyVideo/)	It is produced by U.K.-based Farmers Weekly, a multimedia independent information service for farmers and agricultural businesses. Videos contain information on how to make agri-businesses sustainable, advice on farming careers, and information on different crop inputs (www.foodtank.com).	Farmers, extensionists, agribusinesses	U.K.

(Suchiradipta and Saravanan, 2016)

Youth and ICTs in Agriculture

ICTs for Youth (ICT4Y) in Agriculture

ICTs and youth can prove to be a great catalyst in agriculture. ICTs are the developmental tool kits to empower the rural youth to fight against the odds of poverty, backwardness and illiteracy and come out victorious in their agricultural fields. It is not a mere technological innovation to come and go but application of ICTs in agriculture is a permanent solution to many of the problems of the sector in developing countries. It will also bridge the gap between the present and past making young people the catalyst for the change among their peers. Moreover, forming an opportunity of employment for the rural youth does not only help them but saves the society from many odds. ICTs empower the rural youth and channel their energy in a desired direction.

Of the other direct impact of ICTs on youth in agriculture are forming virtual peer groups to share their experience. The youths from various parts of the world with the Web 2.0 can gain exposure to what is happening in the agricultural front in other parts of the world and thus gain a better understanding of their own situation. This helps them to take better decisions in their own condition. It also gives them the opportunity to give their voice in the policy making process. ICTs increase the productivity and inspire the youth to venture in agripreneurship. ICTs also increase the social status of the farmers and the need for social recognition attracts others to the use of technology in their conditions. It also increases the negotiating power of the farmers and makes them more informed for so as not to be cheated by others.

The role of ICTs, thus, can be felt in every sphere of life of the farm youth. As shown in the Figure 1, it is a lifelong process of learning from their own experiences, from others in the community (both actual and virtual) or from access to the worldwide knowledge repository and also passing that knowledge to those who need it. Not only that, it also helps in their skill formation and proper planning and management of their investments thus ensuring an inclusive individual and community development.

Figure 1. Potential of ICTs in youth and community development (Haddad, 2007)

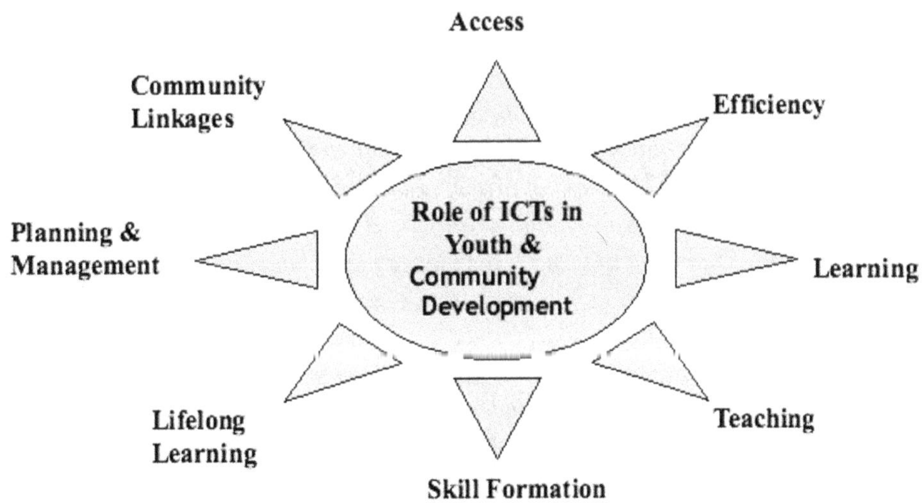

In many countries, mobile phones have reduced the work of farmers allowing they to devote to other works. They do not have to walk long distances in order to transfer money from the bank or get information about market prices. It has also inspired the farm youth to take entrepreneurial steps in innovative ways to make agriculture remunerative and thus reduce the drudgery involved.

Youth for ICTs (Y4ICT)

All the discussions till now have been on how the ICTs can revolutionize agriculture. But like all other technologies, it needs a carrier to take it forward and implement it. ICTs are powerful only as long as they are used to their fullest potential. The youth who form a major portion of the developing world can get the best out of them because of their natural affinity to technology. It is important for the young people to participate in the ICT initiatives and their implementation, as they know best the needs of their community and themselves. It also adds the bottom-up approach to the project implementation making it more realistic and fruitful. The youth can play a big role in bridging the digital divide that exists in developed and developing countries and also the rural and urban areas. They can not only help themselves but also the other community members to get the best out of their land (Suchiradipta and Saravanan, 2013).

Youth and ICTs: Example of a Fruitful Relationship

In North-East India, Jhum cultivation (Slash and Burn Agriculture) is the most predominant form of agriculture. The prevalence of number of tribal dialects makes communication difficult for change agents. Hence, considering the farming situation and difficulty in communication of appropriate farm information, an e-extension project (e-Arik) was initiated since 2007 in East Siang district of Arunachal Pradesh state. To overcome communication difficulty and also to motivate rural farm youth and others to take up profitable farming activities, four educated tribal farm youth were used as 'farmer facilitators' along with ICTs. They were given advanced farm technological training at the constituent college of the Central Agricultural University (CAU) and KVK. The farm youth helped to create general agricultural and rural development awareness among tribal farmers, facilitated eco-friendly and sustainable farm technology dissemination, developed vocational efficiency among farmers, formed farmer groups for self-help, facilitated use of local resources, helped to make timely decisions by the farmers themselves and suggested alternative ways to solve farming and other rural problems in twelve selected villages. By the efforts of farm youth facilitators by following 'farmer to farmer communication' approach and using ICT tools, 44 percent and 92 percent of farmers adopted the information on climate smart practices on paddy (Oriza sativa) and Khasi Mandarin (Citrus reticulata) crops respectively. After three years of project initiation, 55 percent of farmers developed new Khasi Mandarin orchards in their Jhum field, which means they are permanently moving from age old Slash and Burn agriculture to settled cultivation. Even after the completion of the project, trained farm youth are serving as a link between agricultural development departments and tribal farmers for facilitating advanced farm trainings and advisory services and they become 'local knowledge managers' to foster agricultural development in the remote tribal villages" (Saravanan, 2011a & b).

IMPLEMENTATION OF ICTs IN RAS

Appropriateness of ICTs depends on the situation and their use is most successful as a catalyst of development and to do so a few logical steps need to be followed. While the steps may be indicative of the logical delivery of ICT projects, they are not absolute in any terms, but very much flexible depending on the need of the situation and best judgements of the extension organisation, based on detailed need assessment surveys among clientele and other stakeholders (Saravanan et al., 2015a).

1. **Need Assessment:** EAS is most useful and applicable when information and services provided are localized and need based and so for the ICT projects to be successful, the first and foremost action of a host organization should be need assessment of the target community.
2. **Benchmark Survey:** Standards or points of reference are very important for ICT enabled services to meet their objectives and this makes benchmark surveys a necessity. They are also useful as standards of monitoring and evaluation.
3. **Content Development:** Localized and customised content needs to be developed based on the results of need assessment and benchmark surveys to avoid the problem of blanket recommendation.
4. **ICT Selection, Development and Testing:** Based on localised needs, content, and target groups, the appropriate ICT tool needs to be selected, developed and pilot tested for determination of suitability.
5. **Awareness Programs and Registration:** One major drawback in ICT projects is lack of awareness of target users about its existence or benefits and to solve that, innovative awareness campaigns need to be conducted to make them aware of the projects. In the case of subscription based services, it is more important as the users need to register to receive the benefits.

Figure 2. Steps for implementation of ICT enabled RAS (Saravanan et al., 2015a)

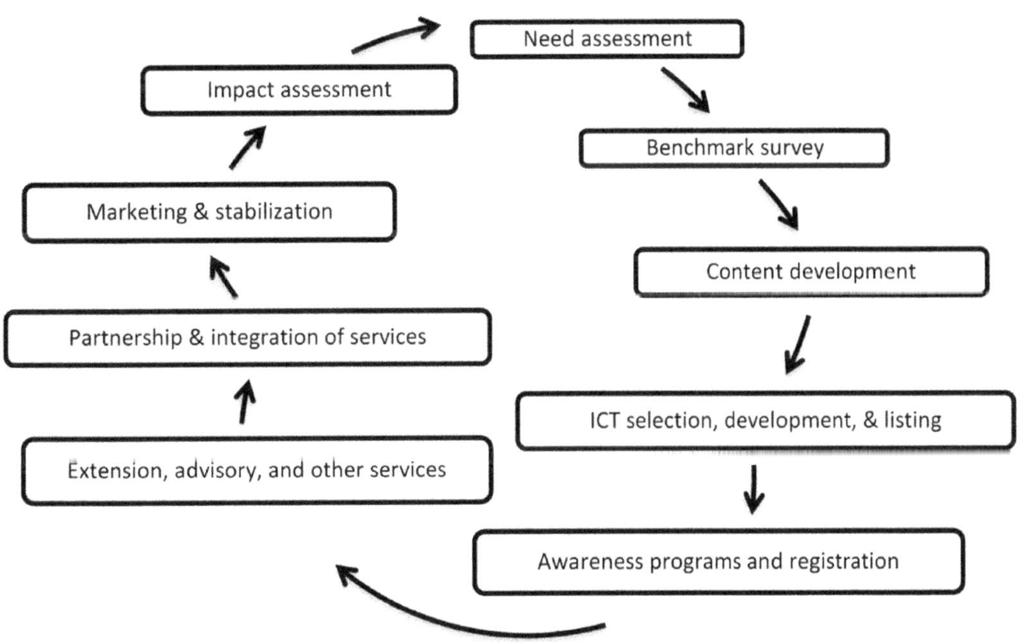

6. **Extension, Advisory and Other Services:** Based on demand and needs of the users, the services are to be provided to the targeted groups.
7. **Partnership and Integration of Services:** Depending on the need of the project and the services provided, collaboration among stakeholders needs to be formed and integration of services are to be determined to provide quality service to the users.
8. **Monitoring and Stabilization:** Continuous monitoring is an important function, especially in the pilot phase, to determine the suitability of the project to target users and accordingly modify the services offered to ultimately scale up the project in a profitable manner.
9. **Impact Assessment:** It remains one of the most important steps in implementation of ICT projects as the impact ultimately decides the degree of success of the projects in bringing the desirable changes in the target group as well as the factors deciding its sustainability for the long run.

SWOC ANALYSIS OF ICTs IN RAS

ICTs no doubt have the immense benefit of reaching the unreached, thus making it one of the most important supplements to RAS. This gives a huge opportunity to reach the BoP farmers and provide them necessary information to commercialize their agricultural practices and become self-sufficient. Then again, ICTs are only useful when the farmers can access the infrastructure, which is oftentimes non-existent in rural parts of the developing world, mostly when they are literate and tech savvy and also can bear the cost of the devices necessary. However, once these challenges of funding of projects, sustainability and acceptance by rural people are overcome, ICTs can become the most important catalyst in development of the rural population in the developing countries. The various strengths, weaknesses, opportunities and challenges of ICTs in RAS are presented in Table 2.

Capacities Required for Using ICTs in Agricultural Development

ICT is an umbrella term used for different types of communication technologies that sometimes require completely different kind of skill set to productively develop, implement, and use them for development purposes. Also, depending on the target user, content modification becomes a very important issue. All

Table 2. Strengths, weaknesses, opportunities, and challenges of ICTs in RAS

Strengths	Weaknesses	Opportunities	Challenges
Better access to services Cost-effective Timely Any time anywhere Supplements the role of extensionists Better research-extension-client system linkages	Success depends on human commitment Lack of personal touch in RAS Lack of ICT skill and competency Lack of institutional ICT policy in RAS Long-term sustainability	Continuous improvement of ICT infrastructure Penetration of high end mobile phones Reducing cost of ICT infrastructure and services Multiple players in RAS services provision using ICTs	Farmer-specific and relevant content Language barriers Low literacy of rural farmers Imparting skill and competence among RAS stakeholders to use ICTs Duplication and contradictory information flow

(Saravanan et al., 2015a)

these require professional capacities in both technology and social sciences. Table 3 below discusses the various capacities needed for successful implementation and ways to develop the major ICT tools predominant in agriculture.

Governance in ICTs for Agricultural Development

In the context of Agricultural Innovation Systems (AIS), to provide a quality extension and advisory services to the users, development and maintenance of the ICT projects needs to be a well thought out and planned measures involving more than one stakeholder working together. And this calls for a well-structured system that involves aptly positioned stakeholders to avoid overlapping roles. Depending on the type of ICT application, the type of governance varies and it is detailed in Table 4.

Table 3. Capacities required in and ways to develop ICT based interventions

ICTs	Capacities Required	Ways to Develop
Social media	Basic knowledge and skill in using internet and social media	Creating interest groups, awareness and training programmes and regular interaction and sharing
Radio, TV, tele-centres	Minimum knowledge of operating the stations and creation of suitable programs for selected medium. RAS professionals should create and deliver appropriate content.	Training in script writing, and content treatment. Need and location specific program creation and inclusion of farmers in the programs
Web portals, expert systems, Mobile apps, MOOCs	Proficiency in IT, expertise in choosing and packaging the content. Knowledge of web content development, content creation, interactive portal management, etc.	Specialized training for developing IT and subject matter experts. Integrating and updating content regularly, interactivity and knowledge base creation

Table 4. Governance of ICT based projects in agriculture

ICTs	Maintenance	Partnership	Stakeholders	Roles
Social media	Individually maintained by extension organizations	Generally does not arise. Groups and communities can be managed in partnership mode.	All agricultural stakeholders can be brought on same platform	Host organization connects to clients and other stakeholder for better coordination and information sharing among the actors in AIS
Radio, TV, tele-centres	Programs recorded by extension organization, TV stations broadcast	PPP model, collaboration among public organizations.	Extension organizations, radio/TV stations or tele-centres, agricultural institutions, farmers	While extension organizations develop content of program, it is recorded by the radio/ TV station or tele-centre crew. Agricultural institutions can act as information resource and farmers are the end users
Web portals, expert systems, Mobile apps, MOOCs	Hosted by extension organization, they provide information to farmers	Depending on stakeholders involved, it can be manned from PPP to profit oriented models.	Extension organization, web content developer/ app developer/course content developer, agricultural institutions (optional), and farmers	Development and management is a collaborative task, considering the involvement of many stakeholders. While the developers need to give constant upgrade, enhancements, maintenance and support, the RAS/ subject matter organization deals with content to be fed and expert service if needed to keep the information up-to-date. Farmers and others are end users.

Costs for Using ICTs for Agricultural Development

Cost of using/ developing ICTs ranges from free to a few million USD depending upon the infrastructure and scale of coverage. Capacity development activities also require considerable amount of money (Table 5).

CRITICAL ANALYSIS OF MAJOR ICTs

It is true that ICTs have transformed communication but in terms of its global reach, there are many more landmarks that need to be crossed. With increased access to the internet and increasing number of communication technologies with diversified working principle, a multifaceted situation is gaining traction which requires immediate attention to details for faster and wider acceptance. This specifically brings attention to the need for critically examining the advantages as well as the drawbacks of all these tools and build upon the information to make them an integral part of rural agricultural advisory systems worldwide. Table 6 deals on the same in details.

Problems

Implementers' Perspective

1. **Connectivity and Infrastructure:** Poor connectivity in rural areas, absence of electricity, transportation, markets and other infrastructural inadequacies restrict the implementers in using ICTs for development.
2. **Technical Workforce:** Technical human resource for ICT projects is very important for their smooth functioning. Functions like regular and timely updating of databases, developing location specific information, web integration and arrangement of links in case of web portals, transforming data/information in accessible formats for all type of users need technical workforce, which needs to be specifically trained in the tasks which is another challenge for the implementing agencies (Saravanan and Suchiradipta, 2015a).

Table 5. Tentative costs of hosting different ICT based initiatives

ICTs	Items	Indicative Cost Involved
Social media	Internet with ICT device	ICT device cost (mobile phone/ or tablet or laptop or computer) and internet charges.
Radio, TV, tele-centres	ICT infrastructure, training for the facilitators, content creation	Depending on air time, the cost can vary from a few hundred to several thousand USDs. However, most of the state/ NGO owned TV, Radio stations can be used freely by the agricultural advisory institutions for extension and advisory services.
Web portals, expert systems, Mobile apps, MOOCs	Hiring a software specialist, training for maintenance, content collation, multimedia content, hosting etc	• Creating a basic website (USD 300 – USD 2,000), CMS integration (USD 2,000 -10,000), simple semantic portals (with 500 nodes of content USD 10,000 – 25,000) and multi-site semantic portals with multiple portlets (USD 25,000 – 60,000) and maintenance of web portals also requires considerable cost. • Expert systems may cost USD 1,000- 10,000 depending upon the design, software, and size of contents. • mApp development can be free of cost to USD 70,000 depending on its architecture and web interface

Table 6. Critical analysis of major ICTs

	Information and Communication Technologies (ICTs)													
	TV		Radio		Mobile Phones (Basic/ Feature)		Computer/ Laptop/ Smart phones							
							Without Internet			With Internet				
	TV Broadcasting	Video with DVD	Radio Broadcasting	Community Radio	Text	Voice	Expert Systems/ DSS/ IMCDs	Digital Video	Animation	Web Site/ Web Portal/ Knowledge Banks/Online Repositories	Tele/ Video Conference	Mobile Apps	e-Learning Platforms (MOOCs)	Social Media
Target groups	Farmers, agripreneur, input dealers, stakeholder in value chain, extensionist	Farmers, agripreneur, extensionist	Farmers, agripreneur, input dealers, stakeholder in value chain, extensionist	Women farmer, agripreneur, stakeholder in value chain, input dealer	Literate farmers, agripreneur, input dealer, stakeholder in marketing channel, extensionist	Farmers, agripreneur, input dealer, stakeholder in marketing channel, extensionist	Literate farmers, extensionist, agripreneur	Farmers, agripreneur, extensionist	Illiterate farmers, extensionist	Literate farmers, agripreneur, input dealer, stakeholder in marketing channel, extensionist, policy maker	Farmer, extensionist	Mostly literate farmers, entrepreneur, input dealer, stakeholder in marketing channel and value addition, extensionist	Extensionist, researcher, academician, farmer, agripreneur, policy maker	Literate farmers, agripreneur, input dealers, stakeholder in marketing channels, extensionist, research and academic institution, experts, policy maker
Major RAS functions	Awareness, advisory, technology transfer, mass advisory, promotional, documenting and sharing of ITK	Advisory, technology transfer, awareness, documenting and sharing ITK	Awareness, advisory, technology transfer, mass advisory, promotional	Awareness, advisory, technology transfer, knowledge sharing, mass advisory, feedback	Awareness, advisory, market information and linking, credit and banking access, mass advisory, promotional, linking with AIS actors	Advisory, market and information linking, mass advisory, feedback, awareness, promotional, linking with AIS actors	Advisory, knowledge sharing, technology transfer, mass advisory	Knowledge sharing, documenting and sharing ITKs, technology transfer, training, knowledge sharing	Advisory, technology transfer, awareness	Advisory, documenting and sharing ITKs, market information and linking, input linking, M&E, enumeration, survey, linking with AIS actors, feedback	Training, advisory, linking with AIS actors, feedback	Awareness, advisory, knowledge sharing, market information and linking, credit and banking access, input linking, feedback, promotional, technology transfer, M&E	Training, education	Awareness, promotional, advisory, knowledge sharing, documenting and sharing ITKs, technology transfer, market information and linking, mass advisory, linking with AIS actors, feedback
Advantage	-Suitable for all stakeholders -Good medium for demonstrations	-Easy way of demonstration -Portability -Popularity in rural community	-Portability -Low cost -Popularity in rural community	-Tool of social learning -Increased women participation	-Low cost -Location and farmer specificity -Timely information	-Low cost -Portability -Easy feedback -Accessible for illiterates -Increased access to public information	-Easy access to expert advice -Location and farmer specific solution -Instant solutions	-Suitable for demonstrations -Low cost -High popularity -Increased women participation	-Suitable for awareness creation -Innovative and interesting	-Integrated information -Location specificity -Interactive -Open access	-Expert advice -Good scope for training	-Interactive -Updated information -Customization of information based on personal needs	-Integrated information -Location specificity -Interactive -Open access	-Interactivity -User generated content -Easy access in multiple devices -Increases dialogue among development practitioners
Limitation	-One way communication -Location specificity -Relevancy of personal context -Popularity	-Availability of playing device with farmers - Relevancy of personal context	-Location specificity -	-Lack of skill -Lack of funding	-Illiteracy -Network coverage - Relevancy of personal context -Popularity in agricultural use	-Network coverage -Relevancy of personal context - Popularity in agricultural use	-Illiteracy -Access to kiosks -Popularity among rural community -Updated database	-Access to player -Very specific type of information -Relevancy in personal context	-Limited scope -Less popular	-Illiteracy -Needs technical expertise -Possibly outdated content -Limited customization for individuals -Limited access by women farmers	-Limited access Limited scope -Unavailability of infrastructure in rural areas -Rarely used	-Illiteracy -Accessible only in smart phones -Information not personalized	-Illiteracy -Needs technical expertise -Possibly outdated content -Limited customization for individuals -Limited access by women farmers	-Illiteracy -Limited access -Costly data charges -Redundancy of information -Unauthenticated information

(Source: Saravanan et al., 2015a; Saravanan et al, 2015 b; Saravanan et al., 2015c; Saravanan and Suchiradipta, 2015c)

3. **Sustainability of Funding:** ICT projects at the initial or pilot phase are more often than not funded by an external agency but after the pilot phase, in most of the cases, funding dries up and so do the projects. Financial sustainability is often addressed too late or never and so, irrespective of the influence of the projects on rural farming, communities have to wrap up, thus disappointing the beneficiaries and making them disinterested in such future attempts.
4. **Effective Participation of Communities:** Lack of proper planning, differences in project objectives and community needs, limited capacity building program of projects, lack of or limited interest of the community in projects due to their temporary nature, uninvolved communities as major stakeholders in projects, etc. (Saravanan, 2012) are some important factors that keep agricultural communities away from the ICT projects.
5. **Lack of Leadership:** To diffuse ICTs and implement them in a large scale in rural areas, effective leadership is needed at both institutional and community level. Opinion leaders in communities often lack interest in them and the younger generation does not have enough followers to effectively diffuse them in the communities, which makes a barrier in increased use of ICTs by rural communities.
6. **Farmers' Access to ICTs:** The rural smallholders often do not have access to any ICTs – personal or otherwise - which makes information delivery through those media or applications redundant.
7. **Institutional Barriers:** Lack of flexibility in content delivery, limited innovation, lack of institutional ownership, limited technical human resource, conflict of interest among project partners, limited research collaboration, red-tapism, and project tourism attitude, etc. (Saravanan, 2012) often take the focus away from the major objective of the projects – development of communities.

End Users' Perspective

1. **Illiteracy:** ICTs, for example mobile phones, social media, web portals, online knowledge repositories, etc. require functional literacy at the least but even that is absent in many rural areas of developing and least developed countries. Furthermore, education illiteracy which is technical illiteracy remains another constraint. And in such cases, the illiterate farmers have very few options in accessing information and so reach is limited.
2. **Infrastructure:** Basic infrastructures in rural areas still have a long way to go. While in some areas telecommunication structures are poor or absent, in many parts of the world, electricity is yet to reach. Added to that is the absence of other infrastructures likes markets, credit facilities, transportation and so on, that limits or discourages the use of ICTs by the rural communities.
3. **Cost of Technology:** Affordability of technology is also a hindrance in its use to many smallholders. While devices like kiosks, computers, video players, etc. can be accessed from telecentres, mobile phones and other such personal devices are often out of reach for many farmers in developing countries. Also, the cost of data and other charges become expensive for many to afford (Suchiradipta, 2012).
4. **Information Explosion:** While a large amount of information is available in the public domain to access by anyone, this information can be confusing to many users. While filtering of important information is done by extensionists to suit the needs of the individual farmer, the farmers may not always be as competent in filtering those which may be more confusing than helpful.

5. **Lack of Awareness and Promotion:** Since ICTs are new introductions in extension services in many rural areas, proper awareness is needed in making the rural communities interested in using them. Also, because of lack of promotion, they often do not participate in the initiatives as active stakeholders.

Evidence of Impact and Potential Scalability of ICTs for Agricultural Development

Impacts of development efforts are generally multifaceted and their impact on social and personal lives of the benefactors is both direct and indirect, making them harder to point out clearly. With a two-sided argument existing around ICTs, mapping out the impact indicators can go a long way in evaluating the impact of the ICT based efforts and clearly state the achievements and limitations. They can also help with the potential scalability of ICT projects across the world, where crossing the hurdle of pilot stage is still a big concern. Table 7 discusses the same in details:

Issues of Sustainability in Using ICTs for Agricultural Development

Major factors that determine the sustainability of ICT use are the ICT tools used, the domain of the service providers (public or private) and the cost incurred in providing the services. Generally, advisory

Table 7. Impact indicators and potential scalability of ICT based projects in agriculture

ICTs	Impact Indicators	Potential Scalability
Social media	Large number of YouTube channels hosted by farmers, popularity of the Twitter handle of AgChat, highly active social media pages in extension like GFRAS, AESA and farmers' pages like Mkulima Young provide the evidence of impact. Participation of the members in the continuous engagement and discussions, creation and sharing of contents, increase in the membership subscription and feedback of members can serve as impact indicators for social media.	Facilitation of social media platforms is key for achieving the growth of audience and scalability is continuous with the increase of members' enrolment which requires high level of participation and engagement of stakeholders
Radio, TV, Tele-centres	Significant knowledge enhancement through radio programs has been noticed among farmers is exposed to radio programs[5]. Before-after comparison of knowledge is an important impact indicator of broadcast services indicator.	Increasing number of programs in the prime time can give the agricultural programs larger audience and increased popularity
Web portals/web sites/ MOOCs	e-Chaupal initiative in India for market price dissemination increased the price of soybean by 1.7 per cent and the result was instantaneous. It increased farmers' share in producers' money and reduced the cost incurred[7]. Access Agriculture, through videos hosted in the web portal, has changed the life of farmers across Asia and Africa, especially of women by taking information to them and organizing and empowering them[8].Time and cost saved, income increased, increased market participation, etc. after use of information provided through web portals, can act as impact indicators for web based information services.	Continuous monitoring and evaluation and based on the findings making necessary changes depending on clients' response is the best way to scale up web portals/ expert systems and MOOCs
Mobile based advisory services	Higher market participation by farmers, reduced wastage, increased transaction, and declined search and transportation costs can serve as indicators of effectiveness of mExtension services. Information through mobile phones have strengthened local livelihoods, natural resources, increased awareness and networking opportunities, especially among women. Time saved, cost saved, income increased after use of information provided through mExtension services can act as impact indicators	Financial sustainability of mobile based services is the major issue in their scaling up and profitable business model is a necessity. Also, development beyond thepilot phase is most important for long term sustainability.

through TV, Radio and community radio, social media are free of cost services. Mobile, web, interactive multimedia CDs, Videos etc. are fee based value added services or free of cost depending upon the host institutions. Governments/ public sector service providers mostly provide free of cost services (Farmers call centres, Front line SMS etc). Private sector actors are generally implementing financially sustainable and profitable models (e-Choupal in India, iCow and Kilimo Salama in Kenya, FCC in Uganda, and IKSL in India etc). However, for the profit oriented services, end users' perception of usefulness of information is the ultimate determinant of the success of the business model.

Sustainability of ICT projects beyond pilot phase has been a major drawback in their long term existence and there is more than one factor that influences their sustainability. The business model is very important factor since finances are the major cause of their cessation beyond the pilot stage. Also, profit oriented or financially sustainable services are more user demand oriented as subscription is important for incoming cash flow and with satisfactory service, users are found to be willing to pay for the services as in case of e-Choupal and IKSL in India, iCow and Kilimo Salama in Kenya, FCC in Uganda among many others. The funding agency also decides the business model to some extent, for example, as with government funding, cost-free services are more common whereas with public or private funding, profit motive is also a major factor along with development. The cost of providing the services impacts the business model to some extent.

Another factor that impacts sustainability of ICT projects are demand based information delivery. Agriculture being location specific, crop specific and time specific, blanket recommendations are helpful to none. Customized demand based information and advisory in ICTs are not choices but a necessity for long term sustainability.

Also, with changing dynamics of agriculture and time, farmers are focusing more on commercialization and so are demanding to be highly engaged to advisory service providers as important and active stakeholders of the system rather than mere beneficiaries of top-down development approach. Applications like social media, mApps, mobile and web based advisory services enable high user engagement and help them to customize the information they retrieve, thus making it personalized and applicable. This is an important feature in sustainability of ICT projects for long term.

CONCLUSION AND RECOMMENDATIONS

With increased technological advancement and its increased use in rural areas, farming communities are to gain a lot from the use of ICTs in information access related to their life and livelihood. Information is more important in agriculture and ICTs have the potential to be an aid in taking the much needed information to the farming communities. Inspite of the potentials, problems are many as discussed earlier and to overcome those, some steps become necessary, as discussed below:

1. **Implementation from Where Users Are:** Information provided to smallholders, irrespective of the means of delivery, needs to be location specific and timely. For any ICT based application, that is an opportunity as well as a challenge as the easier means of communication requires it to be updated regularly. And since access is easier compared to other traditional means, to gain and retain the users' faith in the service, the information needs to be location specific, and if possible, personalized.

2. **Youth Involvement:** The youth are the first to adopt any technological innovation in a social system and their involvement in planning and implementation is very important for overall success of the projects (Suchiradipta and Saravanan, 2013). Moreover, they can be leaders in community and family in diffusion of technologies for agricultural information.
3. **Institutionalization of ICTs:** To make ICTs an integral part of extension, the concerned institutes need to make their use an integral part of the work environment and the employees need to be encouraged about their use. Institutions also need ICT policies to encourage their use along with setting standards and boundaries that define the use of ICTs like social media focusing on the objectives and mandates of the institutions.
4. **Providing Basic Infrastructures:** Use of ICTs can only help the rural communities so far without any basic infrastructures like transportation, markets, credit facilities, etc. to translate the information acquired into practical action. Providing these basic facilities in rural areas is as important as the information, maybe more and so, in collaboration with competent authorities, development of infrastructure also needs to be prioritized.
5. **Appropriate Partnerships:** Appropriate partnerships are very important in the implementation of ICT projects for ensuring a smooth flow of resources (especially the funds to purchase and maintain ICTs), participation of local communities and institutions, regular updating of recent information in timely manner, effective management of stakeholders, and avoiding conflict for smooth functioning.
6. **Partnership with Local Institutions:** ICT applications are best applicable to smallholders when specific to their situations and to develop data and information that is timely and location specific, forging partnership with local research stations, universities, and other local institutions is desirable. Moreover, this makes getting up-to-date data easier and link-ups fruitful (Saravanan and Suchiradipta, 2015a).

REFERENCES

FAO. (2015). *Five agents of change for a sustainable world.* Accessed on 24 September 2015 from, http://www.fao.org/post-2015-mdg/news/detail-news/en/c/281554/

Haddad, W. (2007). *ICTs for Education: A Reference Handbook.* Accessed on October 12, 2008 from, http://www.ictinedtoolkit.org/usere/p_page.php?section_number=0

ITU. (2015). *ICT facts and figures 2015.* Retrieved from http://www.itu.int/ict

Kemp, S. (2015). *Digital, social and mobile worldwide in 2015.* Accessed on August 12, 2015 from, http://wearesocial.net/blog/2015/01/digital-social-mobile-worldwide-2015/

Millennium Project. (2006). *Millennium Development Goals – What they are.* Accessed on September 24 2015 from, http://www.unmillenniumproject.org/goals/

Sadowsky, G. (1996). *The Internet Society and Developing Countries.* Accessed on October 25, 2015 from, http://www.isoc.org/oti/printversions/1196prsadowsky.html

Saravanan, R. (2011a). Tribal Farm Youth for Facilitating Agricultural Advisory Services by ICTs: A Success Story from North-East India. *Abstract Volume of the National Seminar on Attracting Farm Youth to Sustainable Agriculture.*

Saravanan, R. (2011b). *e-Arik: Using ICTs to Facilitate "Climate-Smart Agriculture" among Tribal Farmers of North-East India*. ICTs and Agricultural Adaptation to Climate Change Case Study. Centre for Development Informatics, University of Manchester. Accessed on April 5, 2012 from www.niccd.org/NICCD_AgricAdapt_Case_Study_eArik.pdf

Saravanan, R. (2012). e-Initiative for Agricultural Extension: Browsing for Logout? *Proceedings of the AFITA/WCCA 20012-8th Asian Conference for Information Technology in Agriculture (AFITA) and World Conference on Computer in Agriculture (WCCA)*. Retrieved from http://www.afita.org/graph/web_structure//files/Seminar%20%2807%29-01%281%29.pdf

Saravanan, R. (2013). e-Agriculture prototype for knowledge facilitation among tribal farmers of North-East India: Innovations, impact and lessons. *Journal of Agricultural Education and Extension*, *19*(2), 113–131. doi:10.1080/1389224X.2012.718247

Saravanan, R., & Suchiradipta, B. (2015a). Role of ICTs in family farming: Experiences and way forward. In Family Farming and Rural Economic Development. New India Publishing Agency.

Saravanan, R., & Suchiradipta, B. (2015b). Invest in IT to boost agricultural services. *Smart Agri Post: Empowering Agripreneurs, 1,* 22-24. Retrieved from http://www.smartagripost.com/wp-content/uploads/2015/07/Smart-Agripost-July-20153.pdf

Saravanan, R., & Suchiradipta, B. (2015c). *mExtension – Mobile Phones for Agricultural Advisory Services*. Note 17. GFRAS Good Practice Notes for Extension and Advisory Services. GFRAS: Lindau, Switzerland. Retrieved from www.g-fras.org/en/download.html?download=349:ggp-note-17-mextension-mobile-phones-for-agricultural-advisory-services

Saravanan, R., & Suchiradipta, B. (2015d). *Social media: Shaping the future of agricultural extension and advisory services. GFRAS interest group on ICT4RAS discussion paper*. Lindau, Switzerland: GFRAS.

Saravanan, R., Suchiradipta, B., Chowdhury, A., Hambly Odame, H., & Hall, K. (2015c). *Social Media for Rural Advisory Services*. Note 15. GFRAS Good Practice Notes for Extension and Advisory Services. GFRAS. Retrieved from www.g-fras.org/en/download.html?download=355:ggp-note-15-social-media-for-rural-advisory-services

Saravanan, R., Suchiradipta, B., Meera, S. N., Kathiresan, C., & Anandaraja, N. (2015b). *Web Portals for Agricultural Extension and Advisory Services*. Note 16. GFRAS Good Practice Notes for Extension and Advisory Services. GFRAS. Retrieved from www.g-fras.org/en/download.html?download=356:gfras-ggp-note-16-web-portals-for-agricultural-extension-and-advisory-services

Saravanan, R., Sulaiman, R. V., Davis, K., & Suchiradipta, B. (2015a). *Navigating ICTs for Extension and Advisory Services*. Note 11. GFRAS Good Practice Notes for Extension and Advisory Services. GFRAS. Retrieved from www.g-fras.org/en/download.html?download=351:ggp-note-11-navigating-icts-for-extension-and-advisory-services

Suchiradipta, B. (2012). *Mobiles for mobilizing Agricultural Extension in India*. Credit Seminar presented in School of Social Sciences. *College of Post Graduate Studies, Central Agricultural University, Umiam, Meghalaya*, *793*, 103.

Suchiradipta, B., & Saravanan, R. (2013). Youth and ICTs for Agricultural Development. In K. Narayana Gowda, M. S. Nataraju, & V. Veerabhdraiah (Eds.), *Youth in Agriculture and Rural Development*. New Delhi: New India Publishing Agency.

Suchiradipta, B., & Saravanan, R. (2016). *Social media: Shaping the future of agricultural extension and advisory services*. GFRAS interest group on ICT4RAS discussion paper. GFRAS. Available at www.g-fras.org/en/knowledge/gfras-publications.html?download=414:social-media-shaping-the-future-of-agricultural-extension-and-advisory-services

World Bank. (2011). *ICT in agriculture: Connecting small holders to knowledge, networks, and institutions*. e-Source Book. Report no. 64605. Washington, DC: The World Bank. Available at: http://www.ictinagriculture.org/content/ict-agriculture-sourcebook

Chapter 6
The Temporal and Spatial Development of Organic Agriculture in Turkey

Aylin Yaman Kocadağlı
Istanbul University, Turkey

ABSTRACT

The world's population is increasing rapidly and the food need for the increasing population has been growing also. In the 21st century, efforts for the sustainable use of the natural resources have been raised all around the world. Agriculture has always been on the center of discussions and developments as an activity carried out depending on the nature. Therefore, the sustainable agricultural systems have gained a considerable importance in recent years. This chapter will examine the temporal and spatial development of the organic agriculture in Turkey. The current situation of the agricultural sector in Turkey will be reviewed and the need for a sustainable agricultural system in Turkey be analyzed. The temporal and spatial change of the organic agricultural areas, numbers of producers and products, and organic production amounts in Turkey will be put forward. Also, the organic agricultural potential of Turkey will be evaluated and the reasons for not exploiting this potential will be investigated.

INTRODUCTION

Agriculture is a highly prominent strategic sector for almost all countries as it enables them to produce their basic foodstuff, provides raw materials to other industries, is an area of employment and brings about a foreign currency inflow to the country's economy through exporting activities. In the direction of the world's changing needs and circumstances, in the course of time, some important changes have been seen at the methods and policies applied for the agriculture. As a result of wars bursting out in the last century, and associated food supply problems, application of the technological innovations into agriculture and the rapid increase in the world's population have changed notably the expectations in the agricultural activities. Rapidly increasing population of the world has brought along many necessities, and the need for agricultural products has risen to meet increasing populations. At this stage, the main

DOI: 10.4018/978-1-5225-0942-4.ch006

objective of the agricultural policies has become to raise the productivity and increase the production quantities (Grigg, 1995). Particularly with the policies applied in the 1960-1970s, high effectiveness and productivity have been reached. On the other hand, in obtaining high productivity, the fact that agriculture-related natural resources are limited has been ignored (Badgley & Perfecto, 2007). Agriculture itself has been managed for many years with the understanding that the natural resources are limitless. As a result of these policies, high productivity has been reached and agricultural economic development has been built, but it has been realized that such development is not sustainable any more (Park & Seaton, 1996).

Agriculture is an economic activity dependent on some fundamental natural resources such as soil, water and air. At the end of the 20th century, it was agreed that natural resources, the base of all agricultural production, are restricted and it was realized how important it is to use these natural resources sustainably (Pretty, 1995). At this stage, people have started searching for sustainable agricultural options without damaging the natural resources. Agricultural systems that promote sustainable production without damaging the natural resources as well as policies which support such systems have been brought to the forefront of the world's agenda.

Thanks to the understanding of sustainable agriculture, natural resources are protected and their productivity is guaranteed for a long term. The concept of sustainable agriculture implies a balanced economic, social and ecologic agricultural system. For people's access to sufficient and safe food, the raising of healthy future generations, enhancing rural development, economic improvement and a liveable environment, agricultural areas should be used in a sustainable manner. Today, the concept of sustainable agriculture has great importance. It is so important to create a society that could meet its necessities by not destroying the natural resources and by sustainable use of the existing natural resources especially in light of increasing populations (Grigg, 1995). For this reason, in order to reach a sustainable life and a sustainable agriculture, organic agriculture is gaining importance as an alternative agricultural method around the world.

Organic agriculture is one of several sustainable agricultural systems and it could be generally defined as a production method that is controlled and certified in all of its stages, aiming for protection of the natural resources, having a human and natural friendly understanding and free of synthetic chemical inputs including fertilizers. It is highly possible to protect the natural resources with organic agriculture at one side while to benefit from these resources to the utmost with a sustainable understanding of agriculture (Lotter, 2003). As compared to the conventional agricultural areas, the world's organic agricultural areas are still too small. However, it has been accepted that the world's need for food is increasing rapidly and that the developments happening on the agricultural fields threaten the future of human beings. Because of the awareness, sustainable use of agricultural areas has been raised across the world. Therefore, in the next years, organic agricultural applications will undoubtedly increase.

Turkey is one of the countries where the organic agriculture has gained importance in recent years all around the world. Turkey is an important agriculture country at the present time (Food and Agriculture Organization [FAO], 2014). In reference to the 2014 data, Turkey has the 7th largest agricultural economy of the world. This figure reveals clearly how the agricultural sector is important in Turkey's economic and social structure. Having a sustainable agriculture in Turkey is so important especially in light of the rapidly increasing population. The concept of organic agriculture started to come into the view in Turkey after 1984 and it has improved gradually within the period of time following 2000. Also, the European Union membership process has made a great contribution to the wide spreading organic agriculture in Turkey. Today, Turkey is one of the top ranked countries in the world with regards to organic agriculture areas, number of producers and products, and organic production amounts (Pugliese, Bteich, & Al-Bitar, 2014). Turkey's prominence in the world organic food market scales up day by day.

The aim of this chapter is to examine the temporal and spatial development of organic agriculture in Turkey. Within this framework, the current situation of the agricultural sector in Turkey will be evaluated and the development process of organic agriculture will be reviewed. In addition, the existing challenges of Turkish agriculture and the necessity for sustainable agricultural systems in Turkey, will be studied. Moreover, the temporal and spatial change of the organic agricultural areas, numbers of producers and products, and organic production amounts in Turkey will also be examined.

AGRICULTURE IN TURKEY AND THE POTENTIAL FOR ORGANIC AGRICULTURE

Agricultural Structure of Turkey

Turkey, which is the study area, is a country located at the northern hemisphere between the 25° 40' - 44° 49' eastern longitudes and the 35° 48' - 42° 06' northern latitude (Figure 1). The geographical and special position of the country has brought about different climatic types particularly the Mediterranean climate. Turkey has various characteristics in terms of climate, soil and morphologic features. Such variety has paved the way for a rich agriculture potential and a high importance on the agricultural activities of the country (Doğanay, 2007).

Figure 1. Location map of Turkey

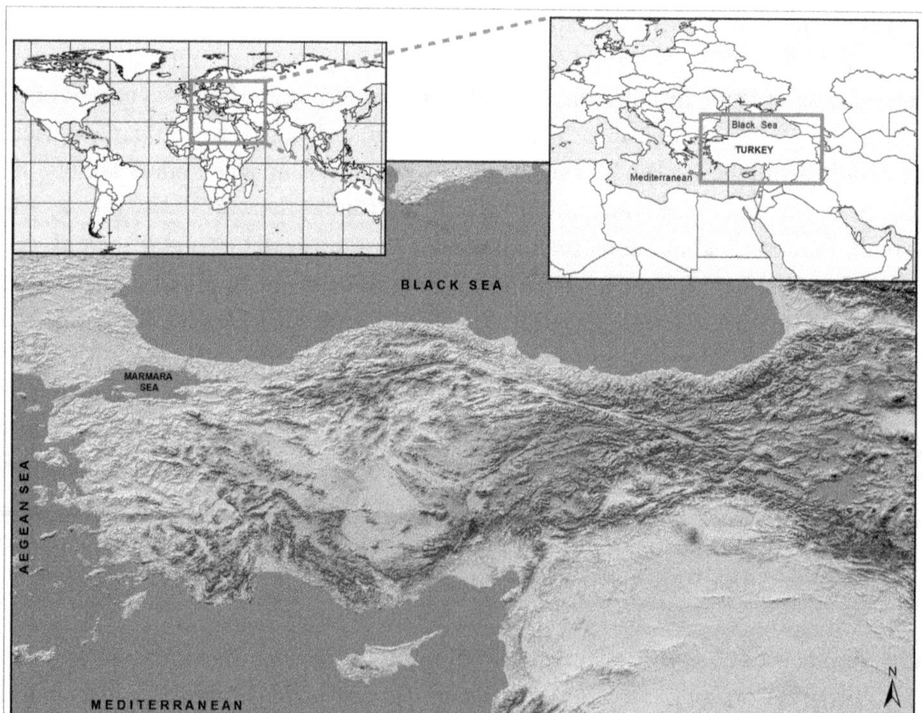

Turkey is one of the countries located on the Mediterranean Basin and the most important economic sector of the Mediterranean Basin is agriculture. At the present time, the agricultural sector uses most of the land and water resources on the Mediterranean Basin and it provides employment for nearly half of the labor force of the basin (Makhzoumi & Pungetti, 1999). It is quite similar in Turkey as well that is located on the Mediterranean Basin and agriculture is one of the most fundamental economic sectors. Agricultural sector has played a very significant role for the economic and social development of Turkey, not only in the past but also in present times.

Agriculture is tightly coupled to the natural environmental features of the space. Natural environmental conditions have influences on the distribution of the agricultural areas, application of the agricultural systems and variety, type and quality of the agricultural products (Knowles & Wareing, 1983). Turkey's structure of land and accordingly changing climatic features has led to different geographic regions and the formation of various micro climates in these regions. Such diversity in the natural environmental features has brought along a rich agricultural potential. There are 7 geographic regions with different ecologic characteristics in Turkey and the production potentials of these regions are considerably different from each other.

Turkey's lands constitute a very old agricultural area. Looking through the historical development of the agricultural land use in Turkey, one can see substantial changes that have happened over time. Depending on the increasing population and developments happening in the agricultural mechanization, a notable increase was detected in the quantity of agricultural areas up to the early 1990s. From the period since 1990 up to our present day, a reduction has been noted in the quantity of agricultural areas and this is due to several different reasons.

Turkey has 78.534.700 hectares of surface area. While in 1928 only 6.6 million hectares was cultivated, it reached to 25.3 million hectares in 1950 and later on nearly 28 million hectares in the early 1990s (Minister of Development, 2014a). Following the end of 1990 the quantity of agricultural areas started to decrease. It dropped to 26.986.548 hectares in 1996 and finally 23.810.672 hectares in 2013. Within the short period of time, 3.175.876 hectares (-11.2 per cent) of agricultural areas have been lost because of different factors. Today, the proportion of the agricultural areas to the country's surface area is 30.3 per cent.

About 20.6 million hectares of the agricultural areas in Turkey consist of arable crops, fallow and vegetable lands in 2013. Nearly 3.2 million hectares of areas are covered with fruit gardens, vineyards and olive groves (Table 1, Figure 2). In about 66 per cent of the agricultural areas, grain products are cultivated while only 3.4 per cent of these areas are allocated for vegetables and in 8.1 per cent of agricultural lands, various fruits are produced. Vineyards constitute approximately 2 per cent of these agricultural areas while 3.4 of them are covered with olive groves. The proportion of fallow lands is 17.4 per cent by 2013, and in the same year, wheat and barley are the mostly cultivated grain products. Wheat is cultivated on 7.773.000 hectares of area and barley is produced on 2.721.000 hectares of area. By 2013, the areas that only wheat and barley are cultivated on, stand for the 44 per cent of the agricultural areas in Turkey.

The agricultural sector maintains its importance as a resource for employment and economic development in Turkey's economy. An examination of the distribution of country's employment to the sectors, revealed some important changes. In the early 1930s, 90 per cent of the working population was employed in the agricultural sector (Göney, 1979). Such proportion dropped back to 75 per cent in 1960 and up to 1990 agricultural sector was dominant over the employment of the country. After 1990, substantial changes took place in the distribution of country's employment to the sectors, (Table 2). In 1991, almost half (48 per cent) of the working population was employed in the agricultural sector.

Table 1. Agricultural land use types in Turkey, 2013

Agricultural Land Use Types	Area (ha)	%
Arable crops	15.618.059	65.59
Fallow land	4.147.587	17.42
Vegetables	808.488	3.40
Fruits	1.937.416	8.14
Grapes	468.792	1.97
Olives	825.827	3.47
Ornamental plants	4.503	0.02
Total	23.810.672	100

(Turkish Statistical Institute [TUIK])

Figure 2. Distribution of agricultural area by land use types, 2013

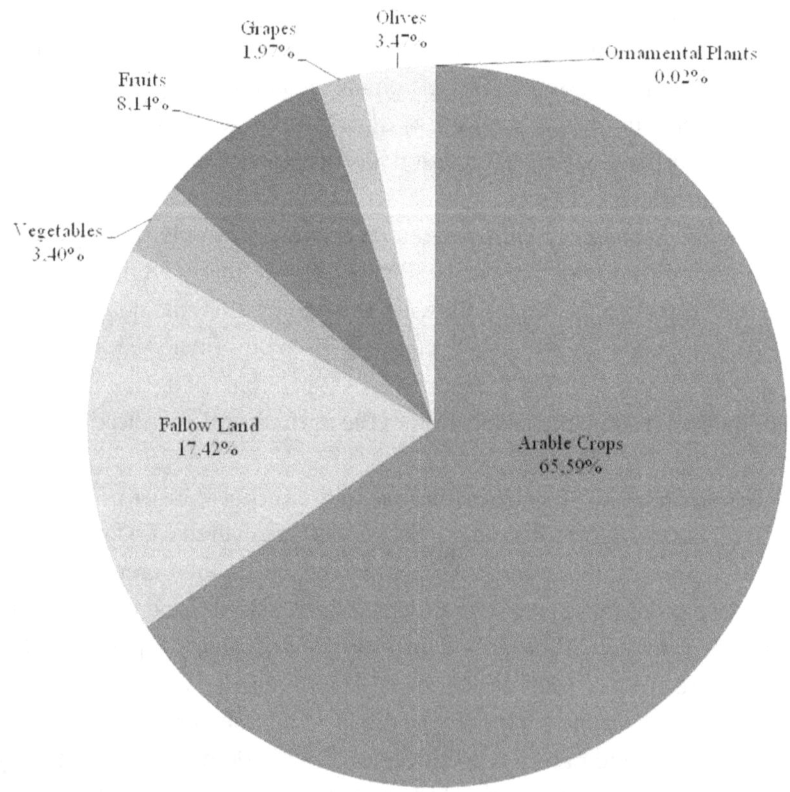

Depending upon the developments happening in the service and industrial sector since 1990 up to our present day, a notable decrease has been seen in the working population employed in the agricultural sector and the percentage of the agricultural sector in the total employment has rapidly reduced. Now, the agricultural sector comprises 21 per cent of the total employment of the country. Although the percentage of the agricultural sector in the total employment has reduced, agricultural sector maintains

Table 2. Employment by sectors

Years	Agriculture		Industry		Services		Total Labor Force
	Employed (thousand)	%	Employed (thousand)	%	Employed (thousand)	%	Employed (thousand)
1991	9.212	48	2.952	15	7.143	37	19.307
1995	9.080	44	3.295	16	8.212	40	20.587
2000	7.769	36	3.810	18	10.001	46	21.580
2005	5.154	26	4.178	21	10.735	53	20.067
2010	5.683	25	4.496	20	12.417	55	22.596
2013	5.204	21	5.101	21	14.297	58	24.602

(TUIK)

its importance in the rural areas as the primary means of employment (FAO, 2014). Nearly 63 per cent of the rural population of the country, in 2011, was employed in the agricultural sector (OECD, 2011). For this reason, to continue the agricultural activities in a sustainable manner is essential for the rural development in the country (Bakırcı, 2007). Apart from these figures, there is illegal employment as well in the agricultural sector of Turkey. In summary, today the agricultural sector maintains its role of being a prominent source of employment for Turkey.

In the development years of Turkey, the agricultural sector contributed a substantial proportion of the national income. In 1970s, the proportion of agricultural sector in Turkey's national income was 30 per cent. Such proportion dropped back to about 10 per cent in 2000, and has decreased to 8 per cent at the present time. Also, the agricultural proportion to total foreign trade has reduced. Agricultural sector constituted 85 per cent of Turkey's foreign trade in 1968 while it was 61 per cent by 1975 (Göney, 1979). The proportion of the agricultural sector in the foreign trade was 6 per cent in 2000 and dropped to 4.7 per cent 2005. In spite of the fact that the proportion of agricultural sector in foreign trade has reduced as compared to other sectors, agricultural sector still maintains its importance in foreign trade. Today, Turkey is one of the world's major agricultural producers and is an exporting country in many prominent agricultural products. In 2012, the agricultural sector in Turkey has 5.2 billion US dollars export volume and it corresponds to 3.4 per cent of country's total exportation.

The agricultural sector is an economic sector which is directly bonded to food production and nutrition of the country's population. Turkey's population is growing rapidly as well. In 1927, country's population was 13.648.270 people and the rural population was more dominant in the country. By 2015, the total population of the country increased to 78.741.053 people, so it means that the total population has increased 5-6 times since 1927. Within the development process of the population, highly important changes have taken place in rural and urban distribution of the population (Table 3). In 1927, 75.8 per cent of the total population lived in the rural areas and the other 24.2 lived in the cities while such percentage has become just the contrary at the present time; in 2012, only 22.8 per cent of the total population lived in the rural areas while the other 77.2 per cent lived in the cities. It is due to the rural-to-urban migration that working population employed in the agricultural sector decreases in spite of the increase in Turkey's total population. Unless such migration is reversed, the decrease in rural population will go on. Keeping the rural population in its place and employment of this population in the agricultural sector are of vital importance for a sustainable agriculture. Even though the rural population has reduced in

Table 3. Turkey population by years

Years	Urban		Rural		Total
	Population	%	Population	%	
1927	3.305.879	24.2	10.342.391	75.8	13.648.270
1940	4.346.249	24.4	13.474.701	75.6	17.820.950
1950	5.244.337	25.0	15.702.851	75.0	20.947.188
1960	8.859.731	31.9	18.895.089	68.1	27.754.820
1970	13.691.101	38.5	21.914.075	61.5	35.605.176
1980	19.645.007	43.9	25.091.950	56.1	44.736.957
1990	33.326.351	59.0	23.146.684	41.0	56.473.035
2000	44.006.274	65.0	23.797.653	35.0	67.803.927
2012	58.448.431	77.2	17.178.953	22.8	75.627.384

(TUIK)

time, as compared to other European countries, the percentage of rural population is still high in Turkey. By 2012, the rural population of Turkey was 17.178.953 people. Today, the unique means of livelihood of Turkey's rural population is agriculture. Therefore, sustainable agriculture is more important for the rural population.

Agriculture plays an important role in providing the food needs of the increasing population. Every country makes efforts to be self-sufficient in the agricultural production. Providing the food needs of the increasing population requires an increase in production at the same time. Soils are used more intensely in order to satisfy the food needs of the increasing population. Agricultural mechanization, developments in the agricultural technologies and increase in use of chemicals and fertilizers brought about high productivity in Turkey, especially between 1970-1990. In this period, the increasing population triggered the motivation to increase production, however the negative effects on the soil and environment of conventional agricultural systems have come to the fore in recent times.

On the other hand, the rapid increase in population has brought along misuse of the lands. The need for new lands due to the pace of urbanization has been answered by opening the agricultural areas to settlement. Needs and tendencies of the rapidly growing cities to enlarge on the agricultural areas gradually threaten the availability of Turkey's fertile agricultural soils. Particularly in the last 15-20 years, the reduction of agricultural areas is highly related to the increase in non-agricultural usage (settlement, industry, tourism, mining and transportation etc.) of these lands. Along with the rapid urbanization in the country, fast growing tourism activities also threaten the agricultural areas as well. The fertile agricultural areas on the coastal regions are almost decimated because of tourism use. Urbanization and tourism activities are intensified generally on the coastal regions and the most productive agricultural areas of Turkey are located on these coastal regions. Data from 2013 show that 41.151.531 people live in the provinces located on Turkey's coastal regions. In these coastal provinces which constitute 28 per cent of Turkey's surface area, 54.4 per cent of country's total population lives. This fact brings into open the population pressure and threat on the productive agricultural areas located on the coastal regions.

Today, one of the other primary problems of Turkey's lands is erosion. Nearly in 79 per cent of the country's lands, there is medium or high level erosion problem. Moreover, there is an erosion problem in 59 per cent of the cultivated agricultural areas and 64 per cent of the pasture areas in Turkey. Such percentages reveal the huge threat to Turkey's soil resources. Erosion decreases the soil depth and removes the fertile upper layer of the soil. About 68 per cent of Turkey's soils are shallow (soil depth is between 0-50 cm). Erosion and consequently shallowness of the soils affect agricultural production. Apart from this, problems seen on the soils such as salinity and pollution threaten Turkey's agriculture more every passing day. In our present time, 4.2 million hectares of area have lost its productivity partially or completely owing to the changing salinity proportions (Minister of Development, 2014b).

Such changes in the physical, chemical and biologic structure of the soil diminish its productivity, and to raise the productivity and maintain an increase in production, much more chemical fertilizers are applied. By 2013, use of chemical fertilizers in Turkey's agriculture is 5.813.612 tons. Within such a short period of time since 2008 up to 2013, a 41 per cent of increase in chemical and fertilizer use has been noted. Within the period from 2012 till 2013, use of chemical fertilizers in the country has increased 9 per cent (Ministry of Food, Agriculture and Livestock, 2014). Although the use of chemical inputs and fertilizers has become more common in Turkey especially for recent years, overall amounts are not so high in comparison to other developed countries (Demiryürek, Stopes, & Güzel, 2008). Particularly in the Central Anatolia, Black Sea, Eastern Anatolia and South Eastern Anatolia regions of Turkey, where the conventional agriculture is more common, the soils have not been polluted so much. Usage of chemical fertilizers is however more intensified in the Marmara, Mediterranean and Aegean regions. The fact that a substantial part of the soils in Turkey is not polluted so much with chemical fertilizers is a definite advantage in its transition to a sustainable agriculture system.

It is highly essential for the countries to provide for the nutritional needs of the increasing population. In Turkey, the population has increased rapidly and the nutritional needs of the increasing population have intensified the pressures on agricultural areas. Population projections set forth that the increase in Turkey's population will continue. According to the estimations, Turkey's population will reach to 88.427.604 people by 2030 and 93.475.575 people by 2050 (TUIK). This will aggravate the pressure on agricultural areas. In order to satisfy the nutritional needs of the increasing population, the sustainable use of agricultural areas is required just the because of the fact that agricultural areas in Turkey has already pushed the limits and creating new agricultural areas is not possible any more. On the contrary, the existing agricultural areas could be lost because of natural and human factors such as erosion and misuse of lands. Consequently, the existing agricultural areas should be used in the most rational way. It is a fact that the agricultural areas in Turkey could not be increased, but it is not impossible to provide the needs of the increasing population through a well-planned and sustainable usage. For this reason, organic agricultural applications have started to gain prominence for the last years in Turkey just as in the other parts of the world.

Organic Agriculture in Turkey

The level that the agriculture sector in the world has come in the recent years shows that agriculture is not sustainable anymore and agricultural innovations are required. Natural resources needed for the agricultural production are not limitless. However, for many years all around the world, agricultural activities have been managed with the principle that these resources as limitless. It is a fact one cannot achieve sustainable development by damaging these limited resources and destroying the natural balance. Therefore, sustainable agricultural systems have become more important in last years. Sustainable agriculture embodies maintenance of production with environmental friendly techniques that protect the natural resources. Sustainable agriculture is defined as production systems that maintain continuous productivity and preserve social benefits (Pretty, 2008).

Organic agriculture is one of the sustainable agricultural systems and it aims for rebuilding the natural balance destroyed because of the wrong applications in the ecologic system, as well as to once more bring the human and environmental friendly production systems to the fore. In this sense, organic agriculture stands for an agricultural system armed with an ecologic, social and economic balance (Topp, Stockdale, Watson, & Rees, 2007). Since it is sensitive to human health and nature, organic agriculture has gained much more prominence. In recent years, organic agriculture has been rapidly spreading across the world and global organic food market has been enlarging every passing day. By 2012 across the globe in 164 countries, 37.5 million hectares is under organic agriculture. Nevertheless, in comparison with the conventional agricultural areas, organic agricultural areas in the world are still a small amount (0.87%).

Turkey is one of the foremost countries that organic agricultural system is performed. Initial activities of the organic agriculture did not become with raising public awareness regarding the sustainable agriculture. Instead, organic agriculture was started completely with commercial motivations in Turkey. Organic agriculture in Turkey started in the 1984-1985 production season upon the organic products request of the European companies from Turkey (Istanbul Chamber Commerce, 2006). The first organic production was carried out with raisins and dried fig in the Aegean Region. Afterwards, these products were increased with the addition of dried apricot and hazelnut production in different regions (Ozkan, 2003). Although organic agriculture started quite late in Turkey, especially for the recent years, an important development has been made. However, this development is based mainly on vegetative production, and organic animal production in Turkey is not so common yet. For this reason, in this study, organic vegetative production of Turkey has been examined. Within time, organic vegetative production in Turkey has made notable progress both in terms of number of products and producers and production areas and quantities. In the following discussion, the temporal and spatial change of the organic agriculture in Turkey will be examined.

TEMPORAL AND SPATIAL DEVELOPMENT

Organic agriculture in Turkey has been subjected to notable changes with regards to both temporal and spatial distribution.

Temporal Development

Turkey has such a rich agricultural potential that many various products could be cultivated in its different geographic regions, territories and climates. Turkey has demonstrated its potential with its organic products variety within the last years. Apart from the arid climate plants, many different plant species including subtropical plants are cultivated in the country. In parallel with the enlargement of the world's organic product market in recent years, organic products cultivated in Turkey have become more diversified. The organic agricultural product range started with traditional products (such as fig, grape, apricot, hazelnut and cotton etc.) has since become varied every passing year. The number of organic products in 1990 was only 8. By 2000, number of organic products rose to 95 and in 2005, it reached to 205. By the year of 2013, the total number of certified organic products in Turkey is 213. As it is seen, within these 23 years, a substantial increase has been noted in the number of organic products.

As consumers' demands for organic agricultural products rise, so has the number of producers adopting organic agriculture. International trade has developed owing to the growth of this demand. All in all, some countries have started to cultivate different organic products demanded from other parts of the world and export these products to these countries in spite of the fact that there is no domestic market or demand for the organic products in these countries. This situation is similarly in Turkey (Rehber & Turhan, 2002). Due to the rise in demand for the organic products in the world, the number of organic agricultural producers has increased.

Looking through the change in number of producers in Turkey, some important changes have been noted (Figure 3). Number of producers in Turkey was only 313 in 1990, and it rose up to 4.039 in 1996 and it reached to 18.385 by 2000. However, after 2000, a decrease has been seen in number of producers in Turkey. The economic crisis which occurred in 2001 in Turkey was responsible for the decrease in number of producers following 2000. On the other hand, after 2008, a substantial increase has been seen in number of producers. While the number of producers in 2008 was 14.926, it rose to 35.565 by 2009. In other words, only within 1 year, the number of producers increased almost 2.5 times. As to 2013, the number of producers reached to 60.797. By the year of 2012, there were 1.927.018 organic agriculture producers all around the world and Turkey became the 6th country with the largest number of producers. Also, Turkish producers now constitute 17 per cent of the total producers in Europe.

Both the temporal distribution of number of producers in Turkey and the spatial distribution of them have been subjected to substantial changes. There are 81 provinces in Turkey as per the civil administration structure. The data of provinces regarding the organic agriculture in Turkey is from 2004 and therefore on this data of 2004 will be used to compare the spatial changes of the organic agriculture.

Figure 3. The number of producers in Turkey

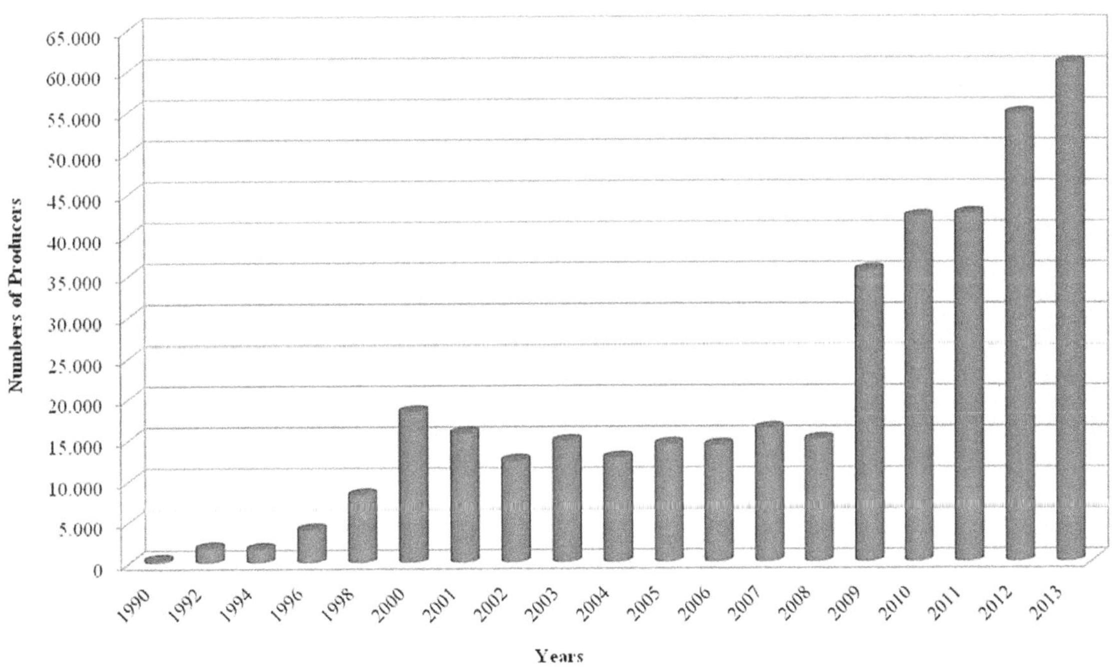

An analysis of maps of 2004 and 2013 demonstrates some important changes in the spatial distribution of the organic agricultural producers. The total number of organic agricultural producers in 2004 was 12.751 and looking at the distribution of producers, no producers can be seen in 28 cities across the whole country. This means that organic agriculture was not practised in 1 of every 3 cities and the cities without any organic agricultural activities are generally located in the Eastern Anatolia and South Eastern Anatolia regions (Figure 4). In 15 cities, number of producers was less than 10. In this way, we conclude that organic agriculture does not matter so much in more than half (43) of the cities. By 2004, the number of cities having more than 1.000 producers was only 6. These cities are respectively İzmir, Aydın, Erzurum, Kastamonu, Konya, and Çorum. As to the cities having 100 and more producers, they were intensified in the Western Anatolia Region and the city having the largest number of producers (1636) was İzmir in 2004.

A look at the map of 2013 shows that with respect to the spatial distribution of the organic agricultural producers, notable changes are seen at first glance as compared to 2004. Total number of producers reached to 60.797 by 2013 and there were only 2 cities (Siirt and Şırnak) without any producers in this year. This shows that organic agriculture has expanded all around the country since 2004 up to our present day. Cities with less producers are mostly concentrated in Central Anatolia Region (Figure 5). The most significant point drawing attention in both of the maps is the great increase in number of organic producers in Eastern Anatolia and Eastern Black Sea regions by 2013, where limited producers were once involved in organic agriculture activities in 2004. The number of cities having more than 1.000 producers rose to 13 in 2013. It is seen that these cities are intensified in Aegean, Eastern Anatolia and Eastern Black Sea regions. There are 5 cities in Turkey that has more than 4.000 producers and these cities are respectively Aydın, Rize, Van, Erzurum, and Kars. Aydın is the city having the largest num-

Figure 4. The distribution map of producer in 2004

Figure 5. The distribution map of producer in 2013

ber of producers (9.417) also in this year. Eventually, it is deduced from the examination of both maps showing the distribution of producers' number in 2004 and 2013, that the number of producers adopting organic agriculture has increased and it has started to spread over the country. Between 2004 and 2013, some notable changes have taken place in the distribution of the producers into the geographic regions. Actually, in all of the regions, the total number of producers has increased between 2004 and 2013. Nevertheless, such increase rate is quite high in some of the regions. After all, the percentage of regions in total number of producers has been changed in time and the most significant change has happened in the Eastern Anatolia Region. While only 13 per cent of the total producers were located in the Eastern Anatolia Region, it has reached to 37 per cent by 2013 (Table 4).

Table 4. Distribution of producer by geographical regions

Geographical Region	2004		2013	
	Numbers of Producers	**%**	**Numbers of Producers**	**%**
Mediterranean Region	744	6	1.183	2
Eastern Anatolia Region	1.627	13	22.179	37
Aegean Region	4.884	38	15.825	26
Central Anatolia Region	1.317	10	2895	5
Black Sea Region	3.093	24	15.451	25
Marmara Region	897	7	1.184	2
Southeastern Anatolia Region	189	2	2.080	3

Upon investigating on the temporal change of the organic agricultural areas in Turkey, we see quite important changes have taken place. Organic agricultural areas have been enlarged ever since the organic agriculture was firstly started in Turkey in 1984. Total area where organic vegetative production is made was solely 1.037 hectares in 1990 and organic agricultural areas were subjected to peaks and troughs between 1990-2008. However, within this period of time, the general tendency of the organic agricultural areas was to enlarge. This enlargement was more rapid after 2008 (Figure 6) and the reason was the fact that organic agricultural policies were getting stronger. By 2012, total organic agricultural areas reached to 523.627 hectares between 2011-2012, a sizeable enlargement was seen. While Greece was the foremost country having the biggest expansion areal enlargement in organic agricultural areas in 2011-2012, Turkey became the fourth country (Willer & Lernoud, 2014). Organic agricultural areas have become enlarged 445 times in Turkey since 1990 up to 2013.

In the evaluation of the organic agricultural areas and natural collection areas in 2013 together, we see that total organic agricultural area in Turkey is 769.014 hectares (Table 5). 60 per cent (461.395 ha.) of this area is vegetative production area while 40 per cent (307.619 ha.) of it is for natural collection. Although the total organic area in Turkey shows fluctuations in some years, generally it tends to grow (Figure 7). In the case that such tendency in growth continues, in Turkey organic agricultural areas will enlarge much more in the coming years. By 2012, Turkey became the 15th country in the world and 8th country in Europe in terms of organic agricultural areas. These qualifications are the sign of Turkey's development in organic agriculture. As compared to the conventional agricultural areas, today organic agricultural areas in Turkey are not so large; they constitute 1.9 per cent of the total agricultural areas in the country. Such percentage shows how the organic agricultural areas are limited as compared to

Figure 6. Organic agricultural lands in Turkey (ha)

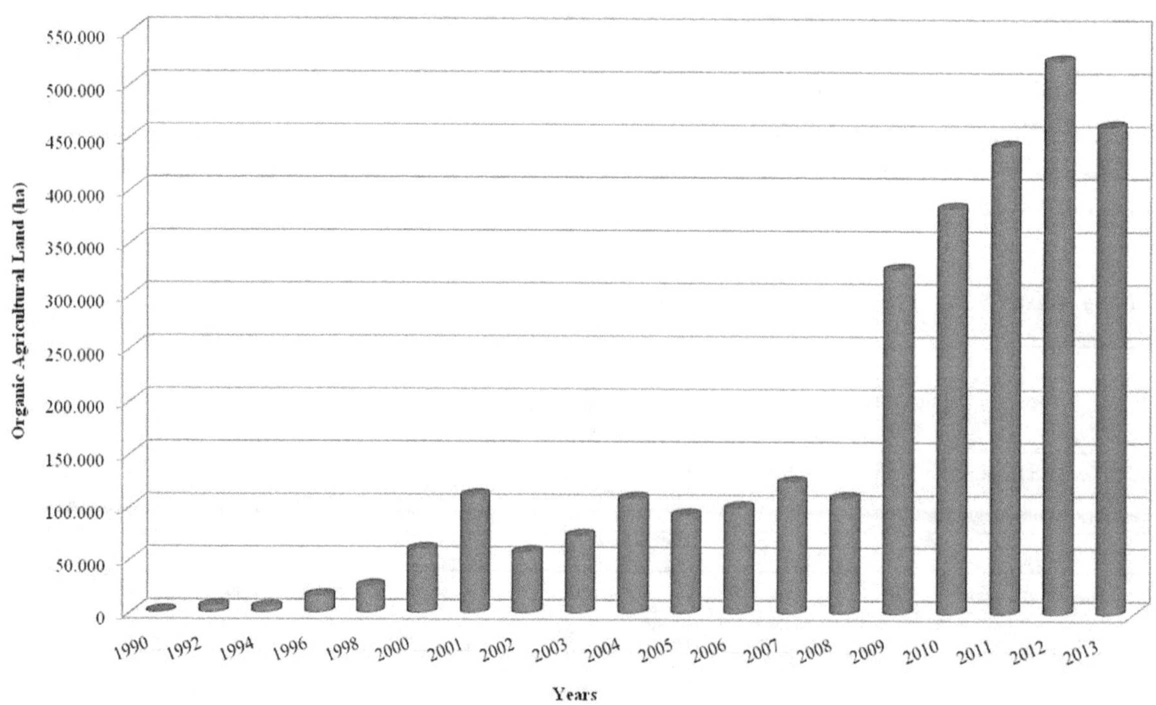

Table 5. Total organic lands in Turkey

Years	Organic Agricultural Land (ha)	%	Non-agricultural Organic Areas (ha) (Mainly wild collection)	%	Total Organic Lands (ha)
2002	57.365	63.9	32.462	36.1	89.827
2003	73.368	64.6	40.253	35.4	113.621
2004	108.598	51.8	100.975	48.2	209.573
2005	93.134	45.7	110.677	54.3	203.811
2006	100.275	52.0	92.514	48.0	192.789
2007	124.263	71.3	50.020	28.7	174.283
2008	109.387	65.5	57.496	34.5	166.883
2009	325.831	65.0	175.810	35.0	501.641
2010	383.782	75.2	126.251	24.8	510.033
2011	442.581	72.0	172.037	28.0	614.618
2012	523.627	74.5	179.282	25.5	702.909
2013	461.395	60.0	307.619	40.0	769.014

(TUIK, 2012; TUIK, 2014)

Figure 7. Total organic lands in Turkey (ha)

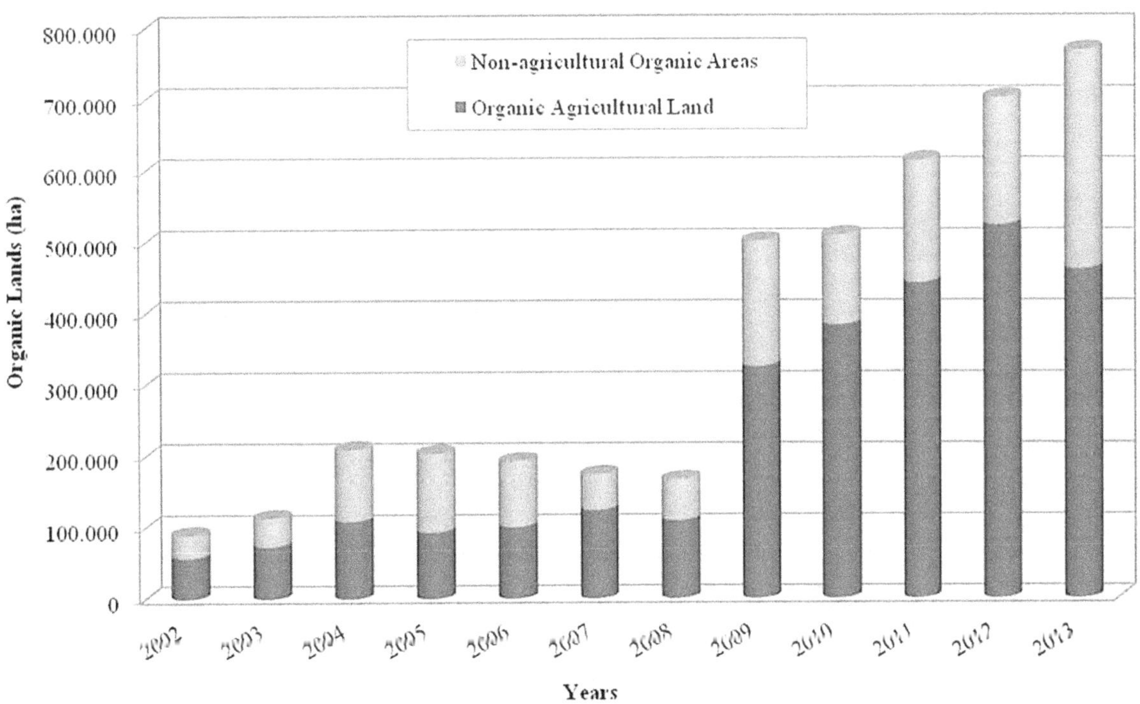

the total agricultural area in the country. On the other hand, in comparison with the past, we see that this percentage has grown. This proportion was 0.2 per cent in 2002, 0.4 per cent in 2008, 1.6 per cent in 2010 and 1.9 per cent at the present time. It is a proof that organic agriculture has been improving gradually within the Turkish agricultural system.

Spatial Distribution

As it has been in the temporal development of the organic agricultural areas, some notable changes have seen in the spatial distribution of these areas. In 2004, the total organic area in Turkey was 209.573 hectares and nearly 52 per cent (108.598 ha) of it was agricultural area while the other 48 per cent (100.975 ha) was for the natural collection area. In the assessment of the map showing the distribution of the organic areas in 2004, it draws attention that there were significant gaps in the country (Figure 8). There was not any organic agricultural areas in 28 cities; mainly in the Eastern Anatolia Region and the eastern parts of the Central Anatolia Region. No data for 15 cities regarding organic land is available that is why no organic land is seen in the cities in the map such as Balıkesir, Bilecik, Erzurum, Eskişehir, Kocaeli, Konya, Nevşehir, Sakarya, Samsun, Tekirdağ, Tokat, Aksaray, Karaman, Iğdır, and Yalova. In 2004, organic production area was less than 1.000 hectares in nearly half (39) of the cities in Turkey and number of cities having a more than 1.000 hectares of organic area was 14. We see that these cities are intensified in the Aegean and Mediterranean regions and only 4 of these cities had more than 10.000 hectares of organic area; these cities are Antalya, Muğla, Şanlıurfa and İzmir. Antalya had 65.900 ha, Muğla had 21.790 ha, Şanlıurfa had 15.981 ha and İzmir had 10.995 ha of organic area. Total organic

Figure 8. The distribution map of organic lands in 2004

production area are in these 4 cities are 114.666 hectares. That is to say, almost 55 per cent of the total organic agricultural areas in Turkey are actually in only 4 cities. This suggests that in 2004 organic agriculture was not distributed equally in the country.

By 2013, there were some major changes in the distribution of thc organic agricultural areas in the country (Figure 9). Total organic area in Turkey rose to 769.014 hectares in this year; 60 per cent (461.395 ha) of this area is the agricultural area while the other 40 per cent (307.619 ha) of it is for natural collection area. While looking at the map showing the distribution of the organic areas in the country in 2013, we see that there 2 cities without any organic agricultural area and these cities are Siirt and Şırnak. However, number of the cities without any organic agricultural areas in 2004 was 28 and the fact that it has decreased to only 2 cities by 2013 demonstrates that organic agriculture has started to spread over the country. (The data of 10 cities have not been obtained and these cities are Bingöl, Edirne, Elazığ, Kırklareli, Kırşehir, Kahramanmaraş, Zonguldak, Aksaray, Karabük and Osmaniye. For this reason, there are not organic any agricultural areas seen in these cities in the map. Nevertheless, it is known that organic agricultural area is quite small in these cities.) By 2013, number of the cities having more than 1.000 hectares of organic area has reached to 41 and it is seen that these cities are distributed almost equally over the country. As to the cities having a more than 10.000 hectares of organic area, the number of them has risen to 16 and these cities are mostly intensified in the Aegean, Eastern Anatolia and Mediterranean regions. The number of the cities with a more than 50.000 hectares of organic area is 6 and these cities are Mersin, Kastamonu, Van, Erzurum, Aydın and Kars. Mersin has 111.868 ha, Kastamonu has 83.714 ha, Van has 67.496 ha, Erzurum has 55.708 ha, Aydın has 52.759 ha and Kars has 52.622 ha of organic area. Between the years of 2004 and 2013, some notable changes have taken place in the distribution of the organic agricultural areas into the geographic regions. In fact, between

Figure 9. The distribution map of organic lands in 2013

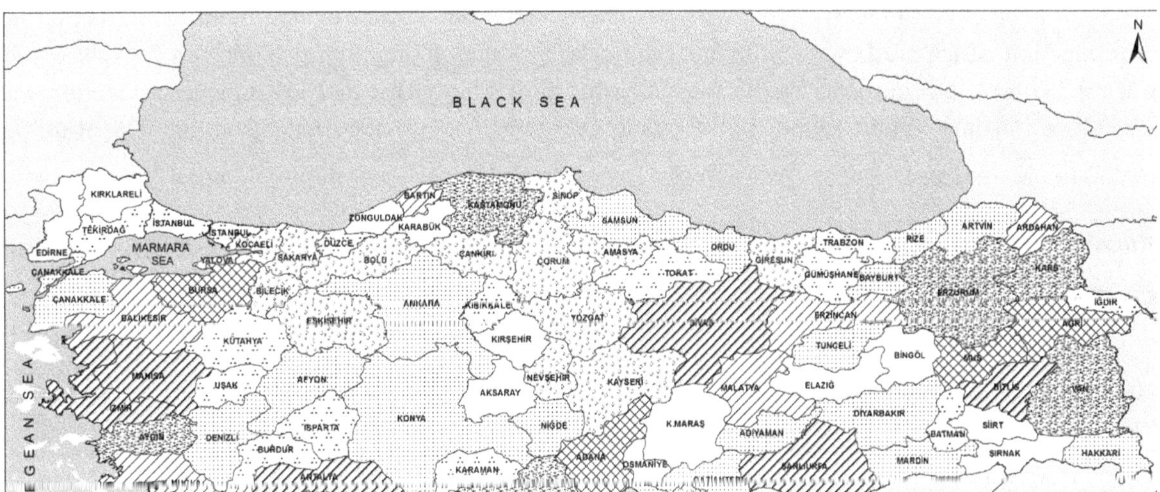

these years, total organic areas has expanded in all the regions but such expansion rate is quite higher in some regions. This has changed the percentages of the regions in the total organic area of the country within time. We see the most substantial changes happening in the Eastern Anatolia Region (Table 6). While only 4 per cent of the total organic area in the country was in the Eastern Anatolia Region; this percentage reached to 38 per cent in 2013. Consequently, an evaluation of the spatial distribution of the organic agricultural areas in Turkey between 2004 and 2013 shows that organic agricultural areas in Turkey have become larger and they have expanded over the country in time.

Depending on the increase in number of farmers and production area, the organic vegetative production quantity in Turkey has gotten bigger. We see some important temporal changes happening in the amount of organic production and the general trend of this change is in the direction of increasing. Organic vegetative production quantity, being only 310.125 tons in 2002, rose to 530.224 tons in 2008. Particularly after 2008, a great increase was noted in the organic production (Figure 10). An increase in the number of producers and production area between 2008-2009 is reflected in the production quantity that reached to 983.715 tons by 2009. In other words, organic production quantity increased by about 85 per cent just within 1 year. Aid given to the organic agricultural producers since 2009 gave a boost to the organic production in forage plants and grains and this led to an overall increase in the total production (The Union of Chambers and Commodity Exchanges of Turkey, 2013). By 2013, the total organic production quantity has reached to 1.620.466 tons. Organic production quantity in Turkey is shaped by the foreign demands. Since Turkey makes export-based organic production, mostly it is the world market that decides the production quantity. However, owing to the rising demands for organic products in the domestic market, the organic production quantity in Turkey is expected to rise as well.

Some notable changes have also taken place in the spatial distribution of the organic production. In 2004, organic production in Turkey was 377.615 tons. Looking at the spatial distribution of production in 2004, one can see 28 cities without any organic production. There were 26 cities having more than 1.000 tons of production and 10 cities having more than 10.000 tons of production and we see that these cities are intensified in the Western Anatolia. There were only 3 cities having more than 20.000 tons of production which are İzmir, Aydın and Şanlıurfa. Production quantity in İzmir was 26.174 tons, it was 39.787 tons in Aydın and 54.198 tons in Şanlıurfa. It shows that in 2004 organic production was intensified in some certain regions of the country (Figure 11). Organic products at that time were the

Table 6. Distribution of organic lands by geographical regions

Geographical Region	2004		2013	
	Total Organic Lands (ha)	%	Total Organic Lands (ha)	%
Mediterranean Region	77.341	47	161.322	21
Eastern Anatolia Region	6.115	4	288.993	38
Aegean Region	49.441	30	98.429	13
Central Anatolia Region	1.297	1	23.459	3
Black Sea Region	7.841	5	110.991	14
Marmara Region	2.718	2	54.296	7
Southeastern Anatolia Region	19.089	11	27.402	4

Figure 10. Organic production (Tons)

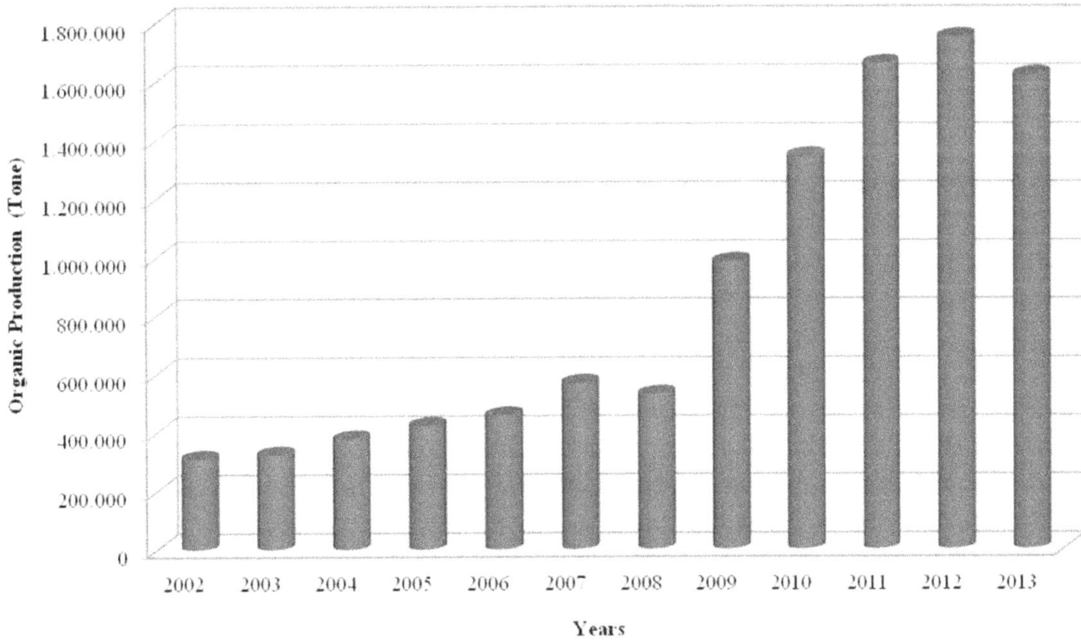

Figure 11. The distribution map of organic production in 2004

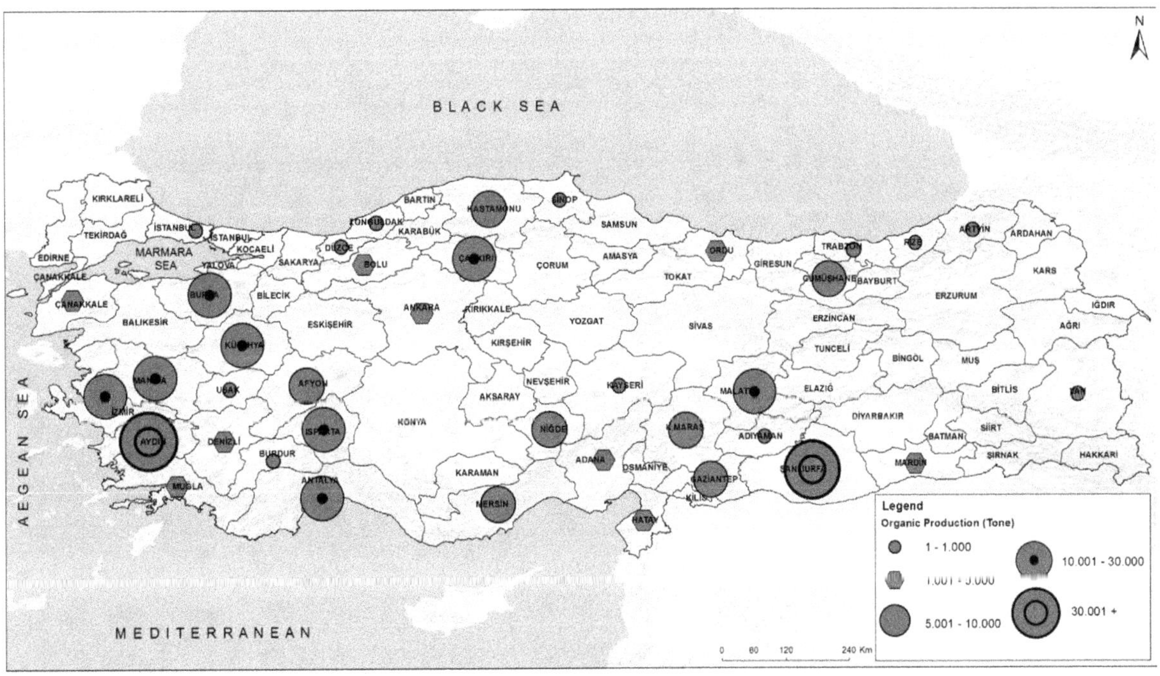

traditional products that were of vital importance in Turkey's agricultural structure; these were grape, fig, apricot, hazelnut, olive, cotton and citrus fruits.

By 2013, the organic production quantity in Turkey rose to 1.620.466 tons and in this year, some substantial changes have been seen in the distribution of the organic production according to the cities (Figure 12). An examination of the map showing the distribution of the organic production into the country in 2013, shows that there are 2 cities without any organic production which are Siirt and Şırnak. (In 2013, we have not accessed to the data of 10 cities regarding the organic production. Therefore, there is no organic production seen in these cities in this map which are Bingöl, Edirne, Elazığ, Kırklareli, Kırşehir, Kahramanmaraş, Zonguldak, Aksaray, Karabük and Osmaniye. However, it is known that there is too small organic production in the aforesaid cities). There are 52 cities having more than 1.000 tons of production. It is possible to conclude that organic production has expanded over the country because organic production is done in most of the cities. The number of the cities having more than 10.000 tons of production is 26 and those having more than 20.000 tons of production are 16. On the contrary, this was limited to 3 cities in 2004. Moreover, there are 12 cities in Turkey having more than 30.000 tons of production and these cities seem to be intensified in the Eastern Anatolia and Aegean regions. In this year, number of the cities having more than 100.000 tons of production is 5 and these cities are respectively Erzurum, Van, Aydın, Kars and Ağrı. The production quantity in Erzurum is 233.408 tons, 198.738 tons in Van, 154.760 tons in Aydın, 137.040 tons in Kars and 135.917 tons in Ağrı. Apart from Aydın, in the rest of the cities, grain production is more dominant. Looking through the maps shows the spatial distribution of the production in 2004 and 2013. At first glance, one can observe that substantial changes have occurred in the spatial distribution of the production within the period of these 10 years,

Figure 12. The distribution map of organic production in 2013

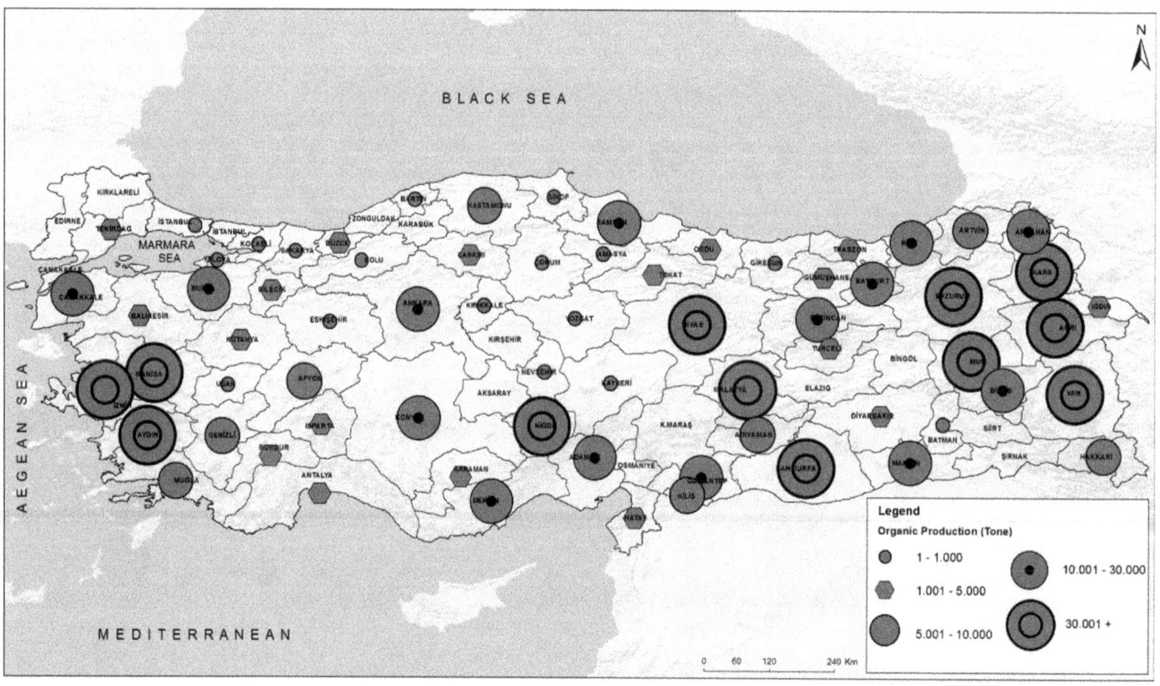

and organic production has expanded all around the country. As to the regional distribution of the organic production, we see a growth in all the regions but there is an extreme growth in the Eastern Anatolia Region (Table 7). In 2004, the organic production in the Eastern Anatolia Region constituted solely 5 per cent of the total production of the country, while it has reached to 57 per cent by 2013. Despite the increase in the organic production in the Aegean Region, the percentage of the Aegean Region within the total production has decreased. A considerable difference can be observed between the Eastern Anatolia and Aegean regions with regards to organic production quantity in 2013. On the other hand, trade volume and organic market value of the products cultivated in the Aegean Region are much more. In the Eastern Anatolia Region, grains are generally cultivated.

Organic agriculture in Turkey develops depending on the export market (Leeuwen, Voort, Sukkel, & Balci, 2008). Almost all the organic production are exported to the developed countries especially EU countries, the USA and Japan. EU countries stand for the most important exportation market of Turkey between these markets and Germany, France and England are the most prominent exportation markets of Turkey within these EU countries (Gök, 2008). A large part of the exported products is dry and dried fruits and later on fresh fruits and field crops. From 19.4 million US dollars in 1998, organic products exportation rose to 26.3 million US dollars in 2005, but it dropped to 15 million US dollars between 2010 and 2011. Such reduction is associated with the global economic crisis in that time. Following the recovery of the markets after the crisis, organic products exportation rose to 24.7 million US dollars in 2012 while it has reached to 46 million US dollars in 2013. Although the exportation values grade up every passing day, the proportion of Turkey in the world organic agricultural market is still too little. In the evaluation of the product groups, it is noted that Turkey has gained the most foreign currency income from organic hazelnut, fig, apricot, grape, strawberry, olive, cotton and fruit juices. The organic agricultural market, which is growing rapidly in the world, was 63.8 billion US dollars in 2012. Considering the total market value of the world's organic agriculture, the revenue that Turkey gains from this market is still very small. Within the 2023 year targets of Turkey, there is an objective as well to raise the organic agricultural exportation to 2.5 billion US dollars. Turkey has the potential to realize this objective as well as to make agricultural the policies needed to stimulate its potential (Ministry of Food, Agriculture and Livestock, 2012).

In order for the organic agriculture to become widespread in Turkey, the domestic market of organic products needs to grow as well. For the time being, Turkey has an export-oriented organic production.

Table 7. Distribution of organic production by geographical regions

Geographical Region	2004		2013	
	Organic Production (Tons)	**%**	**Organic Production (Tons)**	**%**
Mediterranean Region	50.301	17	56.982	4
Eastern Anatolia Region	16.012	5	913.828	57
Aegean Region	111.722	37	303.014	19
Central Anatolia Region	21.501	7	109.879	7
Black Sea Region	20.490	7	72.688	4
Marmara Region	16.210	5	38.351	2
Southeastern Anatolia Region	65.074	22	116.033	7

Consumption of organic products is too small in the domestic market of Turkey. It is estimated that the domestic market of organic products of Turkey is around 14 million US dollars. As compared to other countries, this figure shows how much restricted is the domestic market of organic products in Turkey (Demiryürek, 2011). By 2012, consumption value of the organic food per capita in Turkey was 0.16 US dollar. In other words, consumption value of the organic food per capita is not even one US dollar. On the contrary, in the same year, across the globe, average consumption value of the organic food per capita was 9.08 US dollars. It reveals that in Turkey consumption of organic value is very little and Turkey has a lower consumption value of the organic food in comparison with the other parts of the world. If it is desired for the organic agriculture to become more widespread, domestic consumption should be increased. In the recent years, to increase the organic products consumption in the domestic market, some organic products stores, organic products sections in the supermarkets and organic open-air markets in some days of the week have been opened. Particularly in İstanbul, Ankara and İzmir, organic open-air markets are established at 15 points in the country. There are 7 organic open-air markets in İstanbul. İzmir and Ankara have 2 organic open-air markets and Antalya, Bursa, Eskişehir and Samsun have one organic open-air market each. Domestic consumption has been growing gradually thanks to the recent activities. People have started to hear the concepts of organic agriculture and organic products much more. Along with exports, increased domestic consumption of organic products will make the organic agriculture widespread in the country.

BENEFITS OF ORGANIC AGRICULTURE TO TURKEY

Agriculture is one of the sectors that has always maintained its importance in the economic and social development in Turkey from past to present. Turkey has a very prominent agricultural potential such that apart from arid climate plants, many other plant species could be cultivated including subtropical plants. Making optimal use of Turkey's agricultural potential is possible only with the rational use of its natural resources. Turkey is in need of sustainable agricultural systems that do not decimate its natural resources, pollute or destroy them. Sustainable agricultural production could be applied; systems which do not use up its natural resources and eventually cause to the deterioration of these resources, but rather systems which use these resources effectively so that they could be passed on to the next generations. For this reason, start of organic agriculture and its spread in Turkey in recent years are important steps for sustainable agriculture.

Turkey has made notable progress regarding organic agriculture within the last 20 years. By 2013, Turkey has made a 1.620.466 tons of organic production on a 769.014 hectares of area together with 60.797 producers. Turkey has become the 6th country in the world as having the largest number of producers and the 15th country in terms of its organic agricultural areas. Between 1990-2013, the number of producers has increased 195 times more and organic agricultural area has enlarged 445 times more, and between 2002-2013 organic production quantity has grown 5.2 times more. In spite of all these developments, today Turkey still uses only a very small part of its organic agricultural potential. Turkey is a country with a high organic agriculture potential, but these potentials are not evaluated sufficiently. Here are some of the potentials Turkey enjoys concerning organic agriculture.

- Turkey has different ecologic conditions and such diversity in the ecologic conditions make it possible to cultivate many agricultural products in the country. At one side, arid climate plants are cultivated while mild and sub-tropical plants could be cultivated at the same time. It gives an advantage to Turkey for the variety of products.

- Turkey has a biologic richness and diversity.
- Turkey has adequate and productive agricultural areas. The proportion of the agricultural areas in the surface area of Turkey is 30.3 per cent.
- Turkey has sufficient water resources.
- Turkey has thousands years of agricultural knowledge and experience.
- Turkey has adequate labor force. Researches done in this field have revealed that organic agriculture required much labor force as compared to the conventional agriculture (Turhan, 2005). Having a large, young population in Turkey is an advantage for a cost-effective labor force.
- Another potential is the country's geographical position. Turkey is close to foreign markets and Turkey is especially advantageous due to it being so close to the European Union market.
- Turkey is an important player in terms of transportation systems in the international trade. Apart from the land and air transportation systems, marine transportation is an advantage as well for Turkey because it has outstanding mercantile ports (such as İstanbul, İzmir, Mersin).
- Turkey is the leading producer in some certain products (such as hazelnut, fig, grape, apricot etc.) in the world's agriculture and it has the chance to continue this advantage along with the organic production.
- In comparison with the industrialized countries, the amount of chemical input in the agricultural areas is lower in Turkey. Therefore, chemical pollution is not so dense in Turkey's soils. This will make the transition easier to organic agriculture.
- It is an advantage as well that Turkey has built the fundamental education required for solving the qualified labor force in the organic agriculture. Some 21 Organic Agriculture Departments have been established in 19 universities in order to support the organic agricultural development.
- Having a national legislation for organic agriculture is another advantage. By 2012, Turkey is one of the countries in the world that fulfilled the necessary regulations regarding the organic agriculture (Willer & Lernoud, 2014).

Embracing organic agriculture in the country will provide additional national benefits. Some of these are:

- Since the organic agriculture is an agricultural system that is sensitive to the environment, it will maintain the sustainable usage of natural resources. Turkey will both protect its natural resources and it will get the maximum benefit from these resources as well.
- Erosion intensity, one of the significant problems in Turkey's agriculture, will be diminished.
- Turkey will satisfy the food needs of the increasing population sustainably with organic agriculture.
- Health food supply with organic agriculture will make contribution to raising healthy generations of people.
- Organic agriculture will find solutions for the employment problems of the large rural population to a considerable extent.
- Rural-to-urban migration will be reduced as organic agriculture will contribute to the rural development.
- Wide spreading of the organic agriculture will increase the agricultural employment.
- Producers will be given guarantee thanks to the contracted agriculture.

- Percentage contribution of agriculture in the national income will again increase.
- Turkey will make good use of its potential in the diversity of organic products and will promote its competitive capacity in the international markets.
- Its agricultural exportation revenues will increase. Reduced percentage of the agriculture in the foreign trade will again raise.
- Increasing the foreign trade of the organic agricultural products will provide a substantial foreign currency inflow to the country. It will make contribution as well to the importation-exportation balance.
- High values of the organic products in the market will enable producers to obtain high revenues.
- Use of chemical inputs including fertilizers in the agriculture will be diminished thanks to the wide spread use of organic agriculture, and accordingly farmers will get rid of additional expenses.
- In case of a reduction in the use of the agricultural chemicals and fertilizers, importation expenses of the country will become less since most of them are imported.
- Every year in Turkey, a considerable amount of the agricultural areas remain fallow. Thanks to the organic agriculture, these areas will become smaller, and more agricultural areas could be used for the production.
- It will make a contribution to the regional development and especially to the development of the Eastern and South Eastern Anatolia regions.
- As organic agriculture becomes more common in Turkey, this will contribute positively to the agricultural negotiations in the membership process of the EU (Bakırcı, 2005).

RECOMMENDATIONS

Future agricultural systems will be those that prioritize the principle of sustainability, and the countries having these systems in-house will enjoy sustainable agricultural development. In this sense, if Turkey has expanded the organic agriculture system as soon as possible, it will be a shining example for the future. We have hereby shared our recommendations regarding the sustainable agriculture:

- A sustainable development in the agricultural production will be only succeed with the coherent use of various agricultural technologies. For this reason, Turkey should benefit much more from the agricultural information technologies.
- Geographic Information Systems (GIS) is an effective tool that should be used for the sustainable agriculture. Use of GIS has increased in the sustainable agriculture applications in the world in recent years. GIS helps in forming the sustainable agricultural applications by storing, managing, analyzing, reporting and monitoring agricultural data in huge numbers (Esri, 2008). Therefore, the first priority of Turkey should be establishing agricultural data bases and such data bases should include environmental information, cadastral information, infrastructural and socio-economic information. In Turkey, these data bases should be city-based. The most important factor of the agriculture is soil, but there is not any significant data base of soils yet. For this reason, activities of preparing agricultural data bases should be accelerated.
- Turkey has different ecologic conditions. Therefore, Turkey should consider agro-ecological zoning for sustainable agriculture. By this way, more effective agricultural programs could be pre-

pared for each of the agro-ecologic zones. For an agro-ecological zoning, primary data bases should be developed, that is to say, necessary foundations should be built.

- The most fundamental deficiency concerning the sustainable agriculture is that Turkey lacks for the agricultural data bases. Therefore, information technologies and information management systems should be considered and the relevant studies should be accelerated. Consequently, without sufficient information regarding the physical and human characteristics of the natural environment, a healthy agricultural planning cannot be made.
- Agricultural policies should be based on the concept of sustainability and oriented towards organic agriculture.
- Organic agriculture should be given more priorities in the agricultural development plans.
- The situation of the Turkish farmers should once again be evaluated and in accordance with the local conditions, new agricultural policies should be created.
- Agricultural areas in Turkey are small and partial. Land consolidation of the agricultural areas should be considered and agricultural enterprise areas should be enlarged.
- Support for organic agriculture should be increased.
- Model agricultural applications should be promoted with government aid.
- In order to raise awareness among producers and people, some public service announcements should be prepared.
- Apart from the organic vegetative production, organic animal production should be considered as well and some actions should be taken to improve the production. Large grass and pasture lands (14.617.000 ha) are an advantage that will make it possible an organic animal breeding development in Turkey.
- The entrepreneurs should be encouraged for the organic agriculture and agricultural industrial integration. The number of high value processed organic products should be increased along with the organic raw materials. Thus, the revenue of organic products exports will grow.
- Not only the foreign market but also the domestic markets should be developed.
- An effective and continuous government policy is required for the organic agricultural development in Turkey but the re-shaping of the agricultural policies with every new government will cause instability and slow down organic agriculture development.

CONCLUSION

Nowadays, it has been understood that natural resources are restricted and sustainable use of these resources is highly important. For this reason, future agricultural systems will be those giving priority to the "principle of sustainable agriculture". Organic agriculture, one of the sustainable agricultural systems, has gained considerable importance in the world for recent years. Countries having widespread organic agriculture will enjoy sustainable economic development. Turkey has made substantial progress regarding the organic agriculture for the last 20 years. Moreover, the developments in organic agriculture in the last 5 years are show a lot of promise for the future.

Organic agriculture in Turkey has made notable progress in terms of both number of producers and products and production area and quantity. The number of producers has increased 195 times more and organic agricultural areas have enlarged 445 times more between 1990-2013. Also, organic production quantity has grown 5.2 times more between 2002-2013. Apart from the temporal changes in the organic agriculture of Turkey, some important spatial changes have been noted as well. Intensified in the certain parts of the country at the beginning, the organic agriculture has expanded over the country at the present time. In spite of all these developments, Turkey, today, could not benefit from the potentials and advantages adequately that it has in the organic agriculture. In order for the organic agriculture to be widespread in Turkey, a stable government policy should be developed and applied. Moreover, agricultural information system and management technologies should be considered to aid the sustainable agricultural development in Turkey. Organic agriculture will make great contributions to Turkey's economic development and regional improvement inside the country. Organic agriculture is an important necessity for a sustainable life and a sustainable agriculture. For this reason, it is of essence to spread much wider the organic agriculture in Turkey and all over the world for the sake of bequeathing a livable world to the next generations.

REFERENCES

Badgley, C., & Perfecto, I. (2007). Can organic agriculture feed the world? *Renewable Agriculture and Food Systems*, *22*(2), 80–85. doi:10.1017/S1742170507001986

Bakırcı, M. (2005). The future of organic farming in Turkey and its effects to the agricultural negotiations between Turkey and European Union. *Istanbul University Geographical Review*, *13*, 67–83.

Bakırcı, M. (2007). *Rural development in Turkey: Concepts, policies, practices*. Ankara, Turkey: Nobel Press.

Demiryürek, K. (2011). The concept of organic farming and the situation of organic farming in the world and Turkey. *Journal of Agricultural Faculty of Gaziosmanpasa University*, *28*(1), 27–36.

Demiryürek, K., Stopes, C., & Güzel, C. (2008). Organic agriculture: The case of Turkey. *Outlook on Agriculture*, *37*(4), 261–267. doi:10.5367/000000008787167754

Doğanay, H. (2007). *Agricultural geography*. Erzurum, Turkey: Aktif Press.

Esri. (2008). *GIS for sustainable agriculture* [PDF document]. Retrieved from http://www.esri.com/library/bestpractices/sustainable-agriculture.pdf

Food and Agriculture Organization of the United Nations (FAO). (2014). *Statistical yearbook 2014: Europe and Central Asia food and agriculture*. Retrieved from http://www.fao.org/3/a-i3621e.pdf

Gök, S. A. (2008). *An assessment of trade in Turkish organic products in the expanding market of European Union*. (Expertise thesis). Retrieved from Republic of Turkey Ministry of Food, Agriculture and Livestock expertise thesis database. (50)

Göney, S. (1979). *Geographical essentials of Turkey's agriculture*. Istanbul, Turkey: Istanbul University Press.

Grigg, D. (1995). *An introduction to agricultural geography*. London: Routledge.

Istanbul Chamber Commerce. (2006). *Analysis of organic agriculture and EU market opportunity*. Istanbul, Turkey: Birlik Press.

Knowles, R., & Wareing, J. (1983). *Economic and social geography*. London: Heinemann.

Lotter, D. W. (2003). Organic Agriculture. *Journal of Sustainable Agriculture, 21*(4), 59–128. doi:10.1300/J064v21n04_06

Makhzoumi, J., & Pungetti, G. (1999). *Ecological landscape design and planning: The Mediterranean context*. London: Spon Press. doi:10.4324/9780203223253

Minister of Development. (2014a). *Tenth development plan (2014-2018): Specialization commission report about agricultural structure and food security*. Retrieved from http://www.kalkinma.gov.tr

Minister of Development. (2014b). *Tenth development plan (2014-2018): Working group report about sustainable agricultural land use*. Retrieved from http://www.kalkinma.gov.tr

Ministry of Food, Agriculture, and Livestock. (2012). *Strategic plan for organic farming in Turkey (2012-2016)*. Retrieved from http://www.trakya2023.com/uploads/docs/2806201331nTii.pdf

Ministry of Food, Agriculture, and Livestock. (2014). *Activities of general directorate of vegetative production* [PDF document]. Retrieved from http://www.tarim.gov.tr/sgb/Belgeler/SagMenuVeriler

Organisation for Economic Co-operation and Development (OECD). (2011). *Evaluation of agricultural policy reforms in Turkey*. .10.1787/9789264113220-en

Ozkan, M. (2003). Organic agriculture and national legislation in Turkey. In Organic agriculture: Sustainability, markets and policies (pp. 289-294). Wallingford, UK: CABI Publishing.

Park, J., & Seaton, R. A. F. (1996). Integrative research and sustainable agriculture. *Agricultural Systems, 50*(1), 81–100. doi:10.1016/0308-521X(94)00050-2

Pretty, J. (2008). Agricultural sustainability: Concepts, principles and evidence. *Philosophical Transactions of the Royal Society of London. Series B, Biological Sciences, 363*(1491), 447–465. doi:10.1098/rstb.2007.2163 PMID:17652074

Pretty, J. N. (1995). Participatory learning for sustainable agriculture. *World Development, 23*(8), 1247–1263. doi:10.1016/0305-750X(95)00046-F

Pugliese, P., Bteich, M. R., & Al-Bitar, L. (2014). *Mediterranean organic agriculture: Key features, recent facts, latest figures Report 2014*. Retrieved from http://moan.iamb.it/

Rehber, E., & Turhan, Ş. (2002). Prospects and challenges for developing countries in trade and production of organic food and fibers: The case of Turkey. *British Food Journal, 104*(3/4/5), 371-390. doi:10.1108/00070700210435380

The Union of Chambers and Commodity Exchanges of Turkey. (2013). *Agricultural sector report of Turkey*. Ankara, Turkey: Afşaroğlu Publishing.

Topp, C. F. E., Stockdale, E. A., Watson, C. A., & Rees, R. M. (2007). Estimating resource use efficiencies in organic agriculture: A review of budgeting approaches used. *Journal of the Science of Food and Agriculture, 87*(15), 2782–2790. doi:10.1002/jsfa.3059

Turhan, Ş. (2005). Sustainability in agriculture and organic farming. *Turkish Journal of Agricultural Economics*, *11*(1), 13–24.

Turkish Statistical Institute. (2012). *The summary of agricultural statistics 2011*. Ankara, Turkey: Turkish Statistical Institute Press.

Turkish Statistical Institute. (2014). *Turkey in statistics 2013*. Ankara, Turkey: Turkish Statistical Institute Press.

Turkish Statistical Institute (TUIK) Databases. (n.d.). Retrieved from http://www.turkstat.gov.tr

van Leeuwen, M. A. E., van der Voort, M. P. J., Sukkel, W., & Balci, S. (2008). *Organic agriculture in Turkey: trade opportunities for organic fruit and vegetables*. Retrieved from http://edepot.wur.nl/29925

Willer, H., & Lernoud, J. (Eds.). (2014). *The world of organic agriculture: Statistics and emerging trends 2014*. Retrieved from https://www.fibl.org/fileadmin/documents/shop/1636-organic-world-2014.pdf

Chapter 7
Are GM Crops the Answer to Africa's Critical Food Security Status?
Learning from the Experiences of Developing Countries

Nira Ramachandran
Independent Researcher, India

ABSTRACT

Despite unprecedented technological breakthroughs and rising incomes, the most basic human requirement–food security remains out of reach of many. The countries of Africa are the worst affected, and consistently score lowest on the Global Hunger Index. The downside of technological advance and increasing urbanisation is the decrease in cultivable land and water availability exacerbated by climatic unpredictability. While concerted efforts are on to control the crisis, it is time to take positive action and capture technology to boost food security. Genetically Modified (GM) crops engineered to be drought/flood resistant, reduce pesticide requirements or provide vitamins open up tremendous possibilities for enhancing food and nutrition security and counteracting negative environmental impacts. This paper reviews the current status of GM crops in Africa, the scope for expansion, possible impacts on food security and the validity of existing concerns about bio-safety and farmers' needs in the light of developing country experience.

INTRODUCTION

One of the major challenges that Africa faces today is that of ensuring food security for its population. Over the last few decades, the continent has witnessed severe food shortages, even starvation in one country or the other. This may be attributed to a variety of causes—environmental, economic, strife-related or as a result of pandemics like HIV/AIDS sweeping the continent. The continued heavy dependence on food-aid and cheap food imports also tend to undermine local agriculture and leaves the consumer

DOI: 10.4018/978-1-5225-0942-4.ch007

unprotected and completely vulnerable to global price fluctuations. Food Aid dates back to the late 1950s, but the number of food emergencies have tripled since the mid-1980s (MDG Report, 2013). Climate change which brings in its wake unpredictability of temperature and rainfall as well as an increase in severe climatic events is further exacerbating the situation. In this backdrop, it is encouraging to note that a number of countries in sub-Saharan Africa have improved their food security status in the new millennium. While the decade of the 1990s witnessed extremely slow progress in Africa as against South Asia, there has been more rapid improvement, from 2000 onwards. In fact, the Global Hunger Index (GHI) 2014 finds sub-Saharan Africa outpacing South Asia. The six countries which have made substantive progress reducing their GHI scores by over 50% since 1990 are Angola, Ethiopia, Ghana, Malawi, Niger and Rwanda. Yet even today, 12 countries have GHI scores in the "alarming" or "extremely alarming" categories: Central African Republic, Ethiopia, Comoros, Mozambique, Niger, Sudan, Zambia, Sierra Leone, Madagascar (GHI scores: 20-29.9), and Burundi and Eritrea (GHI scores: 30-39.9) (GHI, 2014). The Global Hunger Index is a composite of three hunger indicators: the proportion of undernourished population, the proportion of underweight children below the age of 5 years and the infant mortality rate, which effectively represent the food security status of a country. As the GHI scores in sub-Saharan Africa are heavily weighted by the proportion of undernourished population unlike in the case of South Asia, where underweight children tend to have more impact on the score, it is not surprising to find that in these countries (with alarming or extremely alarming scores), about one-third of the population does not consume the minimum calorie requirement. In Burundi and Comoros this proportion reaches as high as 65 and 67%, respectively.

The fact that Africa is endowed with vast and underutilised natural resources like huge tracts of cultivable land, water and temperatures suitable for cultivation, makes this chronic shortfall in food production a story of unused potential. With reference to potential land area alone, the FAO (2009) estimates that the unutilised rain-fed area suitable for cultivation could be as much as 700 million ha. It is widely accepted that food shortages in the African context result from the fact that 90% of the cultivators are small and marginal farmers, almost 60% of cultivated area is un-irrigated, and the practice of cultivating a large variety of crops on small farms does not lend itself to the adoption of green revolution technologies, which have been suggested as being more suited to large scale monocultures. In addition, poverty, illiteracy and weak infrastructure preclude access to markets or improved inputs. The Millennium Development Goals Report (2013) states:

In fact, the challenges of the past have grown and become more complex, exacerbating food insecurity. Africa's agriculture is underdeveloped, and people have begun abandoning it for better livelihoods. The effects of climate change are more pronounced and Africa's population is increasing so fast that countries might be unable to produce all the food they need.

An estimated 232 million undernourished people live in Africa (FAO, 2015) of whom almost 95% are in the sub-Saharan region alone. With the population of this region projected to reach 1.5–2 billion by 2050 (FAO, 2009), it becomes imperative to seek out viable solutions to the food security issue. This paper focuses on genetically modified crops and their potential to contribute to a food secure Africa.

Genetically Modified Crops: A Brief History

The selective breeding of plants dates back to the beginnings of settled agriculture. Improved varieties bred for higher yields, changes in colour or flavour or resistance to adverse climatic conditions underlie the development of agriculture in every country and form the basis of indigenous farming knowhow which is handed down through generations of farmers. Agricultural products most in demand in the global market like India's basmati rice, Ghana's cocoa or Hawaii's kona coffee are all the products of selective breeding practices over generations of farmers. Yet, when it comes to the question of genetic modification, there is instant alarm and stubborn resistance to such interventions, which could provide a relatively quicker avenue to improving food security and livelihoods in countries with low availability of food and high levels of malnutrition.

Genetic modification consists of the introduction of genes from a crop with the desired characteristics into the living DNA of another with a view to improving the latter variety. China was the first country to produce genetically modified (GM) crops for the market in the form of a virus resistant tobacco and a GM tomato somewhere in the late 1980s to the early 1990s. Better known is the case of the United States (US), where the first genetically modified food crop was brought into commercial use in 1994 with the *Flavrsavr* transgenic tomato being brought to supermarket shelves in the US. This variety of slow ripening tomato with a long-shelf-life received a positive response with consumers even paying a premium price for the product (Murnaghan, 2014). This breakthrough was quickly followed by the release of three more crops in 1995; The Bt potato, the first pesticide producing crop, Bt corn with insect resistant qualities and herbicide resistant Roundup Ready Soybeans. By 1996, all three crops were also approved in the UK and the European Union (GM Education, 2014).

In 1997, however, the European Commission EC novel Foods Regulation (258/97) came into effect mandating safety assessment for novel and GM foods before release into the market and by 1998 the first GM labelling rules were introduced so that consumers were informed about the use of GM ingredients in the food they were purchasing. These measures signalled the beginning of the disquiet with regard to the possible negative impacts of GM produce (GM Education, 2014).

Yet, the growth of GM crops during the same period was phenomenal. From 1.6 million hectares in 1996, these crops expanded to cover as much as 35 million hectares by 1999. The transfer of GM technology from the developed to the developing world was rapid and fifty two percent of the total area under GM crops is now in developing countries. GM crops cover a total of 148 million hectares (2010) in 29 countries across the world. This near doubling of growth makes GM the fastest crop technology to be adopted in the history of modern agriculture (Adenle, 2011). The largest producers in the developing world are Argentina, Brazil, China and India with Bt cotton dominating GM crop acreage in the Asian and African countries, and herbicide resistant soybean followed by insect resistant corn leading in the countries of Latin America (FAO, 2012).

Despite the growing unease with regard to the environmental impacts of GM crops and the intense debate on the unequal socio-economic benefits of this technology, which are likely to further accentuate the gaps between large and small farmers, as recently as 2010, the European Commission approved *Amflora*, a genetically modified potato for industrial use in paper making and adhesives in the European Union. The EC known for its strong resistance to GM crops also permits the import, but not the cultivation of GM maize for food and feed use (GM Education, 2014).

Do GM Crops Benefit the Small Farmer?

The main argument put forward by opponents of GM crops is that the technology was developed for large scale monocultures quite unsuited to developing countries where agriculture is dominated by small and marginal landholders. The pro-GM camp quotes significant increases in productivity, reduced pesticide costs and impressive growth of farm incomes in both developed and developing countries (Brookes and Barfoot, 2013; Bennett et al. 2004; Adenle, 2011; Morse, Bennet and Ismael, 2013; and Juma, 2014). The opponents, on the other hand, cite falling yields following the initial spurt, the emergence of new pests resistant to even the GM crops and rising costs of inputs for these crops making them unsustainable for the smallholder (Amir, 2014; Kuyek, 2002; Pschorn-Strauss, 2005; Witt, Patel and Schnurr, 2006; and Belay and Guri, 2013). This section aims to adopt an objective approach to the issue by basing its conclusions on available data, rather than rhetoric.

While it is difficult to obtain data on differential impacts of GM crops on large and small farmers, it is possible to compare variations in farm income from GM crops between developed and developing countries. Assuming that over two-thirds of farmers in developing countries have small holdings, this could provide a rough estimate of the extent to which small farmers can benefit from adopting GM technologies. Currently about 52% of the global area under GM crops lies in the developing countries of Asia, Africa and Latin America.

Comparing the economic benefits of GM crops to farmers in developed and developing countries in 2011 (Table 1), it is evident that a little over half (51.2%) of farm income benefits were earned by farmers in developing countries. The largest share of these income gains has been from GM IR (Insect Resistant) cotton and HT (Herbicide Tolerant) soybeans. In the fifteen year period since the introduction of GM crops (1996–2011), the cumulative farm income gain in developing countries was 50.5% of the global figure or USD $49.63 billion (Table 1).

In terms of costs incurred for accessing GM technology, farmers in developing countries incur a lower cost than their counterparts in developed countries (Table 2). For developing country farmers the total cost was equal to 14% of total technology gains as compared to 28% for farmers in developed countries. This difference perhaps reflects factors such as weaker provision and enforcement of intellectual property rights in developing countries and the higher average level of farm income gain on a per hectare basis derived by developing country farmers relative to developed country farmers (ISAAA, 2010).

Table 1. GM crop farm benefits 2011: developing vs. developed countries (million USD)

Crop	Developed Countries	Developing Countries
GM HT soybeans	1,794.2	2,085.0
GM IR maize	5,710.4	1,394.5
GM HT maize	897.1	643.1
GM IR cotton	650.7	5,908.9
GM HT cotton	80.7	86.2
GM HT canola	433.2	0
GM virus resistant papaya and squash and GM HT sugar beet	83.3	0
Total	9649.6	10,117.7

(Brookes and Barfoot, 2013)

Table 2. Cost of accessing GM technology (million $) relative to the total farm income benefits 2011

Crops	Cost of technology: all farmers	Farm income gain: all farmer	Total benefit of technology to farmers and seed supply chain	Cost of technology: developing countries	Farm income gain: developing countries	Total benefit of technology to farmers and seed supply chain: developing countries
GM HT soybeans	1,647.6	3,879.2	3,879.2	5,526.8	606.4	2,691.4
GM IR maize	1,686.6	1,540.2	1,540.2	3,226.8	433.3	1,827.8
GM HT maize	789.6	166.9	166.9	956.5	93.9	737.0
GM IR cotton	758.5	433.2	433.2	1,191.7	536.9	6,445.8
GM HT cotton	313.1	7,104.9	7,104.9	7,418.0	40.5	126.7
GM HT canola	140.2	6,559.6	6,559.6	6,699.8	N/a	0
Others	71.1	83.3	83.3	154.4	N/a	0
Total	5,406.7	19,767.3	19,767.3	25,174	1,711.0	11,828.7

(Source: Brookes and Barfoot, 2013)
N/a = not applicable. Cost of accessing technology based on the seed premia paid by farmers for using GM technology relative to its conventional equivalents

Supporting these findings, the authors (Finger et al. 2011) of a meta-analytical study of 721 single study entries for GM crops derived from a review of 203 publications based on field trials and surveys, discuss the economic benefits of GM crops and conclude:

The results presented here do support the contention that the adoption of GM crops leads on average to a higher economic performance, i.e., benefits, for farmers than conventional (non-GM) crops. This is in agreement with other studies that analyze economic performance of GM crops and is also underlined by the high (and increasing) adoption rates for GM crops in a number of countries. An important finding of the analysis is that the kind and magnitude of benefits are heterogeneous across crops, traits, countries and regions. By and large, the results suggest that the adoption of Bt cotton is particularly (economically) beneficial for the farmers in developing countries while positive effects are present but less significant for developed countries.

COUNTRY EXPERIENCES

Analysing country specific experiences with important GM crops, makes it possible to compare yields, insecticide use and net incomes between farmers adopting GM crops and those cultivating traditional varieties. In India, for example cotton is a major crop accounting for 30% of the country's agricultural GDP. However, prior to the introduction of GM cotton, the country's average cotton yield was 48% lower than the world average with farmers losing 50–60% of their crop to the cotton bollworm (ISAAA, 2012). Bt cotton was introduced in 2002 with 3 hybrids in six states - Andhra Pradesh, Gujarat, Madhya Pradesh, Karnataka, Maharashtra and Tamil Nadu. By 2012, India became the world's largest producer of biotech cotton covering 10.8 million hectares compared to the next largest producer the USA with 4.4 million hectares. The country now has 1097 Bt cotton hybrids approved for planting.

Field studies (Bennet et al. 2004) in the cotton producing state of Maharashtra in India (2002–2003) revealed that:

- Bt cotton yield increased by 45% in 2002, and by 63% in 2003 compared to non-Bt cotton varieties
- Higher revenues were attained by Bt cotton farmers compared to those who planted conventional cotton. The average income difference for 2002 and 2003 were 43% and 63%, respectively.
- The average gross margin (difference of the revenue and variable costs which include seed and insecticide costs) is much higher for Bt growers compared to growers of non-Bt varieties.
- Less insecticide input against cotton bollworms resulted in an average reduction in expenditure per hectare of 72% and 83% in 2002 and 2003, respectively. Savings in spraying were around Rs 2,471 or $54 per hectare on average (at an exchange rate of 1 US Dollar = 46 Indian Rupee[1])

Table 3 compares cotton yields and revenues for farmers adopting Bt cotton and those planting non-Bt cotton.

Table 4 compares the number and cost of insecticide sprays between adopters and non-adopters.

In addition to farm level benefits, the adoption of GM varieties appears to have had a direct impact on household food security. A comparison of households growing GM cotton in India with those that did not do so found that those adopting GM varieties consumed significantly more calories. Juma (2014) stated that: "Each hectare of GM cotton increased total calorie consumption by 74 kcal per adult equivalent per day. Furthermore, a smaller proportion of households were food insecure (7.93% of adopting GM cotton households compared to 19.94% of non-adopting households" The study also showed that GM cotton adoption led to consumption of more nutritious foods – such as fruits, vegetables and animal

Table 3. Cotton yields and revenues

Year	Cotton yield (tons/ha)		Revenue from cotton yield (rupees/ha)		Gross margin (rupees/ha)	
	Adopters	Non-Adopters	Adopters	Non-Adopters	Adopters	Non-Adopters
2002	2.18***	1.50	44,600***	31,078	38,796***	26,005
2003	2.25***	1.38	56,357***	34,597	50,904***	29,279

*** P < 0.001
(Bennett et al. 2004)

Table 4. Number and cost of insecticide sprays against cotton bollworms

Year	Number of bollworm pest sprays per plot		Revenue from cotton yield (rupees/ha)	
	Adopters	Non-Adopters	Adopters	Non-Adopters
2002	1.44***	3.84	692***	2,432
2003	0.71***	3.11	482***	2,881

** P < 0.01; *** P < 0.001
(Bennett et al. 2004)

products. The authors estimate that if the households that do not currently grow GM cotton switched, "the proportion of food insecure households would drop by 15-20%".

Experiences of small farmers from China, South Africa, the Philippines and other developing countries using GM crops clearly show that small farmers can also benefit from the technology. The most consistent observation from these countries is that growing GM crops is a profitable farming endeavour.

A case similar to Bt cotton in India is that of Bt corn in the Philippines, where the Asiatic corn borer caused losses of up to 80% of production, and corn yield levels averaged only 2.8 tons per hectare. The commercial release of Bt corn in 2003 in the Philippines marked the first time that a GM food/feed crop was ever approved for planting in Asia. The initial area under Bt corn covered more than 10,000 hectares. By 2005, it had expanded to 52,000 hectares. Adoption of Bt corn in the Philippines provided small farmers with a 37% increase in yield, reduction in pesticide costs by as much as 60% and an incremental income of up to PhP 10 or USD $0.21, 132 per ha (Yorobe and Quicoy, 2006).

Table 5 compares economic differentials between Bt and non-Bt corn producers in the Philippines

In China, insect-resistant GM rice was approved for food, feed, and cultivation in 2009. Surveys of randomly selected farm households showed that small and poor farm households adopting GM insect-resistant rice obtained yields 6 to 9% higher than conventional varieties and reduced pesticide use by as much as 80% or 16.77 kg/ha, which also had positive health benefits (Huang et al. 2005). The same authors also claim an average reduction in pesticide sprayings for the Asian boll worm from 20 to 6 times per year with the introduction of Bt cotton in China and a fall in cost of production of a kilogram of cotton by 28% as compared the farmers growing non-Bt varieties (Huang, Pray and Rozelle 2002).

In Burkina Faso, almost 100,000 farmers cultivated GM cotton on 260,000 hectares in 2010 with 126 percent increase from 115,000 hectares planted in 2009. In terms of contribution to the economy of Burkina Faso, Adenle (2011) states that:

It was estimated that a benefit of over US$100 million per year could be generated based on almost 30% increase in yield. Additionally, close to a 50% reduction in insecticide sprays may be realised, thereby saving fossil fuels and lowering greenhouse gases emission, hence fighting against climate change.

South Africa, the first and biggest producer of GM crops in Africa, has benefitted substantially from growing GM crops over the past decade. It is the only African country among the five principal coun-

Table 5. Economic indicators: Bt and non-Bt corn farms in the Philippines

Item	Wet Season 2003			Dry Season 2004		
	Bt	Non-Bt	% Diff	Bt	Non-Bt	% Diff
Yield (ton/ha)	4.30	3.31	30.0	4.89	3.79	29.0
Pesticide cost (PhP/ha)	121	159	-23.8	178	410	-56.5
Price received (PhP/kg)	7.20	6.75	6.5	8.95	8.30	8.0
Net cash income (PhP/ha)	8,552	6,249	36.8	23,341	15,489	50.6

(Yorobe and Quicoy 2006)

tries (others include: India, Argentina, Brazil and China) producing GM crops, with 63 million hectares planted in 2010 alone. The economy of South Africa has benefited a great deal from the adoption of GM technology and it has been reported to have enhanced farm income from GM maize, GM soybean and GM cotton by USD $156 million in the period 1998 to 2006, with farm income in 2006 alone estimated to be USD $67 million (James, 2008).

The best known and most often cited case is, however, the success story of small farmers growing Bt cotton in the Mahkhatini flats of South Africa, which soon became a showpiece for Monsanto and paved the way for the widespread cultivation of Bt cotton in the country. The Mahkhatini Flats are the floodplains bordering the Pongola river stretching from just below the Jozini Dam to the confluence of the Pongola and Usuthu Rivers on the Mozambique border covering some13000 hectares to the east of the Umbobo mountains. The region is extremely poor with the majority of the farmers growing mixed crops on small holdings of 2.5–5 hectares; many farms are smaller than 1 hectare. Low incomes and falling employment levels characterised the economy prior to the introduction of Bt cotton.

As a departure from assessing farm level impacts, a study of Bt cotton growers was conducted over the seasons 2003–04 and 2004–05 to assess socio-economic impacts on the livelihoods of the farmers (Morse, Bennet, & Ismael, 2013). Questionnaire based interviews of 100 cotton growers revealed that 88% of the respondents reported higher income from Bt cotton as against conventional seeds and almost 90% increased their asset base, particularly in terms of cultivable land. In addition, some 88% reported a reduction in insecticide related health problems and labour with more time to spend with their families or on other farm activities as spraying was not required.

The study also sought to assess differential impacts on male and female farmers. Gender differentials in benefits from GM crops is particularly relevant in the African context as the region has one of the highest ratios of women to men farmers in the world- almost 50%. The analysis revealed that a larger number of female farmers perceive the beneficial impacts of GM crops in terms of increased income and reduced need to access credit.

A comparison of costs and benefits of GM cultivation for three groups of farmers- those with marginal holdings below 2 hectares, those with small holdings: farm size between 2–3.5 hectares, and larger farmers with holdings above 3.5 hectares reveals that there are no major differences between the three size classes. In fact, farmers with medium sized holdings appear to have the lowest costs and the best yields and revenues, though farmers with over 3.6 hectares have the highest profit margins (Table 6).

The authors conclude that overall the adoption of Bt cotton in Makhathini has not only increased incomes for resource-poor farmers but has also had a positive impact on the livelihoods of these farmers and their households.

On the other hand, the anti-GM group cites data on yields sourced from the South African Cotton Trade industry covering the pre-GM cotton cultivation period (1997 to 2005) when almost 100% of the farmers cultivated GM varieties (Witt, Patel & Schnurr, 2006). This data shows a substantive increase in cultivated area, but little improvement in cotton yields (Table 7). Un-irrigated crops stand more or less at the pre-GM cotton yield of 600 kg seed cotton per hectare as the authors point out, but yields on irrigated lands, though highly erratic do appear to have increased.

The authors suggest that the large scale adoption of Bt cotton is more the result of a structured institutional framework which allows adopters access to seed, credit and support, than any increase in yields. They emphasize that technical interventions like Bt cotton are not inserted into a vacuum, but have to be assessed in the existing ecological, political and economic contexts. Yet on the balance they do find grounds for cautious optimism in the Makhatini area.

Table 6. Farm size differences in the economics of Bt cotton production

	Cotton	2003/2004					2004/2005				
	area	N	Mean	SD	F	Sig.	N	Mean	SD	F	Sig.
Yield (kg/ha)	<2ha	49	1024	290			47	1009	311		
	2 3.5ha	17	1058	153	0.116	ns	25	1140	257	1.811	n3
	>3.5	24	1023	259			24	1075	237		
Revenue (SAR/ha)	<2ha	49	5104	1475			47	5057	1628		
	2–3.5ha	17	5271	960	0.120	ns	25	5767	1351	1.914	ns
	>3.5	24	5068	1428			24	5334	1222		
Total Cost (SAR/ha)	<2ha	49	2759	781			47	2790	976		
	2–3.5ha	17	2922	869	0.791	ns	25	3118	621	1.478	ns
	>3.5	24	3008	893			24	3064	858		
Gross Margin (SAR/ha)	<2ha	49	2345	1159			47	2268	1093		
	2–3.5ha	17	2349	1002	0.538	ns	25	2648	1167	0.962	ns
	>3.5	24	2061	1241			24	2270	1338		

(Morse, Bennet, and Ismael 2013)

Table 7. Cotton production data in KwaZulu-Natal Estimates

Year	Area planted under irrigation (Ha)	Area planted under dryland (Ha)	Yield of cotton planted on irrigated and (Kg seed cotton per Ha)	Yield of cotton planted on dry land (Kg seed cotton per Ha)	Production (200 kg bales cotton lint)	Crop % hand-picked	Crop % ginned
1997–98	1444	6459	2100	600	1130	90	95
1998–99	1502	6785	1800	640	12683	100	80
1999–00	1155	5926	1400	420	7493	100	99
2000–01	528	3587	3000	1150	10419	100	95
2002–03	1403	321	2135	300	5500	91	98
2003–04*	2276	2537	3310	659	14715	58	
2004–05	1173	2242	2860	600	8695	65	

*Estimates
(Cotton SA cited in Witt, Patel and Schnurr, 2006)

The African Experience

GM Crops in Africa: Current Status

The Green Revolution brought about the transformation of agriculture in Asia and Latin America from the late sixties onwards. India in the sixties for instance, was a country with 70% of its population of almost one1 billion living in rural areas and wholly dependent on agriculture for their livelihoods. The huge demand for food had to be met by low yielding traditional food crops and almost complete dependence on rainfed agriculture, the people hovering perpetually on the verge of hunger with food imports, the only insurance against famine. The almost miraculous transformation within a period of barely two decades into a country with massive food surpluses and now a leading exporter of food grain was brought about only through political commitment to ensuring food security and the courage to embark on an unfamiliar path to agricultural development. Admittedly, the negative impacts of green revolution techniques of mono-cropping, excessive use of water, fertilizer and pesticides are beginning to emerge in some areas. There is also evidence that this technological breakthrough benefitted more the large farmers, than their counterparts with small and marginal holdings. Yet the country overall experienced an amazing breakthrough in food production and the fear of hunger is consigned to the past.

The Green Revolution unfortunately bypassed the countries of Africa, perhaps because of unfavourable physical or socio economic conditions or lack of political commitment. There is once again an opportunity for agricultural transformation opening up with the Gene Revolution; can Africa afford to lose this opportunity too?

While in the countries adopting green revolution technology, agricultural growth outpaced population growth from the seventies onwards, in Africa, per capita agricultural output was almost 20% lower in 2005 than in 1970 (Cooke and Downie, 2010). Additionally, with over 70% of the African population engaged in agriculture, Africa still imports a quarter of its food (Paarlberg, 2010).

African interest in biotechnology is reflected in the fact that with the exception of Somalia, all countries in Sub Saharan Africa are signatories to the Convention on Biological Diversity which expressly recognizes the potential of modern biotechnology in the promotion of human well being, particularly in helping meet critical needs for food, agriculture and health care (SCBD, 1992) in addition to ensuring the development of appropriate safety procedures which minimize potential threats to biological diversity and risks to human health. Yet only three countries in the continent — South Africa, Egypt and Burkina Faso have adopted GM technology with the production of food crops being limited to Egypt and South Africa. Egypt was the first North African country to grow GM crops commercially in 2008 with 700 hectares of GM maize planted and the Egyptian government planning to increase the hectarage of GM crops in future (James, 2008).

As far as significant production of GM crops goes, South Africa stands alone. The diversity of environment and socio-economic conditions in Africa is reflected in the varying country responses. The oft cited landmark anti-GM stance of Zambia is contained in the response of President Levy Mwanawasa to the offer of milled GM grains from the US during the food crisis of 2002: "Simply because my people are hungry, that is no justification to give them poison". Three other countries –Zimbabwe, Malawi and Mozambique supported this stance refusing "contaminated food", but eventually accepted food assistance in the form of milled grains. Zambia, however, continued in its refusal and eventually obtained reprieve by the supply of certified non-GM grain from Europe.

The early acceptance and successful progress of GM crops in South Africa and Egypt can be traced to the well-established and adequately funded scientific organisations in both countries. Bt cotton, the largest GM crop in South Africa was approved for commercial release as early as 1997. South Africa is now the eighth-largest producer of biotech crops in the world, growing 2.3 million hectares of genetically modified maize, cotton, and soya in 2011 (ISAAA, 2012). Although GM technology has been in use in South Africa for nearly 17 years and is well accepted in medical applications, its use in agriculture is still a subject of intense debate. While the technology receives strong support from the scientific community, commercial farmers and the government, NGO opposition continues. However, as Cooke and Downie (2010) stated:

In many ways, the debate in South Africa reflects the reality that GMOs are a fact of life and have been for some time. Although opponents continue to question the safety of genetically modified crops and advocate the adoption of the precautionary principle, which argues for a 'when in doubt, don't' approach to GM food, most of them understand such crops are here to stay. Therefore, their efforts center on tightening the regulatory system and preventing the entry of new GMO products into the South African market.

Cooke and Downie (2010) also discuss the rejection of the USAID funded *SpuntaG2* application for a potato modified with a gene to protect against the potato tuber moth, a project given up by Egypt after eight years of field trials and one which would have made South Africa the world's first producer of GM potatoes. They emphasize the fact that the proposal was rejected on economic rather than safety grounds. The fear of losing the two main markets for potatoes and potato seed - Angola and Zambia, the first with no bio-safety laws in place and the second known for its strong anti- GM stance re-emphasizing the importance of trade as a controlling factor.

In the case of Egypt, the realisation of the country's inability to meet its growing demand for food led to an early endorsement of GM technology. Genetic engineering programmes started in 1990. The country has in place a full framework for biotechnology consisting of the Agricultural Research Center, the National Research Center, the Academy of Scientific Research and Technology, the Water Research Center, the Desert Research Center and the Universities and has been intensively researching a variety of GM food crops although none of them have reached the market (Abdallah, 2010). The Agricultural Research Center (ARC) was established as a part of the Ministry of Agriculture in 1983 and brings together 16 research institutes. The crops cover a wide range of foods including staples like salt and drought tolerant wheat, rice with improved protein content and rice varieties resistant to rice blight and stem borers, biotech maize saves on pesticide usage, gives almost 100% protection from stem borers and increases yield by 30–40% over the conventional maize varieties. In 2008, the first approved GM crop, AgeebYG a yellow hybrid maize resistant to three maize borers was launched. A range of vegetables including virus resistant melon, squash and tomatoes are awaiting approval for commercialisation (Abdallah, 2010).

Burkina Faso, the third African country with a significant production of GM crops has a very different history. One among the "Cotton 4" (C4) countries of West Africa (the other three being Mali, Chad and Benin) the country grows and exports a fine variety of cotton. Crop intensification boosted cotton yields from an average of 200 kg per ha in 1963 to a peak of 1400 kg per ha in the mid-eighties. As Vitale et al. (2008) write "Cotton quickly became the region's "white gold", one of the leading sources of export earnings and a major source of employment in rural areas where livelihoods center about cotton, accounting for 60% of household income." In Burkina Faso, over 2.2 million people derive their livelihoods from the cotton industry, with public services like schools, healthcare, roads and agricultural extension being funded through cotton revenues (Vitale et al. 2008). However, the peak yields of the 80s began to decline and the search for a quick solution which would improve productivity in the short term, rather than the long term horizons of pest eradication programmes led to Bt cotton. Burkina Faso's national cotton company, Société Burkinabé des Fibres Textiles (SOFITEX), began a partnership with Monsanto in 2001 to introduce Bt cotton on a commercial basis. Since then the country has emerged as the most progressive in the West African region with reference to biotechnology. After the initial seven year period of field testing, monitoring, and development of bio-safety protocols, Burkina Faso commercially released Bt cotton in 2008. Expansion has been fast paced from an initial 12,000 ha in 2008 to slightly over 125,000 ha in 2009. In 2010, an estimated 400,000 ha of Bt cotton was planted in the country (Vitale et al. 2010).

Expansion of GM Cultivation

While the major GM crop producing countries in Africa have been discussed above, this does not imply that other countries are totally disassociated with bio-technology. Laboratory tests and field trials of a variety of GM crops, both food and non-food are in process across Africa (Tables 8 and 9).

Tables 9 and 10 provide ample evidence that despite opposition and misgivings among certain groups, Africa is well on the way to accepting bio-technology as an option to resolving hunger and malnutrition, particularly in the context of deteriorating environmental conditions.

Table 8. Status of genetically modified (GM) crops in Africa

Application Stage	No. of Countries	Country Names
Commercial production	3	Burkina Faso; Egypt; South Africa
Confined field testing	7	Burkina Faso; Egypt; Kenya; South Africa; Uganda; Nigeria; Malawi
Contained research	At least 15	Burkina Faso; Cameroon; Egypt; Ghana; Kenya; Mali; Mauritius; Namibia; Nigeria; South Africa; Tanzania; Tunisia; Uganda; Zimbabwe; Malawi
Developing capacity for research and development	At least 27	South Africa; Burkina Faso; Egypt; Kenya; Morocco; Senegal; Tanzania; Uganda; Zambia; Zimbabwe; Benin; Cameroon; Ghana; Malawi; Mali; Mauritius; Namibia; Niger; Nigeria; Tunisia; Algeria; Botswana; Ethiopia; Madagascar; Rwanda, Burundi, Sudan

Modified from absafrica.org (NEPAD, 2010)

Table 9. Range of GM crops being developed in Africa

Country	Crop
Kenya	Maize and cotton (insect resistant), cassava, sweet potato (virus resistant)
Uganda	Cotton (insect resistant and herbicide tolerant), banana (nutrient enhanced, fungal resistant), cassava (viral resistant), sweet potato (weevil resistant)
Nigeria	Cassava (nutrient enhanced)
Malawi	Cotton (insect resistant)
Nigeria, Burkina Faso, Ghana	Cow pea (insect resistant)
Kenya, Mozambique, Tanzania, South Africa, Uganda	Maize (drought tolerant)
Burkina Faso, Kenya, South Africa	Sorghum (nutrient enhanced)
South Africa	Maize (herbicide tolerant, insect resistant, virus resistant), Cassava (starch enhanced), Cotton (insect resistant, herbicide tolerant), Potato (insect resistant), sugarcane (alternative sugar)
Egypt	Maize (insect resistant), Cotton (salt tolerant, Wheat (drought tolerant, salt tolerant, fungal resistant), Potato, Banana, Cucumber, Melon, Squash, Tomato (all viral resistant)

(NEPAD, 2010)

DISCUSSION AND EXPERIENCES

Are GM Crops a Feasible Option in Africa?

We have discussed the status of GM crops in Africa and have also assessed whether transgenic crops benefit the small farmer in developing countries. There appears to be substantive evidence that gene technology does improve the livelihoods of small farmers as well as of those with large holdings. Apart from benefits at the farm level, visible improvements have been documented in household asset base and food security in developing countries, where GM crops have been cultivated over a number of seasons. It must be emphasised here that these positive impacts do not deny the emergence of setbacks in the form of lower than expected yields, higher input costs or unintended environmental impacts, which have also been witnessed in some areas and must be given due weight in the decision making process. As Finger et al. (2011) emphasized:

GM crops are designed with particular aims in mind such as insect and herbicide resistance, not with a view to increasing yields, except indirectly through the curtailing of losses. Thus other constraints like infertile soils or drought can only be addressed by varieties designed specifically to overcome these problems. In addition there are always socio-economic problems which GM crops cannot address. Hence, considering GM crops as a panacea for poverty and hunger is groundless.

South Africa and Burkina Faso, the only countries in the African continent where GM crops have been cultivated for over a decade have both registered success on all counts equal to their Asian and Latin American counterparts; yet the opponents of GM crops not only oppose their expansion in the African region on the basis of un-established but nevertheless very real fears of biotechnology, but cite as their explanation the contention that agriculture in Africa being distinct from agriculture in other developing regions is unsuitable for the introduction of advanced technology. In evidence, they emphasize the failure of the Green Revolution in the African context and predict a similar scenario in the case of the Gene Revolution.

Some Constraints to Agriculture

This section will inquire into the constraints that agriculture faces in Africa, why chronic food shortages persist despite abundant natural resources, the reasons behind the failure of the Green Revolution and whether the failure of one technological opportunity necessarily precludes the success of the next.

Africa is known for its diverse agro-ecological regions, abundance of water resources, though unevenly distributed, and almost unlimited land area which could be brought under cultivation — conditions which would be the envy of any agricultural country. Yet, agricultural growth has been slow and has even retrogressed leaving the region perennially short of food, open to droughts, erratic rainfall, attacks by insects and other pests, with survival being assured only by food imports or food assistance during emergencies. Over the past two decades an average of 32 countries in sub-Saharan Africa received emergency food aid each year. The numbers ranged from 22 countries in 1990 to 38 in 2009 (UNDP, 2012).

The issue is further compounded by trade barriers and poor transport systems, which leave food security, particularly in rural areas completely dependent on local production of food staples. It has been stated that "if the food available in sub-Saharan Africa were evenly distributed, all Africans could consume enough calories for their basic functioning" (UNDP, 2012). In fact 75% of the cereals and almost all the roots and tubers, the basic food crops, are grown locally.

In a bid to jump start the development process after gaining independence in the sixties, most African countries focussed on urbanization and industrialisation with the expectation of a trickledown effect to take care of the rest of the economy. The consequent neglect of the agricultural sector and the emerging policy bias, heavy taxation and government control of agriculture over the next two decades are some of the reasons for the stagnation in Africa's food production. At a time, when the developing countries of Asia set upon a trajectory to achieve self-sufficiency in food and free themselves from dependence on food imports, Africa was set on a contradictory mission. In this perhaps, lie the reasons behind the apparent failure of the Green Revolution in the region. With no political backing or supportive measures, technological revolutions are unlikely to succeed in any country and even less so in an agricultural economy characterised by small farmers.

The results of the post-independence policy framework were overvalued exchange rates supporting imports for domestic urban consumers and making agricultural exports less competitive in international markets (Binswanger and Deininger, 1997). Heavy direct and indirect taxes were also levied on agricultural activities. In effect:

these biased policies increased the combined effective tax rate on agriculture (from industrial protection, direct taxation and exchange rate policies) from 5% in the late 1950s to close to 25% in the late 1970s (UNDP, 2012).

Although most staples were unaffected by taxation or support, the food markets came under heavy control. The net outcome was to turn relative prices against farmers.

With the entry of structural adjustment policies, the bias against agriculture was lifted, but the UNDP (2012) states that:

The state retrenchment of the 1980s–1990s disabled the sector as an early driver of growth and retarded the kind of structural transformation needed to increase food security and reduce poverty. Rising population growth compounded the setback, as the need for food outpaced the continent's capacity to produce.

Although there has been a substantial reduction in agricultural tariffs, trade barriers still persist in sub-Saharan Africa slowing the growth of intra-regional trade in food, particularly staples. The 'border effect' or the additional costs of crossing the border between two countries is substantial, and is exacerbated by the price differentials between producing and consuming zones. For example, Brenton, Portugal-Perez and Regolo (2013) find that the effect on relative prices of crossing the Niger–Nigeria border is equivalent to adding an additional 594 km to the distance to markets.

Promoting inter-regional trade in food staples and reducing the costs of bringing food to the market could effectively reduce chronic food insecurity as well as vulnerability to risks. Engel, Jouanjean and Awal (2013) emphasize the importance of facilitating food trade and a more coherent trade policy in the arid and semi arid regions of the Sahel and the Horn of Africa, where the frequency of food shortages is making the population particularly vulnerable.

However, some positive developments are beginning to emerge:

Up until early 2000, the agriculture sector in Sub-Saharan Africa went through policy and budgetary neglect that led to growth stagnation but this has now changed: agricultural GDP in Sub-Saharan Africa as a whole grew by more than 3.5% in 2008, surpassing the population growth rate, and grain production growth rates have been kept above 3% (Ackello Ogutu, Okoruwa, & Bahal, 2012).

Contrasting Experiences with Agricultural Technology in Africa and Asia

While both Africa and Asia faced severe food shortages and the recurring threat of hunger and starvation in the early sixties, development policies in the two regions, as mentioned above, were sharply divergent. Asian governments prioritised food security from the mid-1960s onwards by supporting the agricultural sector to adopt high yielding varieties of food staples known to be highly responsive to fertilizer, water and other inputs. Cereal production increased by 3.4 times during the period 1961– 2010, and per capita availability of cereals also rose sharply by as much as 44%. In Africa, on the other hand, development focussed on industry and export-based cash crops with the expected adverse impacts on food production and the livelihoods of small farmers. Although cereal production in sub-Saharan Africa has tripled over the same period, population growth has outpaced production with the result that per capita output has actually fallen by 13% (UNDP, 2012).

The basic difference between increased cereal production in Asia and Africa is that Asia owes about 44% of its increased production to higher yields, while in the case of Africa, the increase comes in area under cereals, and the exclusive contribution of higher yields is just about one-third of the increase in production. However, this must not be considered a negative factor. It only serves to underline the fact that there is much scope for improving yields in Africa, if the opportunity offered by currently available technologies is grasped. According to the UNDP African Human Development Report (2012):

African farming thus contrasts with the Asian norm of cultivating mainstream staples and cash crops, fertilized and often irrigated, unmixed in a single field on fully settled farms under individual tenure. In sub-Saharan Africa, reliance on many crops and traditional farming practices has resulted in less research and extension (per unit of output and per hectare) than in Asia. More fundamentally, sub-Saharan Africa has not benefited from the farm science–based, input-responsive boost in agricultural productivity that occurred in Asia.

Farming in Africa is characterised by smallholders typically growing a variety of crops, sometimes as many as ten including staples for the home and cash crops for sale. Cultivation is predominantly rain-fed and practised on weathered soils with low fertility and little mechanisation. Low population densities result in poor road network and restricted access to markets. The relative isolation of the African farmer also adversely impacts his access to modern inputs or extension services. "Only 30% of the rural population in sub-Saharan Africa lives within 2 km of an all season road in sharp contrast to South Asia's 58%" (Roberts, Shyam, & Rastogi, 2006).

Across the region (Sub Saharan Africa) just a little over 200 million hectares are cultivated, leaving four-fifths unused. "Excluding protected areas, land covered by forests and regions already settled, and limiting the definition of suitability to five crops, the balance of suitable unused agricultural land is about 200 million hectares, which, if brought into production, would double the amount of cultivated land in sub-Saharan Africa" (Berg and Ostry, 2011). Thus viewed from a different perspective, what appear to be Africa's disadvantages in agriculture, may well prove to be opportunities in the making in the form of unused and underused arable land and unused potential to boost agricultural productivity (UNDP, 2012).

While Africa's agro-ecological zones are more diverse than Asia's, this does not necessarily suggest an impediment to the introduction of improved technologies. The solution would lie in local- specific interventions tailored to meet specific needs and address problems specific to the region. Examples include among others, the development of drought-tolerant varieties of millet and sorghum for farmers in the Sahel and the Horn of Africa (Hunt and Lipton, 2011), GM bananas in Uganda and Bt black-eyed peas in Nigeria (Juma, 2014). Juma and Gordon (2014), while strongly endorsing the technological revolution in Africa, even go so far as to say:

Biotechnology is only a starting point. The introduction of Bt cotton, for example, requires system-wide investments along the entire cotton value chain. This means that the adoption of biotechnology can serve as a trigger for investing in R&D, rural infrastructure, technical training and entrepreneurship (and leapfrog agricultural development in the country.

As aptly summarised in the first African Human Development Report, (UNDP, 2012):

No lack of knowledge, ideas or vision stands between sub-Saharan Africa and this brighter future. And certainly no failure to commit to a new agricultural revolution, as evidenced in African leaders' affirmation of their commitment to agriculture in the Maputo Declaration and the Comprehensive Africa Agriculture Development Programme. But realizing the vision will require policies to increase yields up to the frontier of current science, through faster and wider adoption of better technology, and then policies to expand that frontier so that agricultural yields continue to rise. With rapid growth in agricultural productivity, anchored in rising yields of staple crops, sub-Saharan Africa can reduce mass poverty and ensure food security for most of its people.

CONCLUSION

On all issues discussed above, the evidence is in favour of GM crops, yet most African countries are reluctant to adopt GM technology. A major factor may be financial constraints. As Miller (2007) writes:

The high level of investment needed in GM research and its application has constrained African participation and has led to research that primarily focuses on developed country needs. Transgenic research is very expensive when compared to more traditional biotechnology techniques. For example, the IRMA project is estimated to have cost USD $6 million over 5 years and the transgenic sweet potato research USD $2 million, compared to the average funding of tissue culture and marker technology projects costing on average USD $300,000.

The lack of in-country research has prevented the dissemination of authentic information about GM crops, which would be readily acceptable to the local population. The reluctance to accept information from International Agencies, which are perceived as having their own agenda has created a climate conducive to anti-GM activists. Adenle (2011) writes:

Africa's relatively slow embrace of the "gene revolution", particularly GM technology, has, in part, stemmed from lack of access to GM crops available to beneficiaries such as farmers and scientists. This is partly due to international influence against GMOs.

The communication gap further fuels fears of being held to ransom by seed producing multinational companies, as the terminator gene precludes the traditional practices of seed saving and exchange. Juma (2014) however, adds a rider stating that:

The problem is not simply the presence of large corporations but the low level of the development of domestic seed firms. Such local firms can help focus on indigenous crops. They can also play key roles as business partners with foreign farms.

He further adds that new research shows that in Africa, local start-ups with local scientists are at the forefront of bio-technology research, not multi-nationals.

Yet another fear is that of losing the largest market for agricultural produce- the European Union, which is known to have stringent laws against the import of GM crops. This has led to the adoption of a Go Slow approach by the governments of many African countries (Adenle, 2011). A recent IFPRI (2013) research study concludes:

This European model does not correspond to African needs and realities, however. GM crops can aid farmers by reducing both losses from pests and the costs of weeding. For European countries, in which only a tiny percentage of the population grows crops such as maize, cotton, and soybeans, these advantages are negligible. In Africa, though, where 60% or more of the population depends directly on agriculture for income and subsistence, far more people could benefit directly from biotechnology that includes agricultural GMOs. ....Rather than imitate other regions' regulatory systems, African countries south of the Sahara need to consider and develop their own regulations for biotechnology development that balance risk avoidance (and the accompanying regulation costs) against such technology's net benefits for farmers.

A reasonable conclusion also, is that the fear of short term export losses is also unfounded as the major export crops from Africa are tea, coffee, vegetables and flowers, not staples, which are more likely to be genetically modified.

It must be conceded that these fears are not totally unfounded, but they do not pose insurmountable obstacles to the adoption of this new technology. As Adenle (2011) writes:

Given both the lack of global consensus on GMOs and the growing need to regulate GMO products, the US may offer a good example for Africa and the rest of the world. It is one of the few developed countries to grow GM crops on a commercial basis (more than 70% of foods in the US contain GMOs) and is a leader in the safe application of GMO product.

It appears then, that Africa's path to enhanced food security through the adoption of GM technology is a viable one, given adequate political backing and wholehearted commitment to meeting the needs of the African people.

REFERENCES

Abdallah, N. (2010). GM crops in Africa: challenges in Egypt. *GM Crops*, *1*(3), 116–119. Retrieved 29 May 2014 from, http://www.tandfonline.com/doi/pdf/10.4161/gmcr.1.3.12811

Ackello-Ogutu, C., Okoruwa, V., & Bahal, G. N. (2012). *Long-term challenges to food security and rural livelihoods in Sub-Saharan Africa*. Briefing Paper Number 2, 2012. Washington DC: GDN Agriculture Policy Series. Retrieved 6 June 2014 from, http://www.gdn.int/admin/uploads/editor/files/SSA_2_PolicyBrief_Food_Security.pdf

Adenle, A. A. (2011). Response to issues on GM agriculture in Africa: are transgenic crops safe? *BMC Research Notes*, *4*, 388. Retrieved 1 June 2014 from, http://unu.edu/publications/articles/are-transgenic-crops-safe-gm-agriculture-in-africa.html

Amir, P. (2014). GM crops-no gain for small farmers. *SciDevNet*. Retrieved 30 May 2014 from, http://m.scidev.net/south-asia/agriculture/opinion/gm-crops-no-gain-for-small-farmers.html

Belay, M., & Guri, B. (2013). Why African farmers do not want GMOs. *Truthout*. Retrieved 29 May 2014 from, http://webcache.googleusercontent.com/search?q=cache:http://www.truth-out.org/news/item/20058-why-african-farmers-do-not-want-gmos

Bennett, R., Ismael, Y., Kambhampati, U., & Morse, S. (2004). Economic impact of genetically modified cotton in India, *Agbioforum*, *7*(3), Article 1. Retrieved 2 June 2014 from, http://www.agbioforum.org/v7n3/v7n3a01-morse.htm

Berg, A. G., & Ostry, J. D. (2011). *Inequality and unsustainable growth: two sides of the same coin?* IMF Staff Discussion Note 11/08. Washington, DC: International Monetary Fund. Retrieved 15 July 2014 from, https://www.imf.org/external/pubs/ft/sdn/2011/sdn1108.pdf

Binswanger, H. P., & Deininger, K. (1997). Explaining Agricultural and Agrarian Policies in Developing Countries. *Journal of Economic Literature*, *35*(4), 1958–2005. Retrieved from http://www-wds.worldbank.org/servlet/WDSContentServer/WDSP/IB/1997/05/01/000009265_3971126124251/Rendered/PDF/multi_page.pdf

Brenton, P., Portugal-Perez, A., & Regolo, J. (2013). *Food prices, road infrastructure, and border effects in central and eastern Africa*. Retrieved 8 July 2014 from, https://editorialexpress.com/cgi-in/conference/download.cgi?db_name=CSAE2013&paper_id=919

Brookes, G., & Barfoot, P. (2013). *GM crops: global socio-economic and environmental impacts 1996-2011*. Dorchester, UK: PG Economics Ltd. Retrieved 10 July 2014 from, http://www.europabio.org/sites/default/files/report/2013globalimpactstudyfinalreport.pdf

Cooke, J. G., & Downie, R. (2010). *African perspectives on genetically modified crops: assessing the debate in Zambia, Kenya and South Africa*. A report of the CSIS global food security project. Washington, DC: Centre for Strategic and International Studies. Retrieved 8 June 2014 from, http://csis.org/files/publication/100701_Cooke_AfricaGMOs_WEB.pdf

Education, G. M. (2014). *A brief history of genetic modification*. Retrieved 19 July 2014 from, http://www.gmeducation.org/faqs/p149248-20brief%20history%20of%20genetic%20modification.html

Engel, J., & Jouanjean, M.-A. (2013). *The history, impact and political economy of barriers to food trade in sub-Saharan Africa: an analytical review*. London: ODI. Retrieved 14 June 2014 from, http://www.odi.org/sites/odi.org.uk/files/odi-assets/publications-opinion-files/8803.pdf

FAO. (2009). *High level expert forum, How to feed the world in 2050: The special challenge for sub Saharan Africa*. Retrieved 12 July 2014 from, http://www.fao.org/fileadmin/templates/wsfs/docs/Issues_papers/HLEF2050_Africa.pdf

FAO. (2012). *Statistical Yearbook, Part 4, Genetically Modified crops*. Rome. Retrieved 29 May 2014 from, http://www.fao.org/docrep/015/i2490e/i2490e04d.pdf

FAO. (2015). The State of Food Insecurity in the World 2015. *Meeting the 2015 International Hunger Targets: taking stock of uneven progress*. Rome: FAO. Retrieved 26 January 2016, from http://www.fao.org/3/a-i4646e.pdf

Finger, R., El Benni, N., Kaphengst, T., Evans, C., Herbert, S., Lehmann, B., . . . Stupak, N. (2011). A meta-analysis on farm-level costs and benefits of GM crops. *Sustainability, 3*, 743–62. Retrieved 6 July 2014 from, http://www.mdpi.com/2071-1050/3/5/743

Global Hunger Index (GHI). (2014). *The Challenge of Hidden Hunger*. Welthungerhilfe, IFPRI, Concern Worldwide.

Huang, J., Hu, R., Rozelle, S., & Pray, C. (2005). Insect-resistant GM rice in farmers' fields: Assessing productivity and health effects in China. *Science*, *308*(5722), 688–690. doi:10.1126/science.1108972 PMID:15860626

Huang, J., Pray, C., & Rozelle, S. (2002). *Enhancing the crops to feed the poor. Nature, 418,* 678–84. Retrieved 12 July 2014 from, http://www.nature.com/nature/journal/v418/n6898/pdf/nature01015.pdf

Hunt, D., & Lipton, M. (2011). *Green Revolutions for Sub-Saharan Africa?* Briefing Paper AFP BP 2011/01. Chatham House, Africa Programme. Retrieved 17 July 2014 from, www.chathamhouse.org/sites/default/files/public/Research/Africa/bp0111greenrevolution.pdf

IFPRI. (2013). *Genetically Modified Crops in Africa: Economic and Policy Lessons from Countries South of the Sahara*. International Food Policy Institute Issue Brief 80. Retrieved 12 August 2014 from, http://www.ifpri.org/sites/default/files/publications/ib80.pdf

International Service for the Acquisition of Agri-Biotech Applications (ISAAA). (2010). *Are GM crops Africa's path to food security?* Retrieved 6 June 2014 from, http://www.voanews.com/content/are-gm-crops-africas-path-to-food-security-64567512/157007.html

International Service for the Acquisition of Agri-Biotech Applications (ISAAA). (2012). *S. Africa GM crop area hit record in 2011: Report*. Reuters. Retrieved 12 August 2014 from, http://af.reuters.com/article/InvestingNews/idAFJOE82705320120308

James, C. (2008). *Global status of commercialized biotech/GM crops: 2008* (ISAAA Brief No. 39). Ithaca, NY: International Service for the Acquisition of Agri-biotech Applications. Retrieved 9 July 2014 from, https://isaaa.org/resources/publications/briefs/39/download/isaaa-brief-39-2008.pdf

Juma, C. (2014). Feeding Africa: why biotechnology sceptics are wrong to dismiss GM. *Guardian Professional*. Retrieved 12 June 2014 from, http://www.theguardian.com/global-development-professionals-network/2014/may/27/gm-crops-food-security-calestous-juma-africa

Juma, C., & Gordon, K. (2014). *Leapfrogging African agriculture*. Retrieved 15 June 2014 from, http://www.brookings.edu/~/media/Research/Files/Reports/2014/foresight%20africa%202014/06%20foresight%20african%20agriculture%20juma%20gordon.pdf

Kuyek, D. (2002). The past predicts the future: GM crops and Africa's farmers. *Seedling*. Retrieved 15 July 2014 from, http://www.grain.org/article/entries/344-the-past-predicts-the-future-gm-crops-and-africa-s-farmers

Millennium Development Goals (MDG) Report. (2013). *Section III-food security in Africa-issues, challenges, lessons. Assessing progress in Africa towards the millennium development*. Retrieved 6 June 2014 from, goals.http://www.afdb.org/fileadmin/uploads/afdb/Documents/Publications/Millennium%20Development%20Goals%20%28MDGs%29%20Report%202013%20-%20Section%20III%20-%20Food%20Security%20in%20Africa%20-%20Issues-Challenges-Lessons.pdf

Miller, M. (2007). *Genetically modified crops in Africa*. UNEP. Retrieved 5 May 2014 from http://www.eoearth.org/view/article/152943

Morse, S., Bennet, R., & Ismael, Y. (2013). *Impacts of GM cotton and maize on household livelihoods in South Africa*. University of Reading. Retrieved 9 July 2014, from bch.cbd.int/database/attachment/?id=1575

Murnaghan, I. (2014). *Development and history of GM foods*. Retrieved 10 August 2014, from http://www.geneticallymodifiedfoods.co.uk/development-history-gm-foods.html

New Partnership for Africa's Development (NEPAD). (2010). *Status of crop biotechnology in Africa*. Retrieved 4 July 2014 from, http://www.nepadbiosafety.net/subjects/biotechnology/status-of-crop-biotechnology-in-africa

Paarlberg, R. (2010, November30). GMO foods and crops: Africa's choice. *New Biotechnology*, *27*(5), 609–613. doi:10.1016/j.nbt.2010.07.005 PMID:20637906

Pschorn-Strauss, E. (2005). Bt Cotton in South Africa: the case of the Makhatini farmers. *Seedling*, 13–24. Retrieved 10 July 2014 from, www.grain.org/article/entries/492-bt-cotton-in-south

Roberts, P. K. C., Shyam, & Rastogi, C. (2006). Rural access index: a key development indicator. *Transport Papers 10*. Washington, DC: World Bank. Retrieved 5 July 2014 from, www.worldbank.org/transport/transportresults/headline/rural-access/tp-10-final.pdf

Secretariat of the Convention on Biological Diversity (SCBD). (1992). *The Convention on Biological Diversity*. Montreal, Canada: Text and Annexes.

UNDP. (2012). *Africa human development report: towards a food secure future*. New York: UNDP Regional Bureau for Africa. Retrieved 4 July 2014 from, http://www.undp.org/content/dam/undp/library/corporate/HDR/Africa%20HDR/UNDP-Africa%20HDR-2012-EN.pdf

Vitale, J., Glick, H., Greenplate, J., & Traore, O. (2008). The Economic Impact of Second Generation Bt Cotton in West Africa: Emperical Evidence from Burkina Faso. *International Journal of Biotechnology, 10*(2-3), 167–183. doi:10.1504/IJBT.2008.018352

Vitale, J. D., Vognan, G., Ouattarra, M., & Traore, O. (2010). The Commercial Application of GMO Crops in Africa: Burkina Faso's Decade of Experience with Bt Cotton. *AgBioForum, 13*(4), 320-332. Retrieved 10 August 2014 from, http://soybeaninnovationlab.illinois.edu/sites/soybeaninnovationlab.illinois.edu/files/cotton%20prfotiabilityv13n4a05-vitale.pdf

Witt, H., Patel, R., & Schnurr, M. (2006). Can the poor help GM crops, Technology, representation & cotton in the Makhathini Flats, South Africa. *Review of African Political Economy, 109*, 497–513. Retrieved 12 June 2014 from, http://rajpatel.org/wp-content/uploads/2009/11/Can-the-Poor-help-GM-Crops_final-printed-version.pdf

Yorobe, J. M., Jr., & Quicoy, C. B. (2006). Economic Impact of Bt Corn in the Philippines. *The Philippine Agricultural Scientist, 89*(3), 258–67. Retrieved 5 July 2014 from, https://www.isaaa.org/programs/impact_assessment_of_crop_biotechnology/download/Economic%20impact%20of%20Bt%20corn%20in%20the%20Philippines.pdf

ENDNOTE

1 Applicable in 2002–03.

Chapter 8
Appropriate and Sustainable Plant Biotechnology Applications for Food Security in Developing Economies

Vidya de Gannes
The University of the West Indies – St Augustine, Trinidad and Tobago

Carlos G. Borroto
Yucatan Center for Scientific Research (CICY), Mexico

ABSTRACT

Appropriate plant biotechnology applications could be a major tool in the fight against hunger and poverty, especially in developing economies. Some promising results have already been reported through a range of biotechnology applications, for example the mass propagation of plants through in vitro clonal propagation, use of molecular markers and marker assisted breeding to improve plant breeding, use of bio-products and the application of molecular techniques for quick and accurate diagnosis of plant diseases. Thus, biotechnology may have the potential to deliver solutions to some of the shortcomings of the green revolution such as the limitations of conventional breeding, poor quality and insufficient quantity of planting materials available to farmers and the negative environmental consequences of high usage of inorganic pesticides and fertilizers. This chapter will elaborate the successful application of these technologies for sustainable agriculture, and food security in developing economies. It is envisaged that through partnership between developing and developed economies, and assistance from international organizations and governments, capacity can be developed for developing economies to achieve the full potential of biotechnology and sustainability.

DOI: 10.4018/978-1-5225-0942-4.ch008

INTRODUCTION

The world population may total 8.8 billion by 2050 with 90% of this growth occurring in developing countrics (Botclla-Rodrigucz, 2011). Today, almost a billion people (more than 16% of world's population) live in absolute poverty and suffer from chronic hunger. From the developing economies, most of this population comprises farmers and their families, whose living conditions depend on small plots of poor lands that are mainly affected by drought, salinity, pest and diseases thus resulting in low crop yields (Botella-Rodriguez, 2011). In contrast, there exists an over production of many major crop species and a revolution in biotechnology and associated information technology which aim to improve health and welfare in general in developing economies (Cooper, Spillane, & Hodgkin, 2011). In the future, this food security gap will be even greater taking into account the population growth pattern in both regions. While it is expected that the population in developed countries will not change, in developing countries it will increase to almost 40%. Such was the case during the last century, where green revolution technologies contributed to avoidance of generalized food crisises.E,g., maize cultivation led the modernization process. In 1940, farmers in the United States of America (USA), produced 56 million tons of maize on roughly 31 million hectares, with an average yield of 1.8 t/ha. In 1999, USA farmers produced 240 million tons of maize on roughly 29 million hectares, with an average yield of 8.4 t/ha. This more than four-fold yield increase was due to the impact of modern hybrid seed-fertilizer-weed control technology, promoted as components of the green revolution technological packages.

The green revolution which took place in the late 50s and 60s, was a combination of genetics and technology, the genetics being breeding and the technology the application and availability of petroleum based pesticides and fertilizers and crops that were compatible with these inputs. This resulted in a huge increase in production, but also brought a huge amount of ecological problems (such as eutrophication, high level of water contamination and high amounts of nitrogen oxide) coupled with negative impacts on human health (Pretty, 2008; Tilman, 1999). Of course, this approach is not the sustainable way needed to advance in the future. Just as the food requirements of today's population of 6 billion people could not have been met by the technologies available before the "green revolution", we cannot assume that current practices will feed the increased population of the next 20-30 years. While conventional plant breeding has made some contributions to agriculture it was a slow process with technical limitations. Any further increases in food production should also be based on a sustainable approach. This new sustainable agriculture intensificationcannot depend on a further input from petroleum based products because of the damage that is caused by the applications of these products in terms of contamination of the environment (Brethour & Weersink, 2001) and their negative effects on human health. Thus, it should be almost entirely genetic and using mainly inputs of biological origin (Tilman, 1999). Hence, a major goal will be to redress the ecological imbalance of earlier efforts, by developing genetically diverse new crop varieties, minimizing chemical use with an integrated pest management approach and improving on-farm water efficiency and nutrient management. Diversity of the genetic base of the new plants, the reduction of chemical inputs and the integrated soil, nutrient and water management on-farms are the three pillars of any new sustainable agriculture intensification effort. This can be addressed through biotechnology.

Biotechnology can be defined as any technique that utilizes living organisms or substances derived from organisms to make or modify a product, improve plants or animals or develop microorganisms for specific uses (Fagwalawa, Kutama, & Yakasai, 2013). Too often, the term 'modern biotechnology' refers to transgenic technologies such as with genetically modified crops. However, for the developing world other applications of biotechnology are extremely important. The range includes: the development of new bio-products with better formulation and known action mechanisms (Chandler et al., 2011); mass plant propagation based on tissue culture techniques for rapid multiplication of plants; marker assisted selection to improve the efficacy of plant breeding programs (Cardena, Oropeza, & Zizumbo, 1998) and use of molecular technologies for the diagnosis of plant diseases (Miller & Martin, 1988). From the range of applications of biotechnology available, this chapter will focus on how mass propagation of plants, molecular markers, molecular diagnosis of plant diseases and bio-products have been successfully applied to improve agriculture and food security in developing countries in different scenarios. The goal is to offer these to other developing countries around the world for adoption as part of their agricultural development plans and programmes.

MASS PROPAGATION OF PLANTS THROUGH TISSUE CULTURE TECHNIQUES (MICROPROPAGATION)

Plant tissue culture is the *in vitro* aseptic culture of cells, tissues, organs or whole plant under controlled nutritional and environmental conditions (Thorpe, 2007) mainly to produce clones of the plant. There are different techniques of plant tissue culture with micropropagation being a form of tissue culture which increases the amount of planting material to facilitate distribution and large scale planting. It has been defined as *in vitro* regeneration of plants from organs, tissues, cells or protoplasts (Shankar, Balaji & Sathish, 2014) and "the true-to-type propagation of a genotype of explant" (Debergh & Read, 1991). In this way, thousands of copies of a plant can be produced in a short time. Plants produced through micropropagation have shown to establish more quickly, grow more vigorously, have a shorter and more uniform production cycle, and produce higher yields than conventional propagules. Mass propagation of plants through tissue culture has a significant impact on agriculture development in emerging economies by providing plants to meet the increasing demand through the production and propagation of genetically homogeneous, disease-free plant material (Chatenet et al., 2001).

Planting material available to many small–scale farmers is often of poor quality and insufficient quantity, which pose challenges to potential yields and performance (Ogero, Mburugu, Mwangi, Ngugu & Ombor, 2012). Crops such as yam, cassava, bananas and potatoes belonging to tropical or subtropical agricultural systems, are staple foods in many developing economies and, while being minor crops at the global level, contribute significantly to the food security of rural populations in specific countries and regions (Fujardo, 2010). Production of high quality and healthy planting material propagated from vegetative parts, has thus created new opportunities in global trading for producers, farmers, and nursery owners, and for rural employment. The production of *in vitro* culture crops through micropropagation has several advantages over the traditional methods of propagation (such as through seed, cutting, grafting and air-layering) mainly because it allows a rapid propagation process which can lead to the production of virus free plants (Garcia-Gonzales, Quiroz, Carrasco, & Caligari, 2010). Additionally, plants can be multiplied under a controlled environment regardless of season or weather, on a year round basis.

Generally, the micropropagation process can be summarized in five universally applicable stages according to Murashige (Murashige, 1974).

Stage 0: Pre-propagation stage
Stage I: Initiation of explants *via* surface sterilization and establishment into suitable media
Stage II: Multiplication
Stage III: Shooting and rooting of the explants.
Stage IV: Weaning/Hardening.

Stage 0

The mother plant is selected at this stage and must be pre-screened for diseases and pests, isolated and treated with appropriate fungicides and/ or pesticides (if necessary) to reduce contamination in the *in vitro* cultures. It is extremely important for the pre-screening and treatment process to be carried out in a separate facility from where the clonal propagation process will take place. Clean, enclosed areas or greenhouses will be ideal for this stage to be conducted. Following pre-screening, the explants (plant materials) are selected for clonal propagation and are taken to the facility (laboratory). The explants should be tested further for contamination via standard microbiological techniques (such as plating and gram staining) and or molecular techniques to detect any form of contamination. This is crucial to ensure that the explants are disease free.

Stage I

This stage usually takes place in a lamina flow hood. Most times, explants fail to establish due to contamination, so aseptic conditions in this step are crucial. Before the explant is inoculated into the medium, it should be surface sterilized. A common surface sterilization method involves washing the explant in separate sets of sterile distilled water (3-4 times), rinsing in 70% laboratory grade ethanol and then treatment with chlorine based chemicals such as calcium hypochlorite or sodium hypochlorite based solutions (1-3% (v/v) and then washed again in sterile distilled water. After, they should be blotted in pre-sterilized blotting paper. Materials such as forceps and all other materials used for handling the explant must be pre- sterilized via heat treatment in an autoclave. Mercuric chloride is another common sterilant used, but this should be avoided since it is highly toxic to humans.

Stage II

Stage II is generally considered the propagation stage where the explants are cultured on the appropriate media (chemicals) for shoot induction, shoot multiplication, maintenance and regeneration of roots from multiplied shoots in dark conditions. The main goal is to achieve propagation without losing the genetic stability. The Murashige and Skoog (1962) medium, supplemented with different phytohormones, are commonly used. The containment vessels include test tubes, baby jar flaks or magenta cubes which must be pre- sterilized by autoclaving at 1.06 kg/cm^2 pressure at 121°C for 15 minutes.

Stage III

At this phase, the *in vitro* shoots obtained at Stage II are rooted to produce complete plants. It is important to note that, if multiplied material consists of bud-like structures (e.g. from orchids) or clumps of shoots (e.g. from banana or pineapple), they should be separated after rooting and not before. Many plants can be rooted on the Murashige and Skoog (1962) medium supplemented with a root growth regulator such as auxin. This stage can be considered expensive and has been estimated to account for approximately 35–75% of the total cost of production (Debergh & Read, 1991).

Stage IV

This is the final stage of the plant tissue culture process in which the *in vitro* micropropagated plants are weaned and hardened under greenhouse conditions. The hardening is a gradual process and is achieved using high to low humidity and from low to high light intensity conditions. The plants are usually placed in shade for about 3-6 days, where diffused natural light conditions them to the new environment. Following this, the plants are placed into an appropriate substrate such as sand, peat moss or compost. It should be noted that the transportation of these small micropropagated plants to growers requires special care to avoid plant losses.

During the last thirty years, tissue culture-based plant propagation has emerged as one of the leading global agro-technologies. Between 1986 and 1993, the worldwide production of tissue cultured plants increased by 50%, whereby in 1993, the production was approximately 663 million plants (Fujardo, 2010). By 1997, production had risen to 800 million plants. Since 1995, production has increased by 14% in Asian countries, mainly due to the market entry of China, while the increase in South and Central America was from production in Cuba. More recently, companies from Israel, the USA and UK have shifted their production requirements to emerging economies such as Costa Rica and India where the primary advantage of micropropagation is for the rapid production of high quality, disease-free and uniform planting material. The application of micropropagation through plant tissue culture has revolutionized crop production and many success stories have been reported. Costa Rica was able to produce 600,000 plantlets per year from only two plant tissue culture laboratories (Jaffe & Rojas, 1994) and increased its banana production from 28000ha to 51000ha, thus making it the second largest exporter of banana (Arias, 1999). Also, Cuba has produced over 110 million *in vitro* plants with banana and plantain comprising almost 82%, sugar cane 12% and potato 4% with other crops such as pineapple 2% (Sasson, 2000). Micropropagation techniques exist for several economically important crops such as sugar cane, bananas, pineapple, potato, papaya, citrus, dasheen, cassava (Gomez Kosky et al., 2002; Lorenzo et al., 2001). While many successful studies have been reported, this chapter will focus on potatoes and bananas as these have been well documented and appropriate crops to developing countries' food security needs.

Potato (*Solanum tuberosum* L.) is a major world food crop next in production to only maize, rice and wheat. Potatoes are consumed by over one billion people world over, half of whom live in developing economies. This crop is considered semi-perishable, being susceptible to many diseases and insect pests. A scarcity of good quality seed was recognized as the single most important factor limiting potato production in emerging economies. The availability of plant tissue culture technology, for rapid multiplication of disease-free planting material has facilitated potato seed production to a great extent (Dodds, 1988). For example, 36,000 microtubers were produced from 1,200 culture flasks in a four month period which produced 1,800t potato seed, enough to cultivate 2000 ha of land (Wang & Hu, 1982). Additionally, the possibility of producing 264,500 basic seed tubers from one *in vitro* plant in India has been reported (Naik, 2005). Chung (1989) revealed an increase in the national potato yield in Korea from 11.9 t/ha in

1980 to 20.3 t/ha in 1986 which was due to use of virus-free planting material produced through tissue culture. Such success led to the establishment of an *in vitro* tuberization system in that country (Choi & Kang, 1994). Due to the high success rate of micropropagation, this technique has now become an integral part of potato production in many emerging economies (Donnelly, Coleman & Coleman, 2003).

Banana (*Musa* spp), is another important crop, widely grown in developing countries and is considered the second largest fruit crop in the world after citrus (Madhulatha, Anbalagan, Jayachandaran, & Sakthivel, 2004). Banana cultivars are vegetatively propagated through corms, suckers and sword suckers which usually contain pests (fungi, bacteria, nematodes, viruses, and insects) and diseases and have threatened the banana industry in many countries. Additionally, the rate of multiplication through conventional means was reported to be relatively slow and diseases such as bunchy top virus and banana streak virus were transmitted to the next generation and yield became very poor in many countries. A number of studies has reported on micropropagation of *Musa* spp for mass production and disease free plants with great success (Cronauer & Krikorian, 1984; Kodym & Zapata-Arias, 1999; Muhammed, Hussain, Saqlan, Hamid & Hamid, 2004; Nandwani, Zehr, & Barwale, 2000; Wong, 1986).

During the 1990s, banana bunchy top disease in banana fields of Pakistan (Abdullah, Khan, Siddiqui, 1997), caused a production decline up to 90% in some of the affected areas. Due to heavy loss in production, efforts were made to obtain disease free planting material for re-cultivation. One approach was to import healthy germplasm from other countries, but this approach was unsuccessful as imported germplasm could not adapt well to local soil and environmental conditions. Hence, this led to another option which involved a process to make the existing germplasm disease free and then multiply them at a high rate. Muhammed and colleagues (Muhammed et al., 2004) were able to produce 124 plants from a single shoot tip with great success. Additionally, other researchers (Nandwani et al., 2000) revealed that more than 25,000 banana plantlets, var. "Basrai" were produced through tissue culture and gave 90% survival rate when transferred to the field.

Another success story was with, "The Grand Naine Banana," a high yielding cavendish variety introduced to India from Israel. However, the rate of multiplication through conventional method has been considered to be relatively slow and the variety was also affected by a range of diseases (bunchy top virus, banana streak virus) which resulted in poor yields and thus affected the economy of that country. With the use of *in vitro* mass propagation, survival rate of the plantlets were found to be 84.44% during primary hardening and 100% survival rate when subjected to secondary hardening (Shankar, Balaji & Sathish, 2014) which satisfied the increased demand for disease free and healthy planting materials of banana for commercialization in that country.

Even though there have been many successful reports of *in vitro* micropropagation of crops, there have been some reports questioning the cost effectiveness of this technology; it was considered to be more expensive than the conventional methods of plant propagation. By some, it was also considered a capital-intensive industry, and in some cases the unit cost per plant became unaffordable. The major reasons were the cost of production and the cost of skills acquisition associated with the technique. Furthermore, there were difficulties in selling tissue culture products because the conventional planting material was much cheaper. However, these problems have now been addressed by developing reliable and cost effective tissue culture methods without compromising quality. In Asia and Latin America, where labour is relatively cheaper, consumables such as media, culture containers, and electricity make a comparatively greater contribution to production costs.

MOLECULAR MARKERS AND MARKER-ASSISTED BREEDING TO IMPROVE EFFICIENCY OF PLANT BREEDING

Developments in plant biotechnology have been applied to further increase the scientific base and its application to the plant breeding process (Baenziger, Russel, & Graef, 2006; Jauhar, 2006; Varshney, Hoisington, & Tyagi, 2006.The plant biotechnology era started in the early 1980s with the milestone reports of producing transgenic plants using *Agrobacterium* (Bevan, Flavell, & Chilton, 1983; Fraley et al; 1983). Molecular marker systems for crop plants were developed soon thereafter as a powerful tool to improve the efficiency of plant breeding by detecting important crop traits of interest which could not have been accomplished *via* conventional plant breeding methods (Dekkers & Hospital, 2002; Edwards, Stuber, & Wendel, 1987; Ribaut & Ragot, 2007). Markers used in plant breeding can be classified into two main groups; classical markers and DNA markers. Focus will be placed on DNA markers in this discussion.

Molecular markers are identifiable DNA sequences, found at specific locations of the genome, and which are transmitted by inheritance from one generation to the next (Miller & Martin, 1988). There are various types of molecular markers but those that have been extensively used and have generated promising results from the application to plant breeding are; amplified fragment length polymorphism (AFLP), restriction fragment length polymorphism (RFLP), random amplified polymorphic DNA (RAPD), microsatellites or simple sequence repeat (SSR), and single nucleotide polymorphism (SNP). These marker systems are often linked to genes and present in the simplest form for genetic variation within a population which is known as polymorphism (Ribaut & Ragot, 2007). These become potential genetic markers in plant breeding which then provide the framework for the application of Marker Assisted Selection (MAS). MAS refers to a breeding procedure in which DNA marker detection and selection are integrated into a conventional breeding program. The most extensively used marker- assisted selection is referred to as Marker-assisted or Marker-assisted backcrossing (MABC). MABC has been successfully applied to various types of traits e.g. disease/pest resistance, drought tolerance and quality in many tropical crops such as rice, maize and tomato.

The green revolution saved millions of people with additional food, but it could not prevent hunger and poverty because of certain limitations, such as limitations in conventional breeding due to the lack of resistance genes in crops such as cultivated rice (*Oryza sativa* L) and maize and inadequate understanding of phenotypic (physical expression) variability available to take full advantage of the green revolution. However, the application of biotechnology through the use of molecular markers provided promising results, delivering unique opportunities to overcome these problems. Such results have already been achieved in the development of crop varieties with complex traits and have thus contributed to more sustainable supplies of food and hence could be a major tool in assisting to reduce crop failures and in the fight against hunger and poverty, especially in developing economies. Some application of MAS and success stories are now presented.

It is estimated that the demand for maize in developing countries will grow by 50%, from 558 million tons in 1995 to 837 million tons in 2020, surpassing the demand for wheat and rice Fujardo, 2010). Much of this increased demand will have to be accommodated by increasing developing countries' domestic supply, which will require intensifying production on existing agricultural land (Institute, 2000). Since the majority of tropical maize is grown in the rainy season, the possibilities for alleviation of water stress are limited; as such, it is crucial for farmers to be provided with well-characterized cultivars that can withstand greater levels of drought. Through the marker-assisted approach, MABC, (Ribaut & Rogot

(2007) successfully improved tropical maize grain yield under drought conditions. This was achieved by crossing a drought tolerant line Ac7643, the donor parent with the drought susceptible line CML247, the recurrent parent. This study also demonstrated that mean grain yield of the MABC-derived hybrids were consistently higher than that of control hybrids under severe water stress conditions.

Other successful experiments reported the manipulation of known genes with indirect (linked) markers, including pyramiding of several major resistance genes in rice (Hittalmani, Parco, Mew, Zeigler, & Huang, 2000; Huang et al., 1997; Jiang et al., 2004; Sanchez, Brar, Huang, Li, & Khush, 2000; Sharma, Torii, Takumi, Mori, & Nakamura, 2004; Singh et al., 2001), wheat (XiangYan, ZengYan, LiPu, ZhiYong, & Xiao, 2005), wheat (XiangYan, ZengYan, LiPu, ZhiYong, & Xiao, 2005) and tomato (Barone, Ercolano, Langella, Monti, & Frusciante, 2005; Yang & Francis, 2005) and tomato (Barone, Ercolano, Langella, Monti, & Frusciante, 2005; Yang & Francis, 2005). While rice (*Oryza sativa* L.) is considered to be the most important food crop of the developing world as a staple food, farmers in developing economies do not obtain a significant portion (20-70%) of the potential rice yield because of losses caused by pests and diseases (Datta, 2004). Limitations in conventional breeding arise because of the lack of resistance genes in cultivated rice germplasm and inadequate understanding of phenotypic (physical expression) variability. A major disease of rice in developing countries is bacterial blight caused by *Xanthomonas oryzae* pv. oryzae (Xoo). Researchers have pyramided (stacked genes) three bacterial blight resistance genes (*xa5*, *xa13* and *Xa21*), into cv. PR106, using marker-assisted selection (Singh et al., 2001). Lines of PR106 with pyramided genes were evaluated after inoculation with 17 isolates of the pathogen and genes in combinations were found to provide high levels of resistance to the predominant *Xoo* isolates. Also, some with two and three bacterial blight resistance genes were also evaluated and this combination of genes provided a wider spectrum of resistance to the pathogen populations from the Punjab and the Philippine races of Xoo in PR106 lines.

Tomato, *Solanum lycopersicum* L., is unquestionably one of the most consumed vegetable crop in the world. Most of the commercial cultivars of tomato have been developed through traditional breeding. Some traits such as sweetness and sourness can be related to the sugar and acid content of the fruit respectively (Jones & Scott, 1983), while others such as aroma and texture characteristics are not easily determined by instrumental measures (Causse et al., 2002). Additionally, most of the quality traits exhibit a polygenic inheritance (the inheritance of many genes) and are strongly influenced by environmental conditions. However, with the advent of molecular markers and marker-assisted selection (MAS) technology, tomato genetics and breeding research have entered into an innovative and exciting era. This technology was also utilized for marker-assisted breeding in tomato for several important traits, including sensory quality attributes. For example, in a study, (Causse et al., 2002) identified chromosome regions that gave promising results for improving tomato fruit quality (descriptive profiles of flavor, aroma and texture).

The root crop cassava (*Manihot esculenta*), is grown in over 90 countries and provides food for over 500 million people in emerging economies (Fujardo, 2010). This crop was considered to have a long growth cycle and detailed knowledge about its complex genetic structure was lacking. Through the use of molecular markers, scientists in Colombia were able to develop a molecular genetic map of cassava, the first to be done for a major food crop outside the developed world. This genetic map was able to identify important traits such as increased dry matter content, resistance to cassava blight and reduced postharvest loss (Cockcroft, Herrera-Estrella, & Borroto Nordelo, 2004).

Molecular markers and MAS have impacted the plant breeding paradigm positively. This technology delivered solutions to crops where conventional breeding approaches could not and thus could be considered as a 'powerful' tool for plant breeding. But while the future possibilities and future potential impact of this technology seem promising, there are still some impediments to its use. The availability of sufficient funds have been reported as the key factor for determining the application of molecular markers in crop improvement programmes (Dekkers & Hospital, 2002), followed by poorly trained workers and lack of suitable equipment. The use of markers in selection incurs costs such as for DNA extraction, development of genetic maps, genotyping and sequencing analyses. It is envisaged that through partnerships between emerging and developed economies, public-private sector collaboration and with assistance from international organizations costs can be reduced and capacities can be developed.

BIO-PRODUCTS (BIO-PESTICIDES AND BIO-FERTILIZERS)

Bio-products are utilized as bio-pesticides and bio-fertilizers. In developing countries, notwithstanding the abundance of natural resources, rural poverty and food insecurity have persistently affected more than 55% of the rural population. Losses caused by pests and diseases pose a serious challenge to food security and one way of improving food availability is by management of pests. While synthetic pesticides and fertilizers have helped increase crop yields by nearly 70% in some regions (Pretty, 2008), their use is becoming a significant concern because of issues related to food safety, human health and the environment (Pretty, 2008; Tilman, 1999). Additionally, the excessive and imprudent use of pesticides can result in poor crop growth through pest resurgence, secondary pest problems or the development of heritable resistance. Chemical fertilizers have been used to increase crop productivity. But its excessive use has also resulted in negative effects such as eutrophication, high level of water contamination and high amounts of nitrogen oxide. While there are many definitions for bio-pesticide, it can be defined simply as a mass-produced agent manufactured from a living microorganism or a natural product for the control of plant pests (Chandler et al., 2011). Generally, there are three major groups of bio-pesticides; microbial pesticides, bio-chemicals and semio-chemicals.

Microbial pesticides contain microorganisms (bacteria, fungi, protozoa or virus) as the active agent. The insect pathogenic bacterium, *Bacillus thuringiensis* (Bt), is the most extensively used microbial bio-pesticide. Insecticidal Bt products were first commercialized in France in the late 1930s (Lambert & Peferoen, 1992). Bt has been one of the most consistent and significant bio-pesticides used on crops as an insecticidal spray. It produces a protein crystal (the Bt d-endotoxin) during bacterial spore formation that is capable of causing lysis of the gut cells when consumed by susceptible insects (Carlton & Gonzalez, 1985) causing death in 48 hours (Carozzi, Kramer, Warren, Evola, & Koziel, 1991). Bt sprays have successfully been applied to crops such as maize and cabbage.

Bio-pesticide products based on entomopathogenic fungi have also been developed for use against pests in fruits and field vegetables in both small and large areas of land, most from Central and South America (Choi et al., 2004).The majority of these products are based on the ascomycetes *Beauveria bassiana* or *Metarhizium anisopliae*. Commercial bio-pesticides based on M. *anisopliae* have been successfully used against spittlebugs on approximately 750, 000 ha of sugarcane and 250, 000 ha of grassland on an annual basis (Choi et al., 2005). The bio-pesticide has also been successful for the control of locust and grasshopper pests in Africa and Australia (Choma & Kaplan, 1992) and for locust management (Christou, Capell, Kohli, Gatehouse, & Gatehouse, 2006). In Indian agriculture, bio-pesticides have also been used extensively with great success. For example, the Diamond back moth was controlled by *Bacillus thuringiensis*. Mango hoppers and mealy bugs were controlled by *Beauveria* and rots and wilts in various crops by *Trichoderma*-based products (Gupta & Dikshit, 2010).

A semio-chemical can be defined as a chemical signal produced by one organism that causes a behavioural change in an organism of the same or a different species. Insect sex pheromones are the most extensively used semio-chemical for pest management by monitoring or pest control by mass trapping (Beegle & Yamamoto, 1992), lure-and-kill systems (Gonzalez, Brown, & Carlton, 1982) and mating disruption. Mating disruption has been shown to be successful in orchard crops (Gonzalez & Carlton, 1980 . Bio-pesticides thus have a range of properties that make them good components of Integrated Pest Management (IPM) programmes. Most are selective, produce little or no toxic residue, and development costs are significantly lower than those of conventional synthetic chemical pesticides (Beegle & Yamamoto, 1992).

Biochemical pesticides are naturally occurring substances that control pests. These include substances that interfere with growth or mating, such as plant growth regulators (Gupta & Dikshit, 2010). A commonly used biochemical pesticide is 'azadirachtin' which is extracted from the neem tree (*Azadirachta indica*). This botanical compound has been shown to affect the reproductive and digestive process of a number of important pests. Effective formulations of neem extract have been developed in India and are now commercially produced by other countries. Additionally, extracts from other plants such as papaya, ginger and tobacco have been successfully used to control pests in India (Dutta, 2015). Also, certain microorganisms produce bioactive compounds which can be utilized as a biochemical pesticide. One of the most popular is *Bacillus thuringiensis* which produces specific proteins which are toxic to lepidopterous pests (Gupta & Dikshit, 2010).

Biofertilizers are products which contain specific groups of microorganisms and when applied to seeds, soil or plant surfaces, promote plant growth by increasing the supply or availability of nutrients (Smith, 1992). Microorganisms used belong to a wide array of bacteria and fungi. Some main microorganisms which are important to plant nutrition and are utilized as biofertilizers are: nitrogen fixing bacteria (*Azobacteria* species, *Azospirillum*, *Acetobacter diazotrophicus*, species of *Rhizobium*, *Bradyrhizobium*) and arbuscular mycorrhizal fungi.

Once a suitable strain of bacteria or fungus is selected for use as a biofertilizer, a standardized process of production is applied. For example, the selected microorganism is used to inoculate cheap growth media such as compost or agro-industrial residues enriched with rock phosphate, which then goes through a fermentation process. Another important aspect in the production of biofertilizers is the use of a carrier which helps to deliver an appropriate amount of the biofertilizer in a good state. Carriers can be produced from organic, inorganic, or synthesized from specific molecules such as beads, powder, mineral oils and broth cultures.

The use of biofertilizers in agriculture is compatible with a more sustainable food production system and represents an environmentally acceptable practice. Certain species of *Azobacteria* have the ability to fix atmospheric nitrogen by converting it to ammonia while *Azospirillum* fixes nitrogen compounds through root nodules of gramineous and non-gramineous plants. Brazil is one of the largest producers of sugar cane with an average productivity of 65 tons per hectare with approximately 4 million hectares of sugar cane. The inoculation of nitrogen fixing bacteria, *Rhizobia*, in two suitable varieties of sugar cane has led to low or zero application of nitrogenous fertilizers (Reis, Baldani, Baldani, & Dobereinera, 2000).

The uses of bioproducts offer many benefits to developing economies, but there are still some challenges faced such as selection of suitable microorganisms, development of appropriate formulations and cost of production.

PLANT DISEASE DIAGNOSIS THROUGH MOLECULAR DIAGNOSTICS

Plant diseases are a threat to agriculture worldwide and especially for food security in developing countries. With the increase in human population, the responsibility of protecting food crops from the devastating effects of pests and disease is increasing. It has been noted that approximately 42% of the world's total agricultural crop is destroyed yearly by diseases and pests (Chandrashekara, Raju, & Chandrashekara, 2012). In India, an emerging economy, the crop losses due to pests and diseases have been assessed to be approximately 10-30% of crop production (Kumar, 2014). A number of diseases such as papaya ring spot (papaya ring spot virus), *phytophthora* diseases (*phytophthora parasitica* and *phytophthora palmivora* etc.), sheath blight (of rice/maize), bract mosaic (of banana) and downy mildews (in maize) has threatened agricultural crops (Kumar, 2014), while neck blast and brown plant hopper have caused almost 40% losses in rice in Haryana, India (Kumar, 2014). Also, about 70% of the cocoa produced in Nigeria was lost due to the devastating black pod disease in 1995 (Kutama, Emechebe, & Aliyu, 2011a). Quite often, farmers struggle to find quick and efficient solutions for pest and disease control. If plant diseases and pests are correctly identified and diagnosed in a timely manner, crop losses can be reduced which in turn can contribute to agricultural sustainability and food security. The traditional method of identification of plant pathogens has been through visual means and other methods such as plating, isolation and staining techniques. In most instances, this was only possible after some damage has already been done to the crops; hence treatment was of limited use. While plating, isolation and staining techniques and identification *via* morphology have provided some insights into disease diagnosis, it is a relatively slow process, only a few pathogens are cultivable; meaning that it is not always sensitive and specific and thus accurate diagnosis could be affected. Furthermore, with many bacterial diseases, symptoms do not generally provide a reliable guide to their cause, nor does colony morphology (shape and structure).

Over the last decade, several molecular techniques have been developed and provided several advantages to the conventional methods. These offer a new and exciting future for plant pathogen diagnosis and is possible through DNA based technology such as the polymerase chain reaction (PCR). This includes: Nested PCR, Multiplex PCR, Multiplex nested PCR, Random Amplified Polymorphic DNA (RAPD) and Real Time PCR (RT-PCR). These techniques enable regions of the pathogen's genome to be amplified by several million folds, thus increasing the sensitivity of pathogen detection (Kumar, 2014), improving accuracy in disease diagnosis and thus provide rapid results. Furthermore, many microbial species are closely related and often differ in only a single (single-nucleotide polymorphism [SNP]) to a few bases in such genes. As such, the high degree of specificity of the nucleic acid-based detection techniques, allows detecting such SNPs (Consolandi et al., 2001; Livak, 1999; Papp, Pinsonneault, Cooke, & Sadee, 2003) and this is important because closely related pathogens might have a different host range or exhibit a completely different pathogenicity. Additionally, it is also possible to quantify the amount of pathogens in a sample (through quantification of gene copy numbers) using RT-PCR. While pathogen detection is important, quantification of pathogens is also crucial since it provides valuable foundation for disease management resolutions. Some successes resulting from the use of molecular diagnostics in plant pathology are now presented.

A series of highly sensitive tests utilizing PCR was developed for improvement in identification of a number of plant pathogens in Africa and Asia. For example, the PCR test was used to identify *Ralstonia solanacearum* bacterium (an important pathogen affecting potatoes) in plant material and was found to be particularly useful for screening seed potato tubers for latent infections at export/import borders. It also caused an indirect increase in productivity for farmers who were able to use clean planting material as a result (Mwangi, Nyende, Demo, & Matiru, 2008). Also, *Clavibacter michiganensis* ssp. is a serious pathogen that affects tomato and through PCR this pathogen as well as other species, were identified accurately which enhanced research and extension (Black, Seal, Abubakar, Nono-Womdim, & Swai, 2001). Cassava is a dietary staple and a significant cash crop throughout the tropical world and must maintain its position as a food security crop. The cassava Frog Skin disease is an important disease in Colombia, Venezuela, Costa Rica, Peru and Brazil which caused up to as much as 90% in yield loss. For some time, attempts made to determine the causal agent of this devastating disease were futile. A molecular technique, Amplified Fragment Length Polymorphism (AFLP) was able to identify the causal agent and gave promising results such that, cDNA clones isolated from diseased cassava plants were shown to be similar to the rice Ragged Stunt virus and this allowed for the implementation of a similar management programme.

PCR methods were also useful in disease management such as Sigatoka leaf spot pathogens of banana (*Mycosphaerella* spp.) in Uganda and Kenya. It was believed that spread of the Black Sigatoka (*M. fijiensis- Mf*) was more rapid than distributions of Yellow leaf spot (*Mycosphaerella musicola-Mm*). However, PCR tests revealed the opposite and led to more accurate distribution maps which were beneficial to plant breeders and extension services concerned with management of *Sigatoka* leaf spots.

Plant diseases cause numerous direct and indirect losses such as a reduction in quantity of crops produced, economic loss, threat to the environment and loss of natural resources. Applications of suitable plant biotechnology methods to identify pests and diagnose disease causing organisms based on molecular procedures have already demonstrated great success. Hence, there is no doubt that the application of molecular diagnostics can help alleviate these losses. While molecular laboratories have been developed in some emerging economies, they still lag behind developed countries. There is a need for more funding, substantial growth in human training, technological and infrastructure capacity and enhancement of delivery of diagnostic services to garner the full potential of benefits from molecular diagnostic techniques in plant pathology. Governments have a crucial role to play in supporting diagnostic capacity and sufficient subsidies to support diagnostic capability, but most importantly, molecular diagnostic systems and networks in emerging economies must be sustained.

SUMMARY AND CONCLUSION

The benefits from the application of biotechnology to food security and sustainability are undoubtedly tremendous. The production of *in vitro* culture crops has provided profound implications for developing economies such as, rapid propagation process coupled with the production of disease free plants which can be multiplied under controlled environment on a year round basis. This in turn, can cause an increase in crop productivity, leading to higher incomes for producers and lower prices for consumers. Molecular markers and MAS have provided promising results in the development of crop varieties with new and improved traits (resistance to pests and diseases, resistant to drought or floods and improvement of sensory quality attributes) and hence can enhance food security and increase agricultural sustainability. The uses of bioproducts (biofertilizers and biopesticides) have shown significant benefits to agriculture and the environment by offering an environmentally safe alternative to the use of synthetic pesticides and fertilizers. Plant diseases and pests cause numerous losses to farmers and the application of suitable molecular procedures can certainly provide quick and accurate diagnosis thus leading to better management programmes resulting in a reduction in the loss of crops to farmers.

According to the Nobel Peace Laureate, Norman Bourlag, biotechnology is necessary to boost crop yields to feed the world (Bourlag, 1997).The benefits of the green revolution have already been optimized and hence, such technologies can no longer be the basis for sustained food production and food security in developing economies. Scientific results gained from the application of plant biotechnology to pest and disease diagnosis, plant breeding (through molecular markers and MAS), plant mass production and production of bio-products substantiate the use of biotechnology as a suitable option to meet new challenges for the advancement, sustainability and security of nations' food supply in developing economies. The future potential and impact of biotechnology seem promising for agriculture development and food security, but developing countries still face some major challenges which must be addressed in order for the full potential of biotechnology to be met.

SUGGESTIONS FOR WAY FORWARD

There is still an urgent need for robust and organized outreach services, suitable bodies and infrastructures to facilitate the increased application of biotechnology in developing economies. Governments, research institutions and funding agencies all have major roles to play in committing to sustained support. Graduate programmes from academic institutions must continue to train students in these plant biotechnology applications to equip them to develop, adapt and to help sustain and improve these technologies. The sustainability and improvement of these technologies may incur costs which may be considered to be high but this can be offset by benefits, such as reduced crop losses, mass production of disease free planting material and more ecologically-friendly crop-management practices.

Molecular laboratories have been developed in some emerging and developed economies, but they still lag behind in developing countries. There is a need for more funding, substantial growth in human training, technological and infrastructure capacity and enhancement of delivery of diagnostic services to garner the full potential benefits from molecular diagnostic techniques in plant pathology. Governments have a crucial role to play in supporting diagnostic capacity and providing sufficient subsidies to support diagnostic capability, but most importantly, molecular diagnostic systems and networks in developing economies must be sustained. Also, it is still necessary to conduct some training in the areas of microbial isolation, maintenance and preservation, better formulations of bio-pesticides and its economic utilization.

Additionally, established networks of plant tissue culture techniques, molecular markers and marker assisted selection, the production and use of bio-pesticides and molecular diagnosis of pests and diseases are of great importance. Also, products are tailored largely to the needs of large-scale farmers and industrial processing in the developed world and not sufficiently targeted to small scale farmers in developing countries. It is envisaged that through partnerships among developing, emerging and developed economies, public-private sector collaboration and assistance from international organizations, costs can be reduced, and capacities can be developed.

REFERENCES

Abdullah, K., Khan, I. A., Siddiqui, S. H., Ahmed, M., & Siddiqui, K. A. (1997). *In vitro* culture of indigenous and exotic banana clones for maximum multiplication. *Pakistan Journal of Botany*, *29*(1), 143–150.

Arias, O. (1999). *Addressing market needs through biotechnology: The example of Agribiotecnolog'ıa de Costa Rica.* Paper presented at the Agricultural Biotechnology for Latin America Part 4. Turning Priorities into Feasible Programs.

Baenziger, P. S., Russell, W. K., Graef, G. L., & Campbell, B. T. (2006). Improving lives: 50 years of crop breeding, genetics, and cytology (C-1). *Crop Science, 46*(5), 2230–2244. doi:10.2135/cropsci2005.11.0404gas

Barone, A., Ercolano, M. R., & Langella, R., Monti, & Frusciante, L. (2005). Molecular marker-assisted selection for pyramiding resistance genes in tomato. *Advanced Horticultural Science, 363*(19), 147–152.

Beegle, C. C., & Yamamoto, T. (1992). History of *Bacillus thuringiensis* Berliner research and development. *Canadian Entomologist, 124*(4), 587–616. doi:10.4039/Ent124587-4

Bevan, M. W., Flavell, R. B., & Chilton, M. D. (1983). A chimaeric antibiotic resistance gene as a selectable marker for plant cell transformation. *Nature, 304*(5922), 184–187. doi:10.1038/304184a0

Black, R., Seal, S., Abubakar, Z., Nono-Womdim, R., & Swai, I. (2001). Bacterial spot (*Xanthomonas campestris* pv. *vesicatoria*) of tomato and sweet pepper in Tanzania. *Plant Pathology, 50*(6), 20. doi:10.1046/j.1365-3059.2001.00633.x

Borlaug, N. E. (1997). Feeding a world of 10 billion people: The miracle ahead. *Biotechnology, Biotechnological Equipment, 11*(3-4), 3–13. doi:10.1080/13102818.1997.10818934

Botella-Rodriguez, E. (2011). Cuba's inward-looking development policies: Towards sustainable agriculture (1990-2008). *Historia Agraria, 55*(5), 135–171.

Brethour, C., & Weersink, A. (2001). An economic evaluation of the environmental benefits from pesticide reduction. *Agricultural Economics, 25*(2-3), 219–226. doi:10.1111/j.1574-0862.2001.tb00202.x

Cardeña, R., Oropeza, C., & Zizumbo, D. (1998). Leaf proteins as markers useful in the genetic improvement of coconut palms. *Euphytica, 102*(1), 81–86. doi:10.1023/A:1018392908569

Carlton, B. C., & Gonzalez, J. M. (1985). Plasmids and δ- endotoxin production in different subspecies of *Bacillus thuringiensis*. In J. A. Hoch & P. Setlow (Eds.), *Molecular Biology of Microbial Differentiation* (pp. 246–252). Washington, DC: American Society for Microbiology.

Carozzi, N. B., Kramer, V. C., Warren, G. W., Evola, S., & Koziel, M. G. (1991). Prediction of insecticidal activity of *Bacillus thuringiensis* strains by polymerase chain reaction product profiles. *Applied and Environmental Microbiology, 57*(11), 3057–3061. PMID:1781673

Causse, M., Saliba-Colombani, V., Lecomte, L., Duffé, P., Rousselle, P., & Buret, M. (2002). QTL analysis of fruit quality in fresh market tomato: A few chromosome regions control the variation of sensory and instrumental traits. *Journal of Experimental Botany, 53*(377), 2089–2098. doi:10.1093/jxb/erf058 PMID:12324532

Chandler, D., Bailey, A. S., Tatchell, G. M., Davidson, G., Greaves, J., & Grant, W. P. (2011). The development, regulation and use of biopesticides for integrated pest management. *Philosophical Transactions of the Royal Society of London. Series B, Biological Sciences, 366*(1573), 1987–1998. doi:10.1098/rstb.2010.0390 PMID:21624919

Chandrashekara, K. N., Raju, S. R., & Chandrashekara, C. (2012). Molecular and Biotechnological Approaches in Plant Disease Management. In Ecofriendly Innovative Approaches in Plant Disease Management. International Book Distributors.

Chatenet, M., Delage, C., Ripolles, M., Irey, M., Lockhart, B. L. E., & Rott, P. (2001). Detection of sugarcane yellow leaf virus in quarantine and production of virus-free sugarcane by apical meristem culture. *Plant Disease*, *85*(11), 1177–1180. doi:10.1094/PDIS.2001.85.11.1177

Choi, J. Y., Roh, J. Y., Kang, J. N., Shim, H. J., Woo, S. D., Jin, B. R., Je, Y. H. (2005). Genomic segments cloning and analysis of *Cotesia plutellae* polydnavirus using plasmid capture system. *Biochemical Biophysics Research Communications, 332*, 487-493.

Choi, J. Y., Roh, J. Y., Li, M. S., Shim, H. J., Kang, J. N., Woo, S. D., & Je, Y. H. et al. (2004). Cloning of small plasmids from *Bacillus thuringiensis* subsp. israelensis using plasmid capture system. *International Journal of Industrial Entomology*, *9*(2), 183–186.

Choi, Y. W., Cho, J. L., & Kang, S. M. (1994). Studies on rapid multiplication of microtubers from potatoes (*Solanum tuberosum* L.) *in vitro* and their practical use. IV. Effect of coating and foliar treatments on field performance. *Korean Journal of Horticultural Science*, *35*(2), 323–329.

Choma, C. T., & Kaplan, H. (1992). *Bacillus thuringiensis* crystal protein: Effect of chemical modification of the cysteine and lysine residues. *Journal of Invertebrate Pathology*, *59*(1), 75–80. doi:10.1016/0022-2011(92)90114-J PMID:1541850

Christou, P. T., Capell, A., Kohli, A., Gatehouse, J. A., & Gatehouse, A. M. (2006). Recent developments and future prospects in insect pest control in transgenic crops. *Trends in Plant Science*, *11*(6), 302–308. doi:10.1016/j.tplants.2006.04.001 PMID:16690346

Chung, T. Y. (1989). *Biotechnology in the Republic of Korea*. Paper presented at the FAO Regional Expert Consultation on Biotechnology, Bangkok, Thailand.

Cockcroft, C. E., Herrera-Estrella, L., & Borroto Nordelo, C. G. (2004). *Agricultural Biotechnology in Latin America and the Caribbean. Handbook of Plant Biotechnology* (Vol. 11, pp. 1–10). John Wiley & Sons.

Consolandi, C., Castiglioni, B., Bordoni, R., Busti, E., Battaglia, C., & De Bellis, G. (2001). Development of oligonucleotide arrays to detect mutations and polymorphisms. *Minerva Biotechnology*, *13*(3), 261–268.

Cooper, H. D., Spillane, C., Hodgkin, T., & Cooper, H. (2001). *Broadening the genetic base of crops: an overview. Broadening the genetic base of crop production*. Wallingford, UK: CABI Publishing. doi:10.1079/9780851994116.0001

Cronauer, S. S., & Krikorian, A. D. (1984). Rapid multiplication of bananas and plantains by *in vitro* shoot tip culture. *HortScience*, *19*(2), 353–380.

Datta, K. S. (2004). Rice Biotechnologies: A need for developing countries. *AgBioForum*, *7*(1), 31–35.

Debergh, P. C., & Read, P. E. (1991). *Micropropagation*. Springer Netherlands. doi:10.1007/978-94-009-2075-0_1

Dekkers, J. C. M., & Hospital, F. (2002). The use of molecular genetic in the improvement of agricultural populations. *Nature Reviews. Genetics*, *3*(1), 22–32. doi:10.1038/nrg701 PMID:11823788

Dodds, J. H. (1988). Tissue culture technology: Practical application of sophisticated methods. *American Potato Journal*, *65*(4), 167–180. doi:10.1007/BF02854450

Donnelly, D. J., Coleman, W. K., & Coleman, S. E. (2003). Potato microtuber production and performance: A review. *American Journal of Potato Research*, *80*(2), 103–115. doi:10.1007/BF02870209

Dutta, S. (2015). Biopesticides: An ecofriendly approach for pest control. *World Journal of Pharmacy and Pharmaceutical Sciences*, *4*(6), 250–265.

Edwards, M. D., Stuber, C. W., & Wendel, J. F. (1987). Molecular marker-facilitated investigations of quantitative-trait loci in maize. I. Numbers, genomic distribution and types of gene action. *Genetics*, *116*(1), 113–125. PMID:3596228

Fagwalawa, L. D., Kutama, A. S., & Yakasai, M. T. (2013). Current Issues in Plant Disease Control: Biotechnology and Plant Disease. *Bayero Journal of Pure and Applied Sciences*, *6*(2), 121–126. doi:10.4314/bajopas.v6i2.26

Fraley, R. T., Rogers, S. G., Horsch, R. B., Sanders, P. R., Flick, J. S., Adams, S. P., & Fry, J. S. (1983). Expression of bacterial genes in plant cells. *Proceedings of the National Academy of Sciences of the United States of America*, *80*(15), 4803–4807. doi:10.1073/pnas.80.15.4803 PMID:6308651

Fujardo, J. (2010). Quality Declared Planting Material: Protocols and Standards for Vegetatively Propagated Crops.[). Rome]. *FAO Plant Production and Protection Paper*, *195*, 156.

Garcia-Gonzales, R., Quiroz, K., Carrasco, B., & Caligari, P. (2010). Plant tissue culture: Current status, opportunities and challenges. *Cien. Inv. Agr.*, *37*(3), 5–30. doi:10.4067/S0718-16202010000300001

Gómez Kosky, R., Manuel de Feria, S., Posada Pérez, L., Gilliard, T., Martínez, F. B., Maritza Reyes, V., & Mendoza, E. Q. (2002). Somatic embryogenesis of the banana hybrid cultivar FHIA-18 (AAAB) in liquid medium and scaled-up in a bioreactor. *Plant Cell, Tissue and Organ Culture*, *68*(1), 21–26. doi:10.1023/A:1012905825307

Gonzalez, J. M. J., Brown, B. J., & Carlton, B. C. (1982). Transfer of *Bacillus thuringiensis* plasmids coding for δ- endotoxin among strains of *B. thuringiensis* and *B. cereus*. *Proceedings of the National Academy of Sciences of the United States of America*, *79*(22), 6951–6955. doi:10.1073/pnas.79.22.6951 PMID:6294667

Gonzalez, J. M. J. Jr, & Carlton, B. C. (1980). Patterns of plasmid DNA in crystalliferous and acrystalliferous strains of *Bacillus thuringiensis*. *Plasmid*, *3*(1), 92–98. doi:10.1016/S0147-619X(80)90038-4 PMID:7335824

Gupta, S., & Dikshit, A. K. (2010). Biopesticides: An ecofriendly approach for pest control. *Journal of Biopesticides*, *4*(6), 186–188.

Hittalmani, S., Parco, A., Mew, T. V., Zeigler, R. S., & Huang, N. (2000). Fine mapping and DNA marker-assisted pyramiding of the three major genes for blast resistance in rice. *Theoretical and Applied Genetics*, *100*(7), 1121–1128. doi:10.1007/s001220051395

Huang, N., Angeles, E. R., Domingo, J., Magpantay, G., Singh, S., Zhang, G., & Khush, G. S. (1997). Pyramiding of bacterial blight resistance genes in rice: Marker-assisted selection using RFLP and PCR. *Theoretical and Applied Genetics*, *95*(3), 313–320. doi:10.1007/s001220050565

Institute, I. F. P. R. (2000). 2000 IFPRI Annual Report *Annual report*.Izquierdo, J., & de la Riva, G. A. (2000). Plant biotechnology and food security in Latin America and the Caribbean. *Electronic Journal of Biotechnology*, *3*(1), 1–8.

Jaff'e, W., & Rojas, M. (1994). Abiotic stress and biotechnology in Latin America. *Biotechnological Development Monitor*, *18*(1), 6–7.

Jauhar, P. (2006). Modern biotechnology as an integral supplement to conventional plant breeding: The prospects and challenges. *Crop Science*, *46*(4), 1841–1859. doi:10.2135/cropsci2005.07-0223

Jiang, G. H., Xu, C. G., Tu, J. M., Li, X. H., He, Y. Q., & Zhang, Q. F. (2004). Pyramiding of insect- and disease-resistance genes into an elite indica, cytoplasm male sterile restorer line of rice, '*Minghui 63*'. *Plant Breeding*, *123*(2), 112–116. doi:10.1046/j.1439-0523.2003.00917.x

Jones, R. A., & Scott, S. J. (1983). Improvement of tomato flavor by genetically increasing sugar and acid contents. *Euphytica*, *32*(3), 845–855. doi:10.1007/BF00042166

Kodym, A., & Zapata-Arias, F. A. (1999). Natural light as an alternative light source for the *in vitro* culture of banana (*Musa acuminata* c.v. 'Grande Naine'). *Plant Cell, Tissue and Organ Culture*, *55*(2), 141–145. doi:10.1023/A:1006119114107

Kumar, S. (2014). Plant Disease Management in India: Advances and Challenges. *African Journal of Agricultural Research*, *9*(15), 1207–1217. doi:10.5897/AJAR2014.7311

Kutama, A. S., Emechebe, A. M., & Aliyu, B. S. (2011a). Evaluating the efficacy of seed treatment fungicides in the control of sorghum head smut caused by Sporisorium reilianum, in the Sudan savanna region of Nigeria. *Journal of Phytopathology and Plant Health*, *6*(2), 93–98.

Lambert, B., & Peferoen, M. (1992). Insecticidal promise of Bacillus thuringiensis. Facts and mysteries about a successful biopesticide. *Bioscience*, *42*(2), 112–122. doi:10.2307/1311652

Livak, K. J. (1999). Allelic discrimination using fluorogenic probes and the 52 nuclease assay. *Genetic Analysis*, *14*(5), 143–149. doi:10.1016/S1050-3862(98)00019-9 PMID:10084106

Lorenzo, J. C., Blanco, M., Pel'aez, O., Gonz'alez, A., & Cid, M. (2001). Sugarcane micropropagation and phenolic excretion. *Plant Cell, Tissue and Organ Culture*, *65*(1), 1–8. doi:10.1023/A:1010666115337

Madhulatha, P., Anbalagan, M., Jayachandaran, S., & Sakthivel, N. (2004). Influence of liquid pulse treatment with growth regulators on *in vitro* propagation of banana (*Musa* sp. AAA). *Plant Cell, Tissue and Organ Culture*, *76*(2), 189–192. doi:10.1023/B:TICU.0000007291.31439.6c

Miller, S. A., & Martin, R. R. (1988). Molecular Diagnosis of Plant-disease. *Annual Review of Phytopathology*, *26*(1), 409–432. doi:10.1146/annurev.py.26.090188.002205

Muhammad, A., Hussain, I., Saqlan, S. M., Hamid, R., & Hamid, N. (2004). Banana plantlet production through tissue culture. *Pakistan Journal of Botany*, *36*(3), 617–620.

Murashige, T. (1974). Plant propagation through tissue culture. *Annual Review of Plant Physiology*, *25*(1), 135–136. doi:10.1146/annurev.pp.25.060174.001031

Mwangi, J. K., Nyende, A. B., Demo, P., & Matiru, V. N. (2008). Detection of latent infection by *Ralstonia solanacearum* in potato (*Solanum tuberosum*) using stems instead of tubers. *African Journal of Agricultural Research*, *7*(11), 1644–1649.

Naik, P. S. (2005). Achieving self sufficiency in quality seed production in roots and tuber crops with special reference to potato. In K. L. Chadha, B. S. B.S. Ahloowalia, K. V. Prasad, & S. K. Singh (Eds.), Crop improvement and production technology of horticultural crops (Vol. 1, pp. 243-255). New Delhi: The Horticultural Society of India.

Nandwani, D., Zehr, U., Zehr, B. E., & Barwale, R. B. (2000). Mass propagation and ex vitro survival of banana cv. '*Basrai*' through tissue culture. *Gartenbauwissenschaft*, *65*(6), 237–240.

Ogero, K. O., Mburugu, G. N., Mwangi, M., Ngugi, M. M., & Ombor, O. (2012). Low cost tissue culture technology in the regeneration of sweet potato (*Ipomoea batatas* (L) Lam). *Research Journal of Biology*, *2*(2), 51–58.

Papp, A. C., Pinsonneault, J. K., Cooke, G., & Sadee, W. (2003). Single nucleotide polymorphism genotyping using allele-specific PCR and fluorescence melting curves. *BioTechniques*, *34*(5), 1068–1072. PMID:12765033

Pretty, J. (2008). Agricultural sustainability: Concepts, principles and evidence. *Philosophical Transactions of the Royal Society of London. Series B, Biological Sciences*, *363*(1491), 447–465. doi:10.1098/rstb.2007.2163 PMID:17652074

Reis, V. M., Baldani, J. I., Baldani, V. L. D., & Dobereinera, J. (2000). Biological dinitrogen fixation in *gramineae* and palm trees. *Critical Reviews in Plant Sciences*, *19*(3), 227–247. doi:10.1016/S0735-2689(00)80003-9

Ribaut, J. M., & Ragot, M. (2007). Marker-assisted selection to improve drought adaptation in maize: The backcross approach, perspectives, limitations, and alternative. *Journal of Experimental Botany*, *58*(2), 351–360. doi:10.1093/jxb/erl214 PMID:17158111

Sanchez, A. C., Brar, D. S., Huang, N., Li, Z., & Khush, G. S. (2000). Sequence tagged site marker-assisted selection for three bacterial blight resistance genes in rice. *Crop Science*, *40*(3), 792–797. doi:10.2135/cropsci2000.403792x

Sasson, A. (1993). Biotechnologies in developing countries: present and future. 1. Regional and national survey. *Biotechnologies in developing countries: present and future. 1. Regional and national survey.*

Shankar, S. C., Balaji, P., & Sathish, S. D. (2014). Mass Propagation of Banana (*Musa sp.*) cv. Grand Naine through direct organogenesis by benzyl adenine purine and kinetin. *Journal of Academia and Industrial Research*, *3*(2), 92–97.

Sharma, P. N., Torii, A., Takumi, S., Mori, N., & Nakamura, C. (2004). Marker-assisted pyramiding of brown planthopper (*Nilaparvata lugens Stål*) resistance genes *Bph1* and *Bph2* on rice chromosome 12. *Hereditas*, *140*(1), 61–69. doi:10.1111/j.1601-5223.2004.01726.x PMID:15032948

Singh, S., Sidhu, J. S., Huang, N., Vikal, Y., Brar, Z. L. D. S., Dhaliwal, H. S., & Khush, G. S. (2001). Pyramiding three bacterial blight resistance genes (*xa5, xa13* and *Xa21*) using marker-assisted selection into indica rice cultivar PR106. *Theoretical and Applied Genetics*, *102*(6-7), 1011–1015. doi:10.1007/s001220000495

Smith, R. S. (1992). Legume inoculant formulation and application. *Canadian Journal of Microbiology*, *38*(6), 485–492. doi:10.1139/m92-080

Thorpe, T. (2007). History of plant tissue culture. *Journal of Molecular Biotechnology*, *37*(1), 169–180. doi:10.1007/s12033-007-0031-3 PMID:17914178

Tilman, D. (1999). Global environmental impacts of agricultural expansion: The need for sustainable and efficient practices. *Proceedings of the National Academy of Sciences of the United States of America*, *96*(11), 5995–6000. doi:10.1073/pnas.96.11.5995 PMID:10339530

Varshney, R. V., Hoisington, D. A., & Tyagi, A. K. (2006). Advances in cereal genomics and applications in crop breeding. *Trends in Biotechnology*, *24*(11), 490–499. doi:10.1016/j.tibtech.2006.08.006 PMID:16956681

Wang, P. J., & Hu, C. Y. (1982). *In vitro* mass tuberization and virus-free seed potato production in Taiwan. *American Potato Journal*, *59*(1), 33–39. doi:10.1007/BF02854881

Wong, W. C. (1986). *In vitro* propagation of banana (Musa sp.): Initiation, proliferation and development of shoot-tip cultures of defined media. *Plant Cell, Tissue and Organ Culture*, *6*(2), 159–166. doi:10.1007/BF00180799

XiangYan, Z., ZengYan, Z., LiPu, D., ZhiYong, X., & Xiao, C. (2005). Development of wheat germplasms with multi-resistance to powdery mildew, stripe rust and yellow dwarf virus by molecular markerassisted selection. *Science and sustainable Agriculture, 38*(12), 2380-2386.

Yang, W. C., & Francis, D. M. (2005). Marker-assisted selection for combining resistance to bacterial spot and bacterial speck in tomato. *Journal of the American Society for Horticultural Science*, *130*(5), 716–721.

Chapter 9
Strategic Approaches to Food Security in Developing Countries

Marcus N.A. Ramdwar
The University of Trinidad and Tobago (UTT), Trinidad and Tobago

Narendra Siew
The University of Trinidad and Tobago (UTT), Trinidad and Tobago

ABSTRACT

As the world's population increases, the issue of food security presents a serious challenge. Currently, food security is reliant on a very narrow range of cultivated plant species and is also limited to traditional animal husbandry practices. The use of edible wild plants can be considered to support food security efforts. Such plants have shown to be more resilient compared to traditionally cultivated crops and also have superior nutritional attributes. The issue of protein food security can be addressed through production improvements for local or indigenous chickens, livestock production systems and wildlife to some extent. Food security in developing countries would increasingly become more dependent on widening the biodiversity from which food is selected and utilized and insects, like wild plants, provide another such opportunity. Contemporary reproductive technologies such as lactation induction, embryo transfer and artificial insemination among others can bolster food security efforts in developing countries. Improvements in forage species as well as processing technologies to improve nutritional value of low quality forages can improve overall animal nutrition. Irrespective of the concerns related to genetically modified organisms (GMOs), these sources of food could be beneficial globally. The formation of farmers' groups can be a strategic approach to food security in resource poor developing countries for the purposes of collective action and resource sharing. Sustainable food security requires the integration of several multi-dimensional approaches into a holistic management model to achieve the food security objectives in many developing nations.

DOI: 10.4018/978-1-5225-0942-4.ch009

INTRODUCTION

Global food security objectives, amidst increases in the world's population and environmental changes, would require the implementation of multidimensional strategies to significantly improve the future sustainable production of food. According to Food and Agriculture Organization (FAO, 2009a), the world population is expected to be just over 9 billion by 2050 and most of this population increase is expected to occur in developing countries. While China (1.4 billion) and India (1.3 billion) remain the two largest countries of the world, representing 19 and 18 per cent of the world's population, respectively, Africa has been cited as the fastest growing major area (2.55% annually, from 2010-2015) and is expected to account for more than half of the global population increase between now and 2050 (United Nations, 2015). Africa is followed by Asia, Northern America, Latin America and the Caribbean and Oceania. It is therefore instructive that if population growth is to be concentrated in many of the underdeveloped and developing regions of the world then food production, sustainability and security become a major center piece in policy development for these regions. Rijsberman (2012) in a report related to the future of food indicated that agricultural production is unable to keep up with the global demand for food resulting from an increase in world population levels. The report further suggests that annually there are almost 80 million more mouths to feed. There have been estimates that food production would need to increase by 70% by 2050 (FAO, 2009b). It was recommended that this estimate can be reduced to 60% by drastically reducing huge losses and wastage of food and through the sustainable management of natural resources (Nwanze, Graziano da Silva, Cousin & Frison, 2012). The amount of food globally loss or wasted per year is predicted to be around 1.3 billion tons (Gustavsson, Cederberg, Sonesson, van Otterdijk & Meybeck, 2011). As a result of a myriad of biological, environmental and socio-economic factors, food losses in developing countries are considerably high and urgent intervention is required to address this occurrence.

The urbanization phenomenon has the potential to impact agricultural productivity in the rural regions of developing countries. The pressures of urbanization such as rural to urban migration and rural urbanization would see the removal and reduction of labour and land resources away from agricultural activities. The reality of a shrinking agricultural work force, resulting from employment preferences in other economic sectors, is further negatively impacting agricultural production. The shift of labour from low productivity agriculture sector to the more productive industrialized sectors has increased the level of income derived for Caribbean countries. Notwithstanding the need for industrial development, Lewis (1954) pointed out that "industrial and agrarian revolutions always go together and is the reason why economies in which agriculture is stagnant do not show industrial growth." The industrial development of the Caribbean and other developing countries should not occur with agricultural neglect. It is necessary for governments in the developing regions around the world to place focus on developing and employing sustainable agricultural polices to ensure national food security. Although, urbanization is seemingly a positive driver for economic growth and development within developing countries (Overman & Venables, 2005) there are noticeable effects on the availability, stability, safety and access to food as reported by Matuschke (2009). It should be noted that rapid urbanization may have a long term effect on the sustainable production of food as there would be extreme demands to intensify the agricultural productivity of rural areas beyond their capacity.

The national food security of any country can be improved by either accelerating food production or through the importation of food (Trueblood & Shapouri, 2001). Currently, the agricultural sectors in the Caribbean and many other developing countries have been underperforming resulting in unfulfilled national demands for food. The underperformance of the agricultural sector in such countries has been attributed to constraints on productivity resulting from small scale operations, limited public and private investment and natural disasters (Bourne, 2008). Vehemently outdated farming techniques, improper infrastructure and limited mechanization continue to keep the level of food production well behind the rate of population growth. As a direct result of food production inefficiencies, developing nations throughout the world have grown increasingly dependent on food imports over time (Walters & Keithley, 2012). The gap between domestic consumption and domestic production in the Caribbean is noticeably significant, with consumption being almost two, to four times greater than production (Mendoza & Machado, 2009). The reliance on food imports can have catastrophic consequences in the long run and as such much greater focus must be placed on accelerating food production and exploring alternative food sources within developing countries. It is essential for greater efforts to be directed to insisting, implementing and monitoring the adoption of standard best practices at the farm level if there are to be any significant improvements in the production of food within developing countries.

The unavoidable influence of climate change would undoubtedly exacerbate the challenges already impacting food security aspirations of many developed and developing countries. However, in contrast to the developed countries of the world, the vulnerability of the agricultural sector to climate change in many developing countries is substantially high because of restricted adaptation measures and limited resources (Mendelsohn & Dinar, 1999). There have already been noticeable changes in weather patterns which have impacted rain-fed dependent agriculture in Africa and many other countries in the tropics. Some of the predicted impacts of climate change according to FAO (2005) are increased crop water requirements, increased competition between weed and crops, spread of pests and nematodes and increased salinization of soils. Agricultural practitioners who are prone to climatic vulnerabilities must develop strategies to reduce the impact of climate change. These include and are not limited to the selection of new crop varieties, irrigation technique and management, crop diversification, mixed crop and livestock farming, changing planting and harvesting dates to correspond to the changing pattern of precipitation, soil and water conservation techniques (Juana, Kahaka & Okurut, 2013). There is no escape from the inevitable realities of climate change impacts to food production, however, developing nations must rise to the challenges of averting its undesirable influences to national food security objectives.

Food security in many developing countries requires a comprehensive examination of the many challenges which have been adversely affecting the sustainable production of food. Despite the fact that food security in on the political agendas of many developing nations, the realities of achieving sustainable food security is seemingly bleak. Agricultural practitioners in developing countries with the assistance of government and non-government support must recognize and adopt best practice standard approaches before food production intensification can be worthwhile. The approaches to food security are multidimensional ranging from improving farming techniques to treating the problematic challenges associated with climate change mitigation. The chapter examines several approaches to food security such as the utilization of underutilized indigenous or local plants for food, insects as a protein source, livestock and wildlife production in addition to contemporary approaches to improving livestock productivity. The role of farmers' groups to facilitate a coordinated and collective approach to improving agricultural productivity is also discussed. The strategic approaches discussed in this chapter could be integrated into a holistic model for food security management in developing nations.

UNDERUTILIZED PLANT SPECIES TO SUPPORT FOOD SECURITY

Farmers in the Caribbean and in many developing countries primarily cultivate non-indigenous or exotic vegetable crops many of which are from the brassicaceae (e.g. cabbage, cauliflower, broccoli, pak choi etc) and solanaceae (e.g. sweet pepper, tomatoes etc.) plant families. The availability of these commodities as a source of food, however, is much dependent on certified seed imports and/or on the importation of the commodity itself. In some instances, open-pollinated varieties and heirloom varieties are cultivated. Nevertheless most commercially grown produce are cultivated from imported hybrid seeds. The practice of cultivating such crops of imported origin is attributed to consumers' preferences and demand for high quality vegetable produce. Irrespective of the primary reliance on cultivation of crop plants, the consumption of edible wild plants as a source of food exists. According to Shanley and Luz (2003) billions of people in developing countries use wild plants as a source of food because of their low cost. Traditionally, a vast majority of edible wild plants contributed to the diet among the poor (Bharucha & Pretty 2010) especially those in rural communities. Today, many have now become forgotten or neglected. Petitioning the awareness of edible wild plants and their health benefits can stimulate consumption of this category of food irrespective of social class. Additionally, a unique feature of edible wilds plants is demonstrated by their characteristic resilience during periods of natural or man-made stresses, thus making them a potential source of an emergency food supply. Cultivated crop plants, in contrast to edible wild plants are prone to stresses which can have drastic adverse impacts to food security.

The Food and Agricultural Organization of the United Nations (FAO, 2009) maintains that around 30,000 plant species around the world are edible, but of these, only 7,000 are used as human food resources. The 23,000 plant species unused as human food is a reservoir of unexploited nutrition. In 1996, a FAO global plan of action was adopted for the Conservation and Sustainable Use of Plant Genetic Resources for Food and Agriculture by approximately 150 countries. The plan identified improved conservation and use of neglected and underutilized species as one of its 20 main activities (FAO, 1996). Many of these unused food plants grow wildly in the natural landscape. Campton (2008) indicated that "wild" when applied to plants or plant species refers to those which grow spontaneously in self-maintaining populations in natural or semi-natural ecosystems existing independently of direct human action. Shava (2005) referred to Edible Wild Plants (EWPs) as both indigenous and naturalized exotic plants occurring in the natural environment. There must be caution against the dichotomization of wild versus cultivated food species, since many (if not most) wild species actually fall along a continuum, from wild species under various degrees and types of human management and intervention through to domestication (Bharucha & Pretty, 2010). Many cultivated plants species are in fact relatives of wild plants created through plant breeding and genetic manipulations. Meyer, DuVal and Jensen (2012) pointed to the origins of domesticated food crops as derived from a phylogenetically diverse assemblage of wild ancestors through artificial selection for different traits. In spite of this, it has been suggested that wild food plants are nutritionally superior to some of the cultivated ones (Burlingame, 2000). Edible Wild Plants have been reported to have high contents of photochemical such as phenolic compounds (including flavanoids, which has strong antioxidant properties and which have been implicated in the prevention of age related diseases such as cancer, arteriosclerosis and diabetes (Neudeck et al. 2012).

An inventory of the imported crop diversity and production quantities which contribute to the food supply of developing nations ought to be conducted in order to assess the extent to which these commodities contribute to food security. This would facilitate the formulation of alternative prescriptive strategic approaches aimed at buffering the impact of shortfalls from imported agricultural commodities;

either the commodity itself or the seed planting material. The reliance on imported staples such as rice (*Oryza sativa*) and Irish potatoes (*Solanum tuberosum*) in the Caribbean can be seamlessly substituted with the sustainable production of staple alternatives such as cassava (*Manihot esculenta*), sweet potato (*Ipomoea batatas*), yams (*Dioscorea alata*) and breadfruit (*Artocarpus altilis*). Similarly, the dependence on imported vegetables can be substituted for by using underutilized plant species and selected landrace varieties. The cultivation of underutilized crops would offer some protection against any issues related to the importation of food. The main challenge would be the acceptance of such categories of food by consumers. However, in the event of a food shortage, a source of edible food outweighs the custom of palatability preferences. It is noticeable that countries which are dependent on food imports are vulnerable to unstable global price shocks as well as food restriction polices from exporting countries. The recognition of underutilized and indigenous edible wild plants as a valuable source of nutrition may cushion the shortfalls resulting from a potential food crisis.

Although cultivated domesticated plants species to some degree have evolved from wild relatives, there is growing interest in the identification and development of new crops with the use of neglected and underutilized indigenous "wild" plants species. Undoubtedly, this category of plant species grow naturally in a multitude of ecosystems, and have been found to be valuable potential food sources. There exist an enormous biodiversity of plants which can be used as an easily accessible source of food. The identification of these edible wild plants in developing countries would require accessing the traditional ecological knowledge resource bank of rural communities. In some countries underutilized plants are under threat of extinction as a consequence of urbanization, natural habitat destruction, agricultural intensification, pesticide use and slash and burn agricultural practices. Concomitantly, cultural transformation and the lack of interest of the younger generation in rural communities have led to the loss of plant knowledge, both as a source of food and for medical uses. There is now a trend reversal from underutilized plants as a food source for the poor, to such plants being actively sought after among the affluent and middle class households because of its superior nutritional and medical value especially when compared to conventional crop plants.

It should be noted that some of these underutilized plant foods are considered weeds and are subjected to intense elimination pressure in the agricultural landscape with the use of pesticides and physical means. In Trinidad for example, *Amaranthus dubius* is a popular weed found in many cultivated fields. It is, however, a food source not only consumed in Trinidad but also consumed in other islands in the Caribbean and in many other developing nations in Africa and Asia. The management of *Amaranthus dubius* for food can range from completely unmanaged to intensive formal cultivation. Table 1 shows the comparative volumes of Amaranth traded at a local wholesale market in Trinidad, in relation to other leafy vegetables either imported or grown locally from imported seeds. Because of its low production costs, amaranth is one of the cheapest dark-green leafy vegetables in tropical markets and is often described as the poor man's vegetable (Varalakshmi, 2004). Although the price of *Amaranthus* in Table 2 is relatively close to the cost of pakchoi, the cost of production of pakchoi would be higher. *Amaranthus* which has been traditionally consumed, has been regaining its popularity as a promising food crop for production primarily due to its resistance to heat, drought, diseases and pests, and the high nutritional value of both its seeds and leaves (Wu *et al.*, 2000). Furthermore, it has a high content of essential micronutrients such as beta-carotene, iron, calcium, vitamin C and folic acid (Priya et al. 2007).

The strategy to make available and to incorporate indigenous food into the diet of a population would first require accurate botanical identification and inventory of the edible plant biodiversity which exist in and around farmlands and communities. It would also require a detailed agronomic evaluation of the

Table 1. Annual total volume of leafy vegetable traded for the period 2010-2014 at the Macoya wholesale market, Trinidad, West Indies

Leafy Vegetables	Unit	Annual Total Volume				
		2010	2011	2012	2013	2014
Lettuce (small)	Head	99,257	127,432	90,417	58,371	74,252
Lettuce (medium)	Head	190,700	252,372	175,225	119,750	127,305
Lettuce (Large)	Head	392,994	300,434	199,975	197,628	158,205
Patchoi	Bundle	372,195	267,491	172,939	139,235	143,855
Amaranthus spp (Spinach)	**Bundle**	**219,014**	**203,188**	**146,271**	**138,231**	**120,673**
Cabbage Imported - Green	Kg	608,867	558,217	578,930	528,121	646,185
Cabbage Local- Green	Kg	684,297	786,571	677,918	699,214	618,345
Cabbage Local - White	Kg	28,633	56,016	76,313	73,535	61,983
Cabbage Imported- Purple	Kg	70,088	75,411	88,725	68,487	95,256

Note: Data compiled from the National Agricultural Market Information System (NAMIS)

Table 2. Annual average wholesale prices for leafy vegetable for the period 2010-2014 at the Macoya wholesale market, Trinidad, West Indies

Leafy Vegetables	Unit	Annual Average-Wholesale Prices ($TT*)				
		2010	2011	2012	2013	2014
Lettuce (small)	Head	*3.24*	*3.28*	3.10	3.10	3.55
Lettuce (medium)	Head	*4.43*	*4.42*	4.21	4.26	4.86
Lettuce (Large)	Head	*5.47*	*5.56*	5.29	5.13	6.11
Patchoi	Bundle	*4.89*	*4.84*	4.91	5.49	5.68
Amaranthus spp (Spinach)	**Bundle**	***4.08***	***4.28***	**4.25**	**4.46**	**5.01**
Cabbage Imported- Green	Kg	*7.83*	*7.04*	6.65	7.41	8.01
Cabbage Local- Green	Kg	*8.01*	*7.68*	7.82	8.67	9.82
Cabbage (White)	Kg	*5.25*	*4.89*	4.95	5.35	6.04
Cabbage Imported- Purple	Kg	*10.76*	*9.72*	9.16	10.25	10.30

Note: Data compiled from the National Agricultural Market Information System (NAMIS)
* 6.02 TT equivalent to USD $1.00

"wild" plant to establish the formal cultivation techniques that would enable its successful cultivation as an agricultural crop. Furthermore, active promotion of the nutritional benefits associated with underutilized plants could encourage their utilization and conservation. Farmers who cultivate regular crops should be educated on the existence of wild indigenous plant which can be a source of food and should not entirely eliminate them. The preservation of indigenous plants around farmlands would ensure that they can be made easily available for a source of food in the event of a food crisis natural or otherwise. Moreover, the development and commercialization of underutilized plants can further be valuable to farmers as a low-input crop. Table 3 provides some essential requirements and actions required for the sustainable use of underutilized and indigenous wild plants as a potential source of food.

Table 3. Requiements for the sustainable use of underutilized plants as a source of food

Requirements for Sustainable Underutilized and Indigenous Wild Plant Consumption	Actions Required
1. Identification of underutilized or indigenous wild plants.	• Accessing traditional ecological knowledge from rural communities. • Botanical classification and evaluation.
2. Nutritional and medical benefit profile.	• Accessing traditional medicinal uses. • Nutrient analysis: - proteins, vitamins, minerals • Nutraceutical determination.
3. Agronomic evaluation	Evaluation of determination of specific cultivation requirements.
4. Education	Farmers and public should be educated on the value and potential of these crops as food sources.
5. Research and development	• Research and development in the areas of menu development, palatability studies, preservation, storage and value chain development. • Agronomic research • Analysis of the value chains potential.

The cultivation or preservation of indigenous wild plants is not an attempt to downsize the production of conventional vegetables but to recognize the potential of wild plants in the agro-ecosystem. The development and commercialization of these under-utilized plants as a source of food is essential for sustainable food and nutrition security, particularly for the developing nations across the globe.

LIVESTOCK INTENSIFICATION FOR FOOD SECURITY

Animal production is mainly on a much smaller scale compared to crop cultivation in many developing nations. The lack of space and physical infrastructure along with the need to have costly specialized care inclusive of veterinary services has somewhat stymied farmer's interest in livestock production. Interestingly, in spite of this, the livestock sector is consuming a considerable portion of the world's crops which otherwise can be used as human food. It is essential that livestock production in developing countries not impact the availability of food derived from crop sources. More so, given that animal production is dependent on crop feed sources, it is ever more crucial that sustainable animal production systems are established and maintained to ensure the efficient production of livestock and livestock products. The sustainable development of food derived from livestock must take into consideration the key animal production factors which include animal breeding and genetics, reproduction, nutrition, animal health care and housing. Analytically linking each factor with the type of livestock and the physiological state of the livestock category would enable decisions for optimizing performance of the livestock production system. Additionally, improving the efficiencies in the production cycle of traditional livestock production will contribute directly to an increase in productivity of livestock products at the farm level.

Animal production systems may be broadly categorized as being either intensive or extensive. However, it must be noted that some systems show features of both categories and are hence categorized as semi-intensive. Intensification, as it relates to animal production, may be defined as the rearing of many animals in a single location in close proximity to each other. The intensive livestock production model is essentially a controlled oversight of every stage in the animal production system taking into account the animal production factors. The drastic increase in animal population densities associated with intensification usually results in problems associated with animals' requirements for nutrition and feeding, health and environmental management. In the Caribbean context because of the relatively small sizes of countries, livestock production enterprises are moving towards the adoption of intensive systems for production primarily because of competing interests for land use, praedial larceny, hilly topography and the need to control the environment for animal (Ramdwar, Maharaj & Siew, 2015). The management of livestock using intensive methods of production would ensure the achievement of performance target that will contribute significantly to food security.

Some examples of intensive systems used in the Caribbean region are:

1. Large tunnel ventilated houses for broiler production (with bird density of less than 0.09 m^2 per bird), which utilizes the deep litter system of production.
2. Open side naturally ventilated houses for broiler production using the deep litter system.
3. Slatted floor and deep litter systems for small ruminant (sheep and goats) production.
4. Feedlot systems for cattle and water buffalo production.
5. Intensive swine production facilities.

Local Chickens and Wildlife ("Bushmeat")

Developing countries normally import fertile eggs for commercial layer or broiler production. As a consequence of the importation of poultry for commercial production, the local or indigenous chickens (*Gallus domesticus*) are to some extent neglected especially with respect to the implementation of standard performance targets. These birds are normally not caged and are allowed to range freely and scavenge for food. The production of local chickens is characterized by a low input management system and consequently the productivity of this category of livestock is constrained. The specific constraints associated with the production of local chickens are related to poor housing, poor nutrition, poor reproductive performance, diseases, and predators. In order to increase the productivity of local chickens, there must be greater emphasis directed to improving the overall management of these birds. The overall management approaches directed to improving the productivity of local chickens would include:

- Protection for nesting and brooding sites
- Protection for chicks and adults from adverse weather and predators
- Protection from diseases
- Provision of adequate nutrition
- Genetic improvement and breeding management

Producing indigenous chickens in a more efficient manner would increase the number of eggs produced per bird and also increase the chances of survival of chicks produced per bird. Consequently, the potential number of offspring obtained per bird which can be reared for meat would increase. In essence, housing, nutrition, disease management and reproductive management are essential to fostering a sustainable operational management approach to improving the productive performance targets of local chickens in developing countries. It should be noted also that sustaining the performance improvements of commercial poultry production should be coupled with this additional focus to improving the productivity of local chickens. Increasing the production of local poultry would reduce the reliance on poultry importation and thereby contribute to the inclusion of sustainable local poultry production into the food security agenda of developing nations.

Along with indigenous chickens, there exists a biodiversity of wildlife (sometimes referred to as "bushmeat" or "wild meat") that has historically been used as a source of food. In some rural communities in developing countries "bushmeat" may comprise over 80% of the consumption of animal protein (Adu *et al.*, 1999; Wilkie & Carpenter, 1999). In communities where wildlife is the main source of protein and where greater constraints exist for conventional livestock farming, wildlife farming can be a viable option. Wildlife farming is defined as the rearing of non-domesticated animals for the purpose of captive breeding (Nogueira & Nogueira-Filho, 2011). The farming of wildlife provides a sustainable approach to the conservation of wildlife biodiversity while simultaneously contributing directly and indirectly to food security. Meat from wildlife sources contributes indirectly to food security by providing income from sales or distribution to outside markets (Mainka & Trivedi 2002; de Merode *et al.*, 2004). Brown-Uddenberg *et al.*, (2004) identified several Neo-tropical species in Trinidad and Tobago with the potential for production in micro-livestock enterprises. They included the agouti (*Dasyprocta leporina / D. agouti*); spectacled caiman (*Caiman sclerops*); cocrico (*Ortalis ruficauda*) (this is one of the two national birds of Trinidad and Tobago); red brocket deer (*Mazama americana*); iguana (*Iguana iguana*); lappe/spotted paca (*Agouti paca / Cuniculus paca*); manicou/black eared opossum (*Didelphis marsupialis insularis*); tattoo/nine banded armadillo (*Dasypus novemcinctus*) and wild hog/quenk/collared peccary (*Tayassu tajacu / Pecari tajacu*). Wildlife farming has less negative impacts on the environment, as it reduces the pressure and dependence on wild animal populations (Garcia *et al.,* 2005; Roe 2008; Nogueira & Nogueira-Filho, 2011). Notwithstanding this, the constraints to wildlife farming far exceed the production systems for traditional livestock production systems. The main reasons for this are related to the limited technical knowledge available for the production of wildlife in captivity and lack of funding for the associated high cost of production. Arguably, hunting of wildlife is comparatively much cheaper than wildlife farming itself. In spite of the constraints to wildlife farming, its potential success is highly dependent on having a strong comprehension of the biology and behavioral characteristic of the wildlife being farmed. This is essential in order to map the physiological stages of the animal to the key animal production factors: housing, nutrition, reproduction and animal breeding and genetics. The physiological stage of any animal in production systems refers to the various stages in its growth and development during the production cycle, example oestrus, gestation, lactation, juvenile and maturity stages. Linking the various stages in the growth and developments of an animal in a production system with the specific animal production factors required for each stage would enable the realization of the production goals and performance objectives. This concept is not limited to wildlife farming but for animal production systems in general. The constraint for wildlife farming is the gap which exists between the wildlife species' various physiological stages and the respective requirements for each of the animal production factors. Researchers (Singh et al., 2014; Mollineau *et al.*, 2006; Mollineau *et al.*, 2007[a], 2007[b]) have published extensively on the production aspects of some neo-tropical wildlife in the Caribbean region. Although wildlife farming can be a source of reliable protein for communities in some developing countries it is somewhat inconceivable as a replacement to traditional domestic livestock production in a conventional sense. The introduction or expansion of traditional livestock is more suited for bolstering the sustainable production of food from animal sources rather than solely dependent on wildlife farming and hunting. Nonetheless, wildlife farming has a greater resilience to climate change impacts and provides significant economic opportunities for the rural poor. Additionally, the preservation and conservation of wildlife is essential to the food security aspirations of the developing world since the meat from wildlife is an acceptable protein source for human consumption. While many researchers have suggested the use of neo-tropical species for supporting food security from an animal protein perspective, the magnitude to which these species can be harvested remains a balancing act between overexploitation, conservation and consumption. Potential exist for domestication and production modeling for these species, however, domestication is a time consuming phenomenon and the need for immediate measures to bolster food security from already domesticated species remains pertinent.

Ramdwar, Maharaj and Siew (2015) suggested that world population growth along with several demographic factors inclusive of urbanization have influenced the demand for livestock food products tremendously. In addition to this, economic growth and a concomitant increase in individual incomes and by extension the standard of living; have also contributed to a shift in dietary choices and a subsequent increase in the demand for animal products. Some domesticated animal species that are presently reared on a large scale to provide food for human consumption are outlined in Table 4. A great degree of knowledge on the biology (nutrition, reproduction, genetics etc.) of these animals exist and relevant technologies are been constantly developed to improve their productivity as it relates to the sustained provision of affordable food for human consumption.

Contemporary Reproductive Technologies to Bolster Animal Production

There is growing demand for milk and milk based products, milk production and productivity of dairy animals have either remained constant or declined in many developing countries over the last two decades, resulting in a decrease in per capita milk production and therefore rendering these countries more food insecure (Ramgattie et al., 2014). In Trinidad and Tobago, for example, the Agricultural Services

Table 4. Some of the major domesticated food animal species.

Common Name	Scientific Name	Major Food Produced
Cattle	*Bos taurus* (European Origin) *Bos indicus* (Asian Origin)	Meat (Beef) Milk
Sheep	*Ovis aries*	Meat (Lamb and Mutton) Milk
Goats	*Capra hircus*	Meat (Chevon) Milk
Chicken (fowl)	*Gallus domesticus*	Meat (Broilers) Eggs (Layers)
Pigs (Swine)	*Sus scrofa domestica*	Meat (Pork and Bacon)
Turkey	*Meleagris gallopavo*	Meat
Water Buffalo	*Bubalus bubalis*	Meat Milk
Rabbit	*Oryctolagus cuniculus*	Meat
Ducks	*Anas platyrhynchos* *Cairina moschata*	Meat Eggs
Tilapia	*Oreochromis spp.*	Fish
Atlantic Salmon	*Salmo salar*	Fish

Department of Nestle Trinidad Limited suggested that milk intake declined from around 10 million kg in 1990 to 4 million kg in 2009, with the number of farmers also declining dramatically. The country therefore has to meet the demand for dairy products through importation of dry milk. One of the major reasons for farmers exiting the industry is their inability to turn a profit which directly hinges upon meeting biological efficiencies associated with the dairy animal. Meeting reproductive performance coefficients for parameters such as age at first parturition, calving interval, services per conception, heat detection rate and pregnancy rates are paramount to efficient production and therefore sustainability.

The typical dairy cow is expected to begin producing milk no later than 27 months old (age at first parturition), in many instances producers are unable to get these animals bred on time due to reasons like inadequate nutrition and poor heat detection. Once these animals begin to produce milk they are expected to go through a repeated cycle of a 10 month lactation and 2 month dry period (calving interval of 12 months) until their reproductive life has expired. This reproductively efficient cycle coupled with good nutrition, friendly environment (climatic and disease free) and superior genetics will maximize the productivity of each animal unit. Because producers seldom operate at the desired reproductive efficiency level, novel approaches have been developed to assist the farmer in meeting these targets and improve productivity.

Lactation Induction used to Reduce Non-Productive Days

In a natural situation a dairy animal will commence producing milk at the point of parturition, this more or less suggests that establishment of pregnancy and the production of a neonate is necessary for milk production. The whole process of lactation is therefore dependent on the precise hormonal interaction that commences during pregnancy and continues throughout lactation. As indicated earlier, many dairy farmers suffer tremendous loses because they do not get their dairy cows to initiate lactation at 27 months neither do they get them rebred on time to allow for a calving each year. The end result is extended dry periods (periods of no milk production) or extended lactations (beyond 305 days) with lengthy periods of low milk production. Additionally, poor reproductive performance also denies the producer a calf that can be used to expand the herd, can be selected to improve the genetic composition of the herd, or can be sold to derive additional income (Ramgattie et al. 2014), the net result being reduced revenues and increased expenditures. Naturally, cows that are difficult to breed are culled and replaced, reported culling rates due to reproductive problems ranged from 47% (Allaire, Sterwerf & Ludwick, 1977) to 27% (Nahms, 1996). Generally, because management strategies aim to rebreed a dairy cows when it is in peak lactation under a negative energy balance, the highest milk producers are often the most difficult to rebreed and the farmer is at greatest risk of losing some of his best milking genetics based on reproductive failure. For these animal, lactation induction present a great salvage alternative as they can now be brought back into milk production without becoming pregnant and bearing a calf. Many lactation induction protocols have been evaluated with varying levels of success. Ramgattie et al. (2014) utilized a protocol involving the use of progesterone and estrogen treatment in combination for 7 days (day 1 to 7), quantities were based on the animals' body weight. This was preceded by an intramuscular injection of prostaglandin $F_{2\alpha}$ on days -10 and 0 of the protocol. On day eight, following the progesterone and estrogen treatment, another prostaglandin $F_{2\alpha}$ administration was made and this was followed by a combination of reserpine and dexamethasone from day 9 to 12 after which milking commenced. The authors reported a net profit of approximately US $342.00 per animal treated.

Use of Bovine Somatotropin to Increase Milk Production

Bovine Somatotropin (bST) is a growth hormone (protein) produced by the pituitary gland of cattle. It is used to increase milk production in dairy cows and was approved for commercial use in the U.S by the Food and Drug Administration (FDA) on November 5, 1993. Biotechnology (recombinant DNA technology) has allowed identification of the gene that codes for bST, this gene was inserted into a bacterium and this has led to large amounts of bST being produced under controlled conditions, purified and then made commercially available where cows can be treated with the hormone. The hormone is injected into the cows resulting in significant milk yields. Cows that are treated with bST have to be well fed in order to provide the necessary nutrients for the increased capacity for milk production. Major benefits of bST include better feed conversion efficiency and greater persistency in milk production (arresting of the rate of decline usually observed after peak production. Bauman (1989) suggested that best results are derived from treatment of cows entering mid lactation. Cows that have had more than one calf show a greater increase in milk production than do first lactation heifers (Peel et al., 1989).

Use of Estrus Synchronization and Timed Artificial Insemination Technology

In production of many of food animal species, efficiency in reproduction is greatly dependent on the ability of workers to identify animals in estrus (period of sexual receptivity, also referred to as heat) and present them for breeding. In dairy cattle production, Pregnancy Rate (PR) is a function of Conception Rate (CR) and Heat Detection Rate (HDR), given by (PR= CR*HDR). Poor heat detection rate or missed heat is detrimental to the sustainability of dairy production enterprises. Clunis and Sandy (2008) in a review of the reproductive performance of Eastern Caribbean Institute of Agriculture and Forestry (ECIAF) dairy herd, reported an overall heat detection rate to be 20%, this translated into a PR of 12%, given a 60% non-return rate for artificial insemination. This coincides with the 10% pregnancy rate in dairy cattle in Trinidad and Tobago reported by Hafez (1987). Estrus synchronization reduces the time spent detecting animals in heat while fixed timed artificial insemination protocols almost entirely eliminates the need for detecting estrus as these protocols dictate when the animals will ovulate thus dictating the precise time for insemination. This technology has proven very beneficial in improving pregnancy rates, and as a result the number of offspring produced, the daily quantity of milk produced and ultimately the profitability and sustainability of the enterprise.

Use of Artificial Insemination

In many developing countries one of the greatest hindrances to improving productivity and expanding the scale of production in the livestock industry is the availability of quality breeding stock. Hosein et al. (2013) identified several constraints to the development of the small ruminant industry in the Caribbean, at the top of this list was the unavailability of quality breeding stock. This seem to be a common problem across all livestock subsectors, for example, the average daily milk production per animal (10kg/cow) in the tropics is way below average figures reported for temperate regions (25kg/cow), as such, in a bid to introduce new genetic material, the Government of Trinidad and Tobago embarked on the importation of live dairy cattle to support the struggling dairy industry. The problem with this approach is that the imported animals struggle to adapt to the prevailing climatic, biological/disease and nutritional environment. Simply described, artificial insemination is the breeding of a receptive female without the presence of a male. This technology involves the collection, evaluation, extension and storage of semen that could be used at a later time. Some of the major benefits derived from use of this technology are:

- Rapid genetic improvement
- Widespread use and availability of genetically superior sires (1 bull can breed 500,000 cows in a lifetime).
- After death semen can be used
- Oldest frozen semen 40-45 years old
- Rapid proof of sire (genetic evaluation)
- Progeny testing evaluates offspring for desired traits
- With natural mating will only have hundreds of offspring
- Availability of sires
- Sires anywhere in the world can be used

Use of Sperm Sexing Technology

Production of a male or female offspring may have significantly more economic benefit depending on the enterprise under consideration. In dairy operations the cow is the production unit and therefore the main income generator. Therefore, producing a male calf will have significantly less value than a female as the females are in high demand as replacements and breeding animals. Although bulls are needed, they are needed to a much lesser extent because the majority of dairy cows are artificially inseminated and with artificial insemination the bull to cow ratio is reduced as this technology allows for extension of semen, allowing more cows to be inseminated from a single ejaculate. On the other hand, in a beef production enterprise a bull calf is preferred as the steer is the unit that will earn revenue from beef sales. Each sperm will contain either X or Y chromosome, flow cytometry is used to separate X and Y bearing spermatozoa. The sperm is treated with a dye (DNA stain), because the X bearing sperm has more DNA, more dye is absorbed and therefore more-light is emitted when excited by a laser. The sperm will enter the flow cytometer chamber and pass in single file through a small nozzle where they will be separated in X, Y and dead sperm.

Super Ovulation and Embryo Transfer Technology

Super ovulation involves the stimulation, growth and development, and eventual ovulation of a large number of follicles within a donor animal. The process involves treating (injection) the female with gonadotropins to stimulate follicular growth and development followed by Gonadotropin Releasing Hormone (GnRH) treatment to stimulate ovulation of follicles. The unfertilized eggs can be collected (flushed from the reproductive tract) and fertilized *in vitro*, producing an embryo which can be later transferred into a recipient female. An alternative to this is to produce the embryos *in vivo* by artificially inseminating the females, then flushing the embryos from the reproductive tract of the donor female to be later implanted into a recipient female. The benefit of super ovulation and embryo transfer are:

- It allows rapid multiplication of genetically superior animals
- Allows widespread access to superior genetics through widespread availability of embryos

- It allows continued multiplication of animals with superior genetics which may not be able to carry another pregnancy due to injury.
- It allows use of animals of inferior genetics purely as surrogate for the multiplication of superior genetics

GENETICALLY MODIFIED ORGANISMS AND FOOD SECURITY

WHO (2014) defines genetically modified organisms (GMOs) as "organisms (i.e. plants, animals or microorganisms) in which the genetic material (DNA) has been altered in a way that does not occur naturally by mating and/or natural recombination. The technology is often called 'modern biotechnology' or 'gene technology', sometimes also 'recombinant DNA technology' or 'genetic engineering'. It allows selected individual genes to be transferred from one organism into another, also between nonrelated species". Foods produced from or using GM organisms are often referred to as GM foods. In November 2015, a milestone for agricultural biotechnology was reached when the Food and Drug Administration (FDA) approved the first genetically modified animal for human consumption. Approval was granted for the commercial sale of a strain of salmon (AquAdvantage Salmon) who's DNA has been altered to make it grow faster. Bunge (2015) reported that "the modified Atlantic salmon was developed by Maynard, Mass.-based AquaBounty Technologies, a unit of Intrexon Corp., to grow to market weight twice as fast as wild or farmed versions of that species. It does so by incorporating a gene from a larger, Pacific Ocean species, the Chinook salmon, that helps boost the Atlantic salmon's growth-hormone production". Pollack (2015) reported that "it also contains a genetic switch from the ocean pout, an eel-like creature, this genetic switch keeps the transplanted gene continuously active, whereas the salmon's own growth hormone gene is active only parts of the year". Bunge (2015) also suggested that "AquaBounty reported that the modified fish require 20%-25% less feed to reach the same weight as conventional Atlantic salmon. The company pitches them as a more efficient means to meet growing demand for farmed fish and ease pressure on wild fisheries".

Notwithstanding the benefits and safety level reported by the FDA and producers of the GMO's, consumers are still very reluctant to widely accept these GM foods as common place on supermarket shelves and restaurants, in fact, many food brand distributors have refused to carry these products. Some of the major concerns with the use of GM foods are:

- Possibility of GMO's escaping from the production sites and mixing with the naturally occurring species, out-competing them, becoming preferred mates and putting the natural occurring species at risk of extinction.
- Possible allergenicity associated with the consumption of the GM food.
- Gene transfer from the GMO to naturally occurring organisms.
- Outcrossing
- Susceptibility of non-target species
- Stability of the gene in the medium to long term
- Possibility of the loss in biodiversity in the event of outcrossing
- Movement of resistant genes to other organisms.

The debate on the use of GM foods will continue for a long time, but this technology seems to hold significant potential for addressing food security issues worldwide.

ANIMAL NUTRITION AND ITS POTENTIAL TO IMPROVE FOOD SECURITY

Improved Forage Species

Scientifically meeting the animal nutrient needs remain a challenge in livestock production in the developing world. Given the increasing price of grain on the international market, forage based feeding systems still remain the economically feasible means of meeting the animal's nutrition. In many countries, the quality of the forage remains a major challenge to the improved productivity of ruminant animals. Utilization of improved forage species such as Mulato (*Brachiaria* hybrid CIAT 36061) and Mulato II (*Brachiaria* hybrid CIAT 36087) grass hybrids provides renewed hope for ruminant producers. These species are the result of more than 20 years of research by the International Center for Tropical Agriculture (CIAT) in Columbia. Mulato and Mulato II are *Brachiaria* hybrids of B. *brizantha*, B. *ruzinensis* and in the case of Mulato II, B. *decumbens*. The Mulato grass is highly suitable for use in Trinidad and Tobago seeing that its close relative tanner grass (*Brachiaria arrecta*) grows almost everywhere. Positive attributes of the Mulato grass when compared to tanner grass includes:

- Higher dry matter yield
- Higher leaf to stem ratio
- Higher nutritional value in terms of protein and energy
- Better resistance to pest and diseases
- Good recovery from fire damage
- Comparable tolerance to acid soils

Utilization of By-Products from Local Food Processing and Manufacturing Industry for Feeding Animals

Utilization of by products from processing provides a useful and viable alternative to imported grain for feeding ruminant animals. These could be used in combination to supplement the animals daily nutrient needs. The following represents a selection of commonly used byproducts in Trinidad and Tobago:

- Brewers grain – byproduct from beer production
- Coconut meal – by product from coconut oil production
- Dried citrus pulp – byproduct from citrus processing
- Wheat middling – by product from flour production
- Bakery rejects product e.g. cookies, bread etc.
- Fish meal
- Meat and bone meal
- Poultry offal meal

Utilization of Techniques to Improve the Nutritional Value of Low Quality Forages

The quality of forage material used is influenced by the forage specie, variety, stage of growth or regrowth, crop husbandry practices etc. It is without doubt that in the tropics poor quality forage is a commonplace in animal production because many producers do not pay particular emphasis to recommended forage production practices and the order of the day is to utilize forage where it is available. Technologies exist to assist the producer to enhance the feeding value of these otherwise useless fibrous material, these include:

- Mechanical treatment such as grinding to increase the surface area for rumen microbial action.
- Physical treatment such as temperature and pressure to help in improving digestibility.
- Chemical treatment such as treating with an alkali such as sodium hydroxide or ammonia to bread down resistant lingo-cellulose bonds.

Other Approaches to Improving Nutrient Availability and Utilization

- Improving the nutritional value of cereals – many are low in some of the essential amino acids such as lysine. The use of modern biotechnology is available to produce transgenic cereals with increased essential amino acid content.
- Removal of anti-nutritional factors in forages such as tannins, protease inhibitors, pectins etc. through the use of modern biotechnology.
- Use of enzymes as feed additives to improve digestibility – phytase is a commercial enzyme which is used to bread down phytate and make phosphorus more readily available to the animal.
- Forsberg et al (1986) have suggested the use of "transgenic to produce bacteria with enhanced cellulotic activity, capability to cleave lignohemicellulose complexes, reduced methane production capability decreased proteolytic and/or deaminase activities, increased capability for nitrogen "fixation" and increased ability for microbial production of specific amino acids."

INSECTS AS A SOURCE OF PROTEIN

The consumption of insects, entomophagy, is not novel since many people around the world particularly in Southeast Asia, Oceania, southern Africa, Central America and the northern part of South America have had insects as a component of their regular diet (van Huis et al. 2013). Insects are in fact consumed by at least two billion people in the world and more than 1900 insect species are currently used as food (FAO, 2013). The major group of commonly consumed insects are beetles (Coleoptera), however, other groups of edible insects include, Lepidoptera e.g. caterpillars, Hymenoptera e.g. bees, wasps, Orthoptera e.g. locust, crickets, grasshoppers, Hemiptera e.g. true bugs, cicadas, leafhoppers, plant hoppers, scale insects, Isoptera-termites, Odonata e.g. dragonfly and Diptera e.g. flies (Jansson & Berggren, 2015; Pal & Roy, 2014; FAO, 2013). Although insects are consumed by many people in some developing countries, it is still not widespread enough to have any significant impact on the global food security agenda.

The demands for food especially animal based protein would be increasingly difficult to supply by 2050. In developing countries alone meat consumption is growing at a rate of 5 percent per year (FAO, 2006). The dual pressures of population growth and urbanization would severely reduce the available resources to produce animal based protein. Furthermore the availability of food for human consumption at the global level is heavily impacted by the demands that livestock production places on land and water use (Charlton et al., 2015). Already, livestock production accounts for 70% of all agricultural lands (Steinfeld et al., 2006) and any further increase would not be sustainable. The influence of climate change and environmental effects would further erode the potential to increase animal production to meet the demands of a steadfastly growing world's population. It is imperative that alternative protein sources such as insects be evaluated for inclusion into the human food chain and for animal feed. The replacement of proteins derived from crops with nonconventional protein sources such as insects would redirect the cultivation of arable land solely for food production purposes.

Insects are underutilized as a valuable protein source. The proportion of protein per 100g of various insect species compares favourably with that from mammals, reptiles and fish (Glover and Sexton, 2015). Insects are generally low in cholesterol and fat (Srivastava, Babu & Pandey 2009) and for this reason it is advantageous when compared to the consumption of red meat. It has been reported by Rumpold and Schlüter (2013) that insects are comparable to beef, pork and chicken and contain more polyunsaturated fatty acids and have a higher content of minerals such as iron and zinc. Some insects contain more protein than meat (Johnson, 2010) and can be a better alternative to consuming animal derived proteins. Insects are extremely efficient at converting what they consume into consumable biomass. When compared to cattle the feed conversion efficiency of insects is twenty times higher (Durst & Shono, 2010). Additionally, insects can feed on a broad variety of materials including manure and they have a better feed conversion rate than most other animals, which reduces wastage and costs (Khusro, Andrew & Nicholas, 2012). It is a reality that the cultivation of insects as an alternative source of protein is more sustainable compared to livestock production, considering less land is required and low levels of greenhouse gasses are emitted in producing an even higher quality edible protein.

Western societies are psychologically, culturally and socioeconomically averse to the consumption of insects as a source of food. The taboo associated with insects as a source of food stems from the belief that insects are regarded as scornful pests. It is thus unlikely for the food neophobia associated with insect consumption in western civilization to dissipate, however, the taboo may be somewhat less if insects were used as a feed source for animals. It is expected that soy, maize and fish based ingredients for animal feed formulations would become limited and expensive in the future and as such insects can become a significant protein source for animal feed. There would, however, be a need to select and evaluate suitable insects based on nutritional profile, nutrient bioavailability and animal palatability before establishing sustainable mass production standards.

Although the potentiality of insects as an alternative source of food for humans and feed for animals exist to cope with the emerging problems of food security some major challenges exist. These challenges according to FAO/WUR[1] (2014) include:

- Further awareness raising among the general public to promote insects as healthy food for humans and feed for animals;
- Influencing policy makers to approve insect inclusive food and feed legislations;
- Further research efforts to provide and expand with validated data the available scientific evidence and benefits of using insects in the food and feed chains.

There are currently no international standards, for example, the Codex Alimentarius, governing the regulation of edible insects for human food and animal feed. The absence of such standards would further compound the unwillingness of western societies to include insects as a component of their diet. It is essential for any society to ensure that there are specific food safety guidelines for the consumption of any unconventional food. Hazard Analysis and Critical Control Point (HACCP), a system adopted by the Codex Alimentarius Commission, should be implemented to identify and procedurally regulate to avoid hazards (physical, chemical and microbiological) which may occur in the collection, production and processing of edible insects for food. Tapping into the traditional indigenous knowledge of societies where insects are consumed coupled with sound scientific research must be undertaken before the value of insects to global food security can be fully exploited.

Globally, a lot of emphasis is placed on the elimination of insects rather than exploring their potential for consumption as human food and animal feed. The nutritional value of insects as food and feed diversity has been somewhat overlooked. It would be essential for developing countries, particularly because of rapid population growth, to explore the opportunities for increasing the research, production, preservation and utilization of insects as food and feed. Food security in developing countries would increasingly become more dependent on widening the biodiversity from which food is selected and utilized and insects provides one such essential opportunity.

IMPORTANCE OF FARMERS' GROUPS IN ACHIEVING FOOD SECURITY

In the previous sections emphasis was placed on underutilized plants, local poultry, wildlife and traditional animal production. In this section, focus will be placed on farmers' groups and how they can impact the food security agenda of developing countries. A group is described as a collection of individuals among whom a set of interdependent relationships exist (Windapo & Afolayan, 2005). The terminology 'farmers' groups' and 'farmers' organizations' are commonly used interchangeably when referring to collective action among farmers. There is one distinction however, in that farmers' organizations are considered to be a more formal entity in the form of agricultural cooperatives and associations.

The formation of farmers' groups can be a strategic approach to food security in resource poor developing countries. The arrangement of farmers into groups can be within two interrelated categories namely geographic groups and commodity-specific groups. A geographic farmer's group is a local group within a community that seeks to address general agricultural production matters which are common to all farmers in a specific area. On the other hand, commodity specific groups extend beyond the local community and comprise members who are producers of a specific agricultural commodity. Commodity farmers' groups can be national entities e.g. National Poultry Production Farmer's Group or National Food Crop Farmers' Association.

The formation of farmers' groups enables farmers to come together for the purpose of collective action. Collective action arises when people come together because of constraints and to take joint action and decisions to accomplish an outcome (Sandler, 1992). Despite the potential benefits of farmers forming groups in the pursuit of common goals, there are numerous constraints which impact the longevity and performance of the group. Most of the issues facing farmers' groups are conflict oriented within the group and outside of it. The ideal farmers' group should be one which is democratically governed, has a self-help function and are self-sustaining and should serve the needs of the membership as well as the group as a whole. The purposeful organization of farmers into groups would promote the sustainable production of agricultural commodities. It can avert the issues of oversupply, since not all famers' groups would be engaging in the same type of production activity. Additionally, knowledge and resource sharing among farmers within a group secures the success of its membership and contributes to an increased food production output. Social networks throughout rural communities have been shown to play an indirect role in increasing agricultural productivity through knowledge sharing via networks (Liverpool and Winter-Nelson, 2010). Farmers' groups can provide farmers with many services that are critical to their success in accessing markets. According Stockbridge et al., (2003) these include:

- Marketing services (input supply, output marketing and processing, market information).
- Facilitation of collective production activities.
- Financial services (savings, loans and other forms of credit).
- Technology services (education, extension, research).
- Education services (business skills, health, and general).
- Welfare services, (health, safety nets).
- Policy advocacy.
- Managing common property resources (water, pasture, fisheries, forests).

There is an essential role for farmers' groups in driving the food security initiatives of developing countries. However, the propensity of farmers' groups to be successful is somewhat limited without government support. Government through its public agricultural extension services should encourage farmers to form groups and provide the necessary training to support the development of the groups. Given that individual farm visits are the predominant extension methods used in developing countries, the formation of farmers' groups can be a solution for the high farmer to extension officer ratio in many developing countries. As a consequence, many more farmers could have improved access to extension services, which would build their technical capacities and possibly improve their livelihoods. As such, many developing countries should hasten the formation and development of farmers' groups in a bid to organize their food production activities. Concomitantly, strengthening the technical capacity of extension officers in the areas of group dynamics and group development would improve the effectiveness and efficiency of the extension service in sustainable formation of farmers' groups.

CHAPTER SUMMARY

Food security aspirations of developing countries must adopt a multidimensional approach to ensure the availability of safe food. The first approach to food security would be to ensure that current food production levels are optimized in a sustainable manner. The optimization of current food production would require the implementation of improved farming techniques and the implementation of good agricultural practices. Beyond the optimization of current food production, there are opportunities to explore the use of unexploited or underutilized biodiversity as a food source for humans and for animals. There are many plant species which are either neglected or have not been explored as a source of human food. Developing countries in particular which are dependent on the importation of vegetable seeds or the vegetable commodity itself should attempt to identify local alternatives to augment food supplies. Food from animal sources must also take into consideration overall improvements in animal husbandry practices, improved management for local poultry, wildlife and wildlife farming and even insects. The implementation of contemporary reproductive technologies can improve the performance parameters for livestock enterprises and boost livestock overall productivity. Livestock nutrition is an important contributor to ensuring food security and must be factored into the overall approaches to improved animal husbandry practices. The use of alternative feed sources such as by-products from food processing as well as exploring the use of insects as a protein source for animals can contribute to a reduced reliance on imported feed. Humans and animals are both dependent on arable land for a source of food hence any food alternatives which can reduce this dependence can go a long way towards contributing to overall food security efforts. The coordination of food security efforts in developing countries must involve the agricultural extension officers. Given the high farmer to extension officer ratio in many developing countries, the population of farmers would be most effectively served if there were arranged into groups. The key concepts discussed within this chapter could be incorporated into an integrated food security management model in attempt to achieve some, if not full level of food security in developing countries.

REFERENCES

Adu, E. K., Alhassan, W. S., & Nelson, F. S. (1999). Smallholder farming of the greater cane rat *Thryonomys swinderianus*, Temminck, in southern Ghana: A baseline survey of management practices. *Tropical Animal Health and Production, 31*(4), 223–332. doi:10.1023/A:1005267110830 PMID:10504102

Allaire, F. R., Sterwerf, H. E., & Ludwick, T. M. (1977). Variations in removal reasons and culling rates with age for dairy females. *Dairy Science, 60*(2), 254–267. doi:10.3168/jds.S0022-0302(77)83862-9

Bauman, D. E. (1989). *Biology of Bovine Somatotropin. Advanced Technologies Facing the Dairy Industry: bST.* Cornell University.

Bharucha, Z., & Pretty, J. (2010). The roles and values of wild foods in agricultural systems. *Philosophical Transactions of the Royal Society of London. Series B, Biological Sciences, 365*(1554), 2913–2926. doi:10.1098/rstb.2010.0123 PMID:20713393

Bourne, C. (2008). *Perspectives on Enhancing Sustainable Growth and Development of Caribbean Agriculture.* Keynote Address, 44th Meeting Caribbean Food Crops Society.

Brown-Uddenberg, R. C., Garcia, G. W., Baptiste, Q. S., Counand, T., Adogwa, T. O., & Sampson, T. (2004). *The Agouti (Dasyprocta leporina, D. aguti) Booklet and Producers' Manual.* Trinidad and Tobago: Champs Fleurs.

Bunge, J. (2015, November 19). Food and Drug Administration Approves Genetically Modified Salmon. *The Wall Street Journal.*

Burlingame, B. (2000). Wild nutrition. *Journal of Food Composition and Analysis, 13*(2), 99–100. doi:10.1006/jfca.2000.0897

Campton, R. (2008). *The Economic Value of Wildlife in Eastern and Central Africa.* Research Paper No: 23. Institute of Development Studies (IDS), University of Dar es Salaam, Tanzania.

Charlton, A. J., Dickinson, M., Wakefield, M. E., Fitches, E., Kenis, M., & Han, R. et al. (2015). Exploring the chemical safety of fly larvae as a source of protein for animal feed. *Journal of Insects as Food and Feed, 1*(1), 7–16. doi:10.3920/JIFF2014.0020

Clunis, A., & Sandy, K. (2008). *A review of the reproductive performance of the dairy herd at the Eastern Caribbean Institute of Agriculture and Forestry (ECIAF)*. Project submitted in partial fulfillment of the Diploma in Agriculture.

de Merode, E., Homewood, K., & Cowlishaw, G. (2004). The value of bushmeat and other wild foods to rural households living in extreme poverty in Democratic Republic of Congo. *Biological Conservation, 118*(5), 573–581. doi:10.1016/j.biocon.2003.10.005

Durst, P. B., & Shono, K. (2010). Edible Forest Insects: exploring new Horizons and traditional practices. In P. B. Durst, D. V. Johnson, R. N. Leslie, & K. Shono. (Eds.), *Forest insects as food: humans bite back. Proceedings of a workshop on Asia -Pacific resources and their potential for food development.*

FAO/WUR. (2014). Insects to feed the world: summary report. In P. Vantomme, C. Münke, A. van Huis, J. van Itterbeeck, & A. Hakman (Eds.), *Insects to Feed the World*. Ede, Netherlands: FAO/WUR.

Food and Agriculture Organization (FAO). (1996). *Global Plan of Action for the Conservation and Sustainable Utilization of Plant Genetic Resources for Food and Agriculture*. Declaration adopted by the International Technical Conference on Plant Genetic Resources, Leipzig, Germany. Rome, Italy: FAO, UN.

Food and Agriculture Organization (FAO). (2005). *Impact of Climate Change, Pests and Diseases on Food Security and Poverty Reduction.* Special Event Background Document for the 31st Session of the Committee on World Food Security.

Food and Agriculture Organization (FAO). (2006). *World agriculture: towards 2030/2050. Food and Agriculture Organization of the United Nations*. Rome: FAO.

Food and Agriculture Organization (FAO). (2009a). How to Feed the World in 2050. Rome, Italy, Food and Agriculture Organization.

Food and Agriculture Organization (FAO). (2009b). The State of Food Insecurity in the World. Rome, Italy: Food and Agriculture Organization.

Food and Agriculture Organization (FAO). (2013). *Edible insects: Future prospects for food and feed security*. Forestry paper, 171. FAO.

Forsberg, C. W., Crosby, B., & Thomas, D. Y. (1986). Potential for manipulation of the rumen fermentation through the use of recombinant DNA techniques. *Journal of Animal Science, 63*, 310–325. PMID:3525493

Garcia, G. W., Young, G. M., Amour, K. M., James, D., Lallo, C. H. O., & Mollineau, W. et al. (2005). *The Collared Peccary/Javelina/Sajina/Poco do Monte/Wild Hog/Pakira/ Patira/Taitetu/Catete/Catto/ Quenk (Tayassu tajacu, Pecari tajacu) Booklet and Producers' Manual. GWG Publications.*

Glover, D., & Sexton, A. (2015). *Edible Insects and the Future of Food: A Foresight Scenario Exercise on Entomophagy and Global Food Security. IDS Evidence Report 149*. Brighton, UK: IDS.

Green, C. (1992). An overview of production and supply trends in the U.S. specialty vegetable market. *Acta Horticulturae, 318*(318), 41–45. doi:10.17660/ActaHortic.1992.318.5

Gustavsson, J., Cederberg, C., Sonesson, U., van Otterdijk, R., & Meybeck, A. (2011). Global Food Losses and Food Waste: Extent Causes and Prevention. Rome: Food and Agriculture Organization (FAO) of the United Nations.

Hafez, J. (1987). *Reproduction in Farm Animals* (5th ed.). Lea and Febiger.

Hosein, A., Paul, C., Roach-Benn, C., Borely, J., Blair, T. M., & Fearon, A. (2013). The Small Ruminant Industry in CARICOM countries with particular reference to Jamaica and Trinidad & Tobago. St Augustine, Trinidad and Tobago: Caribbean Agricultural Research and Development Institute.

Jansson, A., & Berggren, A. (2015). Insects as Food – Something for the Future? A report from Future Agriculture. Uppsala, Sweden: Swedish University of Agricultural Sciences (SLU).

Johnson, D. V. (2010). The contribution of edible insects to human nutrition and to forest management: Current status and future potential. In P.B. Durst, D.V. Johnson, R.N. Leslie, & K. Shono (Eds.), *Forest insects as food: Humans bite back. Proceedings of a workshop on Asia -Pacific resources and their potential for development*. Chiang Mai, Thailand. FAO. doi:10.1146/annurev-ento-120811-153704

Juana, J. S., Kahaka, Z., & Okurut, F. N. (2013). Farmers' Perceptions and Adaptations to Climate Change inSub-Sahara Africa: A Synthesis of Empirical Studies and Implications for Public Policy in African Agriculture. *The Journal of Agricultural Science*, *5*(4), 121–135.

Khusro, M., Andrew, N. R., & Nicholas, A. (2012). Insects as poultry feed: A scoping study for poultry production systems in Australia. *World's Poultry Science Journal*, *68*(03), 435–446. doi:10.1017/S0043933912000554

Lewis, W. A. (1954). Economic development with unlimited supplies of labour. *The Manchester School of Economic and Social Studies*, *22*(2), 139–191. doi:10.1111/j.1467-9957.1954.tb00021.x

Liverpool, S. L. O., & Winter-Nelson, A. (2010). *Poverty status and the impact of social networks on smallholder technology adoption in rural Ethiopia*. IFPRI Discussion Paper 970. Washington, DC: IFPRI.

Mainka, S. A., & Trivedi, M. (2002). *Links between Biodiversity Conservation, Livelihoods and Food Security: The Sustainable Use of Wild Species for Meat*. Cambridge, UK: IUCN.

Matuschke, I. (2009). *Rapid urbanization and food security: using food density maps to identify future food security hotspots*. Paper prepared for presentation at the International Association of Agricultural Economists Conference, Beijing, China.

Mendelsohn, R., & Dinar, A. (1999). Climate change, agriculture, and developing countries: Does adaptation matter? *The World Bank Research Observer*, *14*(2), 277–293. doi:10.1093/wbro/14.2.277

Mendoza, A., & Machado, R. (2009). *The Escalation in World Food Prices and Its Implications for the Caribbean* (vol. 2). Economic Commission for Latin America and the Caribbean, Caribbean Development Report.

Meyer, R.S., DuVal, A.E., & Jensen, H.R. (2012). Patterns and processes in crop domestication: an historical review and quantitative analysis of 203 global food crops. *New Phytologist, 196*, 29–48. doi:.2012.04253. x10.1111/j.1469-8137

Mollineau, W. M., Adogwa, A. O., & Garcia, G. W. (2007a). A preliminary technique for electro-ejaculation of the agouti (Dasyprocta leprorina). *Animal Reproduction Science, 108*(1-2), 92–97. doi:10.1016/j.anireprosci.2007.07.017 PMID:17889458

Mollineau, W. M., Adogwa, A. O., & Garcia, G. W. (2007b). Spermatozoal morphologies and fructose and citric acid concentrations in agouti (*Dasyprocta leprorina*) semen. *Animal Reproduction Science, 105*(3-4), 378–383. doi:10.1016/j.anireprosci.2007.08.009 PMID:17919856

Nahms, N. A. (1996). *Dairy management practices*. Washington, DC: USDA.

Neudeck, L., Avelino, L., Bareetseng, P., & Ngwenya, B.N., Teketay, D., & Motsholapheko, M.R. (2012). The contribution of edible wild plants to food security, dietary, diversity and income of households in Shorobe Village, Northern Botswana. *Ethnobotany Research and Applications, 10*, 449–462.

Nogueira, S. S. C., & Nogueira-Filho, S. L. G. (2011). Wildlife farming: An alternative to unsustainable hunting and deforestation in Neotropical forests? *Biodiversity and Conservation, 20*(7), 1385–1397. doi:10.1007/s10531-011-0047-7

Nwanze, K. F., da Silva, J. G., Cousin, E., & Frison, E. (2012). *Opinion: Food Security Must Be On The Table At Rio+20*. Retrieved February 13, from http://www.wfp.org/news/news-release/food-security-must-be-table-rio20

Overman, H. G., & Venables, A. J. (2005). *Cities in the Developing World*. CEP Discussion Paper No 695. Available from: http://cep.lse.ac.uk/pubs/ download/dp0695.pdf

Pal, P., & Roy, S. (2014). Edible Insects: Future of Human Food - A Review. *International Letters of Natural Sciences, 26*, 1–11. doi:10.18052/www.scipress.com/ILNS.26.1

Peel, C. J., Hurd, D. L., Madsen, K. S., & de Kerchove, G. (1989). Monsanto Agricultural Company. In *Proceedings, Monsanto Technical Symposium*. The Monsanto Company.

Pollack, A. (2015, November 20). Genetically Engineered Salmon Approved for Consumption. *Associated Press*.

Priya, V. P., Celine, V. A., Gokulapalan, C., & Rajamony, L. (2007). Screening amaranth genotypes (Amaranthus spp.) for yield and resistance to leaf blight caused by *Rhizoctonia solani* Kuhn. *Plant Genetic Resources Newsletter (Rome, Italy), 147*, 1–4.

Ramdwar, M. N. A., Maharaj, R., & Siew, N. (2015). Sustaining The Environment: Farm and Beyond the Farm. Chapter 7 Pp 213-246. In W.G. Ganpat & W.P. Isaac (Eds.), Sustainable Food Production Practices in the Caribbean (vol. 2). Randle.

Ramgattie, R., Siew, N., Diptee, M., Stoute, V., & Knights. (2014). Effect of mammary stimulation on dairy cows and heifers exposed to a lactation induction protocol. *Open Journal of Animal Sciences, 4*(1). DOI:10.4236/ojas.2014.41001

Rijsberman, F. (2012). *CGIAR: A global research partnership for a food secure future*. Retrieved 25th December 2015, from http://www.cgiar.org/consortium-news/cgiar-global-research-partnership-for-a-food-secure-future/

Roe, D. (2008). *Trading Nature. A report, with case studies, on the contribution of wildlife trade management to sustainable livelihoods and the Millennium Development Goals*. Switzerland: TRAFFIC International and WWF International, Cambridge, UK and Gland.

Rumpold, B. A., & Schlüter, O. K. (2013). Potential and challenges of insects as an innovative source for food and feed production. *Innovative Food Science & Emerging Technologies*, *17*, 1–11. doi:10.1016/j.ifset.2012.11.005

Sandler, T. (1992). *Collective Action: Theory and Applications. University of Michigan Press.*

Shanley, P., & Luz, L. (2003). The impacts of forest degradation on medicinal plant use and implication for health care in Eastern Amazonia. *Bioscience*, *53*(6), 573–584. doi:10.1641/0006-3568(2003)053[0573:TIOFDO]2.0.CO;2

Shava, S. (2005). Research on indigenous knowledge and its application: A case of wild food plants of Zimbabwe. *Southern African Journal of Environment Education*, *22*, 73–86.

Singh, D. M., Adogwa, O. A., Mollineau, W., & Garcia, G. (2014). Gross and microscopic anatomy of the reproductive tract of the female agouti (*Dasyprocta leporina*): A neotropical rodent with potential for food production. *Tropical Agriculture (Trinidad)*, *9*(1), 38–46.

Srivastava, S. K., Babu, N., & Pandey, H. (2009). Traditional insect bioprospecting - As human food and medicine. *Indian Journal of Traditional Knowledge*, *8*(4), 485–494.

Steinfeld, H., Gerber, P., Wassenaar, T., Castel, V., Rosales, M., & De Haan, C. (Eds.). (2006). *Livestock's long shadow. Environmental issues and options*. Rome, Italy: Food and Agriculture Organization of the United Nations.

Stockbridge, M., Dorward, A., & Kydd, J. (2003). *Farmer organizations for market access: A briefing paper.* London: Imperial College.

Trueblood, M., & Shapouri, S. (2001). Implications of Trade Liberalizationon Food Security of Low-income Countries. United States Department of Agriculture. Agriculture Information Bulletin No. 765-5.

United Nations, Department of Economic and Social Affairs, Population Division. (2015). *World Population Prospects: The 2015 Revision, Key Findings and Advance Tables*. Working Paper No. ESA/P/WP.241.

van Huis, A., van Itterbeeck, J., Klunder, H., Mertens, E., Halloran, A., Muir, G., & Vantomme, P. (2013). *Edible Insects: Future Prospects for Food and Feed Security. FAO Forestry Paper 171*. Rome: Food and Agriculture Organization of the United Nations.

Varalakshmi, B. (2004). Characterization and preliminary evaluation of vegetable amaranth (Amaranthus spp.) germplasm. *Plant Genetic Resources Newsletter (Rome, Italy)*, *137*, 55–57.

Walters, L. M., & Keithly, J. G. (2012). *Caribbean Food Import Demand: Influence of the Changing Dynamics of the Caribbean Economy.* Presentation at the Southern Agricultural Economics Association Annual Meetings, Birmingham, AL.

Wilkie, D., & Carpenter, J. F. (1999). Bushmeat hunting in the Congo Basin: As assessment of impacts and options for mitigation. *Biodiversity and Conservation*, *8*(7), 927–955. doi:10.1023/A:1008877309871

Windapo, O., & Afolayan, S. O. (2005). Group Dynamics and Leadership in Agriculture, Extension. In S. F. Adedoyin (Ed.), *Agricultural Extension in Nigeria* (pp. 134–138). Ilorin Agricultural Extension Society of Nigeria.

World Health Organization (WHO). (2014). *Frequently asked questions on genetically modified foods.* Available from http://www.who.int/foodsafety/areas_work/food-technology/Frequently_asked_questions_on_gm_foods.pdf

Wu, H., Sun, M., Yue, S., Sun, H., Cai, Y., & Huang, R. et al. (2000). *Field evaluation of an Amaranthus genetic resource collection in China*. Genetic Resources.

ENDNOTE

1 Wageningen University and Research.

Chapter 10
Horticulture as a Reliable Source of Livelihood in Andaman and Nicobar Islands, India

Shrawan Singh
Indian Agricultural Research Institute, India

D. R. Singh
National Research Centre on Orchids, India

Subhash Chand
National Institute of Agricultural Economics and Policy Research, India

S. Dam Roy
Central Island Agricultural Research Institute, India

ABSTRACT

The tropical region is typically warm to hot and moist year-round with lush vegetation thus is favourable for horticultural diversity. The horticulture sector has shown remarkable progress in production and the shift in consumption patterns towards healthier diets within the principles of sustainability. Plantation crops are major contributors to the farm income and livelihoods of millions of people in the tropical region, while fruits and vegetables have direct roles in improved health through their micronutrient and antioxidants properties. Tuber and root vegetables are staple diets for indigenous tribes and one cannot imagine tribal life without coconut. Plantations crops and spices bring additional farm income and add flavour to foods and life. Medicinal plants have huge potential in the modern pharmaceutical and functional food industries. Research and development efforts have made horticulture an indispensable option for livelihood and nutrition, however, there are several issues which need timely attention for sustainable development.

INTRODUCTION

The potential of tropical horticulture is tremendous. In contrast to temperate regions, environmental conditions in the tropics are defined by less seasonal temperature fluctuations and more by seasonality of precipitation; however, temperature conditions are affected by altitude. The tropics have a rich biodiversity of flora and fauna but also have around 40% of world population which will be going to increase

DOI: 10.4018/978-1-5225-0942-4.ch010

55% by 2050. In addition, anthropogenic and climatic factors have made the region highly vulnerable to sustain the population with decent level of livelihood. The expanding population and shrinking land resources due to rising sea levels are creating further challenges. In the region, industrial development is slow and agriculture and tourism sectors are only source of food and livelihood security. They are also interdependent. In agriculture, horticultural crops, for example plantation crops, tropical fruits and vegetables, spices and medicinal plants are sources of income for the region. Historically, this sector has its roots in existing tribes and settler communities because they can be easily prepared for eating. However, efforts need to be made for sustainable utilization of the regions rich biodiversity and fragile natural resources. In this region, the economies and livelihood are traditionally mainly dependent on a horticultural based farming system. However, a recent trend shows that while the younger generation is opting for other occupations for livelihood and comfort, the horticulture sector still attract a large section of youth from rural and tribal regions. However, for its improvement due attention must be given to promotion of area-specific modern hi-tech technologies, ensuring timely supply of inputs, adopting high value crops, strengthening post harvest technologies, planning proper marketing strategies, effective disease and pest control through quarantine and integrated pest management, creating infrastructure development, crop diversification and integrated farming system, adopting organic farming, rejuvenation of existing old and semi-abandoned plantations with improved dwarf varieties, promotion of low-input-requiring locally adaptable crops, bioprospecting of underutilized horticultural crops, floriculture development through promotion of indigenous orchids and ferns and developing effective plans for germplasm conservation for long-term use and protection from biopiracy incidents. In tropical and coastal regions, the human resource development, effective waste land management and developing horticulture- based tourism will need to be further strengthened with adequate scientific information to present horticulture as a lucrative option for livelihood and thus attract the younger generation. Effective planning and implementation of action plans to address identified issues will certainly contribute to the mitigation of problems associated with employment opportunities and malnutrition, and in-turn improve the livelihood standard without further deteriorating the fragile island ecosystem. Along with this, systematic and robust weather forecasting and associated crop advisories will help to reduce the negative impact of climate change on horticulture in tropical regions. However, there is need for much systematic information on such efforts and strategies for horticultural development in tropical islands. This chapter reviews the efforts and proposes practical strategies for the development of the horticulture sector in tropical regions with special emphasis on Andaman and Nicobar Islands.

HORTICULTURE IN TROPICAL ISLANDS

There are 45 island nations in the world which differ from one another in terms of size, shape, wealth, and natural and economic resources. The tourism, agriculture and fishing sectors are three important sources of livelihood in tropical regions with limited contribution from the industrial sector. An analysis of the agricultural scenario of the islands indicates that agriculture is an important source of income in many of the tropical islands. Nearly 80% of the population of Comoros islands, 70% of Fiji, 65% of Vanuatu and 40% of Dominica are dependent on agriculture. Agriculture also contributes significantly to GDP of the tropical island nations like Solomon Islands (31.70%), Federal State of Micronesia (28.90%) and Madagascar (26.80%) (Singh & Singh, 2014). Horticultural crops occupy large land area with coconut, banana, coffee, black pepper, rubber, cocoa vanilla, and along with tuber crops contribute significantly in their national agricultural output. Island nations with developed horticultural sectors have low unemployment level (Singh, 2011) which signifies the potential of horticulture in generating employment opportunities and in developing the sector as best- livelihood option. They also experience similar constraints to sustainable development arising from small land size, geographical dispersion, vulnerability to natural hazards and disasters, limited terrestrial natural resources, heavy dependence on imports, limited commodities, isolation from markets and many other characteristics and processes. In addressing these well-documented constraints, as well as preparing for new and evolving challenges, small-island nations must continue to draw on their noted capacity for adaptation and innovation and their proven determination and capability to overcome adversity.

The horticulture sector has shown great contribution to the development of many of the countries/ regions in the Mediterranean and temperate regions. Even in the tropics, a number of islands or nations are known for one or more horticultural crops. Notably, tropical regions cover 20% of land surface and can sustain 40% of the world's population mainly because horticulture provides higher returns per unit area as well as employment; even with climate change impacts. Several efforts have been made through international collaborations and regional governments or organizations to promote the horticulture sector in tropical islands. In tropical island regions, though paddy is an important crop in low land areas, horticulture has emerged as best- option for providing food, nutrition and livelihood security. The cultivation of these crops is generating employment opportunities in rural areas while marketing and postharvest management activities are sources of livelihood for youth in urban areas. In many of the tropical islands, established export houses are participating in the international trade of horticultural crops which bring in a lot of foreign exchange and help to reduce the export-import gap. These crops are also used for processing into various products like pickles, preserves, beverages, jam, and squash, which offer employment opportunities to the rural women. The focus on value addition in the horticulture sector is vital for the comprehensive development of the rural economy, which creates jobs and a demand for agro-raw materials. This leads to diversification and commercialization which enhances the income of farmers and creates surpluses for export.

Like other tropical islands, the climate of the Andaman and Nicobar Islands is approximately similar to maritime climate. It has high annual rainfall (3100-3900 mm), high relative humidity (65-92%) and narrow range of minimum and maximum temperatures (20 - 32°C). The islands have two crop seasons as per the occurrence of rainfall pattern, namely, rainy season (May-December) and dry season (January- April). The rainy season is characterized by abundant rainfall (400-600 mm/ month) and average temperature ranges from 23°C to 27°C. The relative humidity also remains much higher (80% or above) during the rainy season. The dry season, is a rain-free period (except some incidental heavy rains during February-March months) and comparatively lower average temperature 21°C – 24°C) and the relative humidity remains low (65 – 75%). A number of horticultural crops or certain varieties in some crops can thus be grown in the islands however, it requires proper analysis with reference to crops, growing periods and suitable technological options.

ANDAMAN AND NICOBAR ISLANDS (A & N)

Andaman and Nicobar Islands, known as the 'Emerald Isles', and form an archipelago of 572 islands spread over 8,249 sq. km. in the Bay of Bengal. The islands are spread between 6° N to 14° N latitude and 92° E to 94°E longitude and bestowed with humid tropical climate (Figure 1). The Andaman Islands

Figure 1. Geographical location of Andaman and Nicobar Islands (Subhash et al. 2015)

and the Nicobar group of Islands are separated by the 10°N Channel (about 90 miles/ 145 km wide), with the Andaman group lying towards the north while the Nicobar group towards the south. The islands are a rich centre of horticultural biodiversity and recognized as one of the hot spots for biodiversity (Myers, 2003). The island biodiversity is situated between two major biodiversity hot spots, namely the Indian sub-continent and the Malaysia- Indonesia region. Notably, the islands manifest biodiversity of extraordinary range with unmatched distribution of plants representing the Indian, Myanmarese, Thai, Malaysian and Indonesian floras. The flora of the Andaman group of islands shows closer affinity to the Indo-Myanmarese-Thai flora, while the Nicobar group of islands is closer to the flora of Malaysia-Indonesia (Balakrishnan & Ellis 1996). The local diversity of horticultural crops has great potential in developing island-specific varieties or the mining of important traits/genes for use in breeding programmes (Srivastava et al. 2009).

Horticultural crops are playing significant role in improving factor productivity, generating employment opportunities, livelihood and economic conditions of the farmers and also ensuring nutritional security to the inhabitants (Singh et al. 2013). Horticulture contributes nearly 24.5% of the national share of agricultural GDP of agriculture in these islands, and is considered as a third pillar of the island economy. The horticultural based farming system occupies about 70% percent of total cropped area (50000 ha), with major horticultural crops being coconut (21768 ha), rice (9081 ha), vegetables (6695 ha), areca nut (4152 ha) and fruits (3160 ha) (Basic Statistics, 2011). Although research and development efforts have been made in the horticulture sector, they are still inadequate to attain the desired results in the islands (Development Report, 2008). Innovative technologies such as improved seeds, good production and protection technologies, efficient tools for land, water and weed management, have huge scope in the island horticulture.

Although a Tsunami in 2004 caused huge losses to the Islands' agriculture, the production of fruits and vegetables rapidly recovered and showed a positive growth again. For this, the financial and input support from state agriculture department and technological back stopping has hastened the pace of their recovery (Chand et al. 2011). Further, the 'demand factor' also played a crucial role in the rapid improvement in vegetable and fruit production. The increase in local population and rise in inflow of floating and tourist populations also contributed to the demand. It was also increased due to a rise in per capita income as a result of post-Tsunami rehabilitation; infrastructure and overall economic activities raised the purchasing power of the islanders in past few years (Chand et al. 2013). Health awareness also played a key role in stimulating the consumption of several underutilized fruits and leafy vegetables (Singh et al. 2014). The rise in consumption was also due to the improvement of transportation infrastructure, particularly the aviation sector and an increase in ferry services between islands (Salam 2010).

Although, the climatic conditions of islands are suitable for various horticultural crops, productivity still remains lower than national average in most of the crops, suggesting the need for major interventions to explore the prospect of island horticulture. The research and development efforts in the horticulture sector during the past three decades showed that substantial improvement in productivity of these crops in the islands is limited by major constraints such as investment in production of disease-free, quality planting materials of improved varieties, promotion of rain water harvesting and drip irrigation, plastic mulching, modules for organic farming and integrated farming, improvement in the marketing and processing sector and human resource development.

Tribes of the Islands

The tribes of Andaman and Nicobar Islands are dependent on the horticulture products and by-products found in the forest areas. The need for food, shelter and indigenous medicines are met from the forest, particularly from the wild horticulture plants. The tribes indigenous to the islands include the Jarawas, the Sentinelese, the Great Andamanese, and the Onges, who reside in the interior forests of Sentinel Islands, Strait Islands and Little Andaman. In 2011, out of a total Andaman and Nicobar population of 384 thousands, the population of indigenous tribes was very meagre and counted in numbers as only 300 Jarawas, 90 Sentinelese, 98 Onges, 48 Greate Andamanese and over 280 Shompens who constitute the primitive tribe (Directorate of Economic & Statistics Census, 2011)).

Agriculture

The agriculture sector provides considerable livelihood support to the local population in A&N Islands. Crops such as coconut, areca nut, and other plantation crops are predominant, mostly in the upper reaches. Vegetables, paddy, spices and fruits are cultivated in valley areas. Pulses and oilseeds are also grown in the rice fallows during the dry season. Coconut occupies more than 50% of the total cultivable area and is the main source of income and livelihood for the farming community. The A&N Islands have a total of 11,349 agriculture holdings which comprises 9,444 individual holdings, 1783 joint holdings and 122 institutional holdings. It is also observed that the number of small holdings has increased over the years

(Chand et al., 2012). In the recent past, the production of perishable food items experienced a spurt because of increased population, inflow of tourists and increased per capita income in the Bay Islands as well as in India. With the research efforts of the Indian Council of Agricultural Research (ICAR) - Central Island Agricultural Research Institute (CIARI), there is a move towards protected cultivation, particularly of fruits, vegetables and flowers.

Animal Husbandry and Fishery Sector

Animal husbandry is an important component of farming in A&N Islands. The livestock population in these islands consists of non-descript cattle, buffaloes, pigs, goats, poultry and ducks. According to the Livestock Census, 2007, these islands possess a total livestock population of 173,970 that include buffaloes, cattle, goats and pigs. Due to continuous inbreeding/ indiscriminate breeding, none of the germ plasm is in pure form resulting in poor productivity. Total milk production in the islands is about 25.28 thousand tonnes (2011). This is however constrained by poor veterinary services, poor quality of animal breeds, and unavailability of quality green fodder during dry phase and dry fodder during rainy season. Other constraints to improved milk production include, worm infestation, high humid and tropical climate, and inadequate attention to minerals and dietary supplements to animals among others (Kundu et al., 2010).

The total potential of marine fisheries is about 0.14 million tonnes in A and N islands. Due to a variety of constraints including insufficient infrastructure and skills, only 20% of the potential is being exploited at present. The capture of fishery is mainly restricted to only 6 nautical miles. The bulk of the potential lies on tuna fisheries which amounts to about 0.65 million tonnes, but the fishermen do not have the skill and required infrastructure to optimally harness the available tuna resources (Chand et al. 2011).

Carrying Capacity of A and N Islands

Available estimates indicate that, given the existing climate, land and water resources as well as technology, the islands can only meet the protein-calorie requirements of a population of about 0.016 million. This puts the maximum carrying capacity with an advanced level of technologies at 0.025 million people. Another estimate which considered the potential of the ocean resources, places the carrying capacity at 0.085 million (Reddy, 2007). However, a criticism associated with this estimate is that it only indicates the upper limit of carrying capacity at a given point of time, but not the one that the islands can sustain over an extended period, environmentally or economically (Dhingra, 2005). It is quite obvious that the current population of 0.038 million have crossed the threshold as per the first estimate noted above and therefore, further expansion needs careful planning and execution of promising programmes. The local demand of fresh vegetables can be met through increasing the area under improved seeds, promotion of suitable protected cultivation technology and efficient cold storage and transportation facilities and so on (Singh et al. 2015[a]).

Tourists Inflow as an Opportunity for Development in A&N Islands

The statistics registered on 1980 by the Andaman and Nicobar administration indicated that only less than 10,000 tourists were visiting every year. Later, with the establishment of adequate infrastructure for accommodation and travelling by the authorities, the tourists' numbers started increasing. International tourists from 162 countries have visited the islands. It could be observed that tourists' arrival from different destinations have shown an increasing trend from 2004-05 to 2010-11 (Rengarajan et al. 2014).

Horticulture Crops in the Andaman Islands

The predominance of horticultural based farming system traced back from colonization of these islands. As stated before, the major horticultural commodities in islands are coconut, areca nut, vegetables, fruits, spices and medicinal plants (Table 1). Commodity-wise area and production- important horticultural crops is given in Figure 3. The horticultural crops are occupying about 70% percent of total cropped area (50000 ha) in the islands. They occupy around 39575 ha area with 180740 MT production (NHB Database, 2014). Since 1991-1992, the horticultural sector showed notable improvement in area, production and the introduction of new crops. This could be due to introduction of varieties having local adaptability, suitable technologies, financial and input support by local administration and assured market with higher prices for horticultural crops (Srivastava et al. 2010; Chand et al. 2013; Singh et al. 2015[a]). However, local production of fruits and vegetables is still unable to meet the demand of island population and tourists. The huge gap between local demand and production of flowers is filled by importing from the mainland resulting in huge money drains from the islands.

Unfortunately, the productivity of horticultural crops on the islands still remains near to stagnant over the years. The comparative analysis with other states in approximately similar climate conditions and national figures also showed a very grim situation of the horticulture sector in the islands (Figure 2; Table 2). This trend of low increase in production is mainly due to low progress in area and productivity front (Table 3). Lack of suitable varieties, appropriate production technologies, poor transport and marketing facilities and high rate of post-harvest losses also contributed in poor trend in this sector. Climatic vagaries, high incidence of insect pest and diseases, weed infestation and poor irrigation facilities in

Table 1. Major horticultural crops grown in the Islands

Category	Crops	Source of Seed/Planting Material	Area (ha)	Production (MT)
Plantation crops	Coconut, areca nut and cashew nut	Local and improved varieties from mainland	27,330	95,710*
Spices	Black pepper, clove, nutmeg, cinnamon	Introduced from mainland India	1,675	3,220
Fruit crops	Banana, mango, pineapple, guava, papaya, citrus,sapota	Local and introduced from mainland India	3,550	29,730
Vegetables	Okra, brinjal, chilli, cowpea, cucurbits, tomato, leafy vegetables, cauliflower, cabbage, beans, drumstick, radish	Local and introduced	6,890	51,790
Tuber crops	Elephant foot yam, tapioca, sweet potato, tannia, arrow root, colocasia, Ariel potato	Local and introduced	410	6940
Flowers and ornamentals	Gerbera, marigold, crossandra and orchids, ferns, Heliconia, jasmine, tuberose, china rose	Introduced from mainland India	130	290
Medicinal plants	Morinda, tulsi, kalmeg, pippli, pudina, sambalu patti *etc.*,	Local	N.A.	N.A.

(NHB Database, 2014)

N.A. – Not available; *coconut in numbers

Figure 2. Major horticultural crops grown in A and N Islands 2013-14 (NHB Database, 2014)

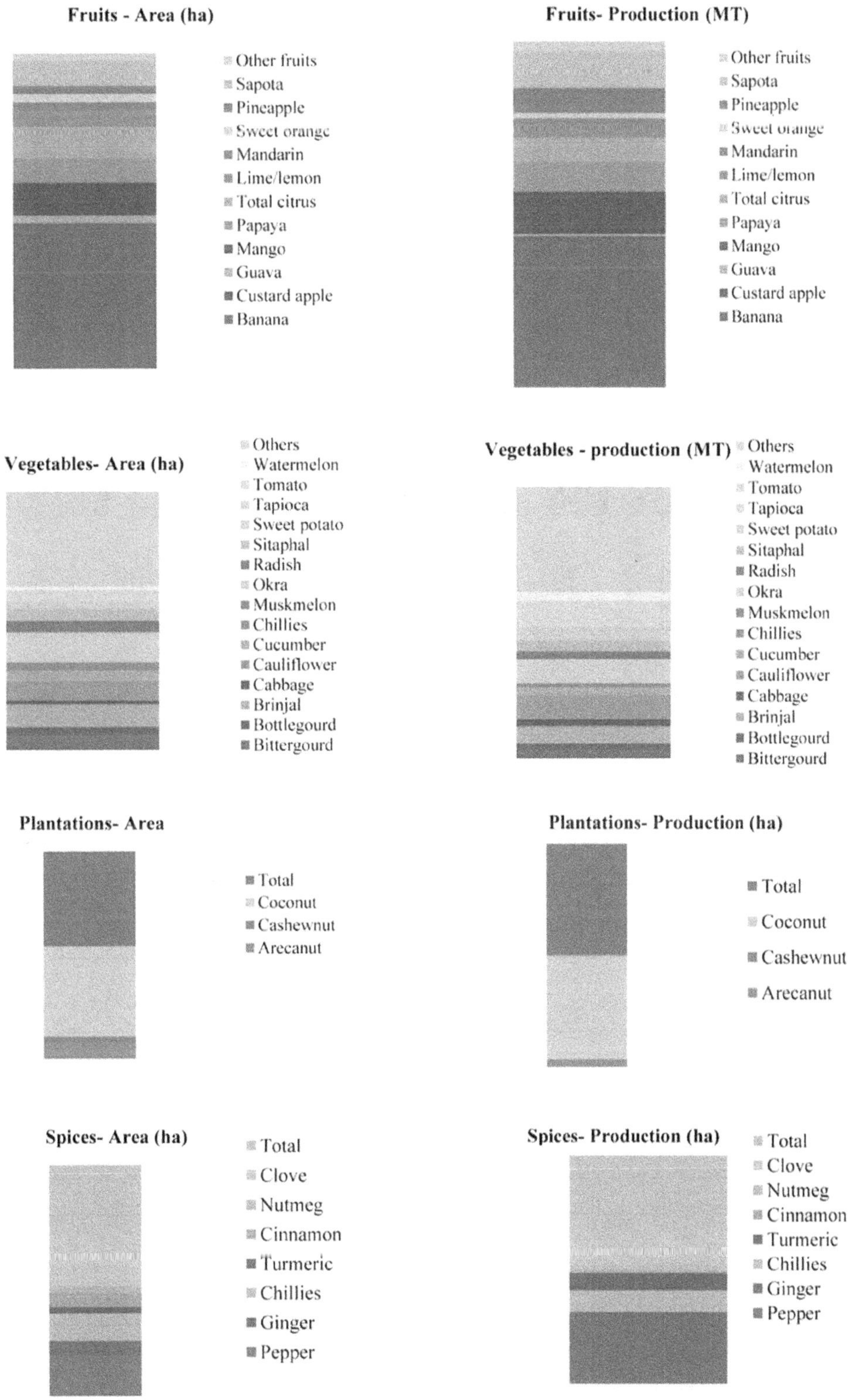

Figure 3. Comparison of productivity (MT/ha) of islands verses southern states in mainland India

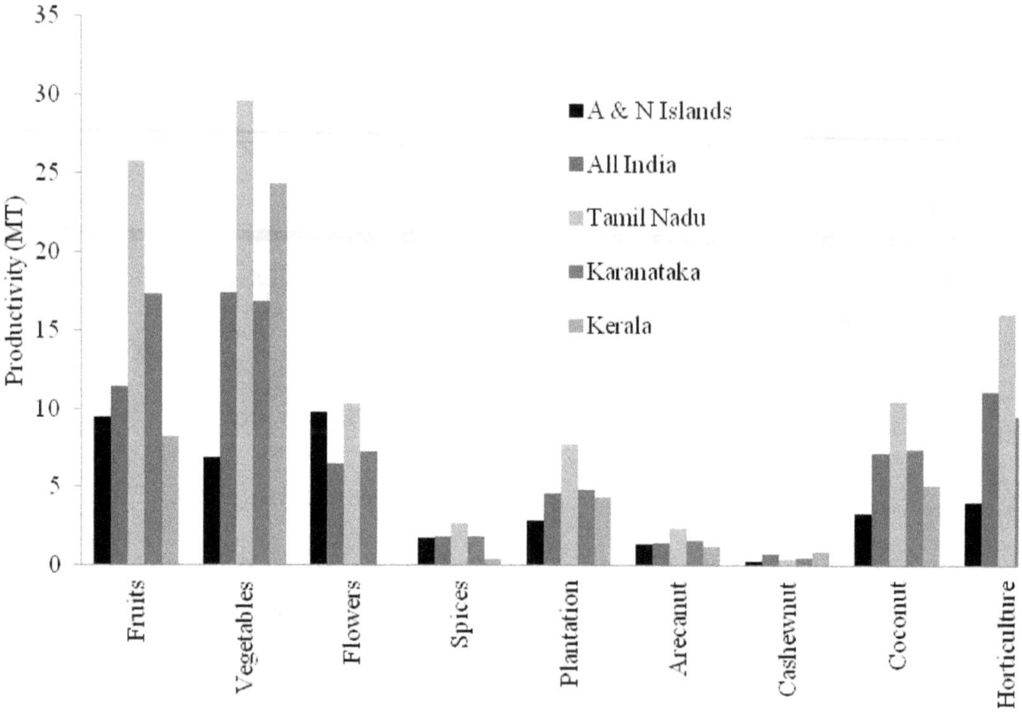

Table 2. Current scenario of horticulture crops in A & N Islands (2013-14)

Crops	Andaman and Nicobar Islands			India			% Difference in Productivity (Island vs Mainland)
	Area (000 ha)	Production (000 MT)	Productivity (MT)	Area (000 ha)	Production (000 MT)	Productivity (MT)	
Banana	1.8	14.0	7.7	802.6	29724.5	37.0	380.11
Custard apple	0.0	0.0	1.3	21.8	165.2	7.6	468.93
Guava	0.1	0.2	2.6	268.2	3667.9	13.7	420.96
Mango	0.4	3.9	9.6	2516.0	18431.3	7.3	-23.89
Papaya	0.3	2.7	8.4	133.4	5639.3	42.3	401.17
Total citrus	0.4	2.3	6.6	1077.7	11147.1	10.3	57.40
Lime/lemon	0.2	1.4	7.9	286.4	2835.0	9.9	24.59
Mandarin	0.1	0.4	4.3	330.0	3431.4	10.4	139.98
Sweet orange	0.1	0.5	6.0	334.9	3886.2	11.6	93.38
Pineapple	0.1	2.3	19.2	109.9	1736.7	15.8	-17.53
Sapota	0.3	3.3	11.8	1345.7	177.0	0.1	-98.88
Other fruits	0.1	1.0	7.1	900.6	5370.6	6.0	-16.51
Total fruits	3.6	29.7	8.4	7216.3	88977.1	12.3	47.23
Bittergourd	0.4	1.7	4.3	78.9	807.5	10.2	140.85
Bottlegourd	0.2	1.1	4.8	103.2	1818.9	17.6	270.96

continued on following page

Table 2. Continued

Crops	Andaman and Nicobar Islands			India			% Difference in Productivity (Island vs Mainland)
	Area (000 ha)	Production (000 MT)	Productivity (MT)	Area (000 ha)	Production (000 MT)	Productivity (MT)	
Brinjal	0.6	3.3	5.8	711.3	13557.8	19.1	230.23
Cabbage	0.1	1.3	15.6	400.1	9039.2	22.6	44.58
Cauliflower	0.5	4.6	9.4	1073.7	433.9	0.4	-95.70
Cucumber	0.3	1.6	5.4	43.3	678.1	15.7	192.00
Chillies	0.2	0.4	1.9	140.0	1687.3	12.0	529.81
Muskmelon	0.0	0.3	15.5	36.7	760.8	20.7	33.76
Okra	0.8	4.6	5.4	532.7	6346.4	11.9	119.00
Radish	0.3	1.5	5.4	173.3	2484.8	14.3	164.08
Sitaphal	0.3	2.0	6.7	19.8	416.1	21.1	212.74
Sweet potato	0.2	2.7	15.8	105.9	1087.9	10.3	-35.06
Tapioca	0.2	4.3	17.7	228.3	8139.4	35.7	101.35
Tomato	0.1	0.7	5.4	882.0	18735.9	21.2	294.49
Watermelon	0.1	1.7	20.6	74.6	1809.8	24.2	17.56
Others	2.5	20.1	8.0	1574.4	19108.2	12.1	52.55
Total	6.9	51.8	7.5	9396.1	162896.9	17.3	130.64
Arecanut	4.2	5.9	1.4	451.9	622.3	1.4	-0.94
Cashewnut	1.2	0.4	0.3	101.7	265.2	2.6	723.28
Coconut	21.9	89.5	4.1	2984.3	465.3	0.2	-96.18
Total	27.3	95.7	3.5	2379.8	461.2	0.2	-94.47
Pepper	0.6	0.1	0.2	32850.8	374579.5	11.4	5162.67
Ginger	0.2	1.9	8.9	65568.3	743519.7	11.3	27.65
Chillies	0.4	0.6	1.5	130058.8	1475892.4	11.3	644.12
Turmeric	0.1	0.5	5.9	259831.2	2948949.7	11.3	93.18
Cinnamon	0.2	0.0	0.3	2.8	5.1	1.8	583.66
Nutmeg	0.1	0.1	0.7	18.9	12.8	0.7	-5.33
Clove	0.2	0.0	0.1	2.1	1.1	0.5	731.07
Total	1.7	3.2	1.9	3163.2	5908.3	1.9	-2.84

(Basic Statistics, A &N Administration, 2. NHB Database 2014)

dry phase are other factors responsible for this grim situation. Although the percent change in area and production is significant in vegetables, fruits and flower, productivity still follows a zig-zag pattern over the years as shown in Table 4. However, the demand has increased manifold due to increased population and tourist inflow. As a result, the islands are facing the scarcity of locally grown horticultural products and largely depend on shipment from the mainland. This results in the peculiar phenomena like ship to mouth, monotonous dietary pattern and huge transportation losses with a resultant high cost of produce.

Table 3. Trend of Area, Production and Productivity of horticultural crops in A&N Islands 1991-92 to 2013-14

Crops	1991-92		2001-02		2005-06		2006-07		2007-08		2008-09		2009-10		2010-11		2011-12		2012-13		2013-14	
	A (ha)	P (MT)	A (ha)	P (MT)	A (ha)	P (MT)	A (ha)	P (MT)	A (ha)	P (MT)	A (ha)	P (MT)	A (ha)	P (MT)	A (ha)	P (MT)	A (ha)	P (MT)	A	P	A	P
Fruits	3300	12900	3700	16700	2900	18500	2800	20800	3000	22500	3005	24950	3116	3160	3160	28772	3240	30500	3245	30730	3550	29730
Spices	-	-	1261	1726	1813	4252	1746	3571	1700	3120	1659	2541	1662	1661	1661	3067	1650	2980	1735	3570	1675	3220
Vegetables	3400	13200	3100	15800	3669	25682	4300	32600	4000	30800	4599	30199	5201	5150	5150	31300	6310	43210	6362	44158	6890	51790
Roots	384	2216	-	-	1701	10257	1314	8077	-	-	866	5411	450	449	449	5193	¥	¥	420	2970	410	6940
Coconut*	24115	844	25200	890	20927	780	21400	812	21600	552	21690	819	21760	21768	21768	950	21800	722.5	21900	89450	21900	89450
Areca nut	3100	4700	4400	4550	4046	3058	4100	5800	4100	5700	4148	5721	4153	4152	4152	5800	4220	5950	4230	5880	4230	5880
Flowers	-	-	10	250	10	254	10	254	30	470	34	335	34	340	34	329	40	340	40	350	130	290

*Production as ('00000 nuts); ¥ - Included in vegetable crops.

(Basic Statistics, A &N Administration, 2. NHB Database, 2014)

Table 4. Percent change in growth trend in horticultural crops in A and N Islands 1991-92 to 2010-11

Crops	Area			Production			Productivity		
	2000-01 (over 1991-92)	2005-06 (over 2000-01)	2010-11 (over 2005-06)	2000-01 (over 1991-92)	2005-06 (over 2000-01)	2010-11 (over 2005-06)	2000-01 (over 1991-92)	2005-06 (over 2000-01)	2010-11 (over 2005-06)
Fruits	12.12	-21.62	8.97	29.46	10.78	55.52	15.46	41.34	42.73
Spices	-	43.74	-8.36	-	146.34	-27.86	-	71.38	-21.28
Vegetables	-8.82	18.35	40.37	19.70	62.54	21.88	31.28	37.34	-13.18
Roots	-	-	-73.60	-	-	-49.37	-	-	91.80
Coconut	4.50	-16.96	4.02	5.45	-12.36	21.79	0.91	5.54	17.09
Areca nut	41.94	-8.04	2.61	-	-	89.67	-	-	84.84
Flowers	-	0.00	236.50	-	1.60	29.64	-	1.60	-61.47

(Basic Statistics, A &N Administration, 2. NHB Database 2014)

These factors have led the islands to a high cost economy and are proving a deterrent to tourist inflow which in turn restricts development. Further, after a Supreme Court order banned the logging of forests, employment opportunities have shrunken.

On a positive note, climatic suitability and assured market indicate a huge potential of horticultural crops in the islands both for generating income and employment through cultivation, postharvest management, processing and marketing. Interestingly, most of the commercial horticultural crops in the islands were introduced from the mainland India with some coming from adjoining regions. Some of them got high acceptance while many were rejected due to lack of acceptance in traditions, taste and cultivation difficulties. The domestication of a few indigenous crops also took place and their commercial cultivation started within recent years but their share is negligible in the island horticulture which is dominated by the commercially introduced horticultural crops.

Coconut and areca nut are predominant plantation crops in the islands which covers around 60% of agricultural land. The coconut and areca nut based farming systems are the dominant systems in the islands' agriculture. They are also integral component of homegardens in the islands, particularly in native tribal homegardens. However, the area under coconut and areca nut in the pre-tsunami period was 25,200 ha and 4,400 ha (in 2001-02) respectively, which reduced to 20,927 ha and 4,046 ha respectively in 2004-05. The decrease in area was primarily due to loss of land by sea water incursion in existing plantations in the coastal areas. It has affected the total production of coconut which came down from 890 (2001-02) to 780 million nuts (2004 05) and areca nut from 4450 MT (2001-02) to 3056 MT (2004-05). However, these production levels are increasing due to financial support and technological interventions through the Coconut Mission (Department of Agriculture, Andaman and Nicobar Islands), ICAR-CIARI, Port Blair and NGOs. The systematic rejuvenation of the old coconut plantations using Noni (*Morinda citriflia* L.) is also undertaken by the local administration with the help of ICAR-CIARI (Singh et al. 2012a). However, no significant improvement was recorded in productivity of coconut and areca nut. This is one of the important reasons for the very meagre contribution of the islands to national coconut (approx.1%) and areca nut (1-1.5%) production. The productivity of coconut is 3,402 nuts per hectare and average yield of the palm in the islands (20 nuts/ palm/ year) is far below than national average of 44 nuts/ palm/ year.

During 1991-2012, the area under fruits showed a marginal increase while a noticeable increase was witnessed in vegetable across the islands. The productivity of fruits was increased due to the adoption of improved production technology for banana, papaya and mango and the supply of quality planting materials by the local administration (Srivastava et al. 2010; Singh and Singh, 2014). The major fruit crops in Andaman and Nicobar Islands are banana, papaya, citrus, mango, pineapple, and to lesser extent underutilized fruits. Though fruits are grown throughout the islands in fragmented plantations, the major growing areas are Neil Islands, Havelock, Hut Bay and North and Middle Andaman. Banana is the most important fruit crop covering about 61% of area and accounting for nearly 59% of total fruit production of the islands. The islands have rich diversity of underutilized fruits which have good extent of inter-specific and intra-specific diversity. These wild or neglected species have great potential in household food items or small scale value added products for tourists and floating population (Singh et al. 2012b). These plants are well adapted to the islands due to favourable climatic conditions. Papaya is ranked as second in area and production after banana. The absence of destructive viral diseases like papaya ring spot virus, favourable climatic conditions and the existence of local types favour cultivation of papaya in the islands. The low production level of other fruits is mainly due to poor availability of quality planting materials and high incidence of diseases and pests. However, the supply of quality planting materials from mainland-based tissue culture laboratories and nurseries by the local administration improved input supply. The adoption of production practices such as drip irrigation, insect-pest management, manuring and fertilizers are other major factors which increased productivity. Some areas have been identified for enhancing vegetable productivity in islands; these include identification/ development of suitable varieties for islands conditions, promotion of indigenous/ underexploited vegetables and the adoption of improved production technologies like broad bed and furrow system, raised bed technology, kitchen/ home gardening etc. Interventions such as protected cultivation, plastic mulching, and micro irrigation also need to enhance productivity of vegetables in the islands.

The major spices in islands are black pepper (*Piper nigrum* L.), nutmeg (*Myristica fragrans* Houtt), cinnamon (*Cinnamomum verum* J.S. Presl), clove [*Syzygium aromaticum* L.) Merrill & Perry] and ginger [*Zingiber officinale* (Willd.)Rosc]. These are being grown on 1,675 ha with total production of 3220 MT (in 2013-14) Table 2. Black pepper is commonly grown in between areca nut and coconut plantations in upland condition (Pandey and Singh, 2010). However, in recent years the response of farmers for adoption of spices in islands was not very encouraging and productivity level also showed declining trend (Table 3). The field surveys by CARI indicated that poor marketing and processing set up for spices was one of the major reasons behind the sluggish growth of the spice sector in the islands. Further, the area and inputs for spices were diverted to other highly remunerative crops like vegetables and flowers which also got full support from the state agriculture department. In the islands, spices like pepper, clove, nutmeg, and cinnamon are grown under multi-tier cropping system and are also important constituents in homegardens (Pandey et al. 2006). Some of the islands produce 'organic spice' which need formal certification and measures to enhance the productivity in organic system. The productivity of spices like black pepper (0.05 t/ha), cinnamon (0.12 t/ha) and clove (0.04 t/ha) is however below the national average of 0.3, 2.1 and 0.5 t/ha, respectively (Table 2).

Flowers have a huge demand in the islands and most of it is being met through shipments from the mainland. In the pre-tsunami period, the commercial cultivation of flowers was negligible but has increased significantly during past few years (Tables 2&3). The immediate returns, good adaptability and secure market attracted farmers towards flower cultivation. The technological backup and financial support from different agencies facilitated the farmers to adopt flower cultivation (Srivastava et al. 2010). As a result, around 34 ha have come up under flower crops in the past 5 years but the level of productivity was not very high. One of the main reasons suggested was the lack of suitable varieties and associated technologies available to the farmers. Secondly, heavy and untimely rains also caused huge losses to the flower crops in the islands.

The islands have been considered as a treasure house of valuable medicinal and aromatic plant species. The Botanical Survey of India (BSI) identified nearly 300 medicinal plants in these Islands, and from them, 14 medicinal plants were selected on priority basis by the Andaman and Nicobar Medicinal Plant Board (ANMPB) for commercial cultivation. These were: Amla (*Emblica officinalis* L.), Ashok [*Saraca asoca* (Roxb.) de Wilde], Arjuna [*Terminalia arjuna* (Roxb.) Wight & Arn.], Bael (*Aegle marmelos* (L.) Corrêa), Brahmi [*Bacopa monnieri* (L.) Wettst], Kalihari (Gloriosa superba),Kalmegh [*Andrographis paniculata* (Burm.f.) Wall. ex Nees], Pippali (*Piper longum* L.), Patcholi [*Pogostemon cablin* (Blanco) Benth], Citronella [*Cymbopogon citrates* (DC.) Stapf.], Lemon grass (*Cymbopogon citratus* (DC.) Stapf.), Tulsi (*Ocimum Sanctum* L.), Stevia (*Stevia rebaudiana* L.) and Vanilla (*Vanilla planifolia* Jacks). Medicinal plants have good potential, in the islands particularly in homegardens, for traditional medicine system as well as commercial farming (Srivastava et al. 2010). The islands are very much suitable for the cultivation of oyster, milky and paddy straw mushroom. Around 98 species of mushroom were documented and promising ones as *Cantherellus cibtarius, Lycopordon pyriformae, Scleroderma verucosum, S. ctrinum Scleroderma verucosum, S. ctrinum* and *Clitocybe floccida* have been identified in the islands and catalogued.

Variations in Island Horticulture

The islands are characterized by variability in several areas which includes climatic variability from the Northern Islands to Southern Islands and crop factors such as input supply, and technological interventions. Such variability results in inter-island variations in cropping pattern and productivity. Neil Island and North Andaman both contribute a large share of total vegetable production, while the share of the southern group of islands to total vegetable production is not only very low but greatly reduced in the post- tsunami period (Figures 4a-b).

Although the climatic conditions are suitable for growing the large number of vegetables, the prolonged rainy season in the May-December period and the moisture deficiency in February to April period reduce the crop growing period. However, during the rainy season only about 700 ha in the uplands is utilized for growing vegetables. This is not sufficient to meet the huge market demand and results in a market deficit. On the contrary, in the dry period (January to April), a large portion of the lowlands and a small portion of uplands are being cultivated for vegetables. This produces enough vegetables to meet the local market demands. However, the prolonged market deficiency for vegetable in the islands creates a monotonous dietary pattern; a 'ship-to-mouth' phenomena for many vegetables, with associated high transportation costs and results in malnourished children and women. It has also led to reduced employment opportunities for islanders. Over and above, this situation also exacerbates climate change effects through increased use of fuel in transporting vegetables from the mainland.

Price Trend of Fruits and Vegetables

The market analysis for vegetables in the islands during 2008 to 2011 showed a volatile behaviour for major vegetables which was governed by crop season, market channels and transport services. The price

Figure 4. (a-b). Spatial distribution (a) area and (b) production of horticultural crop in A & N Islands

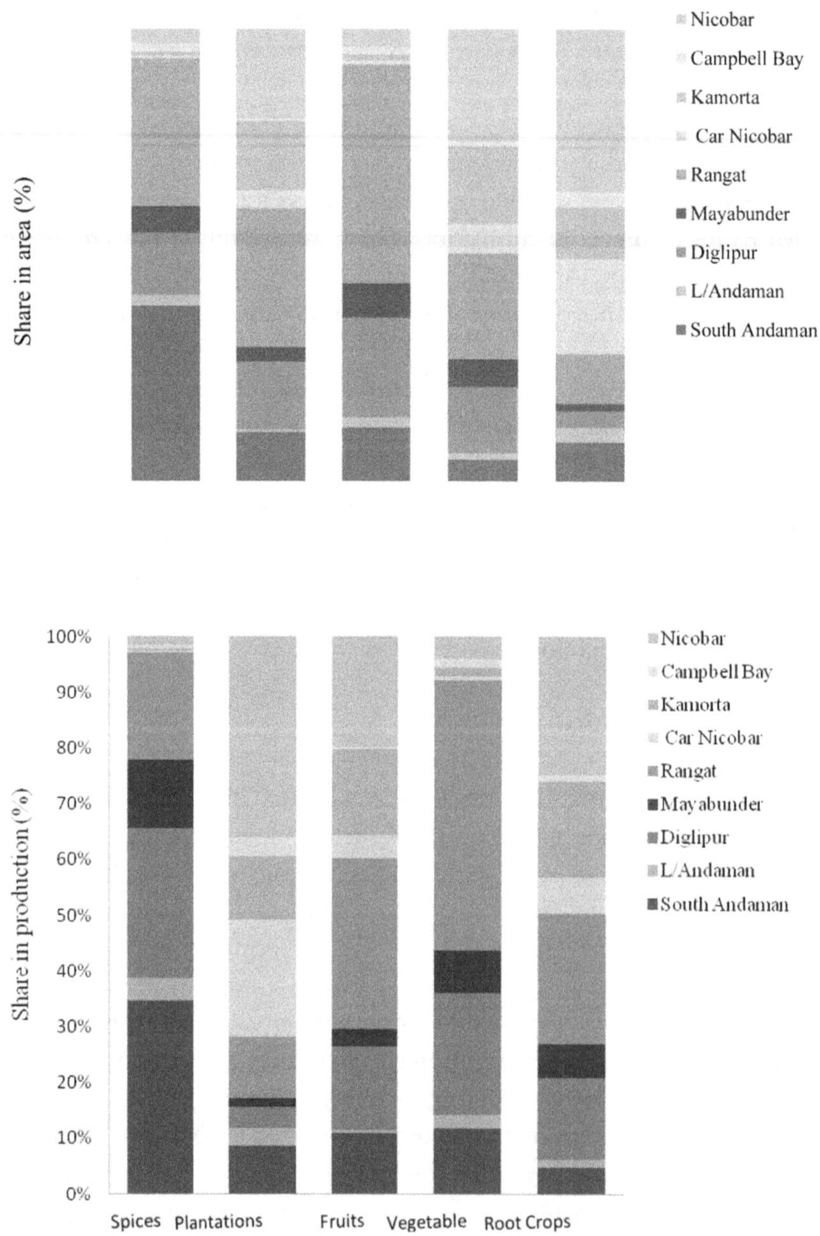

of major vegetables remains low only in Port Blair during the rainy season due to transportation from mainland (Table 5).

However, the sharp decline in the prices during the December to March months was due to increased local production, low price of winter vegetables in mainland India and increased availability of alternate food during this period (Figure 5). However, the prices of capsicum remain very high during whole year. The leafy vegetables show a different trend as their prices go up during March – May months, primarily due to low production. This low production is mainly due to reduced irrigation, no harvest from farm

Table 5. Price (INR) trend analysis of fruits and vegetables in islands

Vegetable	Botanical name	Price (INR/kg) at Mainland India	Range of Market price (INR/kg) in islands		
			Port Blair	Diglipur	Neil Island
Capsicum	*Capsicum annum* L.	15-30	80-150	90-170	90-150
Tomato	*Solanum lycopersicum* L.	5 30	15 40	10 50	5 45
Cucumber	*Cucumis sativus* L.	10-15	20-40	20-30	20-30
Brinjal	*Solanum melongena* L.	10-15	20-30	8-20	8-15
Cauliflower	*Brassica oleracea* var. *botrytis* L.	8-15	40-120	20-130	50-130
Coriander	*Coriandrum sativum* L.	15-25	60-80	50-70	40-60
Potato	*Solanum tuberasum* L.	5-15	15-30	20-40	20-35
Onion	*Allium cepa* L.	5-15	20-40	25-50	25-45
Drumstick	*Moringa oleifera* L.	40-60	80-150	60-80	40-50
Leafy vegetables		10-20	10-30	10-20	10-15
Banana	*Musa × paradisiaca* L.	20-30	25-50	20-30	20-30
Papaya	Carica papaya L.	10-20	20-50	20-40	20-30
Pineapple	*Ananas comosus* (L.) Merr.	20-40	30-80	30-50	30-50
Mango	*Mangifera indica* L.	15-40	40-120	30-50	30-60

Figure 5. Mean monthly price trend of vegetables in Islands (based on2010-11 prices)

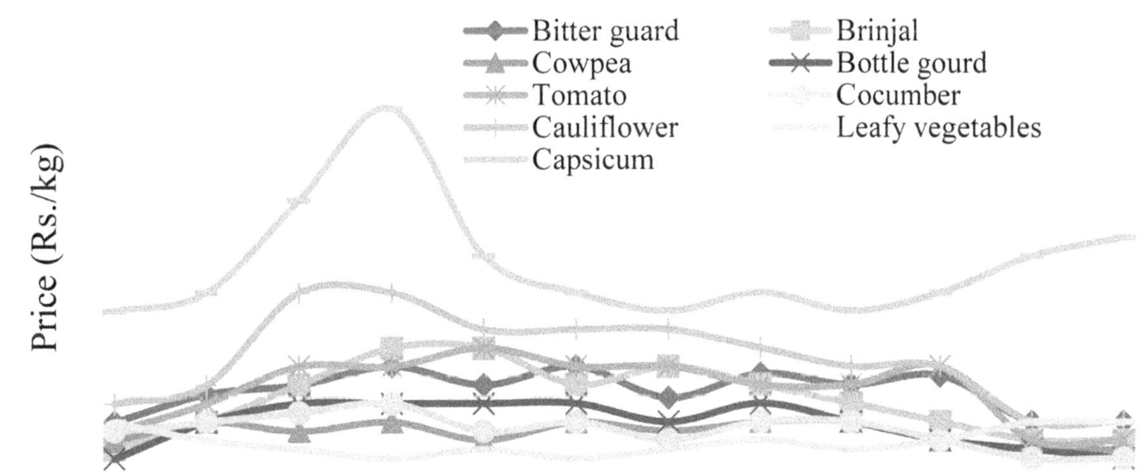

bunds, forests or home gardens and major emphasis on commercial vegetables (Singh et al. 2015a). Further, day-to-day changes in prices were also observed for vegetables in the islands as reported by Myint (2003) in Myanmar and in Sri Lanka by Gunatilke (2003). Usually, the price trend in previous year affects vegetable planting pattern (Shepherd, 2005) but this phenomenon was not found in the islands, mainly due to small size of holdings and huge gap in demand and supply.

Post-Harvest and Marketing

The perishable nature of vegetables, small size holdings, isolated production sites, limited shipping facilities, poor storage and processing and unregulated market structure are the major factors for the slow growth in the vegetable sector in the islands (Singh et al. 2015a). The postharvest losses of vegetables in islands are around 25-40%, Losses are primarily due to uncertainty in shipping schedules and the low importance of vegetables in inter-island shipping services. The Island economy is predominately primary sector based, which needs proper capital investment to engage the increasing population. Currently, the islands lack well established scientific processing units for horticultural commodities. The establishment of these units in the islands will help in value addition and employment generation.

Market Channel and Price Spread

Successful farmers are sometimes those who do the opposite to what is being done by other farmers (FAO, 2005). The existing market channels in the islands take several forms. These are: 1. Grower-Consumer; 2. Grower-Retailer-Consumer; 3. Grower-Forwarding agent-Commission agent-Retailers-Consumer; 4. Grower-Wholesale merchant -Retailers-Consumer; and 5. Grower-Commission agent-Wholesaler-Retailer-Consumer. However, it was noticed that there are mainly three channels which are involved in vegetable marketing for locally produced vegetables. It was found that the Grower-Consumer market channel was more remunerative to the vegetable growers showing a market efficiency about 74% whereas the least remunerative channel was the Grower-Commission agent-Wholesaler-Retailer-Consumer Channel. Therefore, farmers are better off selling directly to the consumer as compared to the channel where middlemen were involved. In the case of vegetables imported from mainland India, the vegetables are brought by the government sponsored. Andaman and Nicobar Islands Development Cooperative (ANIDCO) or private agencies and then channeled to consumers as shown in Figure 6. The vegetables are transported to dif-

Figure 6. Market channels in the islands for vegetables transported from the mainland India

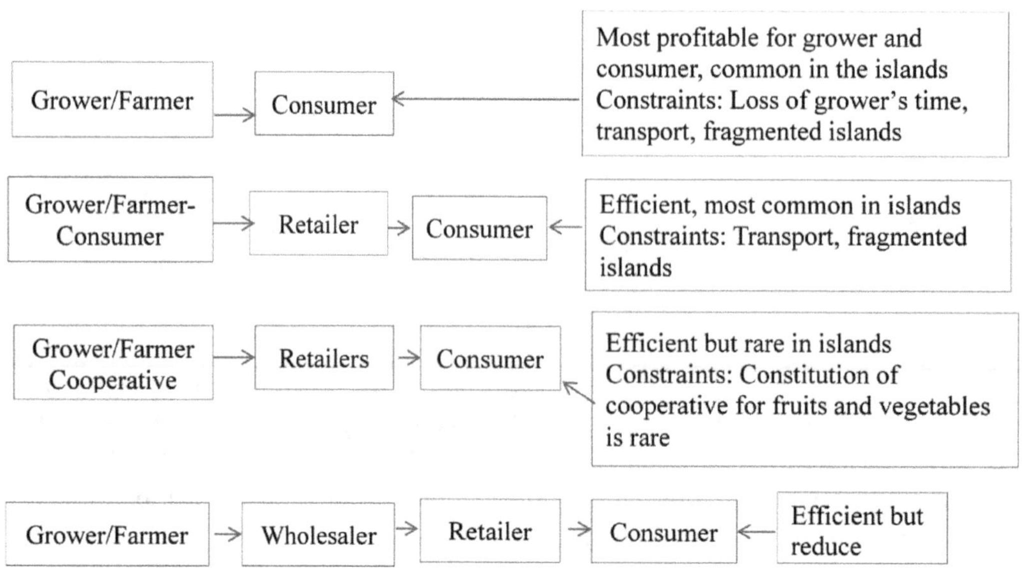

ferent the islands by cargo ships or local ferries from Port Blair and sold to consumers by retailers. The existing system lacks efficiency for price regulation, quality maintenance and ensured supply system to the satisfaction of the consumers. Hence, adequate measures such as a dedicated transportation system to ensure supply, cold storage and refrigerated transport facilities for proper quality and timely supply at affordable prices are the issues which need immediate attention in the vegetable marketing system.

Cultural Linkages through Horticultural Items

The Andaman and Nicobar Islands have six tribal communities and some tribes still follow the barter system for agriculture commodities. The tribal households exchange their surplus fruits, vegetables and animal products among themselves which augurs well for cohesiveness and mutual benefits for their overall socio-economic development. This system may be explored for their development and integration to modern systems. The introduction of modern production technologies to ensure their food and livelihood demands attention among these people. While the sharing of their produce and goods of basic need help them in promoting integration, they need to have a well-developed market which will result in a competitive environment and will ultimately lead to higher agriculture production.

Horticulture in Socio-Economic Development

Socio economic development and cultural association go hand in hand in the Andaman and Nicobar Islands. Therefore, people of these islands may engage horticulture cultivation with the use of modern scientific knowledge for increasing agriculture production. Once people's economic conditions are improved they may opt for specialised horticulture production. Since the tourism sector is emerging in these islands, local fruits and vegetable demand can be met through higher returns per unit of land. For vegetables, these may be for cauliflower, cabbage, French bean, cucumber, okra and Indian spinach. For flowers, it includes marigold, tuberose and gerbera; papaya, guava, mango, banana and sapota are possible fruits; sweet potato and greater yam in tuber crops and Morinda as a medicinal plant. New technologies viz., protected cultivation, drip irrigation, raised bed systems, organic farming, pot culture, value addition and processing activities have great promise for attracting rural youth into the horticulture sector. A number of success stories have been reported with new varieties and technologies in horticultural crops (http://icar-ciari.res.in/). Higher and regular income from horticulture products lifted the status and incomes of farmers and enabled them to spend more money on health, education and basic needs (Chand et al. 2013). However, it is suggested that a systematic impact assessment of the sector on island population be undertaken.

The economic analysis of important agri-horticultural commodities in the islands indicates that vegetables, flowers and fruits are highly remunerative crops due to their higher market prices, comparatively higher productivity levels and more number of crops per year (Table 6). However, level of benefits from these crops can be further increased by improving the productivity through use of suitable high yielding varieties, recommended production technologies and regulating the price fluctuation in the market. These crops are more vulnerable to climatic vagaries and market fluctuations hence adequate technologies, crop insurance or infrastructure support measures are required to ensure their economic viability.

Table 6. Economics of important agri-horticultural crops in Andaman and Nicobar Islands

Crops	Avg. production (MT/ha)	Avg no. of crops/ year	Price (INR/ ton)	Gross returns (INR)	Cost of Cultivation (INR)	Net returns (INR/crop)	Net returns (INR/year)
1	2	3	4	5	6	7	8
Rice	3.04	1	12000	36480	25000	11480	11480
Vegetables	7.6	3	20000	152000	65000	87000	261000
Fruits	8.67	1	25000	216750	95000	121750	121750
Spices	1.95	1	40000	78000	35000	43000	43000
*Coconut**	5.8	1	7000	40600	20000	20600	20600
Flowers	2.3	3	70000	161000	80000	81000	243000
Pulses	0.48	1	65000	31200	17000	14200	14200
Oil seeds	0.35	1	24000	8400	14000	-5600	-5600

*(1000 nuts/ha)

Horticulture in Nutrition

Fruits and vegetables are rich sources of dietary minerals and vitamins and are known as protective foods. It has now been scientifically proven that intake of fruits and vegetables improves the nutrient status of the human body and also improves immunity. The nutrient profiles of indigenous vegetables (Singh et al. 2011[a]) and underutilized fruits (Singh et al. 2011[b]) of the islands revealed their greater potential in meeting the daily need of dietary minerals and vitamins of local tribes and settler communities in rural areas. These plants are also visible occasionally in the weekly markets. Through this channel, these plants also contribute to the dietary needs of both the urban and non-farming population. However, there must be more market and consumer sensitizations to help reduce the micronutrient deficiencies. The cooking processes also affect the dietary minerals and antioxidants (Singh et al. 2015[b]). Further, worm infestation is heavier in rural and tribal areas in the tropical islands due to poor hygienic conditions and the use of tap water which also contributes to iron deficiency. Hence, the right kind of food items with adequate cooking processes and deworming are essential to minimize the incidence of deficiencies of dietary micronutrients.

Indigenous Horticultural Crops in Household Nutrition

The promotion of nutrient-rich indigenous vegetables and fruits in homegardens, kitchen gardens and school gardens is a useful strategy to fight against micronutrient and vitamin deficiencies. This approach has shown positive impact in different regions in African and Asian countries. Researchers have identified a number of fruits and vegetables rich in specific mineral nutrient/antioxidants and plants with overall superior nutrient/antioxidant profiles which can be used to address deficiencies. However, adequate support to ensure supply of seeds or planting materials of identified plants, proper cultivation practices in different growing situations and conditions and health management of plants are issues will have to be addressed. This is because of the absence of a proper seed production system, no documentation of diseases and pests and their management strategies.

The Andaman and Nicobar Islands are situated amidst two biodiversity hot spots and have rich floral and faunal biodiversity. It biodiversity represents both from North-East India and Indonesia-Malaysian regions. For horticultural diversity, this pattern also exists. Around 57 indigenous vegetables and more than 50 underutilized fruit crops have been reported in the islands. Many of the indigenous fruits and vegetables have key roles in the food and nutrition of native tribes and early settlers. The nutritive profiling showed their superiority over many of the commercially grown vegetables and fruits (Singh et al. 2011[a]; 2011[b]). However, some of the underutilized fruits and vegetables are currently seen in local markets or roadside shops but adequate measures for their conservation, improvement and promotion in public are essential to harness their potential in health and nutritional security.

Regarding vegetable crops, the islands are rich in diversity of underutilized legume vegetables such as *Mucuna pruriens* (L.) DC., *M. monosperma* Dc.ex Wigh, *M. gigantea* (Wild) DC., *Canavalia ensiformis* (DC.)Makino, *Canavalia gladiata*(Jacq.) DC, *Canavalia stocksii* Dalzell & A. Gibson, *Parkia roxburghii* G. Don., *Vicia sativa* var. *nigra* L., *V. faba* L., *Atylosia cajanifolia* Haines, *Atylosia scarabaeoides* (Linn.) Baker, *A. grandiflora* Benth. Ex Baker, *A. nivea* Benth., *Cicer microphyllum* Benth., *Abrus precatorius* L, *Acacia andamanica* I.C. Nielsen, *Psophocarpus tetragonolobus* (Stickm.) DC. and *Centrosema pubescens* Benth. Of these *Mucuna* spp., *Vicia* spp., *Dolichus* spp., and *Atylosia* spp. Have their orignins of diversity in the islands, while *Psophocarpus tetragonolobus, Vigna unguiculata, Phaseolus vulgaris* and *Cyamopsis tetragonoloba* were brought by immigrants. Legume vegetables meet the protein requirement in local diets, besides adding nitrogen in the soil thereby improving its fertility and productivity. Tree species having vegetable importance such as *Artocarpus heterophyllus*, *A. incisa*, *Murraya koenigii* and *Tamarindus indica*are are prevalent in home gardens in the ANI. *Cycusrumphi* and *Calamusandamanicus* were also found to be used as leafy vegetables and eaten in the form of stew, soup, pickle, chutney or curry. Root and tuber crops are the third important food crops of mankind after cereals and grain legumes, and constitute either staple or subsidiary food for about one fifth of the world population. In ANI, tubers and aroids are secondary foods for settlers but they constitute the most important part of local diets of Nicobari, Jarawas and Onges tribes. The major edible aroids are represented by two botanical tribes and five genera including Lasioideae (*Crytosperma* and *Amorphophallus*) and Colocasiodeae (*Alocasia*, *Colocasia* and *Xanthosoma*). All parts of the plants including corms, cormels, rhizome, stalks, leaves and flowers are edible. Out of 22 species of *Amorphophallus* in India, five are present in the islands. Of these, *Amorphophallus campanulatus* Blume ex Decne has attained commercial status. The major tuber crops in the islands are *Manihot esculenta* Crantz, *Ipomoea batatas* (L.) Lam., *Dioscorea alata* (L.), *Colocasia esculenta* (L.)Schott, *Xanthosoma sagittifolium* and *Amorphophallus poeniifolius* (Dennst.) Nicholson.

Considering fruit crops, the islands have around 50 species from 25 families. Predominant families are Clusiaceae, Anacardiaceae, Myrtaceace, Moraceae, Musaceae and Rhamnaceae. In the islands, *Dillenia andamanica* C. E. Parkinson, *D. indica* L. and *D. pentagyna* Roxb., *Diospyrous blancoi* (velvet apple), star gooseberry (*Phyllanthus acidus*), *Baccaurea sapida* (sapida), *B. ramiflora* (*khatta phal*), *Haematocarpus validus* (Miers.) Bakh. f. ex. Forman, *Ficus carica* L. and *Ficus racemosa* L. are found in wild habitats. Jackfruit (*Artocarpus heterophyllus*), breadfruit (*A. altilis*), monkey jack (*A. lakoocha* Buch.-Ham.) and cham pedak (*A. chaplasha* Roxb.) are found in the backyard plantations of the Islands. The islands have a rich diversity of *Musa* spp. such as *Musa acuminata* Colla, *Musa balbisiana* Colla, *Musa paradisiaca* L. few other wild relatives of cultivated bananas have also been reported from the Islands. These include *Musa balbisiana* Colla subsp. *andamanica* Singh et al., *Musa textilis* Nees, *Musa sabuana, Musa indandamanensis* L.J. Singh. Some of the wild fruit plants such as *khaariphal* (*Ardisia solanacea* Roxb. and *A. andamanica* Kurz.), *Syzygium cuminii* (jamun) and *S. claviflorum* (wild jamun) are found in the forests of ANI. These have potential as salinity tolerant rootstock. Additionally, Pandanus fruits, also called Kewra fruit, form an essential part of Nicobari diet and are obtained from three species, namely, *Pandanus lerum* Jones ex Fontane var. *lerum*, *Pandanus lerum* var. *andamanensium* (Kurz.) D.C. Stone and *P. tectorius* Soland. *Ziziphus andamanica* Bhandari et Bansali, *Z. mauritiana* Lam., *Z. nummularia* (Burm. F.) Wight et Arn.,*Z. rugosa*, *Z. oenoplia* (L.) Mill. var. *oenoplia*, *Z. oenoplia* (L.) Mill. var. *pallens* Bhandari et Bhansali are the major edible fruit species of Rhamnaceae family in the ANI.Bael (*Aegle marmelos* (L.) Correa), wild lime (*Atlantia spinosa* (Wild) Tanaka), pummelo (*Citrus grandis* L. Osbeck), kafir lime (*C. hystrix* D.C.), citron (*C. medica*), *C. nobilis* Lour.var. *limonellus* Kurz. and*Limonia acidissima* (L.) are used in various culinary preparations.

Home Gardens

Like in many tropical islands, homegardens are very common in the A&N islands. They are important and contribute to ensuring access to healthy diets rich in macro- and micronutrients. Homegardens provide different fruits, vegetables, spices and medicinal plants which help in diversifying the food basket (Pandey et al. 2006). Notably, these gardens are very common in rural areas where people with high vulnerability to hunger and malnutrition live. It is worthy to note that in remote rural areas and in geographically isolated habitats, the efforts of government agencies have limited impacts due to various socio-economic and transportation limitations. In such a situation, the homegarden approach has immense value because of the localized nature of the food production system and ease of access. Further, because the cost of perishable food items such as fruits and vegetables is high, the common man cannot buy to meet family needs. Hence, homegardens are still a vital component of food and nutritional security in these islands. During occasional interactions and visits to rural homegardens in the islands, it was observed that the common crops are plantations (coconut, areca nut), fruits (papaya, banana, guava, mango and such like), vegetables (Indian spinach, chilli, culantro, amaranthus, bottle gourd, pumpkin, brinjal and others), spices and condiments (blackpepper, cinnamon, cardamom among others), medicinal plants (such as ocimum, morinda, and aloe) and some other ornamental plants. The details of these observations of 35 homegardens in different islands are given Table 7. The average number of plants was recorded by the count method. Intensity in homegardens was calculated as: Intensity in homegardens (%) = (No. of homegardens having the particular plant species/Total No. of homegardens) x 100. The contribution to family demand (%) and level of satisfaction with existing plant composition and plant performances are reported based on homegarden owner's direct response. The most common identified interest of the owners were; plants of high yielding adaptable varieties, right quality inputs in adequate quantity in 'homegarden kits' and the right platform to advise on the plant health and cultivation related issues. To ensure supply of quality seeds in adequate quantity, the 'Vegetable Seed Kits' of ICAR-Indian Institute of Horticultural Research, Bengaluru and ICAR-Indian Agricultural Research Institute, New Delhi are quite useful.

Integration of Horticulture in the Ecosystem

The islands have mainly forest and marine based ecosystems which are well preserved by the natural dynamics of biogeographical cycles. But, the introduction or domestication of horticultural crops for commercial cultivation in the islands had a direct influence on the existing island ecosystems. The ma-

Table 7. Horticultural crops in homegardens in Andaman and Nicobar Islands

Group	Crops	Avg. no. of Plants	Replacement of Purchase (%)	Prevalence in Homegardens (%)	Level of Satisfaction with Existing Plants
Plantations	Coconut	1.1	100.0	70.1	Good
	Areca nut	1.4	100.0	72.9	Good
Fruits	Papaya	1.0	30.0	30.6	Good
	Banana	1.5	40.0	41.4	Poor
	Mango	0.8	60.0	28.6	Poor
	Custard apple	0.8	60.0	28.6	Good
	Citrus	0.2	100.0	20.0	Very good
	Guava	0.7	40.0	20.0	Poor
	Others	1.4	35.0	35.7	Poor
Vegetables	Indian spinach	3.9	70.0	70.0	Good
	Amaranthus	50.0	40.0	75.7	Poor
	Chilli	3.8	100.0	60.0	Poor
	Culantro	20.0	100.0	55.7	Good
	Bitter gourd	10.0	60.0	40.0	Good
	Cowpea	20.0	44.0	50.0	Good
	Dolichos bean	10.0	35.0	25.0	Poor
	French bean	35.0	60.0	10.0	Good
	Tomato	5.0	30.0	15.0	Poor
	Brinjal	3.8	40.0	17.1	Poor
	Bottle gourd	5.5	45.0	68.0	Good
	Pumpkin	2.2	65.0	28.6	Poor
	Cucumber	4.0	50.0	30.0	Good
	Moringa	0.5	100.0	40.0	Very good
	Curry leaf	1.0	100.0	38.0	Very good
	Others	2.8	30.0	70.0	Good
Spices	Black pepper	0.4	100.0	40.0	Poor
	Cinammon	0.3	100.0	54.3	Poor
	Cardamom	0.2	100.0	60.0	Good
	Ginger	0.2	25.0	40.0	Poor
	Others	0.2	30.0	25.0	Poor
Medicinal plants	Noni	0.1	100.0	40.0	Good
	Tulsi	0.9	100.0	25.7	Good
	Kalmegh	20.0	8.5	20.0	Good
	Pippali	5.0	15.0	100.0	Good
	Others	0.3	60.0	30.0	Poor
Tuber crops	Taro	55.0	50.0	40.0	Good
	Elephant foot yam	20.0	40.0	25.0	Good
	Sweet potato	35.0	55.0	20.0	Good
	Tapioca	10.0	30.0	35.0	Good

jor change observed was the use of chemical inputs i.e. fertilizers, pesticides and weedicides as well as growth substances, and these have a negative impact on water bodies and soil health. They might have also affected the marine ecosystems but more research is needed to support the hypothesis. Swarnam and Velmurugan (2012) reported pesticide residues in soil and water bodies as a result of vegetable cultivation. In excess, these chemicals may contribute in shift of microbes or insects in the ecosystems and may threaten animal and human life. Consequently, adequate measures are required to minimize the use of such chemical inputs. Heavy rains favour rapid wash-off of chemicals and fertilizers from the crop area to the water bodies which harms the water bodies. However, the Local Government has initiated various activities to ensure sustainable development of the island agriculture without harming the existing natural ecosystems. Some of them are to promote organic farming, integrated plant nutrient management (IPNM), integrated pest management (IPM), and the use of plastic mulching and natural bioagents to counter the challenge pesticide residues pose to the ecosystems in the islands.

Island Horticulture and Climate Change

Developing countries are likely to be more vulnerable to climate change due to projected magnitude and poor adaptive capacity (IPCC, 2001) because of institutional and resource constraints. The contribution of the islands to global climate change is negligible though they bear the brunt of climate associated disasters. This calls for an adaptation-centric climate policy in terms of developmental projects and issues, as compared with the conventional approach in terms of response to an environmental problem. Fussel and Klein (2006) reviewed the evolution of conceptual framework of climate change vulnerability assessments. They called for a fundamental shift in the assessment, by devising a framework of adaptation policy assessment, taking into consideration the adaptive capacity, protective measures, non-climatic drivers of climatic policy. The current investigation was carried out in Nicobar group of islands to determine the extent of vulnerability of these islands to climate change and events associated with climate change, for example, storm surges, primarily from the livelihood angle of the tribal population.

Projections for A&N Horticulture

In the post-tsunami period, a study on the carrying capacity of the Andaman and Nicobar islands was conducted and projections were calculated. It was hypothesized that the islands will be able to meet the local demand of horticultural crops by 2021 and will continue to do so for the ever increasing population and tourist in 2031. For this, an area of around 75 ha and 100 ha under protected cultivation will be needed in 2021 and 2031, respectively. The spatial pattern and seasonality in production are going to be key considerations. Most vegetables and fruit production takes place only in the dry season and is concentrated in a few production areas. Singh et al. (2015[b]) have indicated that around 81% of total vegetable production in the islands occurs during the dry season (January to April) while only 19% occurs in the rainy season (May to December). Major vegetable production areas are Neil Island, North Andaman and Middle Andaman, from where the transportation of vegetables remains a big challenge to local administration and vegetable growers. However, the projections are entirely dependent on population rise, food habits, inflow of tourists and government policies to meet the food demands of the population while keeping in mind the sustainable development of islands. Further, climate change, land use patterns and water scarcity will also affect the production profile of various horticultural crops. As a consequence, the local supply system will also be affected by these factors.

CONCLUSION

Over the years, 17 commodities have been identified for the islands. These include vegetables, fruits, flower, spices, plantations and tuber crops. Some promising production technologies include coconut husk-based raised bed techniques, soil mount technique, raised bed with double row technique, protected cultivation technology for flowers and vegetables. Additionally, other innovations include: year–round production technologies for vegetables, growing media for solanaceous vegetables and ornamentals, grafting techniques for tomato, noni and sapota, organic practices for spices, papaya and flowers. Development of seed germination and micropropagation protocols for newly domesticated crops such as noni (*Morinda citrifolia* L.) and oil yielding species, as well as tissue culture protocols for *Piper sarmentosum* and *Eulophia anadamanesis* have resulted in the production of quality planting materials. Many of these technologies are currently on farmers' fields and are showing impact in increasing production and extending the harvest season. However, horticultural production improvements are faced with; changing climate, shift in food habits, rising trend in tourist and local populations and changing work preferences of rural youth. These all encourage opportunities for the development of more cost and labour efficient technologies to meet these emerging demands and raise employment prospects in the islands.

Over the years, the horticulture sector has developed into a viable option for livelihood and employment generation in the Andaman and Nicobar Islands. However, the burgeoning tourist population attracts immediate opportunities to transform the sector for the production of quality fruits, flowers and vegetable to meet the requirement of the tourism sector as well as local and floating (service class) population. Sooner or later, planners will have to recognize that the island horticulture has huge export potential to meet the demand of adjacent East Asian countries for similar horticultural products. Moreover, technology-enabled production and marketing of the horticultural sector can easily attract rural youth and help in self-employment. It is noted that the demand for tropical underutilized fruits in mainland India is being met through imports from East Asian countries and this can be easily substituted with establishing a production system in Andaman and Nicobar Islands. Even today, India is facing huge loss of money and resources due to crop losses caused by pest or diseases or drought or excess rains. The biodiversity of the islands has the potential to serve as a source for genes/crops which can contribute to the development of stress tolerant breeding programmes for the larger benefit of the Indian agriculture. Interestingly, the literacy level in the islands is 86.27% (Census 2011), highlighting the possibilities of development of highly skilled manpower in hi-tech horticulture. The climatic profile of the islands facilitates the production as a wide range of high value and exportable, underutilized fruits, vegetables and ornamental plants. However, their utilization needs improvement through adequate budgetary allocations and research and development activities. Further, genetic sources or materials for the pharmaceutical industry may be sourced from the islands. The islands also have huge marine resources in the coastal areas and natural flora and fauna in forests which demand planning for eco-friendly agricultural activities to avoid the present practices of indiscriminate use of fertilizers, pesticides and weedicides. Further, the islands have six primitive tribes which use mainly sea food and local (underutilized) fruits, tubers and vegetables for their dietary requirements. Quality planting materials of horticultural crops remains a point of concern in the A&N islands. At present, the demand for seed and planting materials is being met through transports from the mainland by the state department and to some extent by CIARI. However, the urgency of the issue has been recognized and model nurseries are being developed on different islands through national schemes. Finally, there is urgent need to rethink and reorient the research and development programmes.in order to transform the islands' horticulture from present level of unorganized, low level of commercial activities to fully organized and eco-friendly commercial horticulture.

REFERENCES

Balakrishnan, N. P., & Ellis, J. L. (1996). Andaman and Nicobar Islands. In *Flora of India, Part 1*. Calcutta, India: Botanical Survey of India.

Basic Statistics. (2011). Andaman and Nicobar Islands, India.

Census. (2011). *General Census*. Office of the Register General & Census Commissioner, Ministry of Home Affairs, Government of India.

Chand, S., Krishnan, P., & Sethi, S. N. (2013). Economic impact assessment of rehabilitation scheme of Tsunami affected farmers of Bay Islands in India. *Indian Research Journal of Extension Education.*, *13*(1), 56–62.

Chand, S., Singh, D. R., Singh, S., Srivastava, R. C., Birah, A., & Banu, S. (2012). *Vegetables production and role of farmwomen in Andaman and Nicobar Islands, India*. Paper presented at the Global Conference on women in Agriculture at New Delhi.

Chand, S., Singh, S., Parappurathu, S., Roy, S. D., & Kumar, A. (2015). Explaining the status and scope of ecotourism development for livelihood security: Andaman and Nicobar Islands, India. *International Journal of Sustainable Development and World Ecology*, *22*(4), 335–345. doi:10.1080/13504509.2015.1050478

Chand, S., Srivastava, R. C., Krishnan, P., Jaikumar, S., Birah, A., Sethi, S. N., & Ram, N. (2011). *Impact of Tsunami Rehabilitation Programme for Reviving Livelihood of Affected People of Bay Islands, India*. 6th National Extension Education Congress, Old Goa, India.

Development Report. (2008). *Development Report*. Andaman and Nicobar Islands: Planning Commission, Government of India.

Dhingra, K. (2005). *The Andaman and Nicobar Islands in the 20th Century a Gazetteer*. New Delhi: Oxford University Press.

Directorate of Economics and Statistics Census. (2011). *Andaman and Nicobar Islands*. Port Blair, India.

FAO. (2005*).Food and Agriculture Organization of the United Nations.Environment and Natural Resources in Small Island Developing States*. Overview prepared for a Special Ministerial Conference on Agriculture in Small Island Developing States, Rome, Italy.

Fussel, H. M., & Klein, R. J. T. (2006). Climate change vulnerability assessments: An evolution of conceptual thinking. *Climatic Change*, *75*(3), 301–329. doi:10.1007/s10584-006-0329-3

Gunatilke, W. D. (2003). Agricultural marketing system in Sri Lanka. *Agricultural Marketing*, *45*(4), 55.

IPCC. (2001). *Climate Change (2001): The Scientific Basis*. Cambridge, UK: Cambridge University Press.

Kundu, M. S., De, A. K., & Srivastava, R. C. (2010). *Livestock and poultry production policy for Andaman and Nicobar Islands: A scientific perspective*. Port Blair: CARI.

Myers, N. (2003). Biodiversity hotspots revisited. *Bioscience*, *53*(10), 916–917. doi:10.1641/0006-3568(2003)053[0916:BHR]2.0.CO;2

Myint, T. (2003). Agricultural marketing information system in Myanmar. *Agricultural-Marketing, XLV*(4), 16–28.

N.H.B. (2014). *National Horticulture Database, National Horticulture Board*. Harayana, India: Guargaon.

Official website of ICAR-Central Island Agricultural Research Institute, Port Blair, Andaman and Nicobar Islands, India. (n.d.). Retrieved from http://icar-ciari.res.in

Pandey, C. B., & Singh, R. K. (2010). Organic black pepper (*Piper nigrum* L.) cultivation on glyricidia standards in coconut (*Cocos nucifera* L.) plantations in South Andaman: Organic matter production and recycling of nutrients. *Indian Journal of Agricultural Science, 80*(11), 975–980.

Pandey, C. B., Venkatesh, A., & Medhi, R. P. (2006). Diversity and species structure of home gardens in South Andaman. *Tropical Ecology, 47*(2), 251–258.

Reddy, S. (2007). Mega tourism in Andaman and Nicobar Islands: Some concerns, *Journal of. Human Ecology, 21*(3), 231–239.

Rengarajan, S., Veeraragavan, S., Thangaraj, M., & Ramachandran, R. (2014). Geographical analysis of tourism sites in Andaman Archipelago (India) and ecotourism development for Smith Island of North Andaman. *International Journal of Sustainable Development and World Ecology, 21*(5), 449–455. doi:10.1080/13504509.2014.958118

Salam, M. A. (2010). Horticultural sector in Andaman & Nicobar. In *Proceedings of National Workshop cum Seminar on Status and Future strategies for Horticulture Development in A& N Islands*. CARI.

Shepherd, A. W. (2005). *Associations of market traders: Their roles and potential for further development*. AGSF Occasional Paper 7. FAO.

Singh, D. B., Attri, B. L., & Singh, C. (2012b). Underutilized fruit crops of Andaman: Its biodiversity and role in environmental conservation and nutritional security. *Indian Forester, 138*(1), 27–30.

Singh, D. R., & Singh, S. (2014). Horticulture as an indispensable livelihood option in tropical island region. In *Souvenir: Indian Horticulture Congress*. HIS Westville Publishing House.

Singh, D. R., Singh, S., & Roy, S. D. (2012a). *Noni (Morinda citrifolia L.): wonderful medicinal plant*. Port Blair: Published by Director CARI.

Singh, D. R., Singh, S., Salim, M. K., & Srivastava, R. C. (2011b). Estimation of phytochemicals and antioxidant activity of underutilized fruits of Andaman & Nicobar Islands (India). *International Journal of Food Sciences and Nutrition, 63*(4), 446–452. doi:10.3109/09637486.2011.634788 PMID:22080844

Singh, H. P. (2011). Horticulture—a best livelihood option in tropical islands. In R.K. Gautam, M. Sankaran, S. Jaikumar, D. Kumar, S.K. Verma, P. Krishanan, G. George, & S.K. Ahmed (Eds.), *International Conference on Tropical Island Ecosystem—Issues Related to Livelihood, Sustainable Development and Climate Change*. CARI.

Singh, S., & Bainsla, N. K. (2015a). Analysis of climate change impacts and their mitigation strategies on vegetable sector in tropical islands of Andaman and Nicobar Islands, India. *Journal of Horticulture, 2*(1). doi:10.4172/2376-0354.1000126

Singh, S., Sankaran, M., Singh, D.R., & Roy, S.D. (2014, June 22). CIARI Promoting Island horticulture to eradicate micronutrient deficiency. *The Daily Telegrams*.

Singh, S., Singh, D. R., Chand, S., Birah, A., & Roy, S. D. (2015a). Analysis of perspectives of self-sufficiency in vegetable production under tropical conditions. *International Journal of Vegetable Science, 21*(1), 53–68. doi:10.1080/19315260.2013.823585

Singh, S., Singh, D. R., Salim, M. K., Singh, L. B., Srivastava, A., & Srivastava, R. C. (2011a). Estimation of proximate composition, micronutrients and phytochemical compounds in traditional vegetables from Andaman & Nicobar Islands. *International Journal of Food Sciences and Nutrition, 62*(7), 765–773. do i:10.3109/09637486.2011.585961 PMID:21615278

Singh, S., Singh, D. R., Singh, L. B., Chand, S., & Roy, S. D. (2013). Indigenous vegetables for food and nutritional security in Andaman and Nicobar Islands, India. *International Journal of Agriculture and Food Science Technology.*, *4*(5), 503–512.

Singh, S., Swain, S., Singh, D. R., Salim, K. M., Nayak, D., & Roy, S. D. (2015b). Changes in phytochemicals, anti-nutrients and antioxidant activity in leafy vegetables by microwave boiling with normal and 5% NaCl solution. *Food Chemistry, 176*, 244–253. doi:10.1016/j.foodchem.2014.12.068 PMID:25624230

Srivastava, R. C., & Ambast, S. K. (2011). *Water policy of Andaman and Nicobar Islands: a scientific perspective*. Port Blair, India: CARI.

Srivastava, R. C., Singh, D. R., & Singh, S. (2010). Horticultural scenario in Andaman & Nicobar. In *Proceedings of National Workshop cum Seminar on Status and Future strategies for Horticulture Development in A& N Islands*. CARI.

Srivastava, R. C., Singh, D. R., Singh, S., & Srivastava, A. K. (2009). Andaman & Nicobar Islands: Potential Hot Spot for Genetic Resources of Horticultural Crops. Presented and published in 9th Agricultural Science Congress.

Chapter 11
Promoting Agricultural Productivity and Inclusive Growth in Uganda

William Amone
Gulu University, Uganda

Dick Nuwamanya Kamuganga
Uganda Technology and Management University (UTAMU), Uganda

Godswill Makombe
University of Limpopo, South Africa

ABSTRACT

Although many countries in the world including Uganda witnessed high rates of economic growth in the last three decades, the strong growth has failed to holistically deliver the expected prosperity. Amidst Uganda's strong growth of about 7% per annum, of the recent decades, poverty, unemployment and inequality have remained pervasive especially in the rural areas; an indication that the growth process has not been pro-poor and inclusive of the deprived. Agriculture which is the sector employing majority of Uganda's poor has also not developed much; within the same period the sector grew at an average rate of 2% per annum and its productivity has remained considerably low. This chapter reviews the trending conception of inclusive economic growth, and its relevance to Uganda's development process. In its final section, the chapter assesses the factors that affect agricultural productivity and it discusses the most effective means of raising productivity in order to make the growth process of the country more broad-base, pro-poor and inclusive.

INTRODUCTION

Many countries in the world have achieved high rates of economic growth for nearly three decades since 1990, with a few witnessing slowdown in growth following the global financial crisis that began around 2008. As observed by the World Bank (2013), although it is believed that economic growth should generate virtuous circles of prosperity and opportunities, strong growth of the last three decades in many developing countries have not translated into the expected development outcomes.

DOI: 10.4018/978-1-5225-0942-4.ch011

The World Bank (2014) estimated that about 2.4 billion people lived on less than US $2 a day in 2014, and that even after 2015 more than 1 billion people will still live in extreme poverty. Worldwide, more than 200 million people are believed to be unemployed (ILO, 2014). OECD (2012) explains that in most countries, the growth of recent decades has been accompanied by widening inequality. Economic growth in the recent decades did not benefit all groups equally. Similarly, Oxfam (2014) estimates that the richest 85 people in the world own the same amount of wealth as the bottom half of the world's population (3.5 billion people). The record levels of growth seem to have failed to tackle three key global problems of poverty, unemployment, and inequality.

Rising living standards, not economic growth per se, is what societies expect their leaders to deliver. In many developing economies, fairly sustained strong growth has lifted many people out of absolute poverty but improvements in living standards have not kept pace with GDP growth. This experience is most apparent in Eastern Europe and fast-growing emerging Asian economies such as China, India, and Vietnam, and some African economies including Uganda, Zambia and Kenya. Worse still, some countries, including Chile and Mexico have witnessed widening inequality amidst stellar growth (World Economic Forum, 2015).

Literature shows that economic growth is often unequivocally tied to a correlating improvement in employment levels (Felipe & Hasan, 2006). However, over the recent three decades increases in growth have not resulted into expected comparable increases in the numbers of jobs. Felipe & Hasan (2006) estimated that in the 1980s it took 3% GDP growth to generate a 1% increase in employment, but in the 1990s this changed significantly and it then took an 8% percent increase in GDP to get the same result.

The International Labour Organization (ILO) estimates that over 1 billion people are employed in world agriculture, representing 1 in 3 of all workers (ILO, 2015). Agriculture is a significant source of food for citizens and a means of livelihood for the most vulnerable members in most developing countries. According to the World Bank (2014), about 70% of the world's poor live in rural areas, and agriculture is their main source of income and employment. In Africa, agriculture employs about 65% of the labor force and accounts for 32% of gross domestic product. Improvement in agriculture and its productivity is thus fundamental to achieving food security, poverty alleviation and overall sustainable economic development. Raising agricultural productivity must be an important policy goal for governments and development agencies.

Considering that the large share of the working poor is engaged in agriculture, targeting this sector in development may have major impacts on welfare throughout the world. Most literature seems to agree that investments in agriculture and rural development, both private and public, stimulate economic growth and development (World Bank, 2008; IFPRI, 2012). Similarly, good agricultural performance has been very significant in reducing poverty and hunger rates in the more successful countries. The World Bank (2008) also established that agricultural productivity growth has a high poverty reduction pay-off than non-agricultural growth or investments.

For the period from 1990 to 2010, the Ugandan economy achieved high GDP growth rates; real economic growth averaged 7%. However, the record growth has not been able to benefit all sections of the population; significant proportions of the country's population including smallholder farmers still remain impoverished, unemployed or marginalized (World Bank, 2013). Uganda Bureau of Statistics (UBOS, 2015) reported that the high performance of the Ugandan economy has been accompanied by structural changes, with a steadily declining share of agriculture and increasing shares of industry and services. Uganda's high growth process has therefore not been inclusive since the larger proportion of its population (those engaged in agriculture) has been bypassed by the growth process. Most farmers have missed the benefits of growth as indicated by the entrenchment of poverty, unemployment and exclusion.

Recent statistics from the government indicate that Uganda's agriculture employs about 72% of all workers but contributes only about 23% of the GDP (UBOS, 2015). The farmers are primarily rural based, and have a lower standard of living than those working in other sectors. The majority of Uganda's poorest people are smallholder farmers living in remote areas. The country's agriculture is characterized by smallholder farming with hand hoes as the major production tool. The average farm size of the smallholders is 1.2 hectares. The farmers produce various commodities mostly for subsistence.

The major food crops include maize, rice, millet, bananas, sorghum, cassava, and potatoes. The farmers also produce pulses, fruits, and vegetables. Yields are low across all these commodities. Smallholders use traditional farming techniques; they lack capital for modernization, expansion and purchase of vital inputs. They also lack access to financial services to give them the most needed capital. The biggest challenge to Uganda's farmers seems to be low productivity.

Considering that the large share of Uganda's working poor are farmers and that agriculture is the main productive base of the economy, targeting this sector for development may have major impacts on welfare, economic growth and inclusion. This chapter assesses the factors that affect agricultural productivity and proposes some methods of raising productivity, in order to make the growth process of the country more pro-poor and inclusive.

STUDY AREA

Uganda is a landlocked country astride the Equator, about 800 kilometers inland from the Indian Ocean. It lies on the northwestern shores of Lake Victoria, extending from 1 degree south to 4 degrees north latitude and 30 to 35 degrees east longitude. As shown in figure 1, the country is bordered by Tanzania and Rwanda to the south, Democratic Republic of Congo (DRC) to the west, South Sudan to the north, and Kenya to the east. With a land surface of 241,550.7 square kilometers, Uganda occupies most of the Lake Victoria Basin. It has a total population of about 37 million people (UBOS, 2015). The country is blessed with fertile soil and ample rainfall.

Uganda's rural areas account for 85% of the population and 94.4% of the poor, while the urban areas account for 15% of the population but only 5.6% of the poor (World Bank, 2015a). Despite the progress achieved to date in some areas such as security and education, there is little cause for complacency, as most development indicators still remain deficient. A few indicators are shown in table 1.

DEFINING AGRICULTURAL PRODUCTIVITY

Fulginiti and Perrin (1998) define agricultural productivity as the output produced by a given level of input(s) in the agricultural sector of an economy. Similarly, Olayide and Heady (1982) define it as the ratio of the value of total farm outputs to the value of total inputs used in farm production.

Productivity of agriculture is measured as the ratio of final output, in appropriate units, to some measure of inputs. Singh and Dhillion (2000) suggested that "yield per unit" should be considered to indicate agricultural productivity; many scholars have criticized this suggestion pointing out that it considered only land as a factor of production, disregarding other factors. Many scholars believe that agricultural

Figure 1. Map of Uganda showing the major towns and the neighboring countries
(World Atlas, 2015)

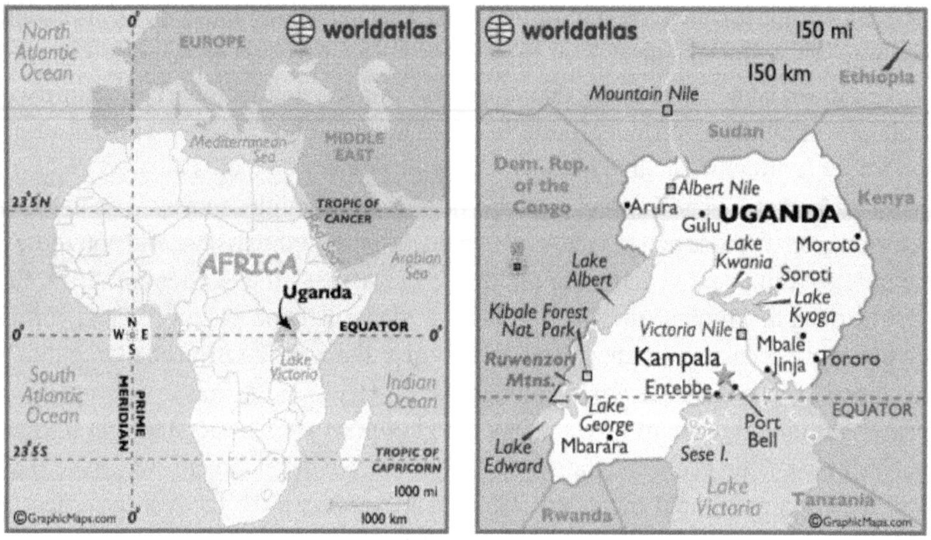

Table 1. Basic information on Uganda

Total population	37 million (2015)
Gross Domestic Product (current US$)	$26.3 (2015)
GNI per capita (atlas method)	$660 (2014)
Agricultural households' population	4.2 million households
Percentage of workers employed in agriculture	72%
Agriculture contribution to total GDP	23.2 (2014)

(UBOS, 2014; UBOS, 2015; World Bank, 2015a)

productivity should contain all the factors of production such as labor, farming experiences, fertilizers, availability and management of water and other biological factors (Dharmasiri, 2009; Zepeda, 2001).

Measures of productivity are categorized as partial or total measures depending on the number of inputs considered as discussed below.

1. **Partial Measures of Productivity:** are the amounts of output per unit of a particular input. Commonly used partial measures are yield (output per unit of land) and labour productivity (output per economically active person or per agricultural person-hour). Yield is commonly used to assess the success of new production practices or technology. Labour productivity can also be used to indicate rural welfare or living standards since it reflects the ability to acquire income through sale of agricultural goods.

As emphasized by Dharmasiri (2009), productivity of land is very important because land is the most permanent and fixed factor among the three inputs of land, labour and capital. Land productivity may be raised by applying input packages consisting of improved seeds, fertilizers, agro-chemicals and labour intensive methods, or by applying crop diversification in a season on the same land.

Productivity of labour is a vital determinant of the income of the population engaged in agriculture. Labour productivity has two important aspects: i) it profoundly affects national prosperity; ii) it principally determines the standard of living of the famers. Agricultural labor productivity may be enhanced through training and increase of incentives or wages. Capital in terms of land purchase or development, drainage, irrigation equipment, livestock, feeds, seeds, agricultural implements and machineries, crop production chemicals should also be given priority as a factor for enhancing agricultural productivity.

Generally, the quality of the three major inputs: land, labour and capital is still very poor in Uganda, and they are not optimally utilized. As a result, both yields and labor productivity are very low. Although the partial measures are used to assess agricultural-production performance, often because they are easy to estimate, the measures normally show higher rates of growth than Total Factor Productivity (TFP).

Output is measured as a function of the conventional inputs (production factors): capital, labour, land and raw materials that are under the control of the farmers and for which markets exist. This is because growth in land and labor productivity may result from more intensive use of inputs including fertilizer and machinery rather than TFP increase. Using the partial measures can be misleading as they ignore the role of other inputs in any observed output changes. Due to this shortcoming, a total measure was developed.

2. **Total Measures of Productivity:** TFP is the ratio of total output (crop and livestock products) to total production inputs (land, labor, capital, and materials). It is the ratio of output value to the value of inputs (IFPRI, 2013). An increase in TFP implies that more output is being produced from a constant amount of resources used in the production process.

According to Zepeda (2001), TFP is also defined as the ratio of an index of agricultural output to an index of agricultural inputs. The index of agricultural output is a value-weighted sum of all agricultural production components. The index of agricultural inputs is the value-weighted sum of conventional agricultural inputs; including land, labour, physical capital, livestock, chemical fertilizers and pesticides. TFP measures are often preferred because they are generic.

As observed by the World Bank (2015b), improving agricultural productivity will play a key role in achieving food security and reducing poverty in Uganda. Achieving high agricultural productivity requires investment in education especially of women, improvement of rural infrastructure including roads, water sources, communication and markets. These must also be complemented by private investment in inputs such as fertilizers, farm machines, and vehicles.

The Inclusive Growth Concept

A key policy challenge preoccupying leaders around the world is expanding social participation and benefits of economic growth. Correspondingly emphasized in the Inclusive Growth and Development Report 2015 (WEF, 2015), major international organizations including the World Bank, IMF, WTO, WHO and OECD are calling upon policy makers to design appropriate policies to strengthen growth and make it more inclusive and sustainable over time based on the perception that strong growth is clearly compatible with social cohesion.

According to DFID (2012), inclusive and sustainable economic growth is crucial for long-term poverty reduction. Inclusive growth should be characterized by people's ability to participate in, benefit from, and create wealth and jobs from growth. Literature on inclusive growth unanimously concludes that for growth to be inclusive equity must be a component of growth and at the same time growth must be pro poor (OECD, 2012; Anand, Mishra & Peiris, 2013; CAFOD, 2014).

The World Economic Forum (WEF) defines inclusive growth as output growth that is sustained over decades, is broad-based across economic sectors, creates productive employment opportunities for a great majority of the country's working age population, and reduces poverty. They reiterate that inclusive growth is about both the pace and pattern of economic growth (WEF, 2015).

Ianchovichina and Lundstrom (2009) stated that that: (i) for growth to be sustained in the long run, it should be *broad-based* across sectors. Issues of structural transformation for economic diversification therefore take a center stage; (ii) it should also be *inclusive* of the large part of the country's labor force, where inclusiveness refers to equality of opportunities in terms of access to markets, resources and unbiased regulatory environment for businesses and individuals. Given the widespread poverty and unemployment in Uganda, leaders and policy makers need to focus on redistributing the benefits of growth.

After witnessing average GDP growth of about 7% for about two decades, Uganda's economic growth has dipped. Beginning 2011 to date, GDP growth has averaged 4%, a rate considerably lower than those recorded by the neighbouring countries of Kenya, Tanzania, and Rwanda (World Bank, 2015b). Uganda's stellar economic growth is therefore not being sustainable in the long term. The record growth process has not been inclusive since a larger proportion of its population (especially farmers) continues to miss the benefits of growth as indicated by the entrenchment of poverty, unemployment and exclusion.

Why the Ugandan Government Should Urgently Embrace Inclusive Growth

1. **Economic Growth Sustainability:** When the poor farmers who have mostly been bypassed by growth and development get to benefit from the growth process, they will become more involved in it. Engaging the farmers will enhance their capacity to better educate their children and undertake investments that will ultimately lead to sustainability of economic growth.
2. **Poverty Reduction:** Inclusive growth is pro-poor, targeting those deprived, commonly based in rural areas. Empowering these segments of the population will significantly lead to poverty reduction. This has already been witnessed in Egypt, Burkina Faso, Ghana, and Zimbabwe (UNDP, 2013a).
3. **Equity:** Involving the disadvantaged sectors or segments of a population in productive economic activities will hopefully augment redistribution of income across the population. Achieving equity is a realistic way of fostering balanced development which can ultimately reduce rural urban migration with all its associated predicaments. Considering that Ugandan smallholder farmers are more unprivileged, targeting them in the process of development is a viable way of ensuring equity.
4. **Enhanced Employment:** By targeting the disadvantaged segments such as the underemployed smallholder farmers and empowering them, an economy is poised to reduce its unemployment level. Those empowered will further gain capacity to create own jobs/employments which can reduce the rate of crimes committed by those formerly unemployed.
5. **Food Security:** In most developing countries especially in sub-Saharan Africa, more than 65% of the populations are engaged in agriculture making the sector a home to more than 70% of the poor. In the same way, agriculture remains an important sector, employing about 72% of all workers in

Uganda (World Bank, 2014). Targeting agriculture and enhancing its productivity will ensure food security for the whole nation.

6. **Resource Utilization:** Just like many other developing countries, Uganda has a number of unutilized natural resources. By economically empowering the poor segment of the country's population, the government will eventually see a number of natural resources exploited. This will probably lead to improved people's income and welfare, and broaden the tax base or revenue of the country.

Literature indicates that growth is always more effective in reducing poverty in countries with low income inequality than in countries with high income inequality (UNDP, 2013b). When growth is inclusive people are more augmented to participate in, benefit from, and create wealth and jobs from the growth. Inclusive growth is inextricably linked with the contemporary definition of economic development. Therefore, to achieve economic development, developing countries should unanimously promote inclusive growth. In the case of Uganda this may be achieved through increasing agricultural productivity of the smallholder famers who are more disadvantaged. As reported by UBOS (2014), production by smallholders constitutes over 75% of both food crop and cash crop aggregates.

Major Constraints to Agricultural Productivity in Uganda

Despite several attempts to raise agricultural productivity in Uganda, success has been significantly low. The modest increases in productivity since 1990 are attributed to expansion of cultivated land rather than improvement in productivity per unit area of land (World Bank, 2008). Nabbumba and Bahiigwa (2003) observed that expansion of cultivated land is becoming unsustainable since access to land is increasingly constrained by the high population growth.

Growth of agricultural productivity in Uganda is constrained by several factors. In this chapter, the key factors were identified based growth accounting exercise. Based on the work of Barro (1998), growth accounting can be defined as a technique for explaining the factors that determine growth; it breaks down economic growth into components associated with changes in factor inputs and Solow residual. It provides an understanding on the sources of growth hinging on the Cobb-Douglas production function. Other determinants of productivity were ascertained by disaggregating the key variables captured in the main growth accounting equation (equation 1).

$$G_t^Y = G_t^A + \alpha G_t^K + (1 - \alpha) G_t^L \tag{1}$$

Equation (1) implies that the growth rate of output equals the growth rate of the technology term, plus a weighted average of capital growth and labour growth; where the weight is determined by the parameter $\boldsymbol{\alpha}$. The equation provides estimates of how much agricultural productivity growth over a certain period comes from growth in the number of workers, how much comes from growth in the stock of capital and how much comes from improvements in TFP. The major factors identified in this chapter are discussed below.

- **Capital:** Since most smallholder farmers are poor, lack of capital prevents them from acquiring modern equipments, using hybrid seeds, or expansion of their farms. The farmers often fail to cultivate in time, cannot afford to buy fertilizers and cannot spray their crops or treat their livestock in case of diseases. In most cases, the farmers are forced to use family labour. They plough small areas and spend a lot of time doing worthless work.

- **Pests and Diseases:** Pests and diseases are serious problems to nearly all farmers in Uganda, causing yield losses of up to 50% sometimes. Maize, beans, bananas, simsim, cassava, millet and coffee are extremely susceptible to diseases and pests. The possibility of chemical control of pests and diseases by smallholder farmers is limited by finances and availability of quality pesticides. Most farmers resort to resistant varieties which are generally low yielding.
- **Farming Skills/Knowledge:** Over 70% of the Uganda's smallholders have never received formal training in farming or agriculture; most farmers lack the required technical skills in agriculture.
- **Land Problems:** Uganda's rapid population increase and the analogous increase in the number of agricultural households have increased land pressure. Average agricultural household land holding is as low as 2.4 hectares (6 acres). However, in the northern part of the country some households own more than 4 hectares of land, while their counterparts in the central and western Uganda own less than 1.6 hectares. Land limitation forces farmers to undertake small scale farming activities, and the repeated use of a farmland depletes its fertility making the land less productive. Since most smallholder farmers lack the technical farming skills, they do not know how to replenish soil nutrients. Recent economic update by the World Bank (2015b) shows that most land in Uganda is not registered; less than 20% of Uganda's arable and pastoral land is registered (having land titles), with over 90% of rural land being undocumented.
- **Variable Climate and Weather; Drought, Flood and Wind:** Due to reliance on natural climatic factors, Ugandan farmers have become more vulnerable to the impacts of climatic change. Further still, poverty cannot allow the farmers to undertake irrigation. Extreme weather events including drought, flood and wind are expected to severely affect farmers further due to climate change. Less than 2% of the farmers apply irrigation. Most farmers in the country have also failed to control flood and extreme wind, except in western Uganda where some farmers practice terracing.
- **Labour Problems:** Availability and the quality of labour are serious constraints to farm productivity in Uganda. The high cost of labour or their limited availability makes planting, weeding, and harvesting often a serious problem. Shortages of labour causes delay in planting, weeding or harvesting; bad timing in any of these activities leads to major losses to the farmers.
- **Limited Application of Modern Inputs:** The key intermediate inputs include fertilizers, pesticides, energy, feeds, seeds and parent/breeding livestock. The application of modern intermediate inputs is not widespread in Uganda and most smallholders do not use them. Due to limited knowledge, poverty and conventionality, most farmers grow their crops or keep their livestock without using modern inputs. Although some farmers lack the money to buy them, others believe that they can do without the inputs perhaps because their land is still fertile or because they prefer to practice organic farming.
- **Poor Infrastructure:** Uganda has the poorest infrastructure in the region; most roads are basic and they are hardly maintained. There are almost no functional irrigation facilities. The poor state of infrastructure has had long-term detrimental effects on productivity growth. Policy makers need to reconsider their budget priorities in order to mitigate the deleterious impact of budget cuts on future growth. Under funding of agriculture is now a big problem (UBOS, 2014).
- **Lack of Government Support in the Form of Subsidies, Loans and Extensions:** Over the last twenty years, the agricultural sector has been receiving less than 5% of Uganda government

budget allocations. Based on Annual National Budgets (MFPED, 2015), the government allocates more money to security (army), education, health and transport sectors. Under financing of the agricultural sector implies support to the smallholder farmers in form of subsidies, loans or extensions/trainings are limited.

Although the government has put in place the National Agricultural Advisory Program (NAADS) to support farmers, only about 11% of the farmers received support. The supports given are often too small to improve the living standard of the farmers. Supports extended included seeds, animals (goats and cows), fertilizers and specialized trainings. In most cases supports are given to farming groups.

- **Fluctuation in Prices:** Uganda experiences fluctuating food prices; prices increase during and after the long dry spell (between December and June) and are low during the rainy seasons. The main reason for this fluctuation is the reliance on natural factors. Immediately after harvest periods, prices for most food stuff are extremely low due to very high supply, and the prices will rise in times of scarcity. However, even during scarcity the smallholder farmers do not gain much from soaring prices since they do not store their products to sell later at higher prices.
- **Political Instability and Conflicts:** Uganda has witnessed several episodes of war over time that caused various atrocities and seriously ravaged the country. In the last fifteen years however, most parts of the country were peaceful. To some extent, the various episodes of wars and conflicts have destroyed or lessened the infrastructure and other systems needed for the development of agriculture, thereby limiting agricultural productivity.
- **Post-Harvest Losses:** According to Alliance for a Green Revolution in Africa (AGRA, 2013), crop post-harvest losses in Uganda have been estimated at 5-15% for cereals and legumes, 20-25% for root and tubers and over 35% for fruits and vegetables. These losses are mostly attributed to poor storage conditions, damp floors, inadequate sunshine and high humidity. Ugandan farmers' need additional training to enable them mitigate continuous post-harvest losses.

Ways of Improving Agricultural Productivity

For national development strategies to be successful, food security must be part of the national agenda, and the promotion of rural development should be a key component of such strategies. Achieving rural development requires sustainable strengthening of agricultural productivity and competitiveness (World Bank, 2008). Based on these views, for growth to be effective in reducing poverty, unemployment and income inequality, the poor must contribute to deliver growth and they must also benefit from the process. Enhancing agricultural productivity must be a prime target in any policy design for sustainable growth and development in any poor agriculture based country. The subsequent section presents the possible ways of raising agricultural productivity in Uganda:

- **Resource Support for Farmers:** Agricultural growth depends on whether Uganda's scarce resources, including farm land, labour, irrigation water, farm equipments and other capital, can move freely between farms to higher value uses. Efficient resource reallocation through structural adjustment or otherwise is an important productivity driver at industry level. As previously indicated, a major constraint limiting agricultural productivity is lack of capital for acquiring modern equipments, buying hybrid seeds, or expanding the enterprise. Unfortunately, the unrelenting underfunding of the agricultural sector implies that the little public support extended to the farmers cannot adequately foster improvement in their livelihood. In case of natural hazards, assistance programs such as drought/flood assistance may also be introduced to provide farming incentives and to act as a safety-net to the victims.

- **Access to Assets:** There is need to improve access by the smallholders to various assets and also to involve them in decision making processes, especially on issues that influence their own lives. The government should provide low interest loans to farmers with minimum requirements since most smallholders do not have the required collaterals. Given that most farm land is not registered, the government should subsidize land registration and ensure faster processes of formal land documentation. Government should also invest in expensive farm equipment including tractors and avail them for hire at more affordable cost at every sub-county. These would enable timely land opening and will foster cheaper farm expansion by the smallholders.
- **Invest in Research, Development and Extension System:** Investing in rural R&D is critical to increasing productivity, sustainability and resilience of rural farm activities. Public investment in R &D, and effective and timely extension activities will foster adoption of such innovations on the latest technologies by farmers. Investments by the Ugandan governments in rural R&D and extension have not had a significant effect on agricultural productivity. Most rural farmers are not informed about new research findings from the National Agricultural Research Organization (NARO) because little extension services actually reach them. Since most farmers have had low level of education and less than 30% received formal training in agriculture, rural R&D and extension are significant in promoting productivity. Several studies which examined the return to public investments in R&D and extension in agriculture have shown high rates of return (Mullen, 2007; World Bank, 2008).

Although both domestic and international R&D findings should be used, domestic findings lead to more tailored and better suited policies to the local conditions. R &D should particularly focus on high yielding varieties, and pests and disease control. Generally, future agricultural productivity growth will depend on the capacity of rural RD&E systems to holistically supply innovations to a diverse section of rural Uganda.

- **Rigorous Use of Modern Inputs Including Hybrid Seeds, Fertilizers, Pesticides and Feeds:** Several scholars have recognized intermediate inputs as the major embodiment of modern scientific farming (Donovan, 2012). As outlined previously, smallholder farmers in Uganda still underutilize modern inputs despite their positive correlation with productivity. Addressing the low use of modern inputs may require an integrated approach. The government should subsidize inputs so as to make them more affordable to the farmers. Subsidies can help to overcome temporary market failures, reduce risks and offset fixed costs. Furthermore, the inputs should be made available in the rural areas and their benefits should be well communicated to the farmers.
- **Rural Infrastructural Development:** The magnitude of smallholder supply response to trade and price changes depends on, among other factors, rural infrastructure (irrigation, roads, transport, power, and telecommunications), markets, rural finance, and research. Rural infrastructure is still very poor in Uganda. The significance of rural infrastructure and their unrelenting deficiency implies government must invest in them in order to reduce the cost of doing business and to take advantage of trade reforms.

Private sector competitiveness may be significantly constrained by infrastructure deficiencies such as unreliable electricity and inadequate rail or road networks. The Ugandan government has recognized this and it has increased public investment to finance infrastructure projects in energy and roads in the budget allocations of 2015/16 with the aim of reducing the cost of doing business (MFPED, 2015). Nevertheless, there is need to monitor the implementation of the projects. The government should open and maintain rural roads, install irrigation facilities, extend electricity, and improve communication services.

CONCLUSION AND RECOMMENDATIONS

Although the Ugandan government expects that economic growth should generate virtuous circles of prosperity and opportunities, strong growth of the recent decades has not translated into the expected development outcomes. As indicated in several reports (World Bank, 2011; World Bank, 2015a; MFPED, 2015), the distress of poverty, unemployment and inequality continue to acutely affect people, especially the rural farmers. Uganda's growth process has thus not been pro-poor and inclusive since the living standards of most farmers have not improved.

The biggest challenge of Uganda's agriculture seems to be low productivity (IFPRI, 2008; Nabbumba & Bahiigwa, 2003). The country's agricultural productivity is constrained by lack of capital, pests and diseases, and limited farming skills. Land limitation, bad climate and weather, and limited use of modern inputs also limit productivity. Other factors that constraint productivity include poor rural infrastructure, poor government support to farmers, price fluctuation, and post-harvest losses.

Augmenting agricultural productivity will require increasing resource support for farmers; increasing access to various assets; and investing in research, development and extension system. The government should also foster rigorous use of modern inputs including hybrid seeds, fertilizers, pesticides and feeds; and it should develop rural infrastructure. Further researches should consider careful decomposition of Uganda's economic growth (growth accounting) and also ascertain the country's binding constraints to inclusive growth in order to foster formulation of well-tailored policies that coherently offer generic solutions to poverty, unemployment, and inequality problems.

REFERENCES

AGRA. (2013). *Establishing the Status of Post-harvest Losses and Storage for Major Staple Crops in Eleven African Countries (phase I)*. Retrieved July 23, 2015, from file:///C:/Users/Bond/Downloads/establishing-the-status-of-postharvest-losses-and-storage-for-major-staple-crops%20(4).pdf

Anand, R., Mishra, S., & Peiris, S. J. (2013). *Inclusive Growth: Measurement and Determinants*. Washington, DC: International Monetary Fund.

Barro, R. (1998). Notes on growth accounting. *Journal of Economic Growth*, *4*(2), 119–137. doi:10.1023/A:1009828704275

CAFOD. (2014). *What is "inclusive growth"?* CAFOD Discussion Paper.

DFID. (2012). *Growth Building Jobs and Prosperity in Developing Countries*. Retrieved July 4, 2015, from http://www.oecd.org/derec/unitedkingdom/40700982.pdf

Dharmasiri, L. M. (2009). Measuring agricultural productivity using average productivity index. *Sri Lanka Journal of Advanced Social Studies*, *2*(1), 25–44.

Donovan, K. (2012). *Agricultural risk, intermediate inputs and cross-country Productivity Differences*. Job Market Paper.

Felipe, J., & Hasan, R. (2006). *The Challenge of job creation in Asia.* Asian Development Bank: ERD Policy Brief No. 44.

Fulginiti, L., & Perrin, R. (1998). Agricultural productivity in developing countries. *Agricultural Economics*, *19*(1-2), 45–51. doi:10.1016/S0169-5150(98)00045-0

Hausmann, R., Rodrik, D., & Velasco, A. (2005). *Growth Diagnostics.* Inter-American Development Bank. Retrieved October 9, 2013, from http://www.hks.harvard.edu/fs/rhausma/new/growthdiag.pdf

Ianchovichina, E., & Lundstrom, S. G. (2009). *Inclusive Growth analytics framework and application.* Policy Research Working Paper 4851. World Bank.

IFPRI. (2008). *Land Tenure & Agricultural Productivity in Uganda*. Kampala: IFPRI.

IFPRI. (2012). *Rural-Urban Transformation in Uganda.* Kampala: IFPRI. Retrieved December 10, 2014, from http://www.ifpri.org/sites/default/files/publications/usspwp10.pdf

IFPRI. (2013). *Total and Partial Factor Productivity.* Retrieved March 4, 2015, from http://www.ifpri.org/sites/default/files/gfpr2012_indicators_tfp.pdf

International Labor Organization (ILO). (2014). *Global Employment Trends 2014: The risk of a Jobless Recovery*. Accessed March 30, 2015, from http://www.ilo.org/global/research/global-reports/global-employment-trends/2014/WCMS_234107/lang--en/index.htm

International Labor Organization (ILO). (2015). *Agriculture; Plantations; other rural sectors.* Accessed March 10, 2015, from http://ilo.org/global/industries-and-sectors/agriculture-plantations-other-rural-sectors/lang--en/index.htm

MFPED. (2015). *Maintaining infrastructure investment and promoting excellence in public service delivery. Background to the Budget, 2015/16.* Kampala: MFPED.

Mullen, J. D. (2007). Productivity growth and returns from public investments in R & D in Australian broad-care agriculture. *The Australian Journal of Agricultural and Resource Economics*, *18*(1), 18–29.

Nabbumba, R., & Bahiigwa, G. (2003). Agricultural Productivity Constraints in Uganda: Implications for Investments. Kampala: Uganda.

OECD. (2012). *Promoting Inclusive Growth: Challenges and Policies*. Paris: OECD.

Olayide, S. O., & Heady, E. (1982). *Introduction to Agricultural Production Economics* (1st ed.). Ibadan: Ibadan University Press.

Oxfam. (2014). *Oxfam: 85 richest people as wealthy as poorest half of the world.* World Economic Analysis. Accessed July 4, 2015, from http://www.theguardian.com/business/2014/jan/20/oxfam-85-richest-people-half-of-the-world

Singh, J., & Dhillion, S. S. (2000). *Agricultural Geography* (2nd ed.). New Delhi: McGraw Hill.

UBOS. (2014). *2014 Statistical Abstract*. Kampala: Uganda Bureau of Statistics.

UBOS. (2015). *2015 Statistical Abstract*. Kampala: Uganda Bureau of Statistics.

UNDP. (2013a). *Assessing Progress in Africa toward the Millennium Development Goals. MDG Report 2013*. New York: UNDP.

UNDP. (2013b). *Human Development Report 2013: The Rise of the South*. New York: UNDP.

Worldatlas. (2015). *Uganda*. Retrieved August 4, 2015, from http://www.worldatlas.com/webimage/countrys/africa/ug.htm

World Bank. (2008). *Agriculture for development. World Development Report 2008*. Washington, DC: World Bank.

World Bank. (2011). *Uganda: Growing Out of Poverty. Poverty Reduction and equity*. Accessed July 20, 2015, from http://web.worldbank.org/WBSITE/EXTERNAL/TOPICS/EXTPOVERTY/EXTPA/0,contentMDK:20204772~menuPK:435735~pagePK:148956~piPK:216618~theSitePK:430367,00.html

World Bank. (2013). *End Extreme Poverty 2030 Promote Shared PROSPERITY. Annual Poverty Report 2013*. Washington, DC: World Bank.

World Bank. (2014). *World Bank and Agriculture in Africa*. Accessed June 20, 2014, from http://web.worldbank.org/WBSITE/EXTERNAL/COUNTRIES/AFRICAEXT/0,contentMDK:21935583~pagePK:146736~piPK:146830~theSitePK:258644,00.html

World Bank. (2015a). Agriculture and rural development. *World Development Indicators*. Accessed July 1, 2015, from http://data.worldbank.org/topic/agriculture-and-rural-development

World Bank. (2015b). *Can Uganda's land support its prosperity drive?* World Bank Report No 99060. Washington, DC: World Bank.

World Economic Forum. (WEF). (2015). *The Inclusive Growth and Development Report 2015*. Geneva: World Economic Forum.

Zepeda, L. (2001). *Agricultural Investment, Production Capacity and Productivity*. Geneva: FAO.

KEY TERMS AND DEFINITIONS

Agricultural Productivity: The ratio of the value of total farm outputs to the value of total inputs used in farm production. It is the output produced by a given level of input(s) in the agricultural sector of an economy.

Economic Development: The process of improving the quality of all human lives and capabilities by raising people's standards of living, self esteem and freedom.

Economic Growth: The continuous improvement in the capacity to satisfy the demand for goods and services, resulting from increased production scale, and improved productivity (innovations in products and processes). It is measured using GDP.

Inclusive Growth: Output growth that is sustained over decades, is broad-based across economic sectors, creates productive employment opportunities for a great majority of the country's working age population, and reduces poverty. It involves growth in both income and equity or income distribution.

Poverty: The lack of basic needs and services such as food, clothing, beddings, shelter, healthcare, education, information and communication.

Unemployment: All the individuals who are without work/jobs: they may either be in the process of moving to new jobs or actively seeking work.

Chapter 12
Integrating Spatial Technologies in Urban Environments for Food Security:
A Vision for Economic, Environmental, and Social Responsibility in South Bend, Indiana

Edwin Joseph
Indiana University – South Bend, USA

Elizabeth O'Dea
Indiana University – South Bend, USA

ABSTRACT

Food security for the urban poor has been an important topic for both developed and developing countries over the last 15 years. Although South Bend Indiana is a city in a developed country, declining economic circumstances have caused the city to show significant urban decay somewhat similar to some cities in developing countries. In this chapter, we explore South Bend's history and economic development strategies, and review practices aimed at strengthening food security for the urban poor. The chapter documents how numerous disparate organizations have been trying to help alleviate urban poverty and hunger, and reviews previous strategies used to foster sustainable growth and development. The integration of spatial technologies will become a key factor for promoting community social networks, participatory planning, and collaboration. The case is presented for the use of Geographic Information Systems (GIS) and associated technologies to help organizations, community leaders, local organizations, city planners, higher education institutions and the urban poor, work together to alleviate poverty and malnutrition through networking and sustainable urban agriculture.

DOI: 10.4018/978-1-5225-0942-4.ch012

INTRODUCTION

The World Food Summit of 1996 defined food security as *"when all people, at all times, have physical, social and economic access to sufficient, safe and nutritious food which meets their dietary needs and food preferences for an active and healthy life"*. Food insecurity is defined as, *"a household-level economic and social condition of limited or uncertain access to adequate food"* (World Health Organization, 2015). The United States Department of Agriculture (USDA)'s definition of food security is, *"access by all people at all times to enough food for an active, healthy life* (USDA, 2015).*"*

ECONOMIC PROFILE—SOUTH BEND (2015)

The city of South Bend is considered the economic and cultural hub of the greater seven-county region known as Michiana, Indiana in the United States of America. The Chamber of Commerce of St. Joseph County, Indiana defines Michiana as St. Joseph County and "counties that contribute at least 500 inbound commuting workers to St. Joseph County daily." Those counties include Elkhart, La Porte, Marshall, St. Joseph, and Starke in Indiana, and Berrien and Cass in Michigan (Wikipedia, 2016). South Bend is the fourth largest city in the state, with a population of 101,190 (Indiana Demographics, 2016).

The 2015 unemployment rate in St Joseph County was 4.8%. The top three employers came from education [Notre Dame University], health care [Beacon Health System], and K-12 Public Schools [South Bend Community School Corporation] (South Bend Region Economic Development, 2016). There are concerns over the ecological unsustainability within some areas of the city. There is also growing inequality among the urban poor. The big issue is to identify possible remedies for inner city poverty in South Bend, knowing that the gap in the wealth among American households is more extreme than at any time since the Great Depression (Matthews, 2014). Seeking solutions for these problems may take many forms. One way is to explore pathways towards economic growth through urban sustainable agriculture. It is also instructive to review trends that are currently occurring in developing countries.

POVERTY AND SUSTAINABLE DEVELOPMENT WORLDWIDE

There seems to be much programmatic interest about living standards and health among neighborhoods in developing countries, but very little in developed countries. Researchers have indicated that population growth will be concentrated in cities and towns by 2030 (Montgomery & Hewett, 2005). This statistic is supported by the United Nations Department of Economic and Social Affairs world population and distribution research data, which shows that 66% of the world's population would be urban, with North America currently having 82%.

At the 2012 Rio+20 meetings, political leaders acknowledged that there were many challenges to sustainable development, and reiterated that many of the world's poor depend on rapidly disappearing and fragile biodiverse ecosystems (Chappell, et al., 2013). The United Nations department of Economic and Social Affairs stated that there are more than 1 billion people living in extreme poverty and income inequality worldwide. As a result, inadequate and unsustainable consumption and production patterns exist, but at huge economic and social costs (DESA, 2013).

URBAN FOOD SYSTEMS

The term "food system" describes the complex and interconnected activities of agricultural food production, processing, marketing, consumption, and disposal. It is an emerging discipline that brings together scientists and humanists to help address the cultural, social, ecological, physical, ethical and political aspects of food (Berman, 2011). According to Berman (2011), there is no single accepted definition for local food systems, but it is often defined as an area within a 400-mile radius of food origin. The scale of urban agriculture can vary from less than an acre of land worked by a handful of people to several acres.

Simply put, urban agriculture is growing or producing food in a city or heavily populated town or municipality. Producers vary from individuals, friends, non-profits, neighborhood groups and so forth. Reasons for production also vary from sales to restaurants, farmers' markets and donations to local food kitchens and churches. Urban agriculture increases access to locally grown food, and reinforces the importance of community and culture. In short, it is the front line of the food system (Greensgrow, 2016)

Urban Poverty in America: A Brief Overview

In his book about the causes of urban poverty in 1987, Harvard Professor, William Julius Wilson argued that changes in the structure of the U.S. economy were the primary drivers of increased social and economic dislocation of the urban poor. As the American economy shifted from predominantly manufacturing to service-driven, the urban underclass was severely impacted. According to Wilson, joblessness, low socioeconomic status, social isolation, impoverished neighborhoods, low education, lack of community safeguards, resources, and support were typical features of African-American urban communities. It wasn't a cultural problem, but a problem of isolation from community and necessary resources. Discrimination facilitated the persistence of an urban "underclass." (Wilson, 1987).

Urban Sustainable Agriculture: Examples across the US

Agriculture in the United States has been in social and economic turmoil for over two decades while Urban Sustainable Agriculture and Community Supported Agriculture (CSA) has been gaining in popularity. We can trace CSAs from the 1960s in Japan, when women who were frustrated with the pesticide laden, process prepared, imported food formed an alliance with farmers to supplement their food needs from locally grown, nutritious, organic food. The alliance was known as Teikei, or partnership—a more detailed/literal Japanese translation is "food with the farmer's face on it." Teikei therefore was born out of the rejection of conventional agriculture, on grounds of social, environmental, and economic justice in addition to a simple desire for fresh, quality food (Schnell, 2007) (Henderson & Van En, 2007). The concept eventually made its way to the United States.

Farmers, consumers, and activists in the sustainable agriculture community in Pennsylvania learned to exploit ruptures in the conventional food system and developed new forms of food provisioning in local economies. The partnership has aided in local economic development and poverty alleviation, and demonstrated that this alternative model of food production and consumption has encouraged the rise of urban gardens all over the nation (Trauger, 2009).

In May 2000, Greenpoint Manufacturing and Design Center (GMDC) embarked on a $6 million development of The Greenpoint Wood Exchange—a rehabilitation of an 80,000 square foot former bowling alley building at 810 Humboldt Street Brooklyn New York, and built a rooftop garden. The rooftop accommodates a 15,000 square-foot hydroponic greenhouse that produces over 100 tons of food per year and supplies year-round produce to nearby grocery stores and restaurants (GMDC, 2015).

The Will Allen organization, Growing Power, designed a system of greenhouses and farm locations around Milwaukee that divert and compost more than 44 million pounds of food waste per year from nearby households, beer breweries, coffee shops, and even corporations like Walmart. The organization also produces one million pounds of organically-grown food each year (Kumar, 2015).

In 2011, Noah Link and Alex Bryan built an urban farm in Detroit Michigan known as Food Field that offers a CSA (Community-Supported Agriculture) service. Their goal is to create an alternative to the corporate food system. Local residents enjoy the sustainable produce in weekly CSA boxes containing salad greens, mulberries, and other food items. Food Field now has an aquaponics system for raising Catfish and Blue Gill. The farm also raises chickens and ducks for eggs (FoodTank.com, 2013).

The Distributed Urban Farming Initiative (DUFI) in Bryant Texas transformed vacant lots to produce healthy food. The goal is to build gardens in empty spaces, inspire Bryan residents to eat healthy food, and drive entrepreneurship and tourism. The initiative has been able to successfully grow broccoli, cauliflower, cabbage, and lettuce on raised beds and pallet gardens. DUFI has been working hard to reconnect the Bryan community with the food system (DUFI, 2015).

The Chicago Council on Global Affairs published a report highlighting the importance of urban food security, and offers an action plan for addressing urban food security and nutrition (Lee, 2013).

What is known about many urban centers in the US is that healthy, safe, nutritious food at affordable prices is difficult to find (The Chicago Council on Global Affairs, 2013). In many ways, the City of South Bend Indiana is no different from other cities in its quest to feed its growing population. Although over the past few years there have been efforts to address poverty and malnutrition through community gardens, much more collaborative work is needed to make these efforts sustainable.

URBAN SUSTAINABLE AGRICULTURE IN SOUTH BEND, INDIANA

A report published by the American Planning Association (APA), provides guidance for supporting and expanding urban agriculture. Like Chicago, Cleveland, Detroit, Missouri, Milwaukee, Philadelphia, among others, South Bend is an old industrial city, and possess acres of vacant abandoned city lots. These areas are potential centers for urban agriculture initiatives (Hodgson, Campbell, & Bailkey, 2011). Currently, the most popular form of urban agriculture in the city is community gardening, which contributes significantly to developing social connections, capacity building, and community empowerment. Our investigation seeks to find out how much collaboration exists to support important social networks to help in closing the gap in the wealth among households, and in the satisfaction of food needs.

There is a growing awareness about the role of urban sustainable agriculture in the context of food security and poverty alleviation in South Bend Indiana; however, its implementation is very informal. The local government verbally supports the grassroots efforts of the urban agricultural movement in South Bend. However, the city has yet to incorporate its support into the City Plan (Radican, August 2010). This makes urban agriculture vulnerable, and may also jeopardize its sustainability. Nonetheless, urban sustainable agriculture in the city seems to be steadily increasing. There is also a growing demand for a greener city, social justice, job security, and increased wildlife restoration.

Currently, there is a public listing of active community gardens and food services for individuals and families at risk in South Bend. The original listing included contact persons, telephone numbers, and emails (these have been removed from the Table 1.). In order to view the distribution of the gardens and food services within a Geographic Information System (GIS), we geocoded the addresses by assigning geographic coordinates to each location (latitude/longitude). This made it convenient to view other spatial entities such as roads, parks and schools, to understand trends, patterns and distribution in relation to each other.

Table 1. Distribution of community gardens in South Bend IN

Food Entity	Address	Zip	Latitude	Longitude	Description
Corvilla - Bulla	18443 Bulla Road	46637	41.7043715	-86.2273586	Garden
Corvilla - Roelke	1318 E Roelke Dr	46614	41.617287	-86.233396	Garden
Corvilla Office Unity Garden	3231 Sugar Maple Business St.	46628	41.720718	-86.2975079	Garden
El Campito Unity Garden	1024 Thomas St	46601	41.6741565	-86.2676142	Garden
Garden of St Therese	915 N Olive Street	46628	41.6873507	-86.2858679	Garden
Haircrafters Unity Garden	602 Lincoln Way East	46601	41.6702888	-86.2449995	Garden
IHC Unity Garden	1901 W. Western Ave	46619	41.6731319	-86.2800307	Garden
IUSB Unity Garden	1700 Mishawaka Ave	46615	41.6800027	-86.2401604	Garden
Jewish Federation	3202 Shalom Way	46615	41.678144	-86.205174	Garden
JJC Unity Garden	1001 S. Michigan St	46601	41.6640819	-86.2522822	Garden
Kroc Center Unity Garden	900 W. Western Ave.	46601	41.671611	-86.2644002	Garden
LaSalle Square	3701 Prast Blvd	46628	41.6863911	-86.3027314	Garden
Madison Square Unity Garden	1230 E Madison Street	46617	41.6806026	-86.2337549	Garden
McKinley School	228 E McKinley Ave	46615	41.6804295	-86.1814233	Garden
Monroe Park	608 Carroll St.	46601	41.6693708	-86.2492739	Garden
Morningside Unity Garden	333 W Colfax Ave.	46601	41.6781427	-86.2571334	Garden
Muessel Primary Center	1020 Blaine Ave.	46628	41.6883679	-86.2696123	Garden
Ravina Park Unity Garden	1511 Dale Ave	46613	41.6596109	-86.239353	Garden
Red Cross Refuge Unity Garden	3220 E Jefferson Blvd	46615	41.6724718	-86.2039691	Garden
Robinson Center	921 North Eddy Street	46617	41.6880438	-86.2370864	Garden
Safe Station Unity Garden	1322 LincolnWay East	46614	41.6597928	-86.2316855	Garden
Smallest Unity Garden	528 E Colfax Ave	46617	41.6776081	-86.245224	Garden
St Peter's UCC Unity Garden	915 N Ironwood Street	46615	41.687646	-86.2184267	Garden
St Vincents Unity Garden	520 Crescent Ave	46617	41.6838757	-86.2460485	Garden
Unity in God Garden	646 LaPorte Ave	46628	41.6831259	-86.2739902	Garden
Zion Unity Garden	224 E. Jefferson Blvd	46617	41.6731176	-86.1806002	Garden
Notre Dame Ctr.	1045 W. Washington St.	46601	41.6762414	-86.2673326	Garden
Beacon Triangle Unity Garden	1345 N Michigan Street	46617	41.6924393	-86.2528915	Garden
Eddy St Eberhart Enviro Garden	317 S. Eddy Street	46615	41.6730586	-86.2372438	Garden

continued on following page

Table 1. Continued

Food Entity	Address	Zip	Latitude	Longitude	Description
Walnut Street	421 S. Walnut Street	46619	41.6716617	-86.2726509	Garden
Walnut Street	423 S. Walnut Street	46619	41.6716618	-86.2726732	Garden
Hear Our Tears	2421 Bertrand St	46628	41.6815726	-86.2860361	Garden
Hope Ministries Rescue Mission	432 S Lafayette Blvd	46634	41.6712481	-86.2555248	Soup Kitchen
Center for the Homeless	813 S. Michigan Street	46601	41.6659525	-86.2527165	Soup Kitchen
St. Vincent de Paul	3408 Ardmore Trail	46628	41.6885417	-86.2990941	Food Asst.
Berean Community Service	601 W Colfax Ave	46601	41.677674	-86.2600018	Food Pantry
Food Bank Of Northern Indiana	702 S Chapin St	46601	41.6679634	-86.2625769	Food Bank
Community Food Pantry	702 S Chapin St	46601	41.6671644	-86.263539	Food Pantry
Downtown Soup Kitchen	333 N Main St	46601	41.6800648	-86.2546771	Soup Kitchen
Northeast Neighborhood Center	803 N. Notre Dame Avenue	46617	41.685896	-86.2401508	Food Pantry
Catholic Charities Food Pantry	1817 Miami St.	46613	41.6554923	-86.2372697	Food Pantry
La Casa De Amistad	746 S Meade St	46619	41.6667298	-86.28753	Food Pantry
River Park United Methodist	920 S 23rd St	46615	41.6653659	-86.2157277	Food Pantry
Little Flower Catholic Church	5419 N Ironwood Drive	46635	41.6972864	-86.2186849	Food Pantry
Needs for Neighbors	54301 Maple Lane	46635	41.7041949	-86.2159955	Food Pantry
Christ Temple Apostolic	55454 N Lexington Ave.	46628	41.6869748	-86.3148238	Food Pantry
Share Your Blessing	3001 W. Cleveland Rd.	46628	41.7250602	-86.2932771	Food Pantry
Clay United Methodist Church	17646 Cleveland Rd.	46637	41.7229223	-86.2758369	Food Pantry

A review was undertaken of other aspects of urban agriculture that are being used to raise awareness, and inform neighborhoods, decision makers, and planners about the different types of urban agriculture that have the potential to increase household incomes, promote healthy lifestyles, and generate a distribution chain through strong community networks. The importance of GIS cannot be understated since it can transform the way people network, communicate, and collaborate to make decisions—in this case, improve the standard of living for the urban poor. The significance of GIS technology to boost ongoing initiatives by key players and stakeholders in South Bend to improve the livelihoods of the urban poor must be recognized and factored into solutions.

HISTORY AND ITS EFFECTS ON AFRICAN-AMERICANS IN SOUTH BEND

African-American historical narratives not only document the lives of enslaved people, but offer invaluable perspectives of first-hand experience and mistreatment of people of color throughout an important period of American history. After the Civil War, former slaves continued documenting their experiences to ensure social and economic progress as affirmed by the nation's leaders. The African-American History narrative in South Bend mirrors some of the characteristics of post-war slavery, but is unique in many ways. For this reason, this should not be a truncated, incomplete, or untold story. It is a narrative that has the "potential to evoke pride and add a level of complexity" to South Bend's black heritage and history (Andrews & Adams, 2016).

Despite its rich history, little is known about the African-American experience in South Bend, or the efforts that were made to improve the quality of life within the black community; and enhance educational and economic opportunities. The South Bend History Museum documents the life of John Charles Bryant—a descendant of one of South Bend's first African American families with portraits dating back to the 1800s. For Bryant, portraits of his grandparents and great-grand parents have special meaning because they tell the story of how against all odds the Bryant-Powell family were able to escape the south, and settle in Indiana.

John Charles Bryant's grandfather worked at Studebaker, his father worked the tables at the Old LaSalle Hotel, his mother Georgia Bryant, was an organist for 42 years at the Olivet AME church, South Bend's first established African American church. This and other narratives are intertwined with the fabric of American History .

A brief review of South Bend's history and economy reveals that like most of America, African Americans experienced a measure of discrimination in the workplace and in the community. South Bend was a hub of manufacturing fueled by the presence of the Studebaker car company and other significant industries. The 1883 South Bend City Directory lists several African American men employed by local industries such as Wilson Brothers Shirt Factory, Birdsell Manufacturing Company, Singer Sewing Machine Company and the F.P. Box Company, The Oliver Chilled Plow Works (The History Museum, 2016).

Between 1870 and 1910, the African American population increased from 73 to 640; and in 1903 when Booker T. Washington visited South Bend, he found additional opportunities for employment for African Americans at Studebaker. Unfortunately, employment for African-Americans became difficult after the end of World War I when they had to compete with European immigrants for jobs. In the early 1920s the population of South Bend Indiana was 70,983, occupying a land area of 15.4 square miles, and an average population density of 4,609 per square mile. The city was ranked 100 out of the 100 largest urban places in America (U.S. Bureau of Census, 1998).

The population of African-Americans in South Bend was estimated around 1,269 during the 1920s, and increased to 3,431 in 1930 (Robinson, 2015). Studebaker employed more than 400 African-Americans, and at times up to 700. This was considered unique for the period, as several other local companies did not employ African Americans until the 1940s (WNDU, 2010). However, Studebaker underwent financial problems and the plant ceased production in 1963. As a result, the city lost about one-third of its population, and experienced severe growth and economic problems.

During World War II, South Bend became part of the wartime employment boom. Many African-Americans were employed as factory workers, but by the early 1950s, employment for African-Americans was again on the downturn in South Bend. In spite of the unemployment challenges, the 1950s Great Migration of African-Americans from the South caused an increase in the African-American population. And by the mid-1950s, the African-American population exceeded the foreign-born population (South Bend Tribune, 2015).

Based on archived reports, there was a concerted effort to end racial policies in South Bend. For example, in 1963, South Bend Mayor Frank J. Bruggner appointed a 34-man Bi-racial Committee to study the problems of racial discrimination in South Bend. The committee met with community leaders to help eliminate, by voluntary measures, the cause of racial tension (Mayor's Biracial Committee Progress Report, 1963). South Bend also passed an ordinance (4091) known as the South Bend Fair

Employment Practice Ordinance which prohibited discrimination in employment because of race, color etc. (City of South Bend Common Council, 1965).

In 1968, the South Bend Human Relations and Fair Employment Practices Commission endorsed open housing, thus raising the possibility of an open housing ordinance at the county level (The Reformer Newspaper Vol. 1, No. 29, 1968). In spite of these efforts, there were many examples of discrimination in the sale, rental, and financing of private housing in South Bend. One important example is the public hearing on March 19, 1963, in Notre Dame Law School Auditorium (Citizen's Fair Housing Committee of South Bend-Mishawaka, 1963)

South Bend Today

In 2011, Newsweek put South Bend on its list of America's Dying Cities, underscoring the 2.5% decline in the number of young people over the previous decade. (Newsweek, 2011). In 2014 South Bend was listed among the least happy regions in the United States by a Harvard and University of British Columbia study, taking the number 3 spot (UBC News, 2014). This was based on a survey of 300,000 people in 2009. The 2013 Gallup-Healthways Well-Being Index also lists South Bend as one of the "most unhappy" regions in the nation. The degree of unhappiness was based on a number of factors such as physical and emotional health, healthy behaviors, work environment, social and community factors, financial security and access to necessities like food, shelter and health care (South Bend Tribune, 2016).

Moreover, economists have established volumes of literature on the relationship between food insecurity and household incomes (classified by the poverty line). Researchers have concluded that food insecurity has increased dramatically in recent years, and is one of the most important nutrition-related public health issues in the U.S. (Gundersen, Kreider, & Pepper, 2011)

In 2015, South Bend's poverty rate stood at 23.9%, with almost 70% of students in the public school system eligible for free or reduced-cost lunch. South Bend also has double the national average rate of families living in poverty (Buttigieg, 2015). Crime risk statistics show South Bend to be above the US average for personal crime risk, murder risk, rape risk, robbery risk, property crime risk, burglary risk, larceny risk, and automotive theft risk. Assault risk, however, is lower than the US average (Point2Homes, 2016). The mayor of South Bend acknowledged that there were many excellent organizations working on problems, but needed partnership with the city.

In spite of these troubling statistics, the city of South Bend was ranked third in the nation for America's Best Complete Streets Policies of 2015 with a score of 98.2/100. This is an integrated transportation system that supports safe travel for pedestrians, cyclists, transit riders, motorists and others, within the city. The policy is one milestone encouraged in the Mayors' Challenge, a U.S. Department of Transportation initiative in which South Bend is a participant, committed to improving sidewalks, bike lanes, public transportation stops, frequent and safe crossing opportunities, median islands, accessible pedestrian signals, curb extensions, narrower travel lanes, and roundabouts (Smart Growth America, 2016).

A recent article by the South Bend Tribune discussed how it may be possible to develop, attract and keep talent so that South Bend area can compete in the modern economy. The Tribune acknowledges that at its heart, the region is still a manufacturing center, and manufacturing accounts for more than one out of every four jobs. (South Bend Tribune, 2014). Young, talented people are looking for quality of life, and will choose a community with a rich lifestyle over a good job elsewhere. The city knows that young people are more likely to locate near "livable" communities with bike trails, public transportation, and vibrant downtowns among others. South Bend agreed to work together with institutions of higher learning to include significant grass-roots involvement in visioning such a future.

As we fast-forward to the year 2016, we believe it is important to review a 30-year old problem, and carefully examine the issue of how food and other essential needs can be better satisfied among the urban poor in South Bend Indiana, and how we can use Geographic Information Systems Technology (GIS) to analyze proximities, adjacencies, intersections, and containment issues to help solve urban poverty.

For example, one may wish to know how close a community garden might be to a landfill in an urban area; or city planners may want to identify roads that pass within a certain distance of a city park. These are proximity issues. Questions such as, what is the shortest street network route to all community gardens are important proximity questions for understanding distance relationships, and are applicable for both developing and developed countries. Adjacency is generally about objects that are zero distance away from each other—in short, they share at least one common boundary.

An intersection analysis calculates the geometric intersection of features on a landscape. A simple analysis could be which river or stream crosses a major road—at what specific location; or which trail passes through a particular parcel in the city. The analysis does not limit itself to linear objects, but may extend to point on point analysis. In other words, where are the houses (point location) that have gardens on the lot (another point location) - that is considered an intersection of two points. Containment analysis answers questions such as which gardens are within which neighborhoods. These are very important spatial questions that we can query on-the-fly, and use for decision-making related to urban poverty.

Facing the Realities in South Bend: There is Hope

Many organizations work towards the goal of a sustainable South Bend. They include the Food Bank of Northern Indiana, Unity Gardens, Young Professionals Against Poverty (YPAP), City of South Bend Division of Sustainability, Center for a Sustainable Future, Links Incorporated (South Bend Chapter), The Civil Rights Heritage Center, United Way, 100 Black Men, and a few others can be located using the University of Notre Dame's link http://green.nd.edu/sustainability-in-action/community/.

These disparate groups tend to work independently, but would be more effective if there are stronger collaborations. For example, the Food Bank of Northern Indiana works in partnership with the community to feed the hungry, increase awareness of the effects of hunger and lead programs designed to alleviate hunger (Food Bank of Northern Indiana, 2016). The Food Bank describes that the rate of hunger insecurity in Northern Indiana as a silent epidemic, and sees the challenge for the city of South Bend as addressing food and nutritional security in a sustainable way given very high poverty levels (Foodbank of Northern Indiana, 2015). Their programs include Healthy Choices Market, Senior Bag Program, Mobile Food Pantry program, and Food 4 Kids.

The Unity Gardens, a 501(c)3 Non-profit in South Bend, is committed to improving community health by increasing accessibility to fruits and vegetables by developing a local food system. The goal of Unity Gardens is community building through a "framework of gardens". The gardens would not only provide healthy food for those in need, but also educate the community on how to grow, and utilize healthy food. The overall mission is to improve community health which includes access to locally grown fruits and vegetables, education on food preparation and consumption, increasing social capital, and opportunities for the disadvantaged (Unity Gardens, 2009).

Young Professionals Against Poverty (YPAP) was founded in 2006 by Steve Camilleri and Matt Money. They wanted to raise awareness about poverty in South Bend, and brought young professionals together to help organize events and raise money geared towards improving the quality of life. Steve Camilleri is also the Executive Director of the Center for the Homeless which has been endorsed by the US Department of Housing and Urban Development (HUD). The Center has provided more than 700,000 safe nights and over 1.5 million meals to more than 44,000 men, women, and children since opening in December 1988. The focus is on providing life-changing services to their guests (Center for the Homeless, 1988).

South Bend's chapter of Links Incorporated is an organization whose mission is to promote and engage in educational, civic, and intercultural activities in order to enrich the lives, and to work together toward achieving community goals. The organization's vision is to eliminate disparities by reducing barriers to services through advocacy, education, and service. Their goal is to empower communities and increase the number of sustainable and measurable programs; expand collaborative partnerships; and identify communities with greatest needs (The Links Incorporated, 2015).

The Civil Rights Heritage Center (CRHC) was established on the campus of Indiana University South Bend for the study and documentation of civil rights history. The Center engages in scholarly inquiry, student research, and community education, and is dedicated to creating partnerships with the state and local communities and offer leadership in creative solutions for 21st-century problems. (Civil Rights Heritage Center, 2015). The center sponsors group discussions related to equity and social justice, wages and income, access to healthy food and barriers to employment, among others.

In 2016 the center hosted a number of significant events. Some included community lectures on asset building, and civil rights. Other events were partnerships, for example, the center partnered with the University of Notre Dame to host a Wikipedia Edit-A-Thon to teach people how to become a Wikipedia editor and use proper guidelines for writing articles and finding resources for researching and writing about South Bend's history.

The Center also celebrated the history and experience of African Americans in South Bend through stories from the Black LGBT experience, and explored the significance of open-access, digital resources using The Black Press Research Collective (BPRC). It hosted a scanning event to preserve and share photographs and documents that tell about black history and the black experience. Guests were encouraged to bring artifacts to the Civil Rights Heritage for scanning and inclusion in Michiana Memory, the free, online archive (http://michianamemory.sjcpl.org/). In celebration of Women's History Month, the Civil Rights Heritage Center part sponsored a public talk on the topic "A Band of Sisters: Fighting for Employment Equality from the Assembly Line"—the story of a small group of blue-collar women who, in a class action lawsuit, uncovered endemic gender discrimination in South Bend.

South Bend Reconnecting People through Sustainable Programs

Recently the city of South Bend has been working in conjunction with the Environmental Protection Agency (EPA) and participating in their program called the Food Recovery Program (FRP), which helps prevent food waste by diverting leftover food from local restaurants to impoverished people in need (Rivest, 2015). There is also renewed interest in traditional food systems in South Bend. In an attempt to reconnect land and people through sustainable gardening, Indiana University South Bend (IUSB) is currently conducting research on urban sustainable agriculture. IUSB has developed agricultural courses focusing on sustainable local food production techniques, and is well on the way to developing a Sustainability Showcase—an area on campus dedicated to demonstrating urban sustainable practices.

Thc Showcase would contain a plaza/entryway featuring native plants, a zero impact home with the capacity to continue to showcase new home building design and products. It would also serve as a site for constructing Tiny Homes, and invite community participation for the regular construction of this new concept to increase accessibility to low cost housing in South Bend. The showcase will be open to area schools, neighborhood groups, non-profits, and so forth; all working together to solve social, economic, and environmental problems in the city. It will also serve as an open living laboratory to compare best practices for urban gardening. The goal is to share information and strengthen community relationships and networking.

TECHNOLOGY: A PATHWAY TO ECONOMIC RECOVERY IN SOUTH BEND

We believe that introducing new technology may be an appropriate pathway to economic recovery and progress in South Bend. Information and spatial technologies have the potential to provide a platform for people with similar interests to solve problems in the community. These technologies can provide a technical framework for food production and act as a communications platform even in low-income communities. New technologies can be separated into two categories—sustaining, and disruptive. A sustaining technology relies on incremental improvements to an already established technology, while a disruptive technology may not yet have a proven practical application, but has the potential to significantly alter the way organizations operate (Christensen, 1997).

Geographic Information Systems (GIS) technology falls into the category of a disruptive technology. It is a location-based technology that has made people more aware of the dynamic world in which we live. It has improved our ability to interpret our environment, and understand spatial interactions. Therefore, by nature, GIS technology is adaptive, and suited for dynamic, fast-paced innovations. Many entrepreneurs view GIS as a survival imperative that creates revenue opportunities. Some companies see it as a tool to expand their markets, new products or services. Dynamic leaders are curious, and welcome a challenging future, and are constantly looking for institutional changes that adapt to "urgent" disruptive technologies. GIS technology, therefore, encourages a growth mindset—essentially welcoming change instead of fearing it. Yet, GIS technology has not been fully embraced and exploited for solving urban design and environmental problems in South Bend. Integrating GIS with urban sustainable land use planning that includes gardening is crucial for realizing the aim of food security in South Bend Indiana. The city is able to explore and exploit the potential of GIS technologies not only for monitoring the urban infrastructure, but also as an information technology which when shared at the grassroots level through participatory planning, has the potential to promote sustainable gardening and responsible land use. No other technology can provide a powerful combination of digital, mobile, social, geographic and video innovations as tools for investigating different land use options as with GIS. In the case of South Bend Indiana, existing community collaborations provide a solid network of private organizations and individuals, community leaders, government institutions, and educators, to develop urban food networks and related economic activities that fit local realities.

GIS in South Bend, Indiana

Developed in the late 1960s, very few planning departments could afford a GIS because of the prohibitive cost of hardware and the limited capabilities of the software. But by the 1980s, GIS technology was considered the formalized computer-based information system for effective decision-making in urban planning (Han & Kim, 1989). South Bend's GIS department has been using ArcGIS software since 1999 to manage geographic data, and generate maps for different city projects. The city uses GIS to provide services such as; hot spot mapping of crime, tracking of basement flooding, utility inventory and similar others. The city also makes data and information available via an on-line map server (ESRI, 2011). Currently, the system provides basic information about property within the city, and maps include aerial photographs (orthophotos), parcel boundaries containing parcel identification number, property address, and property owner information (City of South Bend Indiana, 2016).

Among the geographic layers that are available for download from the City of South Bend's website are:

- Demographic Data
- Vacant and abandoned properties
- Street centerlines
- Public facilities
- Social services location
- Parks
- Businesses

These information layers are valuable for environmental research, public safety, marketing, etc. (Government Technology, 2016). It is now possible for organizations in the city to invest their time to perform tasks that were previously impractical or impossible without the current datasets. For example, modeling, predictions, and analysis can now be done through dynamic map overlays of physical, cultural, and economic features within the city landscape. GIS technology will enhance public participation, promote greater collaboration among public agencies, and would also be useful to city planners and educators who focus on geographic literacy/education and GIS mapping.

In the quest to acquire data for spatial analysis, several problems were encountered. Important metadata for datasets were missing or incomplete. Online map servers run by the city allowed minimum analyses. Data export was also limited to graphic formats. When GIS data was available for download, they were usually from other sources such as the Census Bureau which, because of scale, did not satisfy the desired unit of analysis.

EVALUATING POPULATION AND INCOME DISTRIBUTION IN SOUTH BEND

South Bend is the largest city in St Joseph County, Indiana, and occupies 457.9 square miles. South Bend's population distribution per square mile was 584.4; with a total population of 101,190 in 2014. This is a 0.2% rise in population from 2010 (STATSINDIANA, 2015). Table 2 shows that the 2015 South Bend population dropped to 101,168 with a race distribution of 60% white, 26% African American, 12% Hispanic or Latino. These three races account for more than 98% of the population. Source shows small variations in population distribution[1].

Table 2. 2015 South Bend population by race

Race	Population	% of Total
Total Population	***101,168***	***100***
White	61,199	60
Black or African American	26,906	26
Hispanic or Latino	13,116	12
Asian	1,318	1
American Indian	478	Below 1
Three or more races	313	Below 1

Source: suburbanstats.org

The 2015 median age in South Bend was 33, with average income for people within the age group 25-44 being $36,620 (Suburban Stats, 2016). Family income distribution when compared with the State of Indiana and the U.S show that above $50,000 family income in South Bend is lower than the State of Indiana, and the U.S (USA.com, 2016). However, when we compare family income below $50,000, South Bend is higher (see Figure 1).

Poverty by Zip codes

Sixteen zip codes are included in the city of south Bend; 46601, 46613, 46614, 46615, 46616, 46617, 46619, 46624, 46626, 46628, 46634, 46635, 46637, 46660, 46680, 46699 with some zip code routes intersecting the city boundary (zipcode.org, 2016). Population, race, and income characteristics for Whites, Blacks, and Hispanics (the largest three groups) are presented in Table 3 below, for each zip code[2]. Zipcodes 46601, 46613, 46616, 46617, and 46619 show the highest poverty rates.

Figure 1. Family income distribution

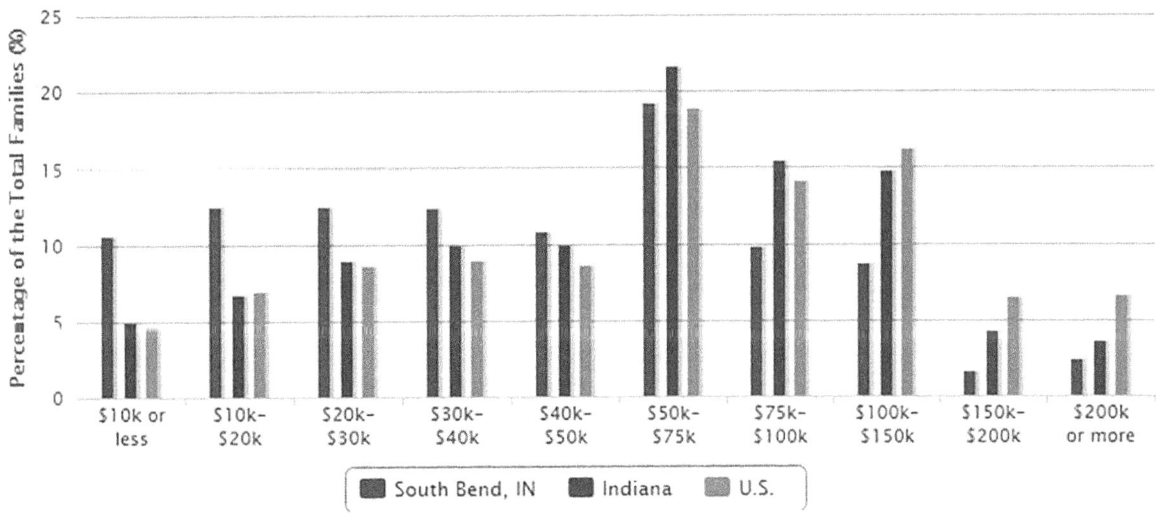

*Table 3. Poverty level data was obtained from zipatlas.com * (NA=Not Available)*

Zipcode	Average House Value USD$	Median Age	Income per H/hold USD$	% Below Poverty Level	% White Pop.	% Black Pop.	% Hispanic Pop.	% of Total Pop.
46601	70,300	33.1	17,434	39.98	43.25	48.49	7.29	92
46613	48,300	30.10	30,717	25.46	64.05	22.21	14.18	100
46614	93,300	41.80	43,160	8.06	91.33	4.89	2.42	99
46615	70,200	34.10	33,557	15.65	82.87	9.43	4.03	13
46616	62,800	32.80	33,449	26.82	66.49	24.95	6.91	98
46617	85,900	34.30	39,517	21.84	67.28	26.52	3.81	98
46619	60,100	33.80	33,463	20.09	59.72	24.86	18.48	103
46624	N/A	N/A	N/A	N/A	N/A	N/A	N/A	N/A
46626	N/A	N/A	N/A	N/A	N/A	N/A	N/A	N/A
46628	69,900	36.80	36,951	17.96	63.67	30.64	4.44	99
46634	N/A	N/A	N/A	N/A	N/A	N/A	N/A	N/A
46635	105,400	44.70	54,296	4.69	90.93	4.49	2.24	98
46637	95,700	37.80	43,225	13.82	89.06	5.45	2.61	97
46660	N/A	N/A	N/A	N/A	N/A	N/A	N/A	N/A
46680	N/A	N/A	N/A	N/A	N/A	N/A	N/A	N/A
46699	N/A	N/A	N/A	N/A	N/A	N/A	N/A	N/A
46616	62,800	32.80	33,449	26.82	3,940	1,583	460	402
46617	85,900	34.30	39,517	21.84	7,010	2,338	610	626
46619	60,100	33.80	33,463	20.09	12,250	5,789	5,911	4,284
46624	N/A	N/A	N/A	N/A	N/A	N/A	N/A	N/A
46626	N/A	N/A	N/A	N/A	N/A	N/A	N/A	N/A
46628	69,900	36.80	36,951	17.96	16,079	8,678	1,975	1,821
46634	N/A	N/A	N/A	N/A	N/A	N/A	N/A	N/A
46635	105,400	44.70	54,296	4.69	5,617	800	318	429
46637	95,700	37.80	43,225	13.82	13,345	1,402	627	933
46660	N/A	N/A	N/A	N/A	N/A	N/A	N/A	N/A
46680	N/A	N/A	N/A	N/A	N/A	N/A	N/A	N/A
46699	N/A	N/A	N/A	N/A	N/A	N/A	N/A	N/A

AARP Livability Index

The AARP (American Association of Retired Persons) Public Policy Institute developed a Livability Index to measure community livability. The scores are based on affordability and access to housing, access to live, work and play, safe and convenient options for transportation, clean air and water, health, civic and social involvement, opportunities for inclusion. The 2015 livability index[3] for South Bend Indiana, was 51 out of 100p (AARP, 2016). Areavibes.com has also developed livability indexes based on their opinion of important metrics. They include amenities, cost of living, crime rates, education, employment, housing and weather. The individual category scores for AARP are presented in Tables 4 – 17.

Figure 2. Zipcode map of South Bend Indiana

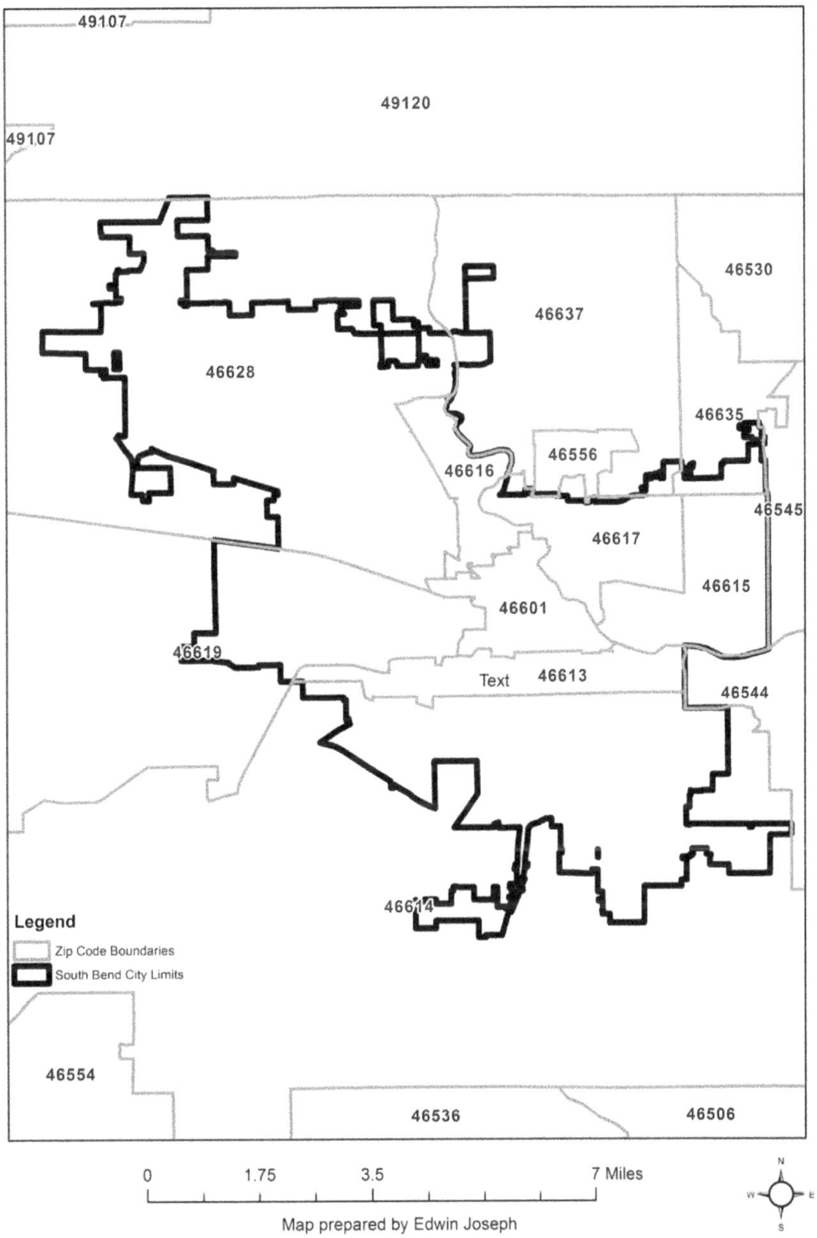

Areavibes.com Livability Index

Areavibes.com (2016) assigns South Bend a livability index of 71. This was derived from a distribution and number of important amenities such as grocery stores, other food and drink establishments, shopping malls, coffee shops, schools, parks, libraries, entertainment, fitness. The spatial distribution for these amenities is shown in Figures 3 – 12.

Table 4. AARP livability index for South Bend Indiana

Category	Index
Housing—affordability and access	58
Neighborhood—access to life, work, and play	49
Transportation—safe and convenient options	55
Environment—clean air and water	50
Health—prevention, access, and quality	48
Engagement—civic and social involvement	49
Opportunity—inclusion and possibilities	47
Livability Index	**51**

Table 5. AARP livability index for zip code 46601

Category	Index
Housing—affordability and access	69
Neighborhood—access to life, work, and play	52
Transportation—safe and convenient options	58
Environment—clean air and water	49
Health—prevention, access, and quality	44
Engagement—civic and social involvement	53
Opportunity—inclusion and possibilities	44
Livability Index	**53**

Table 6. AARP livability index for zip code 46613

Category	Index
Housing—affordability and access	**57**
Neighborhood—access to life, work, and play	48
Transportation—safe and convenient options	58
Environment—clean air and water	51
Health—prevention, access, and quality	44
Engagement—civic and social involvement	51
Opportunity—inclusion and possibilities	44
Livability Index	**53**

Table 7. AARP livability index for zip code 46614

Category	Index
Housing—affordability and access	55
Neighborhood—access to life, work, and play	42
Transportation—safe and convenient options	47
Environment—clean air and water	45
Health—prevention, access, and quality	50
Engagement—civic and social involvement	50
Opportunity—inclusion and possibilities	46
Livability Index	**48**

Table 8. AARP Livability Index for zip code 46615

Category	Index
Housing—affordability and access	58
Neighborhood—access to life, work, and play	52
Transportation—safe and convenient options	57
Environment—clean air and water	52
Health—prevention, access, and quality	53
Engagement—civic and social involvement	44
Opportunity—inclusion and possibilities	50
Livability Index	**52**

Table 9. AARP Livability Index for zip code 46616

Category	Index
Housing—affordability and access	58
Neighborhood—access to life, work, and play	47
Transportation—safe and convenient options	58
Environment—clean air and water	51
Health—prevention, access, and quality	50
Engagement—civic and social involvement	45
Opportunity—inclusion and possibilities	46
Livability Index	**51**

Table 10. AARP livability index for zip code 46617

Category	Index
Housing—affordability and access	56
Neighborhood—access to life, work, and play	48
Transportation—safe and convenient options	57
Environment—clean air and water	48
Health—prevention, access, and quality	49
Engagement—civic and social involvement	49
Opportunity—inclusion and possibilities	47
Livability Index	**51**

Table 11. AARP livability index for zip code 46619

Category	Index
Housing—affordability and access	59
Neighborhood—access to life, work, and play	46
Transportation—safe and convenient options	49
Environment—clean air and water	48
Health—prevention, access, and quality	47
Engagement—civic and social involvement	46
Opportunity—inclusion and possibilities	48
Livability Index	**49**

Table 12. AARP livability index for zip code 46624, 46626, 46628 & 46634

Category	Index
Housing—affordability and access	69
Neighborhood—access to life, work, and play	52
Transportation—safe and convenient options	58
Environment—clean air and water	49
Health—prevention, access, and quality	44
Engagement—civic and social involvement	53
Opportunity—inclusion and possibilities	44
Livability Index	**53**

Table 13. AARP livability index for zip code 46635

Category	Index
Housing—affordability and access	54
Neighborhood—access to life, work, and play	46
Transportation—safe and convenient options	44
Environment—clean air and water	52
Health—prevention, access, and quality	53
Engagement—civic and social involvement	53
Opportunity—inclusion and possibilities	44
Livability Index	**49**

Table 14. AARP livability index for zip code 46637

Category	Index
Housing—affordability and access	63
Neighborhood—access to life, work, and play	42
Transportation—safe and convenient options	45
Environment—clean air and water	49
Health—prevention, access, and quality	53
Engagement—civic and social involvement	50
Opportunity—inclusion and possibilities	41
Livability Index	**49**

Table 15. AARP livability index for zip code 46660

Category	Index
Housing—affordability and access	58
Neighborhood—access to life, work, and play	52
Transportation—safe and convenient options	57
Environment—clean air and water	52
Health—prevention, access, and quality	53
Engagement—civic and social involvement	44
Opportunity—inclusion and possibilities	50
Livability Index	**52**

Table 16. AARP livability index for zip code 46680

Category	Index
Housing—affordability and access	63
Neighborhood—access to life, work, and play	42
Transportation—safe and convenient options	45
Environment—clean air and water	49
Health—prevention, access, and quality	53
Engagement—civic and social involvement	50
Opportunity—inclusion and possibilities	41
Livability Index	**49**

Table 17. AARP livability index for the State of Indiana

Category	Index
Housing—affordability and access	55
Neighborhood—access to life, work, and play	45
Transportation—safe and convenient options	48
Environment—clean air and water	60
Health—prevention, access, and quality	41
Engagement—civic and social involvement	47
Opportunity—inclusion and possibilities	62
Livability Index	**51**

Figure 3. Location of grocery stores
Grocery Stores—20 locations

Figure 4. Locations of food & drink establishments
Food and Drink—20 locations

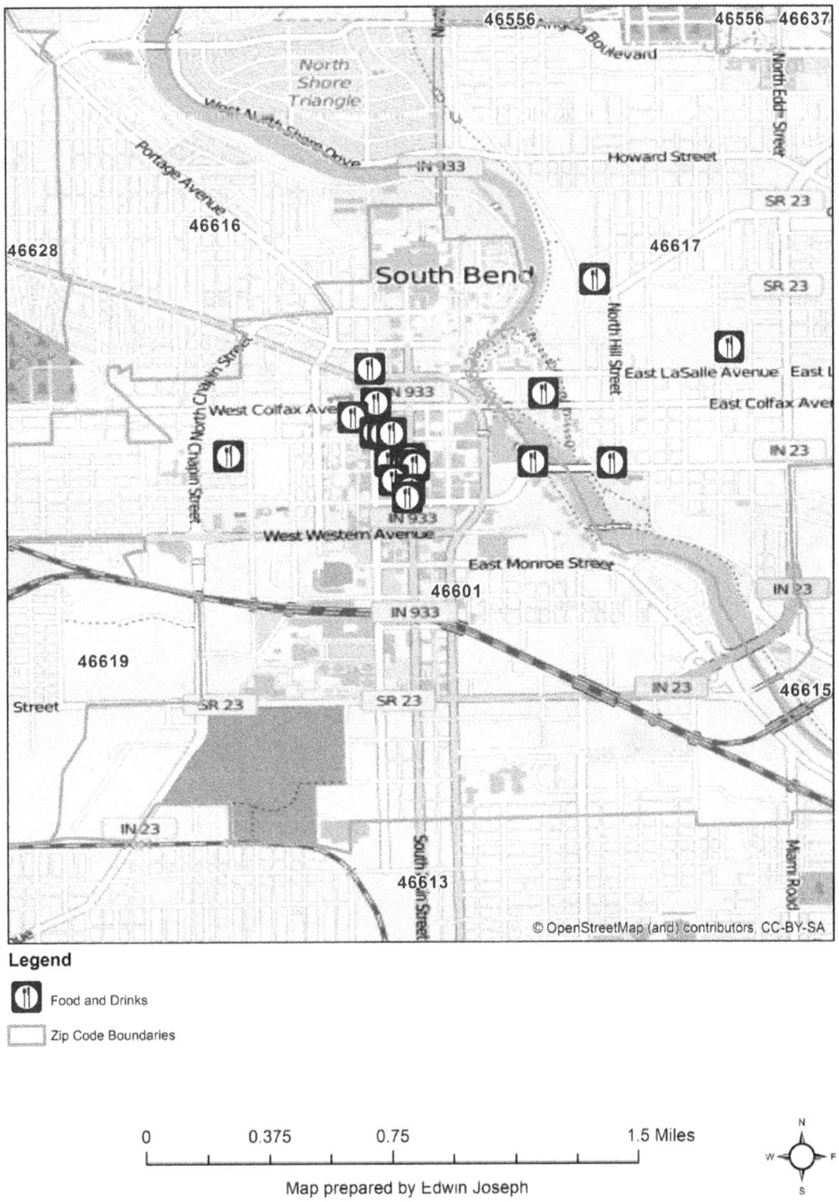

The City of South Bend Department of Community and Economic Development, with assistance from the Area Plan Commission of St. Joseph County, gathered data from the 2010 census and other recent census products and compiled maps for displaying population, race, housing units, and income. This community profile provides information about demographic trends in St. Joseph County. Much of this data is presented at the census tract level. The data shows that St. Joseph County became more racially diverse with a decrease in the white population. The number of residents who are African American,

Figure 5. Shopping locations
Shopping—20 locations

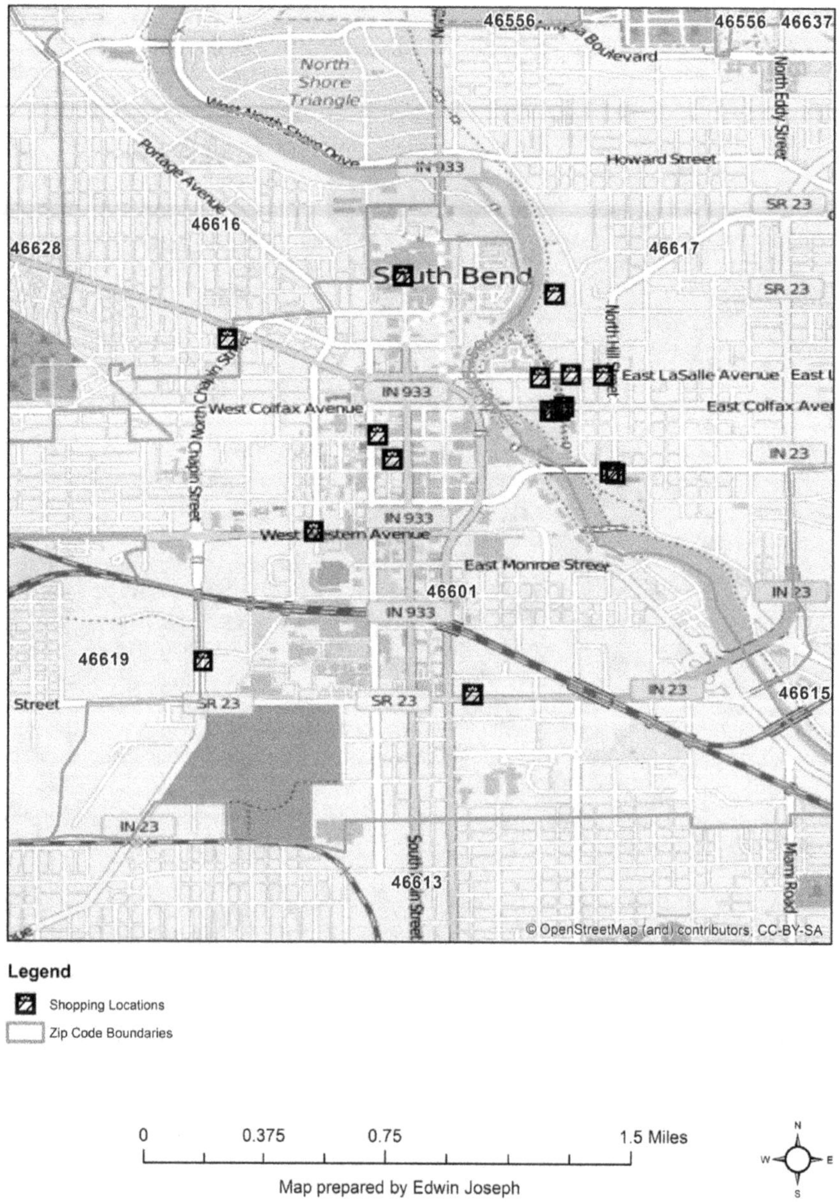

another race, or of two or more races all increased in absolute and percentage terms. Minority groups remained concentrated within South Bend, where the vacancy rate was relatively high—particularly on the near south and west sides of town. (Department of Community & Economic Development, 2011)

A recent review of the status of the urban poor in South Bend (based on 2000-2013 census block group data) revealed that 16.7% of residents had incomes below the poverty level in 2013 while that for the whole state was 9.5% (CDW Services, 2015).

Figure 6. Locations of cafe
Coffee—7 locations

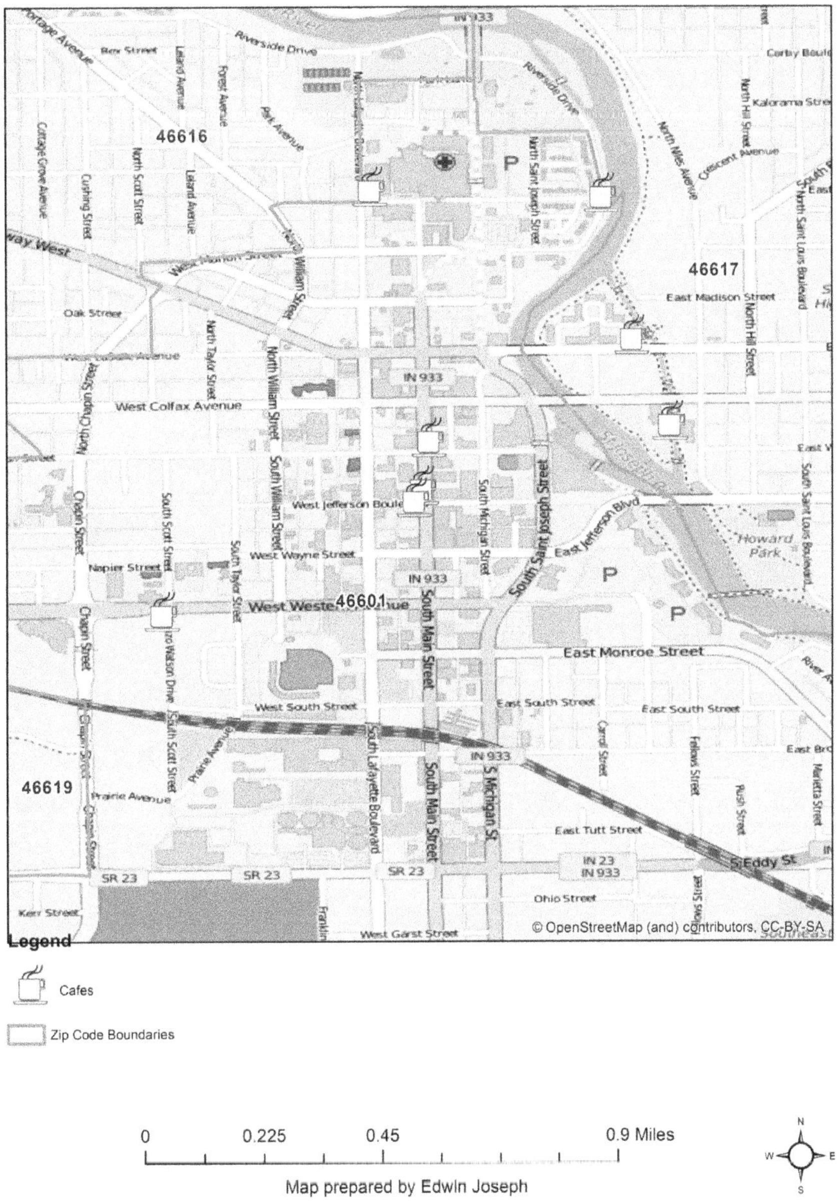

The spatial distribution of poverty shown at a larger scale identifies 4 census blocks with the highest poverty rates (Figures 13 – 16). These blocks are located either south or west of the city center.

Given the spatial distribution of poverty in South Bend, it is important to understand the use of GIS technology to seek solutions for solving environmental, social, and economic problems. Some critical areas range from public safety, to health, and to transportation (Government Technology, 2016). South

Figure 7. Location of schools
Schools—20 locations

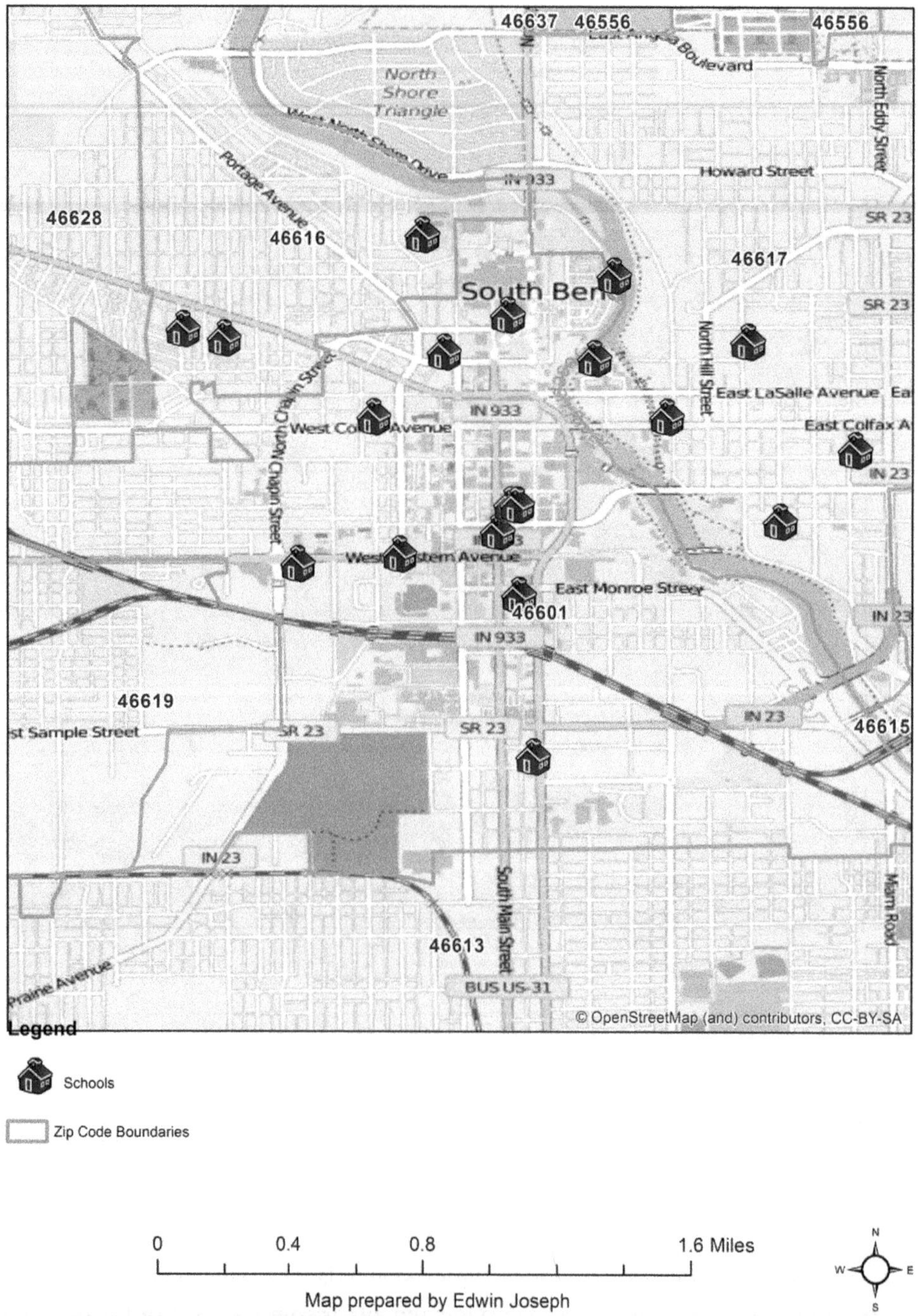

Bend city planners have invested much time in regulating land use, monitoring infrastructure, and traffic management among other things. And like many other cities, GIS is now used for performing spatial analyses. Planners also realize that the spatial and temporal aspects of GIS not only allow for modeling and predictions, but also useful for in-depth analysis.

Figure 8. Distribution of parks
Parks—11 locations

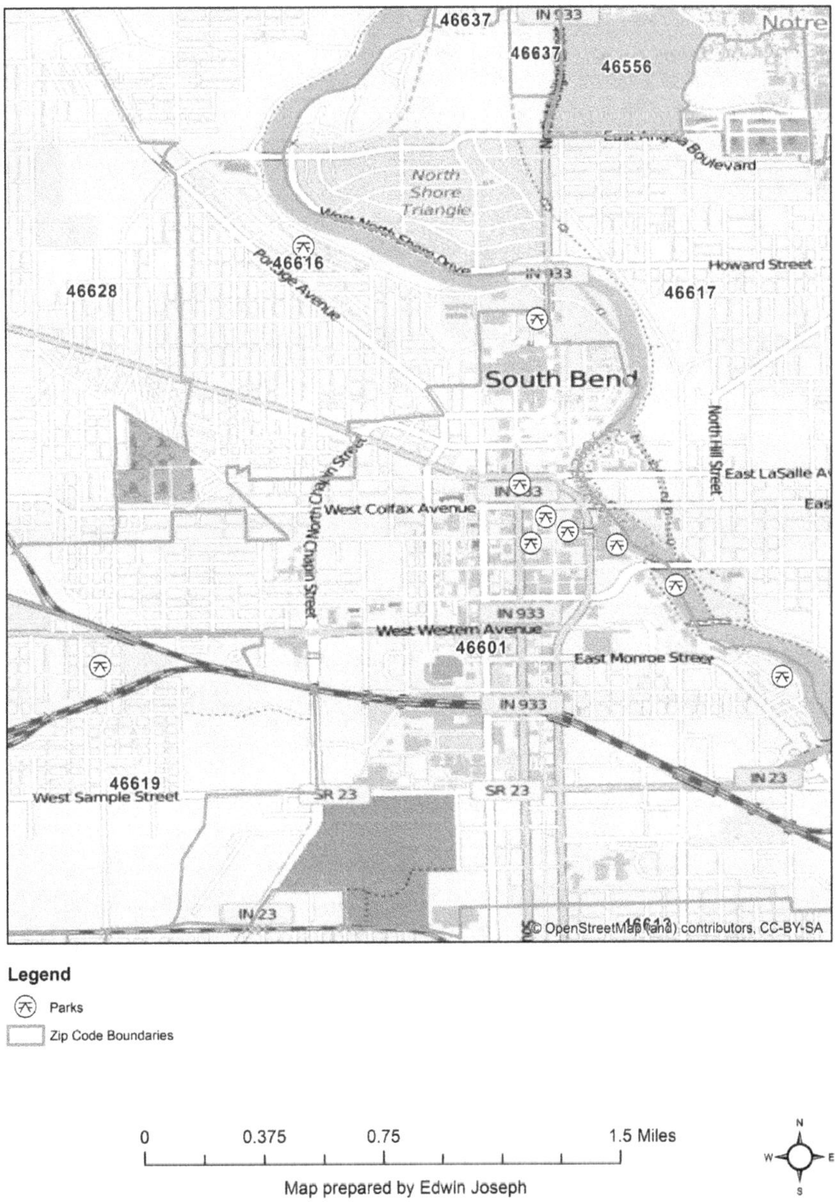

HOW GIS CAN HELP URBAN AGRICULTURE AND INCREASE FOOD SECURITY IN SOUTH BEND

There are many strategic advantages for integrating GIS in Sustainable Urban Agriculture. GIS is a location-based system and is ideal for environmental monitoring, flood management, soil classification, pest and disease management, strategic planning, among others. For the average gardener, planning is generally some combination of guesswork and luck, but integrating GIS allows for more spatial organi-

Figure 9. Distribution of libraries
Libraries—11 locations

zation. Factors that can affect urban agriculture are sunlight, shade, soil moisture content, pH, soil type, artificial materials, dust and lead; all of which can affect the quality and quantity of crops.

A combination of GIS, GPS and Multimedia mapping will provide continuous data for sustainable food production in the city. Surveying new areas to plant crops, or examining soil and air quality will be an enormous challenge without spatial technologies. These technologies lend themselves to crowd sourcing projects, and greatly assist in managing food production and distribution. The World Food Program,

Figure 10. Entertainment locations
Entertainment—11 locations

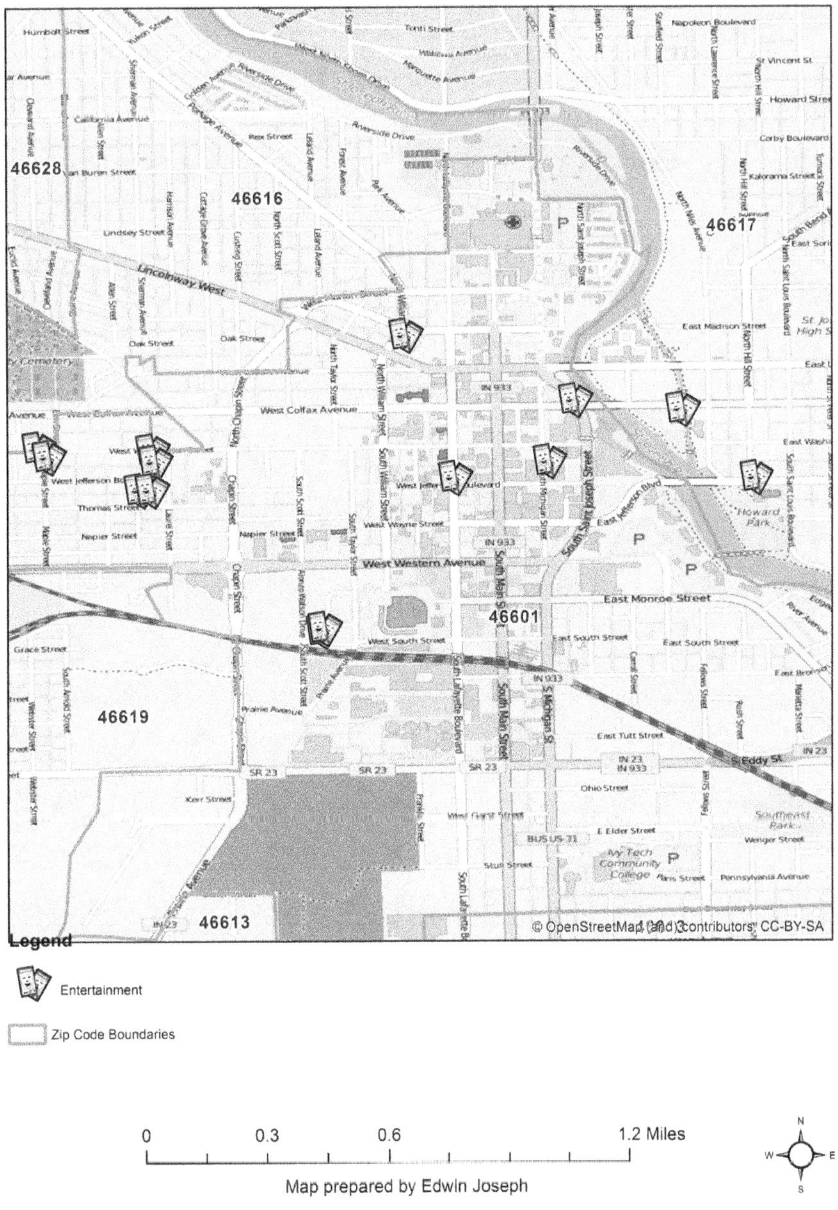

the division of the United Nations recognizes the importance of GIS in food production and is one of the biggest users of GIS data (ArcNews, 2014). The program recognizes the importance of safeguarding food sources, and engages communities in activities such as land use planning, biodiversity education, and soil fertility management. The World Food Program also monitors changes in the environment, and uses remote sensing to identify disaster prone regions.

Figure 11. Distribution of fitness centers
Fitness—6 locations

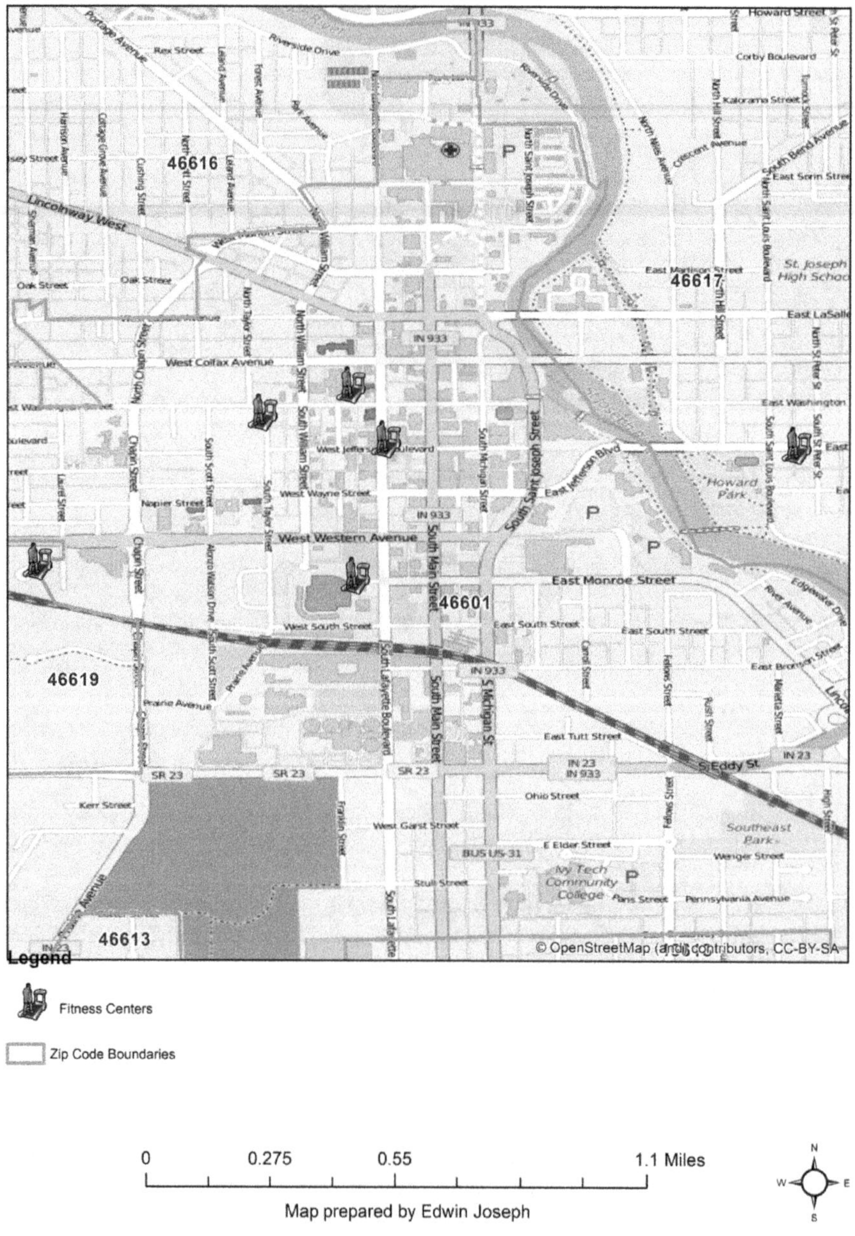

In South Bend it is possible to use drone mapping technology to capture high resolution imagery of urban gardens to monitor production. The addition of high resolution orthoimagery (photographic maps) provides information necessary to adapt sustainable agriculture practices. Orthoimagery from drone mapping creates 3-D models of urban spaces that are critical for landscape analysis. Thus, generating models for potential gardens will ensure better planning for future generations.

Figure 12. Distribution of all amenities
All amenities

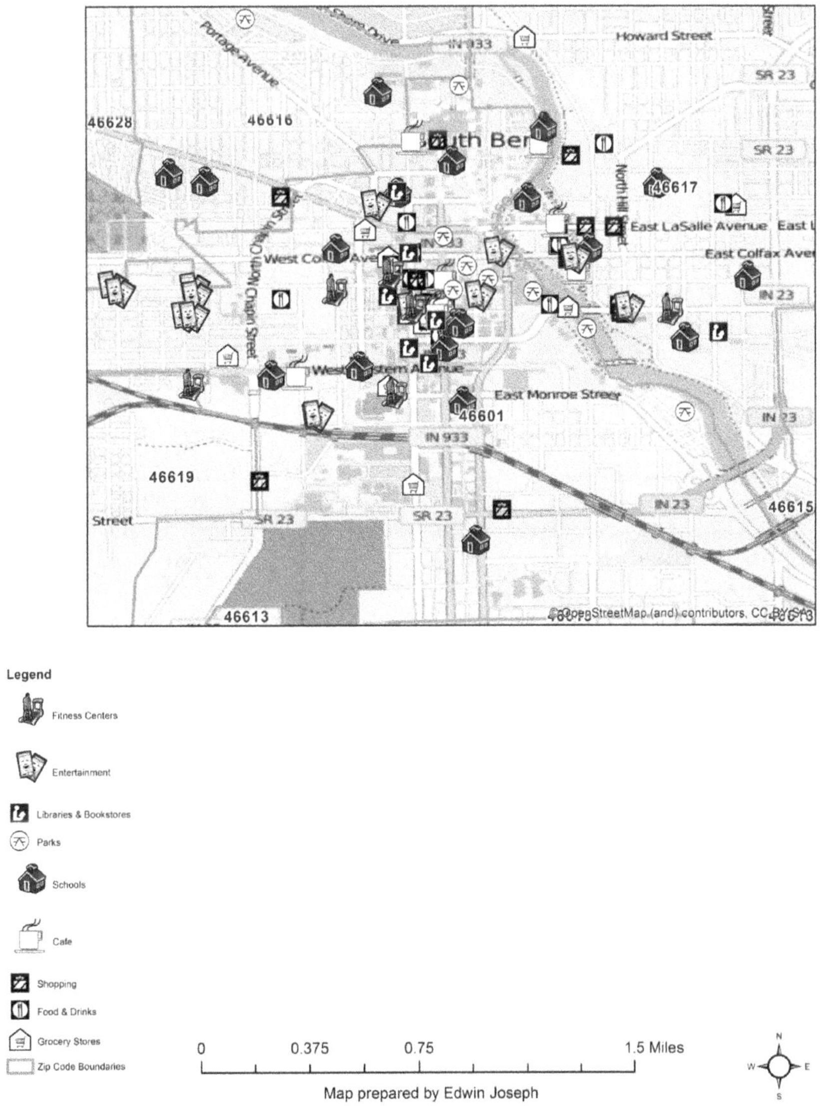

Urban Sustainable Agriculture has the potential to become a dynamic economic sector in South Bend. City planners need to establish a proper program for developing the social, economic and ecological framework for urban agriculture. They must work closely with the urban poor, and as partners, develop a sustainable and resilient city that is socially inclusive, safe, and environmentally-healthy. The challenge for South Bend is putting together a conducive policy for urban agriculture, and urban land use. The city also has to provide easy access to vacant open urban spaces and partner with higher education institutions to develop formal urban garden programs that include technical workshops, participatory action research, and specific skills field training. Potential urban gardeners should not only have access

Figure 13. Residents with income below the poverty level (74.5%): Located roughly along West Sample St., & South Michigan St., zip code 46601 & 46613 (74.1%-76% poverty).

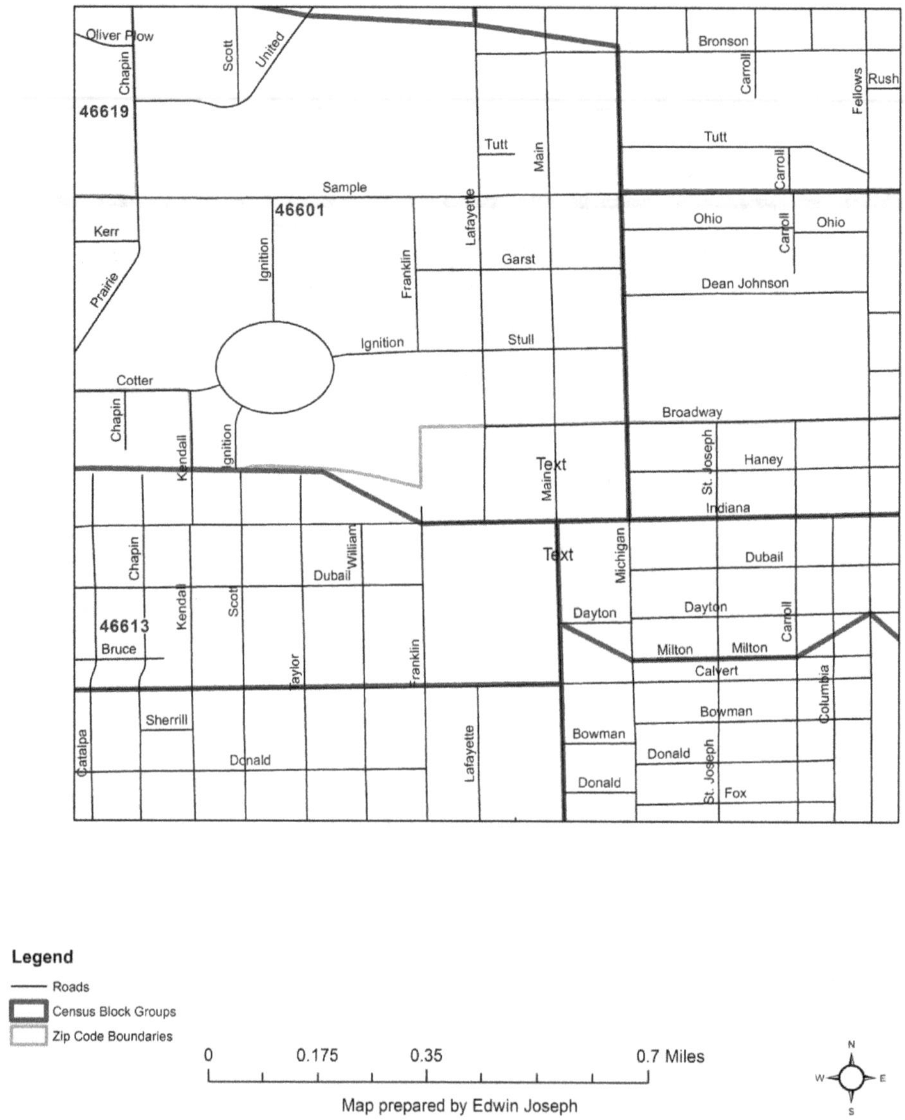

to technical advice, but also credit (in the case where they seek agri-business opportunities within the city) for activities such as urban beekeeping, hydroponics, aquaponics, egg production and small stock (rabbits) production.

NEW DIRECTIONS FOR A FUTURE SOUTH BEND

At the 2016 State of the City address, Mayor Pete Buttigieg declared that South Bend was at the midpoint of a decade that would go down in history as the most transformative period for the city in modern times

Figure 14. Residents with income below the poverty level (64%): Located roughly along Lincoln Way West, and Ardmore Trail, zip code 46628 (62.7%-64.6% poverty)

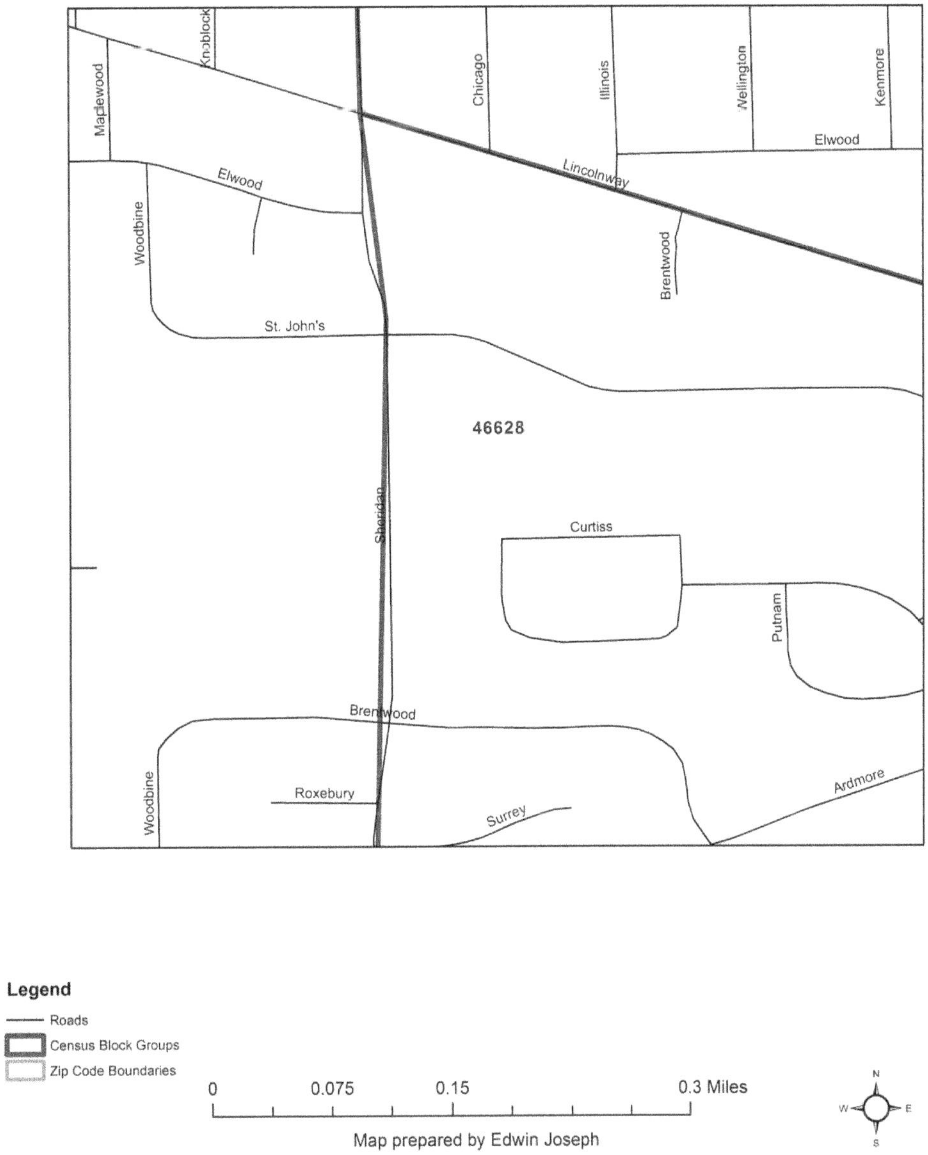

(Catanzarite, 2016). The Mayor reported that in 2015, four hundred and thirty-three houses were repaired through a combination of code enforcement and strong community partnerships. The Mayor identified community partners, for example, social service providers, neighborhood representatives, respected pastors, nonprofit stakeholders, police officers and local, county, and federal law enforcement officials working toward stopping violence.

The 2016 address also promised funding ($700 million) to improve the community, empower people, and allow individuals to live the life of their choosing. The Mayor stated that investing in people and places meant investing in neighborhoods, and outlined the quality of life enhancements such as curbs

Figure 15. Residents with income below the poverty level (59.4%): Located roughly along West Washington St., & South William St., zip code 46601 (58.9%-60.8% poverty).

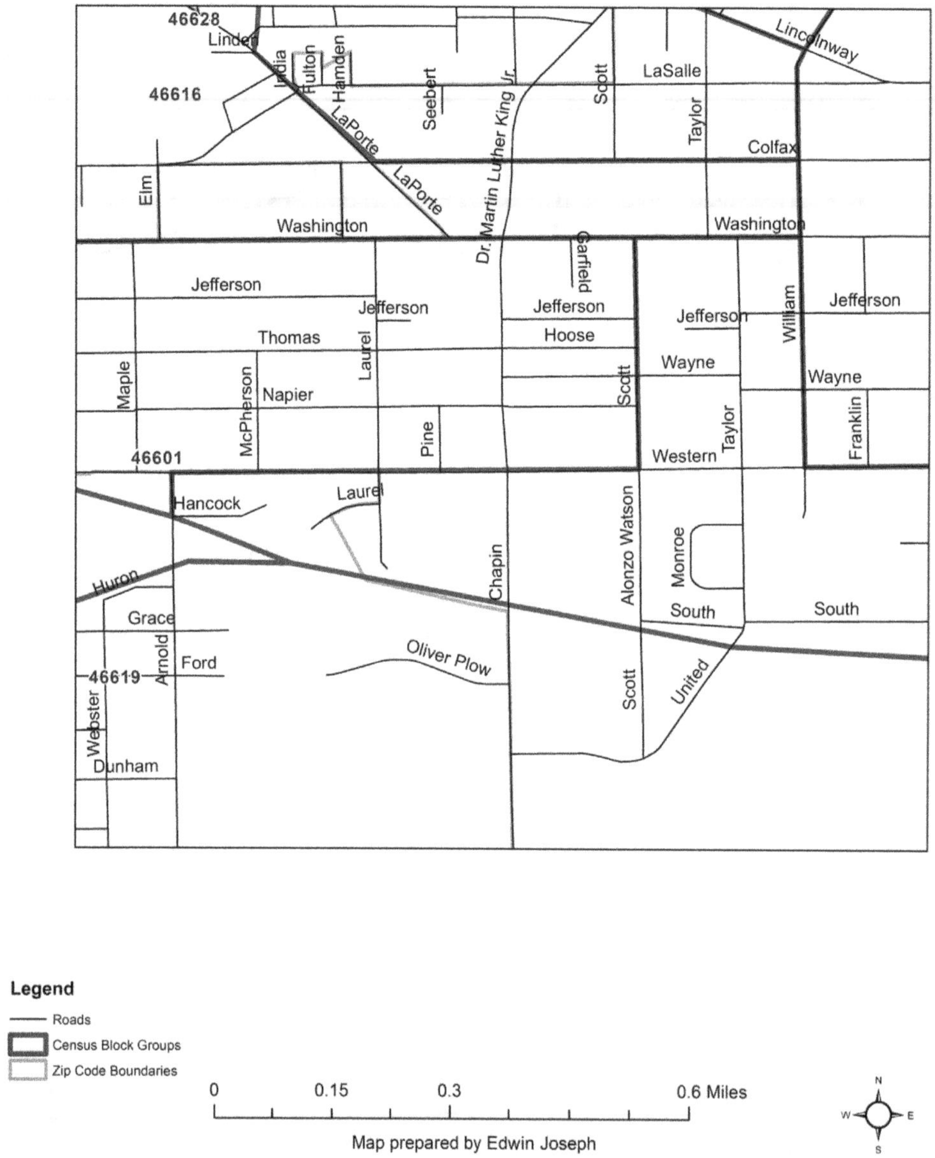

and sidewalks, ramps, intersections, and wheelchair access completed. The city also invested in parks and recreation, and planted approximately 821 trees in 2015. A common theme in the address was the importance of including diverse community voices and perspectives, and focusing on unlimited imagination. In some ways, the mayor encourages imagination and sees it as a precursor to innovation.

The city's interest in gardening was apparent in the Mayor's address when he spoke about his summer 2015 tour of a vegetable garden with a hundred 4th-graders. He also publically thanked the leadership of the St. Joseph County Health Department and the Reducing Obesity Coalition of St. Joseph County

Figure 16. Residents with income below the poverty level (49%): Located roughly within Lindsey St., Lincoln Way West, & North William St., zip code 46601 & 46616

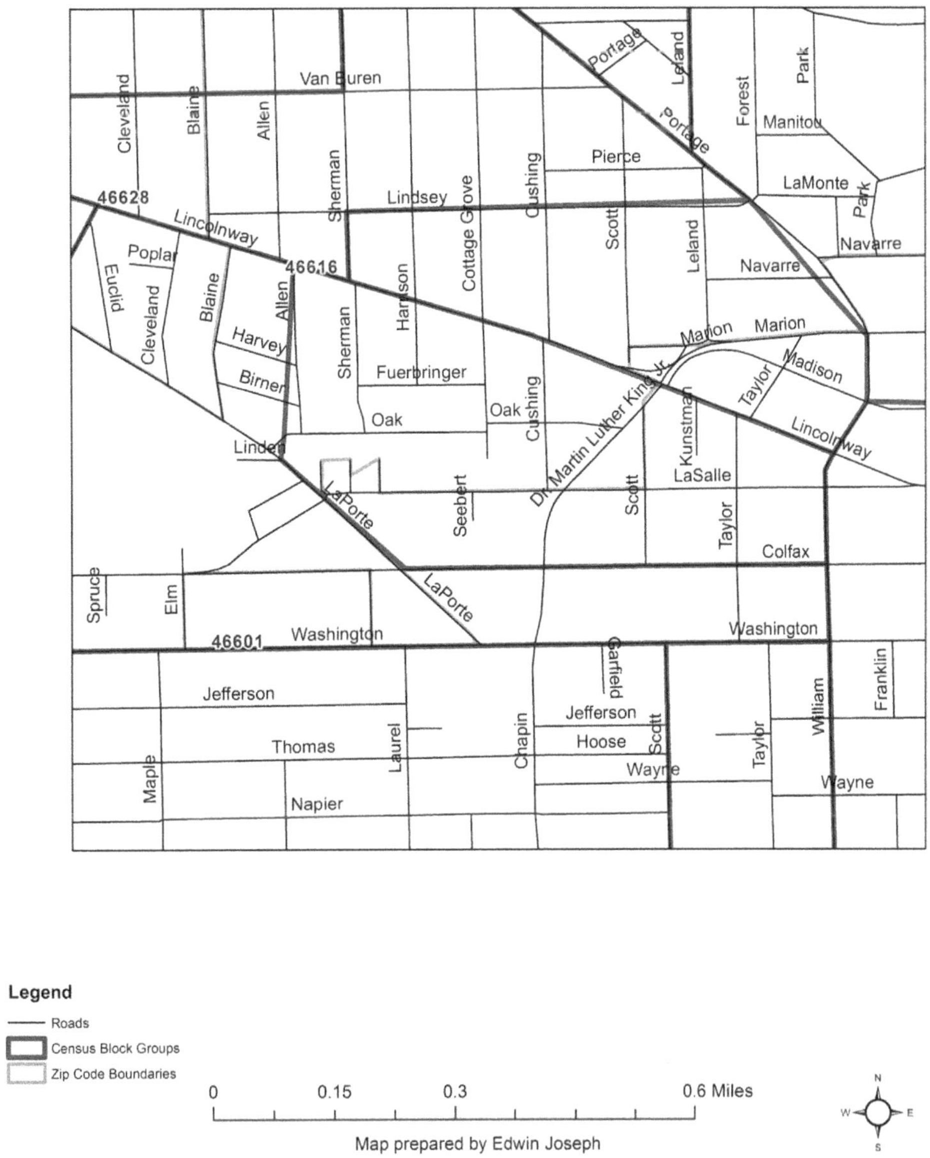

because South Bend was recognized by First Lady Michelle Obama as the first city in Indiana to meet all five goals of the "Let's Move!" initiative for healthy kids.

The city passed an innovative food truck ordinance, and created an online tracker that shows on a dynamic map where the food trucks are actually located, and is now working on an online tool to visualize vacant lots, so residents and potential investors or nonprofits could find ways to research and evaluate vacant land. These two initiatives demonstrate the need for expanding the use of GIS because the obvious benefits of GIS is the location factor. Knowing where things are, and how they change in space and time can be viewed in real time. This, however, requires dedicated programming, time, and effort.

DISCUSSION AND CONCLUSION

GIS and GPS technology are indispensable but neglected tools. These spatial technologies have the potential to generate partnerships among food producers, volunteers, government and other institutions. Developing geographic awareness through collaborative mapping projects will help foster learning communities that can work together to develop sustainable urban garden initiatives—and may include the use of Free and Open Source Software (FOSS) as valuable tools to help satisfy an overwhelming demand for spatial information.

In South Bend, spatial technologies will help maximize food production because of the ability to visualize spatial relationships and patterns, which will contribute to building a more sustainable environment (ArcUser, 2010). Through spatial analysis, we will understand the relationship among factors such as, soil type, pH, soil moisture, and nutrient deficiency. We can generate maps identifying where things are, how they move, and other important relationships.

Spatial technologies will also connect community gardeners by introducing visual components to aid in site selection, planning, and modeling; and enable timely responses to issues that affect production. Integrating spatial technologies with urban gardening will provide an ideal platform for networking. As South Bend thrives to become a leader in sustainability, it has been shown how community participation will strengthen social relationships. South Bend needs to concentrate on growing more of its vegetables within the city. This will ensure healthier lifestyles and stronger communities. The aim is not just to improve health and decrease obesity, but create a more cohesive society with an equal distribution of incomes (Wilkinson & Pickett, 2009).

GIS and GPS technology has proven to be a quick and efficient approach for acquiring field data. GIS therefore, has the potential to become the hub or major planning tool for urban sustainability in South Bend. When a GIS is combined with GPS, multimedia, and internet technologies, it provides a complete range of quantitative and qualitative research opportunities for future planning. Adopting these technologies would likely foster worthwhile sustainable investments, and would keep the momentum of GIS moving. However, the social, political, economic, and institutional climate must be supportive (North, 1990). A typical example of GIS in agriculture that is immediately applicable is pest and disease monitoring. Cell phones are now equipped with GPS, and a picture taken of a disease in the field can be crowdsourced and mapped in real time. Gardeners can seek information/solutions by uploading their images via community map servers where they can be viewed by experts.

There are many examples of communities transforming desolate landscapes into food oases. At Indiana University South Bend (IUSB), we recognize the importance of effective and sustainable urban land management practices, and have utilized geospatial technologies to map and monitor urban gardens in the city. As we look into the future we believe that the next wave of mapping technology would include AUVs (Aerial Unmanned Vehicles) for capturing landscape details that are otherwise unavailable to community gardeners. We believe that high resolution orthophotos (photographic maps) as overlays in Google Earth would help urban gardeners see the potential for online networking and collaboration. AUVs fitted with high-resolution visible-spectrum cameras and integrated GPS (Global Positioning System) are capable of acquiring overlapping aerial photos and elevations at garden sites. These would be invaluable for 3-D visualization, collaborative planning, and viewing temporal changes in the landscape.

It is hoped that in the near future, there would be more open access to city data so researchers would be able to share a platform that is built on collaboration. The current review of the spatial distribution of poverty in South Bend reveals a spatial pattern, but in-depth analysis was difficult because of incomplete and sometimes incorrect data entry from the data sources. Most of the analysis were limited to using on-line GIS resources that could not integrate new/external data, but simply reclassified in-house data, and generated outputs as static maps.

The extent of inner city agriculture can be easily mapped and monitored using crowdsourcing techniques. Crowdsourcing is an important data mining activity that taps into the collective intelligence of the public. Developing countries can take advantage of this platform to acquire data through user-generated media for example, blogs, Tweets, Wikipedia, Wikimapia, Facebook, and YouTube. GIS can be easily integrated onto these platforms. The integration of these services will enable the promotion of a network of food producers who can work together to alleviate poverty and malnutrition through community mapping and sustainable urban agriculture.

REFERENCES

AARP. (2016). *AARP Public policy Institute*. Retrieved from http://livabilityindex.aarp.org/search#South+Bend+IN+USA

ABC57 News. (2015, May 22). *Meet a descendant of one of South Bend's first African American families*. Retrieved from abc57 News: http://www.abc57.com/story/29135949/meet-a-decedent-of-one-of-south-bends-first-african-american-families

Andrews, W. L., & Adams, E. M. (2016). *North American Slave Narratives*. Retrieved from Documenting the American South: http://docsouth.unc.edu/neh/intro.html

ArcNews. (2014). *GIS Supports World Food Programme's Food Security Program*. Retrieved from http://www.esri.com/esri-news/arcnews/spring14articles/gis-supports-world-food-programmes-food-security-program

ArcUser. (2010). *The Next Generation of GIS. Dealing with design and space-time*. The Magazine for ESRI Software Users.

Areavibes. (2016). *South Bend IN, Livability*. Retrieved from http://www.areavibes.com/south+bend-in/livability/

Berman, E. (2011). *Local Food Systems: Selected Resources*. Retrieved from Science and Technology Resources on the Internet: http://www.istl.org/11-summer/internet.html

Buttigieg, P. (2015). *South Bend's New Economic Direction*. Retrieved from Pete Buttigieg, Mayor: http://www.peteforsouthbend.com/economicplan

Catanzarite, M. (2016, February 24). *Mayor Buttigieg delivers "State of the City" address*. Retrieved from http://www.wndu.com/content/news/Mayor-Buttigieg-delivers-State-of-the-City-addresss-369918351.html

CDW Services. (2015). *City-Data.com*. Retrieved from South Bend, Indiana (IN) Poverty Rate: http://city-data.com/poverty/poverty-South-Bend-Indiana.html

Center for the Homeless. (1988). Retrieved from Center for the Homeless: http://cfh.net/

Chappell, M. J., Wittman, H., Bacon, C. M., Ferguson, B. G., Barrios, L. G., Barrios, R. G., & Perfecto, I. (2013). Food sovereignty: An alternative paradigm for poverty reduction and biodiversity conservation in Latin America. *F1000 Research*, *2*, 235. doi:10.12688/f1000research.2-235.v1 PMID:24555109

Christensen, C. (1997). *The Innovator's Dilemma*. Harvard Business Review Press.

Citizen's Fair Housing Committee of South Bend-Mishawaka. (1963). *Transcript of public hearing concerning discrimination in the sale, rental and financing of private housing in South Bend and Mishawaka: held March 19, 1963 in Notre Dame Law School Auditorium. Discrimination in housing--Indiana--South Bend*. South Bend, IN: Citizen's Fair Housing Committee of South Bend-Mishawaka.

City of South Bend Common Council. (1965). *South Bend Fair Employment Practice Ordinance. City Ordinance No. 4091 prohibiting discrimination in employment, providing for the creation of the South Bend Fair Employment Practice Commission*. South Bend, IN: City of South Bend Common Council.

City of South Bend Indiana. (2016). *Interactive Mapping-GIS*. Retrieved January 2016, from City of South Bend: http://southbendin.gov/government/content/interactive-mapping-gis-0

Civil Rights Heritage Center. (2015). *Civil Rights Heritage Center*. Retrieved from https://www.iusb.edu/civil-rights/about-us/index.php

Department of Community & Economic Development. (2011). *2010 Census Report: South Bend-St. Joseph County Community Overview*. South Bend, IN: City of South Bend Department of Community & Economic Development Division of Community Development. Retrieved from https://www.southbendin.gov/sites/default/files/files/CI_Census2010Report.pdf

DESA. (2013). World Economic and Social Survey 2013: Sustainable Development Challenges. New York: United Nations Publication.

DUFI. (2015, August 15). Retrieved from Distributed Urban Farm Initiative: A community and economic growth project by Advent GX: http://dufi.adventgx.com/

ESRI. (2011, Spring). *South Bend, Indiana, Uses GIS for Brownfields Inventory*. Retrieved from ArcNews: http://www.esri.com/news/arcnews/spring11articles/south-bend-indiana-uses-gis-for-brownfields-inventory.html

Food Bank of Northern Indiana. (2016). *Foodbank of Northern Indiana*. Retrieved from http://www.feedindiana.org/mission

Foodbank of Northern Indiana. (2015). *About the Food Bank*. Retrieved from Foodbank of Northern Indiana: http://feedindiana.org/

FoodTank.com. (2013, October 12). Retrieved from Five Examples of Successful Urban Agriculture Done Differently Around the World: http://foodtank.com/news/2013/10/five-different-examples-of-urban-agriculture-from-around-the-world

GMDC. (2015, September 10). *Greenpoint Manufacturing and Design Center*. Retrieved from http://www.gmdconline.org/buildings/810-humboldt-street

Government Technology. (2016). *Disruptive Technology that Could Transform Government-Citizen Relationships*. Retrieved from Government Technology: http://www.govtech.com/budget-finance/The-Disruption-Pipeline-Government-Citizen-Relationships.html

Greensgrow. (2016). *What is Urban Farming*. Retrieved from http://www.greensgrow.org/: http://www.greensgrow.org/urban-farm/what-is-urban-farming/

Gundersen, C., Kreider, B., & Pepper, J. (2011). The Economics of Food Insecurity in the United States. *Applied Economic Perspectives and Policy*, *33*(3), 281–303. doi:10.1093/aepp/ppr022

Han, S. Y., & Kim, T. J. (1989). Can expert systems help with planning? *Journal of the American Planning Association*, *55*(3), 296–308. doi:10.1080/01944368908975417

Henderson, E., & Van En, R. (2007). *Sharing the harvest: a citizen's guide to Community Supported Agriculture*. Chelsea Green Publishing.

Hodgson, K., Campbell, M. C., & Bailkey, M. (2011). *Urban Agriculture: Growing Healthy, Sustainable Places*. APA Planning Advisory Service.

Indiana Demographics. (2016). *Indiana Cities by Population*. Retrieved from http://www.indiana-demographics.com/cities_by_population

Kumar, R. (2015, October 5). *How Urban Agriculture Is Revitalizing Local Economies*. Retrieved from The Blog: http://www.growingpower.org/

Lee, S. (2013, June 19). *The Chicago Council on Global Affairs*. Retrieved June 25th, 2015, from Global Food For Thought: http://globalfoodforthought.typepad.com/global-food-for-thought/2013/06/tackling-urban-food-security.html

Matthews, C. (2014, October 31). *Wealth inequality in America: It's worse than you think*. Retrieved from Fortune: http://fortune.com/2014/10/31/inequality-wealth-income-us/

Mayor's Biracial Committee Progress Report. (1963). *Progress report, July-October 1963, Mayor's Biracial Committee.* Retrieved from Michiana Memory: http://michianamemory.sjcpl.org/cdm/compoundobject/collection/p16827coll4/id/2286/rec/9

Montgomery, M. R., & Hewett, P. C. (2005). Urban poverty and health in developing countries: Household and neighborhood Effects. *Demography*, *42*(3), 397–425. doi:10.1353/dem.2005.0020 PMID:16235606

Newsweek. (2011, January 21). *America's Dying Cities*. Retrieved from http://www.newsweek.com/americas-dying-cities-66873

North, D. C. (1990). *Institutions, Institutional Change and Economic Performance*. Cambridge University Press. doi:10.1017/CBO9780511808678

Point2Homes. (2016). *South Bend*. Retrieved from pointhomes.com: http://www.point2homes.com/US/Neighborhood/IN/South-Bend-Demographics.html

Radican, E. A. (2010). *Growing Food Where it is Needed.* Indiana University, South Bend. South Bend Indiana: Center for a Sustainable Future (CSF). Retrieved November 10, 2015, from https://www.iusb.edu/csfuture/docs/WHITE%20PAPER%20NO.%202.pdf

Rivest, S. (2015, November 5). *ABC57 News*. Retrieved from www.abc57.com: http://www.abc57.com/story/30451531/local-businesses-work-to-combat-food-insecurity

Robinson, G. (2015). *Better Homes of South Bend: An American Story of Courage*. Charleston, SC: The History Press.

Schnell, S. M. (2007). *Food with a Farmer's Face: Community Supported Agriculture in the United States*. Retrieved from Researchgate.net: https://www.researchgate.net/publication/227701611_Food_with_a_Farmer's_Face_Community_Supported_Agriculture_in_the_United_States

Smart Growth America. (2016). *The Best Complete Streets Policies of 2015*. Retrieved from smartgrowthamerica.org: http://www.smartgrowthamerica.org/best-complete-streets-policies-of-2015

South Bend Region Economic Development. (2016). *Economic Profile*. Retrieved from southbendregion.com: http://southbendregion.com/index.php?src=gendocs&ref=EconomicProfile&category=Main

South Bend Tribune. (2014). *Can We Dream Big?* Retrieved from http://www.southbendtribune.com/app/economy/

South Bend Tribune. (2015, May 17). *South Bend's History Seen in Five Periods of Settlement*. Retrieved from http://www.southbendtribune.com/news/local/history/sb150/south-bend-s-history-seen-in-five-periods-of-settlement/article_79cb9dc6-f99e-11e4-b2c4-cbe41056da99.html

South Bend Tribune. (2016). *Did South Bend eserve the Harvard Ranking?* Retrieved from http://www.southbendtribune.com/news/local/did-south-bend-deserve-the-harvard-ranking/article_2000116e-5218-5175-8299-1d3a732165b2.html

STATSINDIANA. (2015, November). *STATSINDIANA Indiana's Public Data Utility*. Retrieved from Indiana Indepth Profile: http://www.stats.indiana.edu/profiles/profiles.asp?scope_choice=a&county_changer=18141

Suburban Stats. (2016). *Current South Bend, Indiana Population, Demographics and stats in 2016, 2015*. Retrieved from suburbanstats.org: https://suburbanstats.org/population/indiana/how-many-people-live-in-south-bend

The Chicago Council on Global Affairs. (2013). *Feeding an Urban World: A Call to Action*. Chicago: The Chicago Council on Global Affairs.

The History Museum. (2016). *Local African-American History*. Retrieved from historymuseusb.org: http://historymuseumsb.org/local-african-american-history/#

Links Incorporated. (2015). *The South Bend Area (IN) Chapter of The Links, Incorporated*. Retrieved from The Links Incorporated: http://southbendarealinksinc.org/About_Us.php

Human Relations Commission Praises County Housing Position. (1968, March 10). *The Reformer Newspaper*.

Trauger, A. (2009). Social agency and networked spatial relations in sustainable agriculture. *Area*, *41*(2), 117–128. doi:10.1111/j.1475-4762.2008.00866.x

U.S. Bureau of Census. (1998, June 15). *Population of the 100 Largest Urban Places: 1920*. Retrieved from https://www.census.gov: https://www.census.gov/population/www/documentation/twps0027/tab15.txt

UBC News. (2014). *Study Reveals Unhappiest Cities in the US*. Retrieved from http://news.ubc.ca/2014/07/21/unhappy-cities/

Unity Gardens. (2009). *Unity Mission/Story*. Retrieved from Unity Gardens: http://theunitygardens.blogspot.com/p/unity-mission-story.html

USA.com. (2016). *South Bend, IN Income and Careers*. Retrieved from USA.com: http://www.usa.com/south-bend-in-income-and-careers.htm#Poverty-Level

USDA. (2015, September 8). *USDA ERS-Food Security in the US*. Retrieved from United States Department of Agriculture Economic Research Service: http://www.ers.usda.gov/topics/food-nutrition-assistance/food-security-in-the-us.aspx

Wikipedia. (2016, March 19). *Michiana*. Retrieved from https://en.wikipedia.org/wiki/Michiana

Wilkinson, R., & Pickett, K. (2009). *The Spirit Level: Why More Equal Societies Almost Always Do Better*. Allen Lane.

Wilson, W. J. (1987). *The Truly Disadvantaged*. Chicago, IL: Chicago Press.

WNDU. (2010, February 1). *Delve into South Bend's past on this Black History Month*. Retrieved from WNDU 16 South Bend Indiana: http://www.wndu.com/home/headlines/83231437.html

World Health Organization. (2015). *Trade, foreign policy, diplomacy and health*. Retrieved from WHO Food Security: who.int/trade/glossary/story028/en

zipcode.org. (2016). *City of South Bend, IN Zip Codes*. Retrieved from zipcode.org: zipcode.org/city/IN/SOUTHBEND

ENDNOTES

1 The data contained in this product are based on the Census 2000 sample. The data are estimates of the actual figures that would have been obtained from a complete count. Estimates derived from a sample are expected to be different from the 100 percent figures because they are subject to sampling and nonsampling errors.

2 The data are based on the Census 2000 sample. The data are estimates of the actual figures that would have been obtained from a complete count. Estimates derived from a sample are expected to be different from the 100-percent figures because they are subject to sampling and nonsampling errors.

3 The Livability Index is a signature initiative of the AARP Public Policy Institute to measure the quality of life in American communities across multiple dimensions: housing, transportation, neighborhood characteristics, environment, health, opportunity, and civic and social engagement (http://livabilityindex.aarp.org/?cmp=LVABLIDX_MAR25_015).

Chapter 13
Farm Security for Food Security:
Dealing with Farm theft in the Caribbean Region

Wendy-Ann Isaac
The University of the West Indies – St. Augustine, Trinidad and Tobago

Wayne Ganpat
The University of the West Indies – St. Augustine, Trinidad and Tobago

Michael Joseph
The University of Trinidad and Tobago, Trinidad and Tobago

ABSTRACT

Agricultural production in the Caribbean is being threatened by many factors such as decreasing availability of arable land, climate change effects such as increased incidences of flooding and drought, labour shortages, and competition from importers. However, one of the most important threats to agricultural production is the often under-recognised and under-reported area of farm theft (referred to as praedial larceny in the Caribbean). It involves the theft of agricultural produce (crops, livestock and fisheries) and farm equipment. One of the main reasons why this threat is so important is that theft of this type is very hard to prove. If indeed perpetrators are caught, and prosecuted successfully, the penalty is practically negligible. This paper examines the current status of farm theft in the Caribbean region, explores some of the main factors influencing farm theft, reviews some of the strategies attempted in the Caribbean and other places around the world and makes several suggestions to create a more secure food region. While the discussion calls on food producers to take several best practice actions to mitigate losses to praedial larceny, it emphasises that the primary responsibility is with government-led actions in the areas of modernised policies, updated laws and enhanced enforcement efforts.

DOI: 10.4018/978-1-5225-0942-4.ch013

INTRODUCTION

Nobody wannu plant the corn
Everybody want to raid the barn
Who yuh a guh blame it on
When is a next man yuh a depend pon
Well yuh wrong
Nobody wannu plant the corn
Everybody want to raid the barn
Haffi sing yuh owna song
Can't compete with careless John.....

These lyrics, written by Anthony B, aptly reflect the unremitting challenge facing the typical farmer in the Caribbean region. Referred to as praedial larceny throughout the Caribbean, farm theft is one of the most extensive among all crimes committed in the region in terms of the number of persons and families affected. Praedial larceny refers to the theft of agricultural produce such as crops, livestock, and fisheries. This term also covers the theft of agricultural equipment such as spray cans, brush cutters, water pumps and other irrigation equipment. It can also be extended to the theft of agriculture inputs and secondary products such as feed and fodder. Praedial larceny has been cited as one of the major challenges impacting the growth of the agricultural sector throughout the Caribbean and has been described as the "Achilles heel" of the agricultural sector across the world. Estimates indicate that the Caribbean region is losing over USD $321 million annually to praedial larceny and it has now become one of the most pervasive and entrenched crimes in business and livelihoods (Caricom, 2011).

A Caricom 2010 study reported that among regional stakeholders there is general consensus (more than 90%) that praedial larceny was the single most discouraging and serious disincentive threatening food security in the region. Farmers throughout the region have been clamouring their Governments to introduce legislation to combat this increasing threat. However, there are concerns by the farming community throughout the region that praedial larceny is not being treated as a very serious crime. Praedial larceny is the only crime at a regional level that consistently trends upwards (Little, 2011). Praedial larceny has moved from the theft of small amounts to large amounts of produce involving in some instances truckloads of bananas in or an entire field of pineapples or other vegetable and root crops or even the entire harvest of a freshwater fish pond. Thieves may also pose serious dangers to farm families and farm workers as many cases of threats on lives of farmers or even homicide have been reported. Some farmers have abandoned their entire enterprise due to heavy losses and the high cost paid for security.

According to Smith (2010), this type of criminal activity typically occurs in a changing social landscape, affected by demographic changes, trends and by the introduction of new policing practices. No longer can the notion of the stereotypical praedial larcenist be viewed as a rural criminal or as a "piper" as many refer to them. The activity has now moved to the status of organized crime, extending away from rural areas into more urban settlements where agriculture is practised. This organized activity now requires a different set of skills and practices for policing the urban landscape (Smith, 2010)

This chapter provides an overview of the problem of praedial larceny facing the Caribbean region, tracing the historical roots of the problem. The chapter examines the drivers and nature of this, now organised criminal activity, focusing on the opportunistic and professional nature of praedial larceny. It finally explores the various strategies being used throughout the Caribbean to address this growing scourge to move toward farm security for food security.

OVERVIEW OF FARM THEFT (PRAEDIAL LARCENY) IN THE CARIBBEAN

A study titled "An Analysis of the State of Praedial Larceny in Member States of CARICOM" conducted in collaboration with the Caribbean Disaster Emergency Management Agency (CDEMA) under the umbrella of the Agricultural Sector Disaster Risk Management Committee (ASSC/TMAC) found that in terms of the number of persons and families affected by praedial larceny, it is the most extensive among all crimes committed in the Caribbean (Food and Agriculture Organization, 2013) Farm theft was for decades, not taken seriously and perceived as petty by many, but has however, transformed itself into a very serious and complicated crime, with now organized overtones, impacting food security and vulnerable populations in the region (Beckford & Campbell, 2012). Considered a "rural crime", its neglect has been attributed to social disorganisation. This theory begins with the fundamental assumption that places with high levels of expressions of cohesion and solidarity have lower rates of crime, however, places which display less order and more disorganisation tend to have higher rates of crime (Donnermeyer & Scott, 2013).

It has been recognised as a problem not only in the Caribbean, and there is a growing body of academic studies conducted in America, Australia, Britain and Scotland which relate specifically to this rural crime (Sugden, 1999; Barclay, 2001; Donnermeyer & Barclay, 2005; Yarwood & Gardener, 2005; Jones, 2008; Spore, 2009; Smith, 2010; Little, 2011). These studies all show that farm theft is regarded as one of the major deterrents to agricultural production. In many cases, this crime has been one of the major contributing factors to the exodus of farmers from the sector. Little (2011) indicated that it has been recognized by the highest level of leadership in the Caribbean as one of the major constraints to sustainable agricultural development in the region. These losses from praedial larceny can be classified under the categories of financial, physical or prevention/solution (Little, 2011). According to a 2010 survey carried out among regional stakeholders, more than 90% agreed that praedial larceny was the single most discouraging aspect of agriculture and has become a disincentive to investment in the sector and a threat to livelihoods in farming and fishing communities (Little, 2011). The report further indicated that on average, 82% of farmers and fishers affected are commercial or semi-commercial producers. The report also revealed that 18% of the value of farm output regionally is taken by thieves, resulting in the loss of over USD $321 million annually (Caricom, 2011), while worldwide losses are estimated as high as USD $5 billion on an annual basis (Swanson et al., 2000). In some countries in Europe, the loss is between 6 and 18% of agricultural output.

The CARICOM (2011) study revealed the following estimated country losses.

- **Trinidad and Tobago:** Losses of USD $22.6 million annually.
- **Jamaica:** Losses in excess of USD $55 million annually.
- **Belize:** Losses in excess of USD $300,000 annually.
- **St. Vincent and the Grenadines:** Losses of USD $2.3 million annually.
- **Bahamas:** Losses to its marine fish industry of US $16 million annually.

Losses due to praedial larceny can be so devastating that many farmers quit farming altogether. As thieves become better organized and armed, farmers not only suffer loss of their agricultural products and income, but also risk their lives trying to scare off thieves or defend their crops and livestock. Many of the countries spend millions in praedial larceny prevention programmes, some of which will be described later in this chapter. Indeed, St. Lucia is spending in excess of USD $400,000 annually on piloting activities to prevent praedial larceny.

CHANGING ECOLOGY OF FARM THEFT IN THE CARIBBEAN

As a complicated crime, it is extremely difficult to prove and therefore difficult to arrest and convict the perpetrators (FAO, 2013). With rising crime in many Caribbean territories, Governments focus on what Dingwall and Moody (1999) refer to as "mean streets myopia", where they focus their "criminological gaze" on urban criminality, such as drug and gang related crimes (Smith, 2010). Dingwall and Moody (1999) argue that criminologists have long viewed crimes against agricultural operations as unreal and as such it is not of interest to the police or to criminology scholars and thus has been neglected (Swanson, Chamelin & Territo, 2000; Smith, 2010). Many farmers refuse to report this crime due to frustration with inaction by police, who oft times dismiss farm related crime in rural areas as trivial or motiveless. According to Little (2010), only 45% of incidences are reported to the police regionally and very often farmers take matters into their own hands and often end up on the other side of the law. Many farmers are afraid to report for fear of reprisals. Data have also shown that just over 35% of produce stolen at the regional level is used for household food and other needs in the home while around 27% is disposed of in the trafficker (higgler/huckster) trade which is dominated by rural women, most of whom are single, vulnerable household heads and with no other source of income (FAO, 2013).

An FAO brief on praedial larceny in the Caribbean explains that offenders operate in a complex, social and economic environment encompassing varied groups of individuals who have become organized and entrepreneurial by developing livelihoods and businesses from the stealing of agriculture produce, equipment, and materials (Food and Agriculture Organization of the United Nations, 2013). There is a growing body of literature on examining the changing ecology of agricultural crime, criminality and policing (Donnermeyer & Scott, 2013; Smith et al., 2013; Smith & McElwee, 2013). The basic ideology, described by Smith, Laing and McElwee (2013) and Smith and McElwee (2013), around which the policing of rurality is governed posits that urban areas are the natural habitat for the serious and organised criminal. There is less crime in rural areas - known as the rural idyll thesis, with the socially constructed perception of the rural criminal as a loveable, rural rogue and small time thief. Furthermore, the policing of rurality suggests that the majority of rural crime is committed by urban criminals - known as urban marauder thesis; rural crime is somehow less serious than urban crime and therefore requires less of a policing presence. These ideologies are however changing, and as Smith (2010) explains, rural crime is now organized, semi-organized or committed by outsiders or sometimes committed by "rogue farmers" or "exploitative farmers" themselves (Wilkinson, Craig & Gaus 2010). The Caribbean region has in the last 20 years or so noted increasingly, this changing ecology of praedial larceny in terms of frequency of incidences, the large volumes of crop, livestock and fish stolen, and the highly organized and often violent behaviours of the thieves. The significance of the loss resulting has become a disincentive to investment in agriculture and fishing and a danger to farmers, fisher folks, aquaculturists, their employees and families (Little, 2010).

Each criminal group has developed its own dynamics in a distribution chain where the crime enters undetected into the normal processes of legitimate industry, especially of domestic fresh food distribution (FAO, 2013). Smith (2010) and McElwee (2009) refers to this as "illegal rural enterprise"; the most pervasive and hidden type of entrepreneurial criminality. In some cases it involves insider knowledge, thus adding to its complexity of the complicity of rural communities (Smith, 2010). Similar rural criminal activity has been described by Sergi and Lavorgna (2012) and Smith et al. (2013) in Europe. Here there has been in the last decade, an escalation in rural crimes such as the theft of farm machinery and tools, and into unregulated butchery practices. These are all evidence of the danger that serious and organised crime groups can pose to rural areas when they seek to expand their criminal activities in the current economic recession.

Table 1. describes the typology of most farm related crimes throughout the Caribbean (adapted from Smith, 2010). Smith (2010) categorises farm crimes as situation-specific and context-bound, wildlife crime as predatory (carried out by organised gangs of urban criminals), illegal rural enterprise (symbiotic/entrepreneurial) and village crime (opportunistic or context-bound).

Further to this, there is a hidden cost to agriculture production, productivity, and food security when farmers decide to leave the sector or when high quality genetic breeds of livestock and crop varieties are stolen from breeding stations and agriculture research facilities and sold as food (FAO, 2013). Further, there are the likely consequences for public health and subsequent industry fallout should tainted, uncertified produce gain entry into the domestic food chain, which has been the case where crops which have been sprayed with chemicals are stolen by thieves (George Ramtahal, Personal Communication, November 2015). Although not officially reported, because of the difficulty of detection by farm owners, many farms may be the location for clandestine drug operations, especially marijuana as in the case in St. Vincent and the Grenadines, where, after the loss of preferential market access for their banana in the 1990s, it is alleged that several farmers got into the illegal production of marijuana in the more remote mountainous parts of the country. Several authors have described the growing trend of this type of illegal activity globally (Muhammad, 2002; Weisheit & Fuller, 2004; Barclay & Donnermeyer, 2011; Garriott, 2011).

On average 82% of farmers and fishers affected are commercial or semi-commercial producers between the ages of 45 and 50 years. Poor and vulnerable farming populations are also targeted victims. Ninety-eight percent of all producers surveyed have experienced loss of produce from theft - more than 332,000 fisher folk families, and well over 1,000,000 crop and livestock farm families. In addition, farmers have

Table 1. A typology of rural related farm crimes in the Caribbean adapted from Smith (2010)

Farm Crime (Predatory/Organised/ Context Bound)	Wildlife Crime (Predatory)	Illegal Rural Enterprise (Symbiotic/ Entrepreneurial)	Village Crime (Opportunistic/ Context Bound)
Theft – farm equipment (irrigation lines, tractors, diesel, water pumps, weed whackers, fertilizer and pesticides etc.) Vandalism – to farm and buildings and fences, damage to crops Fire-raising – buildings and farms Cruelty to livestock	Hunting out of season for wildlife – agouti, manicou, lappe, deer Illegal fishing for tilapia etc. Trapping Theft of turtle eggs etc.	Theft of livestock/sheep, goat, cattle, poultry and small stock Theft of seedlings, fruits and vegetables Squatting on private lands Drug cultivation – cannabis farming	Petty theft (pipers) – generally opportunistic Illegal trespassers and shooters Breaking and entering Dumping of rubbish on farmland

been killed and others have been known to have received threats and physical attacks by persons who have returned to the communities after serving time in prison for praedial larceny. Farmers' crops have also been damaged by these same persons.

The crime, therefore, has potentially high, though undetermined, social costs to welfare in farming communities, livelihoods, and household food security. Concomitant with this, more than 90% of producers regionally agree that it is the single greatest disincentive to investment in the sector.

Policing programmes to combat praedial larceny have now become an integral consideration in determining how agriculture producers, processors, and distributors function in the region in an effort to secure their on-farm investments. For example in Trinidad and Tobago, the praedial larceny squad was established to manage theft in many farming areas. However, in this country, the burden of proof appears to lie with the Police Service. It is required that they produce evidence (the allegedly stolen goods). The problem is: these goods are usually highly perishable. Proper storage is required but often not available. Compounding matters is the fact that the judiciary is burdened by huge backlogs of cases resulting in untimely delays. By the time the cases reach the courts, there is no evidence; crops rot or animals die.

Many governments in the region have also been strengthening policies in order to build strategies which will secure the economic gains from public investment in the sector.

HISTORICAL PERSPECTIVES ON FARM THEFT IN THE CARIBBEAN

Farm theft, better known as praedial larceny, can be traced historically from the Plantation economies of many of the Caribbean islands (Engledow, 1945; Shepherd, 1945; Eisner, 1961; Grossman, 1997 and Bryan, 2000). Eisner (1961) suggested that the problem of praedial larceny, like that of illegitimacy, has its origin in slavery and both have been held responsible for poverty. Bryan (2000) noted that the crime was considered a 'typically black perversion' reflective of the structural problems in the countryside, however, he mentioned that one observer referred to it as the 'rotten and unscientific state of relations between labour and capital', and this is what lay at the root of the problem. Grossman (1997) cited the seriousness of this long-standing problem in Engledow (1945) supplement to the Moyne Commission Report:

No circumstance affecting home-grown food production is more widely or more forcibly brought to notice than what in these colonies is known as praedial larceny. It affects agriculture in many ways..... Praedial larceny is, indeed, a profound handicap to agricultural development.

A report on agriculture in St. Vincent in 1953 similarly mentions *the wave of praedial larceny which prevails and so discourages peasants from growing more food.*

In St. Vincent, Shepherd (1945) reported that *bitter complaints volunteered by many settlers indicate that the losses are serious and that the prevalence of theft is a deterrent to the more extensive planting of food crops.* He also revealed that the problem of theft was more severe with food crops than with export crops. This probably accounts for the reason why farmers up to today still focus on the production of banana over extensive food crops, in what can be described as a culture of dependency (Isaac et al. 2012). According to Shepherd (1945):

Food crops are usually confined to land least suited to the staple cash crop and as these areas are frequently the most remote from the cultivator's home, they are seldom under continuous observation. Food crops can be consumed by the thief, and even if sold are unlikely to arouse suspicion. But cash crops, such as sugar-cane, seed cotton and arrowroot rhizomes, require processing: the number of buyers is limited, and in most cases they are registered, and required to keep an account of their transactions: thus the thief runs a greater risk of detection.

Eisner (1961) noted that this crime was ranked equally important with technical inefficiency as the "drawback to the system of peasant proprietorship". Reported as a criminal offence since the 1850s, statistics show that in many Caribbean islands it was compounded by land tenure status. This was because many rural farmers did not always own or cultivate the parcel of land. Landlords were also away from their estates for long periods of time. These situations facilitated praedial larceny as detection rates were very low and as Eisner (1961) suggested, "detections bear no proportion to commission". Eisner (1961) reported that in Jamaica in 1905, "the average value assigned to the stolen property in 2,349 cases brought into court amounted to only 1.03 shillings". Given the high cost associated with going to court for farm theft related cases, it was not worth the time for small farmers to pursue cases and hence the actual reports and costs associated with this crime were underreported. According to Eisner (1961), in Jamaica in 1883, it was claimed that only one-twentieth of all cases were brought into the court, and in spite of the various measures adopted, the number of convictions for this crime continued to fall. Eisner (1961) also reported that a Baptist minister offering evidence to the 1897 Royal Commission, explained that the problem was not as serious as the complainers suggested, citing the large acreages of canes, banana and coconuts left unprotected in all parts of the island, and yet rarely touched by the poor.

Bryan (2000) citing the reasons for praedial larceny, reported that it was sometimes attributed to ignorance or to hunger. Supporting the hunger thesis, he points to the decline of praedial larceny when crops were good, and the rise of praedial larceny during periods of severe drought. Bryan (2000) also stated that there was evidence pointing to the marketability of stolen produce as a major incentive to praedial larceny, including larceny of citrus, logwood and pimento, all export commodities in Jamaica. He cited an example of a planter, who reported that his grapefruits were never touched until a market for them appeared in the United States.

Bryan (2000) cited Justice Gibbons on the relationship between logwood stealing and local merchants. According to Gibbons:

Since I require the production of the wharf books in a case of logwood stealing at May Pen, when I was informed by an overseer of Denbeigh, that the Lord Penrhyn had been robbed of 60 tons (£300) during the then season, I have observed that no prosecutions have been instituted. There is no disguising the fact that Mr. W. C. the Custos, is a large shipper of logwood.........I have observed a suspicious connection between larceny of produce, particularly logwood and pimento from a small cultivator tried before me, there was no doubt of the thief having sold the stolen pimento to a local buyer who was himself the employer of the buyer and the party who ultimately profited by the theft of pimento....So, too, in several cases of logwood stealing it has appeared that the stolen logwood was readily purchased at the local wharf belonging in almost every case to a magistrate. These facts which have only casually come to my notice are doubtless well known to the bulk of the people (Bryan, 2000, pp. 25).

In Jamaica, praedial larceny was consistently regarded as an offence typically committed by members of the working classes. Thus, Act 32 of 1889 disallowed casual roadside buying and selling of produce, although this 'tended to restrict legitimate trade'. Law 37 of 1896 (the Produce Protection Law) tried to limit the offence from the receiving end (Bryan, 2000).

The punishment for this crime was corporal punishment or flogging (with the cat-o'-nine-tails in Jamaica and Trinidad and Tobago), an important method which was used to intimidate and terrorise the black population into obedience up to the end of the nineteenth century according to Bryan (2000). After the abolishment of this method of punishment in 1886 however, it was observed that there was no increase in the offence. A report in 1896 describes a wealthy banana producer and shopkeeper, advocating, "Flog the Negro; that is the only thing they are afraid of………" The same report mentions a boy receiving twelve lashes for stealing one orange.

Bryan (2000) mentioned Robert Love's anger over the legislation which proposed the continuation of flogging for praedial larceny in 1896. Love's anger was triggered from his belief that flogging was another example of class legislation since it was well known that only blacks were likely to be victims of the lash. In an editorial Love declared:

Whilst a thief who lifts a shop, or robs a bank, may be punished by a few months imprisonment, the poor wretch who steals a 2lb yam, or a stick of logwood, must, besides imprisonment, be lacerated by a cat-o'-nine tails….We hate this law because we are convinced that it was intended to operate on a class only. How many white men will be whipped for praedial larceny? Not one. Why? Because the circumstances are such that they are not exposed to the commission of the offence. We hate this law because, while its penalty will brutalise the offender, it will leave him no less an offender…… By a strange species of reasoning, they propose to drive out of us by the lash, the evil which they had driven into us, by the lash, for nearly 300 years.

In Trinidad, Brereton (2010) informed that in ordinance 6 of 1868—just thirty years after the end of slavery—corporal punishment was allowed as the punishment for praedial larceny and the practice of Obeah. An earlier Brereton (1979) document reported that in 1882,

One person was sentenced to four floggings of 36 lashes each by a Couva magistrate, and it was the governor who remitted three of them. A petition organised by a Methodist minister in 1883 to abolish flogging (—the great and disgusting vice of slavery in the brutal practice of flogging to blood the labouring population for petty larceny and other crimes) failed completely; indeed, an 1893 Ordinance actually extended the practice by making flogging mandatory for a second conviction for praedial larceny.

The situation in St. Vincent and other Eastern Caribbean countries showed changes, in that advanced plantation production technologies served as a deterrent to the praedial larcenist. The British and EU governments, the main importing countries of banana, imposed pressures for improved fruit quality, thus mandating technological innovations in harvesting and packing (Grossman, 1997). These technological innovative practices provided banana production another advantage over local food crop production that is related to the perennial problem of crop theft according to Grossman (1997). This movement from plantation crops, however, led to an emphasis on crop theft in other food crops and now this acted as a deterrent for many farmers getting into production (Shepherd 1945; Rubenstein 1975, 1987). According to Grossman (1997),

A thief could easily go into someone else's food garden under cover of night, dig up provisions, put them in a sack, and carry them to Kingstown the next day to sell to hucksters without being discovered. In contrast, technological innovations in harvesting and packing banana have made them less susceptible to theft than food crops. St. Vincent and the other Windward Islands have been forced to adopt a succession of complicated, intricate, labour-intensive processing methods-field packing, paco pack, and, most recently, cluster packing. Given the increasingly intricate, detailed steps required in these systems, stealing banana at night, processing and packing them according to specifications, and then selling them to the St. Vincent Banana Growers Association (SVBGA) the next day are extremely difficult - yet one more advantage of banana production over local food crop production.

In a historical review from colonial Zanzibar, Grassman (2011) describes a situation where crop theft was considered a 'hobby' (Makinde) or a means of economic sustenance since squatters who lived on large plantations harvested whatever they wanted from the landlord's plantation anytime he was absent. In the 1920s, crop theft was the most common form of property crime. These squatters were often in possession, with the landlord's consent, of large surpluses of coconut or clove, which they sold in the market as they wished (Grassman, 2011).

INFLUENCING FACTORS FOR FARM THEFT

There has been little attention paid to the study of patterns and causes of agricultural crimes in the Caribbean. However, there are several factors that contribute to, and encourage the growing spread of this crime. Several studies have however been conducted within recent years on farm crime in Australia, the United Kingdom, and the United States. These studies have consistently found that most farm crime is property-related, including the theft of livestock, spare parts, tractors and other machinery, fuel, tools, agrichemicals, and farm produce and the destruction of property (vandalism) (Barclay, 2001; Anderson & McCall, 2005; Donnermeyer & Barclay, 2005; Mears et al., 2007a; Jones, 2008; Barclay & Donnermeyer, 2011; Bunei et al., 2013).

A review of literature on agricultural crime found that crime against farms is widespread and costly to farmers (Barclay, 2001; Grassman, 2011; Little, 2011; FAO, 2013). Studies have shown that: (1) larger farms tend to experience more victimization than smaller ones in more developed countries (McCall, 2003; Mears et al., 2007b). However, this is not the case in the Caribbean where, regardless of size, farms are not immune to victimization (Little, 2011; FAO, 2013); (2) farmers use few security measures and are reluctant to report crimes (Jones, 2008; Little, 2011; Dickson, 2013); and (3) although farms are isolated, farm property can be easily accessed from public roads and have many portable commodities and items which are easy to steal (Barclay, et al., 2001). Studies have also indicated that employee theft can be a problem (Swanson et al., 2000; Donnermeyer & Barclay, 2005), especially for farm operations that hire part-time or immigrant or low-paid workers (Anderson & McCall, 2005; Little, 2011; FAO, 2013).

Most of the literature from these studies shows that farm crime is linked to the specific nature of physical, social, geographic, and cultural environments of farming communities and areas (Barclay, 2001; Barclay et al,. 2001; Barclay & Donnermeyer, 2002; Barclay et al., 2004; Anderson & McCall, 2005; Mears et al., 2007a; Jones, 2008; Barclay & Donnermeyer, 2011; Grassman, 2011). According to Barclay et al. (2001), unemployment and drug abuse are the most common social problems linked to farm crime. A study conducted in Australia by Anderson and McCall (2005) showed that isolated farmlands, larger farms with higher incomes, and proximity to urban centres, were the greatest predictors of being a victim of various types of farm crime. Although isolated farmlands and proximity to urban centres seem to be contradictory correlates of agricultural crime, they are not. The isolation refers to the distance from one farm to another, which affects guardianship. Hence, there is a large flow of non-local people (especially by road) in farm areas near towns. Additionally, a farm family's house may be too far from storage buildings, supplies, livestock and other valuable property to be seen, and farm neighbours can still be far enough away from each other so as not to notice intruders and thieves.

Similar studies conducted by Mears et al. (2007a) in the United States found that farm properties which are highly attractive, portable and have value, such as fruits and nuts, were more likely to experience theft. Further, proximity in terms of target and offender has been shown by several studies to be linked to high rates of farm crime victimization. Mears et al. (2007a) and Barclay (2001) found that farm theft was related to the number of employees on the farm, with some farm workers being responsible for crime directly, or by passing information to criminals for a fee. Further, Swanson et al. (2000) argued that farmers themselves constitute a source of support for farm theft and may readily purchase stolen commodities at a bargain price. The fact is that property crime on farms is highly situational, with certain factors being strongly associated with certain types of farm crimes (Barclay et al., 2001).

Even though most empirical studies on agricultural crime have been conducted in Australia, the United Kingdom, and the United States (Anderson & McCall, 2005; Barclay, 2001; Barclay et al. 2001; Jones, 2008; Mears et al., 2007b), factors that explain the various types of crime in developing countries such as the Caribbean region, have not been thoroughly investigated. Indeed, farm crime presents major problems to farm economies in these countries. Thus, such crimes must be given the due attention they deserve if agricultural theft is to be effectively reduced and managed in a growing and increasingly globalised economy.

Cohen and Felson (1979) described the routine activity theory, one of the main theories of environmental criminology, which is derived from an earlier Hawley (1950) theory of human ecology. The routine activity theory explores the temporal aspects of human behaviour in community environments. It includes perspectives such as situational crime prevention and the crime prevention through environmental design model (CPTED) and Newman's (1972) notion of defensible space. This theory suggests that criminal offences are directly related to patterns of daily social interaction of both victims and offenders in time and space, which define in part the situation or context under which crime occurs (Cohen & Felson, 1979). Tittle (2000); Felson (2002); Brantingham and Brantingham (1995) and Barclay et al. (2001), all describe the concept of place to the study of crime. They argue that in order for a crime to occur, three elements must converge. These three elements are (1) a motivated offender, (2) the absence of a capable guardian, and (3) the presence of a suitable target. As the authors explain, they involve proactive environmental techniques such as target hardening, controlled access and effective surveillance in an attempt to dissuade offenders and reduce the opportunistic potential for the crime. In this theory, a motivated offender is anyone who has the tendency or inclination to commit the crime, whereas the target refers to the object, person, place or property against which the crime occurs. Capable guardianship according to Cohen and Felson (1979) includes anyone or any object (that is, forms of physical security) which can limit the chances of an offender committing a crime. The routine activity theory indicates that offenders are motivated to make their decisions based on the characteristics of the target. The choice of the target is based on its value, accessibility, visibility, concealability, removability and disposability (Bunei et al., 2013). Bursick and Grasmik (1993) discussed the work of Cohen and Felson (1979) by referring to guardianship as both a human (human presence or physical guardianship) and a non-human (for example, locks, alarms) phenomenon. Guardianship is the availability of others who may prevent crimes by their mere presence or by offering assistance to ward off an attack. Capable guardianship can include neighbours, friends, relatives, passers-by, plus physical measures for example locks, alarms, and remote cameras; all of which can act as substantial obstacles to offenders (Clarke & Felson, 1993; Clarke, 1995). Barclay and Donnermeyer (2011) in a later study challenged these mainstream criminological theories and concepts of social disorganisation and collective efficiency. They rationalise on the importance of developing perspectives for place-based or ecological theories of crime. The aim is to develop a critical criminology of agriculture and food, exploring studies of rural 'others'. In addition, giving greater analytic attention to divisions and marginalities of peoples living in smaller and more isolated places based on gender, race, and lifestyles, among other factors (Barclay & Donnermeyer, 2011).

According to Bunei et al. (2013), the nature of rural communities together with transformations in their socio-economic and cultural makeup may tend to increase the number of offenders and reduce guardianship, thus exposing the farm property to greater risks. Mears et al. (2007b) describe the vulnerability of farms to criminal activity relative to their remoteness and isolation, increasing chances of theft from perpetrators including neighbours and employees. These authors argue that there is a traditional reluctance by farmers to seek help from law enforcement, which may serve to increase the opportunities for farm crime victimization from reprisals. This, Muhammad (2002) lamented, is due to the tendency for many rural communities to keep community problems within the community.

The routine activity theory stresses the importance of the exposure due to the isolation of property as a key ingredient in assessing vulnerability. Guardianship on a large farm where a great deal of the property is far from the location where the owner lives, tends to be minimal and is not effective in reducing the vulnerability of property (Mears et al., 2007b). Agricultural crimes occur at specific places on agricultural operations and these places reflect the visibility of property from a road by other people (opportunistic) as well as from the place where the farmer lives (guardianship).

In a later publication, Barclay and Donnermeyer (2011) noted that improvements in roads, the increasing cost of farm machinery and farm inputs, increasing reliance on transient or seasonal workers, and encroachment of urbanization into formerly rural and remote areas, have increased visibility, attractiveness and accessibility of farm properties. A study conducted in Kenya, found farm theft to be associated with high market integration and market availability (Omiti et al., 2007). This phenomenon somehow resembles the situation in many Caribbean countries. From the perspective of routine activity theory, this represents an increase in the attractiveness of the target. There is oft times little or no interest in the legitimacy of the source of the produce based on market availability and as a result, there is an imbalance in their daily demand and supply of fresh food and the buyer is often clueless. According to the FAO (2013), one of the major drivers for farm theft in the Caribbean lies in maintaining the supplier/producer relationship, in which, reliability of supplies, freshness, smaller amounts not requiring storage and an acceptable price necessitates what can be referred to as an organized trade. Praedial larcenys' most readily identified business feature is, therefore, the ability of the supplier to combine large volumes, timely delivery and a level of determination to ensure his delivery, that does not rule out the use of violence (FAO, 2013). Hence, sometimes overnight, a farmer's entire crop can be harvested to maintain this supplier/buyer relationship. Residing near major transportation routes would also increase a farmer's vulnerability to farm theft because potential offenders can easily navigate between urban and rural areas.

Barclay et al. (2001) examined ecological factors in conjunction with crimes occurring on farms over a two year period to identify those factors that influenced a farm's vulnerability to victimization. The ecological factors they identified included; the terrain of the property, the cover, the layout of the property, the visibility between the house and shed on the property and the distance of the property from a public road or a highway and the nearest town or service centre. These were all drivers to increase the risk of farm crime.

In a rather interesting study on "illegal pluriactivity" in the farming community by Smith and McElwee (2013), the authors describe a case study in which the minority of farmers are presented as "rogue-farmers" or criminal-entrepreneurs operating across legal and illegal domains and an anti-authority attitude is embedded in their *modus operandi*. Here, the farmer is part of a wider entrepreneurial network which exploits an alternative shadow work economy. A similar situation is now growing throughout the Caribbean. These criminal acts are all unified by the concept of illegal pluriactivity because they provide alternative income generation strategies which supplement their legitimate incomes.

HHB & Associates Limited, Media and Market Research Consultants (2007) examined the nature and extent of praedial larceny in Trinidad and Tobago by interviewing a random sample of 750 farmers – 700 from Trinidad and 50 from Tobago. They determined that the hardest hit areas were Wallerfield and Sangre Grande; 21% - 30% of victims were located in these areas. They also found out that within a six month period, the incidence of theft can be as frequent as one to four times for equipment, two to four times for livestock and two to six times for crops. The incidence of praedial larceny was highest among vegetable farmers (17%) and lowest among those with large animals (for example, Cattle - 2%). The time of attacks was significant during both day and night, although attacks at night were notably higher. They also found that praedial larceny is under-reported to the police with reports ranging from 19% for root crops and 90% for fertilizers. The study discovered that farms with road frontage, as well as those with easy access, were more likely to fall victim to praedial larceny. Another factor which contributed to the increased incidence of praedial larceny was whether or not non-family members were employed. The research showed that farms which employed non-family members were more likely to fall victim to praedial larceny than those which employed family members only. The most frequently cited reason for praedial larceny, however, was the ready market for stolen goods and the farmers' recommendations for combating praedial larceny were to have stiffer penalties for praedial larceny and more police patrols.

Risk management of praedial larceny according to FAO (2013), therefore requires significant inputs from the criminal justice system, traceability systems for proof of ownership under the law, information sharing for deterrence and for capacity building, networking and intelligence gathering on the nature and reduction of the risk.

The major weaknesses and deficiencies in the legislation and the criminal justice system related to praedial larceny (FAO, 2013) include:

1. Deficiencies in the social and economic consequences among the critical players, including the police and judiciary, resulting in the failure to treat praedial larceny as a serious offence. Fines and penalties awarded under the law have not proven to be a deterrent to the thieves. The low priority given to the hearing of praedial larceny cases by the Courts and the ease with which bail is granted to repeat offenders undermines efforts to reduce praedial larceny.
2. Unfamiliarity and underutilization by the police of relevant legislative acts that can be used for prosecution, including the Trespass Act and the Unlawful Possession Act, which may be applied when the source of the produce is not identified in a timely manner.
3. Lack of reporting due to frustration with inaction by authorities means only 45% of incidences are reported to the police regionally. Thieves are therefore fairly confident they will face limited or no consequences.

4. Insufficient proof of ownership. Data management for traceability of produce is inefficient at the farm level. Ineffective registration processes do not permit produce to be followed from the source to the consumer and the monitoring of movement of produce within countries is not enforced
5. Information sharing on praedial larceny is absent at the community level, so the data and information are inadequate to provide guidance to the Police in critical areas such as hotspots, preferred subsectors and commodities or to build intelligence on the habits of suspicious or known perpetrators.
6. Farmer organizations and farmers are not sufficiently vigilant in information sharing and networking and do not adequately support the receipt book system including proper record keeping and the reporting of incidences or suspicious activities.

STRATEGIES FOR MITIGATING FARM THEFT

It is accepted that farm theft is a worldwide phenomenon. Many countries have established different methods to combat this problem. Since 1983, The United Nations Food and Agriculture Organization Praedial Larceny Prevention Act of 1983 granted agricultural wardens the power to detect offenses and arrest suspects in the Caribbean. The routine activities theory suggests that reducing criminal opportunities serves a key role in reducing the prevalence of crime (Brantingham & Brantingham, 1993; Felson, 2002; Barclay et al., 2001).

Early methods of dealing with praedial larceny in many Caribbean islands included a traditional folk belief that if the maljo bean (*Canavalia ensiformis*) is planted around crops it prevents petty theft (Winer, 2009). This bean is also reportedly planted in West African gardens in order to protect (them) from praedial larceny.

Globally, law enforcement agencies refer to the 3D's when attempting to mitigate the impacts of crime. Craig (2011) explained that these include: deterrence, detection, and delay. Deterrence involves the use of lighting, gates, fencing, no trespass signs, security systems and dogs. Detection includes systems to alert you when someone enters your property, which may include the use of motion lights, cameras, sensors and detectors, visual surveillance by your employees or neighbours. Delay is part of the strategy to slow access to your property or equipment and may include the use of cables across field lanes, fencing, locking doors on equipment storage facilities, parking equipment away from public viewing when left in fields overnight (Craig, 2011).

Strategies have also been recommended for farmers to review their operation from a thief's viewpoint and ask, what would be the easiest method to enter and steal items. Farmers should be cognizant of potential targets and conceal them as thieves have many strategies when identifying potential targets. One such strategy they use is calling addresses to see if anyone is home. It is recommended not putting your name on your mailbox which can provide easy access to phone numbers. They may even scout areas, observing farms to establish patterns of use. Altering daily routines can act as a deterrent, making movements unpredictable to thieves.

Asking for the credentials of any unknown or unexpected visitors is also important.

Record keeping is important, especially for keeping tabs on employees who manage your locks and keys on the farm. Records should be kept of all locks, the location of each lock and the number of keys that exist for each lock. Keep a list of any employee that has each key and inventory all keys periodically. It is important not to issue keys for convenience to employees and only provide a key when necessary and then follow through on proper return (Craig, 2011). It's also important that farmers take an inventory of all their equipment, taking pictures where necessary of distinguishing marks and recording serial numbers that can assist in recovery in the case of theft.

Locks should be of the highest security quality, with keys that are not easily reproduced and pick resistant, marking all keys with "Do Not Duplicate". It is important also to use locks with an added security measure of hasps that fold over, which will prevent access to mounting screws.

Precaution must be taken when leaving large equipment such as tractors in the fields by removing keys, locking doors if possible and never leaving equipment within easy access to roads. Use of a lockable fuel cap is also recommended as reports have been made of fuel theft from parked farm vehicles.

In 2007 at The Public Consultation on Food Prices held in *Trinidad*, the Association of Professional Agricultural Scientists of Trinidad and Tobago (APASTT) presented a novel proposal designed primarily to secure food production and also mitigate praedial larceny. The proposal included the following features:

- The establishment of Designated Agricultural Zones. These zones would be provided with the necessary physical infrastructure: roads, electricity, and water. There would be controlled access to the zones.
- From the outset, the zones must be established with inputs from the Police service regarding security concerns;
- There must be close collaboration between the entities in the zone and the Police Service;
- Farmers would be provided with performance leases for the lands within the zones;
- Farmers would be encouraged to form co-operatives as economic units within the zones;
- Each economic enterprise must be registered, and have its own labels and bar-code;
- All vehicles transporting produce out of the zones must have the appropriate signage;
- All produce leaving the zones must be labeled;
- Each zone must be mapped along with all relevant data to produce Geographical Information Systems (GIS). This system should be made available to the Police Service;
- The officers of the Praedial Larceny Squad would be provided with hand-held devices linked to the GIS. This would facilitate verification of information during stop and search exercises.

The *Jamaican* Government addressed the problem of praedial larceny by establishing a receipt book system which was used by farmers and agricultural traders to demonstrate legal proof of sale or purchase of agricultural products. They also initiated a Praedial Larceny Public Education Program which aimed at increasing awareness about unacceptable levels of praedial larceny. In addition, all persons involved in agricultural transactions had to register with the Rural Agricultural Development Authority (RADA) and Island Special Constabulary Force (ISCF) officers would be assigned to various parishes to preside over the implementation of the Praedial Larceny Program.

In *St. Lucia*, the Agriculture Ministry attempted to address the praedial larceny problem by implementing a "four-pronged strategy". The first arm of this strategy involved the enactment of stronger legislation to deal with praedial larceny and to regulate the sale of agricultural produce. The second arm aimed at a national identification program and licensing of bona fide farmers and traders. Thirdly, the government set upon working with Local Government and sought funding through the European Union Social Recovery Program to re-introduce the Rural Constabulary. Finally, the government embarked on an intensive public sensitization program to educate the public on the seriousness of praedial larceny.

A similar situation exists in *Antigua/Barbuda* where "the agriculture sector continues to lose millions of dollars through the effects of praedial larceny" (Hilson Baptiste, Minister of Agriculture, Lands,

Housing and the Environment, Antigua/Barbuda, 2009). This Government also introduced an agricultural receipt book system and farmers were encouraged to register their business or form cooperatives. A public education program was also established to increase awareness of praedial larceny and to educate farmers to use prompt action in detecting and reporting praedial larceny.

When dealing with praedial larceny there are several considerations that must be addressed through a multisectoral and multifaceted approach to ensure appropriate integration of policy issues. These may include (FAO, 2013):

- Issues of land tenure and proof of ownership of land and produce under the law. Programs for alternative livelihoods to manage the perceived relationship between vulnerable and poor households and praedial larceny in rural populations and along rural/ urban livelihood chains.
- Public education and communication strategies that empower and build resilience in farmers and fisher folks and their organizations at all levels - regional, national and community - in praedial larceny prevention and reduction.
- Information systems that provide clarity in the relationships in the legitimate businesses in the trafficker (higgler/huckster) trade between islands.

Some specific actions should revolve around:

- Policy and legislative frameworks to support the work of the police and the judiciary through the criminal justice system and the Agriculture Ministry.
- Mainstreaming of praedial larceny into the work of Agriculture Ministries and technical partners for implementation of a Plan of Action for praedial larceny prevention.
- Integration of praedial larceny prevention into the planning and monitoring systems of the National Agriculture Strategy and the national strategies for crime prevention.
- Information generation and knowledge management at the Ministries of Agriculture and partners including baseline data and mechanisms for knowledge-based planning and evaluation such as mapping praedial larceny hotspots.
- Capacity building for praedial larceny prevention among all stakeholders, Ministries of Agriculture in collaboration with regional and national farmers and fisher-folks and their organizations.
- Strategic partnerships for development and implementation of the Plan of Action for praedial larceny prevention.
- Use of the Comprehensive Disaster Management Strategy framework to provide additional co-ordination for praedial larceny at the regional level. The processes of CDM already provide a mechanism for management and sharing.
- Research to identify solutions to praedial larceny – The University of the West Indies to lead.

CONCLUSION

Praedial larceny can be labelled as a scourge for food producers; it is a very complex and serious problem affecting the agricultural sector worldwide. It involves a wide range of agricultural produce and equipment and its complexity lies in the fact that it is very difficult to prove that a crime has been committed. In the past, the problem of praedial larceny was not given the attention it deserved, but as the global problem of food availability and food security comes to the forefront, more attention is being paid to this problem. In addition, farmers are now being shown more respect and their voices are being heard- they are saying that they are fed up with this problem. Steps are currently being made to mitigate this serious crime as can be seen by the various programs being set up by governments of different countries throughout the region to combat or reduce the prevalence of praedial larceny.

In developing countries, food producers generally struggle to keep financially afloat, producing food under disadvantageous conditions which include: unfavourable weather conditions, frequent pests and disease outbreaks, limited or no support from the government, trade barriers and unfair regulations. Most farmers barely make ends meet, far less generate profits which can be reinvested to move them further along the continuum to full commercial operation. It is in these circumstances that any theft of farm produce or equipment becomes a serious blow to farmers, and if it is a regular occurrence it has the effect of causing the farmer to become impoverished; unable to feed one's family or frustrates movement away from subsistence to commercial farming or make profits to have a better standard of living.

Praedial larcenists do an injustice to farmers. While farmers can do somethings at the farm level to restrict the level of praedial larceny, actions at the higher level can be more effective. This chapter reviewed several actions that have been tried which may be useful for countries struggling with this scourge to consider. Moreover, several suggestions are put forward; actions at government level which brings agriculture, legal and enforcement agencies to the table can be most productive.

One bag of hot pepper taken by a thief may not be much in the eyes of the general public, but to a poor farmer it may determine whether his children will get all their school books in the new school term!

REFERENCES

Anderson, K. M., & McCall, M. (2005). *Farm Crime in Australia*. Canberra, Australia: Australian Institute of Criminology. Retrieved on October 5th, 2015 from www.aic.gov.au/publications/current%20 series/cfi-120/cfi119.aspx

Anderson, S. (1997). Crime and social change in Scotland. In G. Dingwall & S. Moody (Eds.), *Crime and conflict in the countryside* (pp. 24–41). Cardiff, UK: University of Wales Press.

Barclay, E. M., & Donnermeyer, J. F. (2002). Property Crime and Crime Prevention on Farms in Australia. *Crime Prevention and Community Safety: An International Journal.*, *4*(4), 47–61. doi:10.1057/palgrave.cpcs.8140169

Barclay, E. M., & Donnermeyer, J. F. (2011). Crime and Security on Agricultural Operations. *Security Journal*, *24*(1), 1–18. doi:10.1057/sj.2008.23

Barclay, E. M., Donnermeyer, J. F., Doyle, B. P., & Talary, D. (2001). *Property Crime Victimization and Crime Prevention on Farms. Report to the New South Wales Attorney General's Crime Prevention Division*. Armidale, Australia: The Institute for Rural Futures, Incorporating the Former Rural Development Centre. University of New England.

Barclay, E. M., Donnermeyer, J. F., & Jobes, P. C. (2004). The dark side of gemeinschaft: Criminality within rural communities. *Crime Prevention and Community Safety: An International Journal*, *6*(3), 7–22. doi:10.1057/palgrave.cpcs.8140191

Barclay, L. (2001). *A review of the literature on agricultural crime: report to the criminology research council*. Institute for Rural Futures, University of New England. Retrieved on December 19th, 2015 from: www.criminologyresearchcouncil.gov.au/reports/barclay.pdf

Beckford, C. L., & Campbell, D. R. (2012). *Domestic Food Production and Food Security in the Caribbean: Building Capacity and Strengthening Local Food Production Systems*. Palgrave Macmillan.

Bekele, I., Singh, R., Mustapha, N., & Ramsaroop, M. (1998). Incidence of Praedial Larceny in Food Crop Production in Trinidad: Major Highlights. *Proceedings of UWI, Ag. 50*.

Bouffard, A. L., & Muftíc, R. L. (2006). The "Rural Mystique": Social disorganization and Violence beyond Urban Communities. *Western Criminology Review, 7*, 56–66.

Brantingham, P., & Brantingham, P. (1981). *Environmental Criminology*. Prospect Heights: Waveland Press.

Brantingham, P. & Brantingham, P. (1995). Criminality of Place: Crime Generators and Crime Attractors. *European Journal on Criminal Policy and Research, 3*(3).

Brereton, B. (1979). *Race Relations in Colonial Trinidad 1870-1900*. Cambridge, UK: Cambridge University Press.

Brereton, B. (2010). The Historical Background to the Culture of Violence in Trinidad and Tobago. *Caribbean Review of Gender Studies, A Journal of Caribbean Perspectives on Gender and Feminism, 4*, 1-16.

Bryan, P. (2000). *The Jamaican People 1880-1902, Race, Class and Social Control*. The University Press.

Bunei, E. K., Rono, J. K., & Chessa, S. R. (2013). Factors Influencing Farm Crime in Kenya: Opinions and Experiences of Farmers. *International Journal of Rural Criminology, 2*(1), 75–100.

Bursik, R. J., & Grasmik, H. G. (1993). *Neighborhoods and Crime: Dimensions of Effective Community Control*. New York: Lexington Books.

Caricom. (2011). Food Security in Caricom. *Caricom View*. Retrieved from http://www.caricom.org/jsp/communications/caricom_online_pubs/caricom_view_jul_2011.pdf

Clarke, R.V. (1995). Situational Crime Prevention. *Crime and Justice, 19*, 91-150.

Clarke, R. V., & Felson, M. (Eds.). (1993). *Routine activity and rational choice*. New Brunswick, NJ: Transaction Publishers, Inc.

Cohen, L. E., & Felson, M. (1979). Social change and crime rates trends: A routine activities approach. *American Sociological Review, 44*(4), 588–608. doi:10.2307/2094589

Craig, P. H. (2011). *Protecting you Farm from Theft. PennState Extension, Cumberland County*. Retrieved on July 7th, 2011 from: http://extension.psu.edu/cumberland/news/2011/protecting-your-farm-from-theft

Dickson, D. (2013). Praedial Larceny - A Serious Crime. Trinidad and Tobago Chamber of Industry and Commerce. *Feature, 13*(1), 26–27.

Donnermeyer, J., & Scott, J. (2013). How Rural Criminology Informs Critical Thinking. *International Journal for Crime, Justice and Social Democracy.* Retrieved from http://eprints.qut.edu.au/68038/2/C1_Donnermeyer_Rural_Criminology_Pub_Paper.pdf

Donnermeyer, J. F., & Barclay, E. M. (2005). The policing of farm crime. *Police Practice and Justice Research, 6*(1), 3–17. doi:10.1080/15614260500046913

Eisner, G. (1961). *Jamaica 1830 - 1930, A Study in Economic Growth.* Manchester, UK: Manchester University Press.

Engledow, F. L. (1945). *West India Royal Commission: Report on Agriculture, Fisheries, Forestry and Veterinary Matters. Cmd. 6608.* London: His Majesty's Stationery Office.

Felson, M. (2002). *Crime and everyday life* (3rd ed.). London: Sage.

Food & Agriculture Organization (2013). *Praedial Larceny in the Caribbean.* Issue Brief #3.

Garriott, W. (2011). *Policing methamphetamine: Narcopolitics in rural America.* New York University Press.

Grassman, J. (2011). *War of Words, War of Stories: Rural Thought and Violence in Colonial Zanzibar.* Indiana University Press.

Grossman, L. S. (1997). *The Political Ecology of Bananas: Contract Farming, Peasants and Agrarian Change in the Eastern Caribbean.* Chapel Hill, NC: The University of North Carolina Press.

Hawley, A. (1950). *Human Ecology: A Theory of Community Structure.* New York: Ronald Press.

Isaac, W., Joseph, M., Ganpat, W., Wilson, M., & Brathwaite, R. (2012). The Caribbean's Windward Islands banana industry: A heritage of dependency. *The Journal of Rural and Community Development, 7*(2), 98–117.

Jones, J. (2008). *Farm crime on Anglesey: Local partner's and organisations.* Views on the Issue, second report, January 2008. Retrieved on December 12th, 2015 from http://www.aber.ac.uk/en/media/jane-jones---second-report.pdf

Little, D. (2011). *Praedial Larceny: Its Consequences For Caribbean Agriculture.* Caricom View. Retrieved on June 3rd, 2015 from http://www.caricom.org/jsp/communications/caricom_online_pubs/caricom_view_jul_2011.pdf

Marshall, B., & Shane, J. (2005). *Crime in rural areas: A review of the literature for the Rural and Evidence Research Centre.* London: University College London.

Mawby, R. I. (2004). Myth and reality in rural policing: Perceptions of the police in a rural country of England. *Policing: An International Journal of Police Strategies & Management, 27*(3), 431–446. doi:10.1108/13639510410553158

McCall, M. (2003). *Results from the 2001-02 National Farm Crime Survey.* Australian Institute of Criminology, Trends and Issues in Crime and Criminal Justice, No.266.

McElwee, G. (2006). Farmer's as entrepreneurs: Developing competitive skills. *Journal of Developmental Entrepreneurship, 11*(3), 187–206. doi:10.1142/S1084946706000398

McElwee, G., & Vik, J. (2011). Diversification and entrepreneurial motivations of farmers in Norway. *Journal of Small Business, 49*(3), 390–410. doi:10.1111/j.1540-627X.2011.00327.x

Mears, D. P., Scott, M. L., & Bhati, A. S. (2007a). Opportunity theory and agricultural crime victimization. *Rural Sociology, 72*(2), 151–184. doi:10.1526/003601107781170044

Mears, D. P., Scott, M. L., & Bhati, A. S. (2007b). *Policy, theory and research from an evaluation of an agricultural crime prevention program*. Washington, DC: The Urban Institute.

Muhammad, B.L. (2002). *Rural Crime and Rural Policing Practices (Multicultural Law Enforcement)*. An applied research project submitted to the Department of Interdisciplinary Technology as part of the School of Police Staff and Command Programme.

Newman, O. (1972). *Defensible Space*. New York: Macmillan.

Omiti, J. (2007). *Participatory prioritization of issues in smallholder agricultural commercialization in Kenya*. Nairobi: KIPPRA.

Rubenstein, H. (1975). The Utilization of Arable Land in an Eastern Caribbean Valley. *Canadian Journal of Sociology, 1*(2), 157–167. doi:10.2307/3339805

Rubenstein, H. (1987). Remittances and Rural Underdevelopment in the English Speaking Caribbean. *Human Organization, 42*(4), 295–306. doi:10.17730/humo.42.4.0709108414538471

Saugeres, L. (2002). Of Tractors and Men: Masculinity, Technology and Power in a French Farming Community. *Sociologia Ruralis, 42*(2), 143–159. doi:10.1111/1467-9523.00207

Sergi, A., & Lavorgna, A. (2012, September). Trade Secrets: Italian Mafia expands its illicit business. *Janes Intelligence Review*, 44-47.

Shephard, C. Y. (1945). *Peasant Agriculture in the Leeward and Windward Islands*. Trinidad: Imperial College of Tropical Agriculture.

Sherman, L. W., Gartin, P. R., & Buerger, M. E. (1989). Hot spots of predatory crime: Routine activities and the criminology of place. *Criminology, 27*(1), 27–56. doi:10.1111/j.1745-9125.1989.tb00862.x

Smith, R. (2010). Policing the Changing Landscape of Rural Crime: A Case Study from Scotland. *International Journal of Police Science & Management, 12*(3), 18–30. doi:10.1350/ijps.2010.12.3.171

Smith, R., Laing, A., & Mcelwee, G. (2013). The Rise of Illicit Rural Enterprise within the Farming Industry. *International Journal of Agricultural Management, 2*(4), 185–188. doi:10.5836/ijam/2013-04-01

Smith, R., & McElwee, G. (2013). Confronting Social Constructions of Rural Criminality: A Case Story on "Illegal Pluriactivity" in the Farming Community. *Sociologia Ruralis, 53*(1), 112–134. doi:10.1111/j.1467-9523.2012.00580.x

Spore Magazine. (2009). *Theft: Tactics for battling crime*. Technical Centre for Agricultural Land Rural Cooperation (CTA). Retrieved on March 21st, 2016 from: http://spore.cta.int/images/stories/pdf/SE139-web.pdf

Sugden, G. (1999). Farm crime: out of sight, out of mind: a study of crime on farms in the county of Rutland, England. *Crime Prevention and Community Safety: An International Journal*, *1*(3), 29–36. doi:10.1057/palgrave.cpcs.8140023

Swanson, C. R., Chamelin, N. C., & Territo, L. (2000). *Criminal investigation*. Boston: McGraw Hill.

Swanson, C. R., & Territo, L. (1980). Agricultural Crime: It's Extent, Prevention and Control. *FBI Law Enforcement Bulletin*, *49*(5), 8–12.

Weisheit, R. A., Falcone, D. N., & Wells, L. E. (2006). Crime and policing in rural and small-town America (3rd ed.). Waveland Press.

Wilcox, P., Land, K. C., & Hunt, S. (2003). *Criminal Circumstance: A Dynamic Multicontextual Criminal Opportunity Theory*. New York: Walter de Gruyter.

Wilkinson, M., Craig, G., & Gaus, A. (2010). Forced Labour in the UK and Gangmaster Licensing Authority. Report for the Contemporary Research Centre, University of Hull.

Winer, L. (2009). *Dictionary of the English / Creole of Trinidad and Tobago*. McGill-Queen's University Press.

Yarwood, R., & Gardener, G. (2005). Fear of Crime, Cultural Threat and the Countryside. *Area*, *32*(4), 403–411. doi:10.1111/j.1475-4762.2000.tb00156.x

Compilation of References

AARP. (2016). *AARP Public policy Institute*. Retrieved from http://livabilityindex.aarp.org/search#South+Bend+IN+USA

Abarche, J. B., Dickerson, A., & Green, F. (2004). Trade liberalization and wages in developing countries. *The Economic Journal, 114*(493), 73–96. doi:10.1111/j.0013-0133.2004.00188.x

Abbott, P. (2003). *Towards more socially responsible cocoa trade.* International Agricultural Trade Research Consortium, working paper 03-3. Retrieved June 20, 2015, from http://nbcgib.uesc.br/cicacau/arquivos/producao_tecnico_cientifica/wp03-03.pdf

ABC57 News. (2015, May 22). *Meet a descendant of one of South Bend's first African American families*. Retrieved from abc57 News: http://www.abc57.com/story/29135949/meet-a-decedent-of-one-of-south-bends-first-african-american-families

Abdallah, N. (2010). GM crops in Africa: challenges in Egypt. *GM Crops, 1*(3), 116–119. Retrieved 29 May 2014 from, http://www.tandfonline.com/doi/pdf/10.4161/gmcr.1.3.12811

Abdullah, K., Khan, I. A., Siddiqui, S. H., Ahmed, M., & Siddiqui, K. A. (1997). *In vitro* culture of indigenous and exotic banana clones for maximum multiplication. *Pakistan Journal of Botany, 29*(1), 143–150.

Ackello-Ogutu, C., Okoruwa, V., & Bahal, G. N. (2012*). Long-term challenges to food security and rural livelihoods in Sub-Saharan Africa*. Briefing Paper Number 2, 2012. Washington DC: GDN Agriculture Policy Series. Retrieved 6 June 2014 from, http://www.gdn.int/admin/uploads/editor/files/SSA_2_PolicyBrief_Food_Security.pdf

Adedeji, T.O., Nosiru, M.O., Akinsulu, A.A., Ewebiyi, I.O., Abiona, B.G. & Jimoh, T.S. (2013). Adoption of new rice for Africa (NERICA) technology in Ogun State, Nigeria. *Journal of Development and Agricultural Economics, 5*(9), 365-371.

Adenle, A. A. (2011). Response to issues on GM agriculture in Africa: are transgenic crops safe? *BMC Research Notes, 4*, 388. Retrieved 1 June 2014 from, http://unu.edu/publications/articles/are-transgenic-crops-safe-gm-agriculture-in-africa.html

Adu, E. K., Alhassan, W. S., & Nelson, F. S. (1999). Smallholder farming of the greater cane rat *Thryonomys swinderianus*, Temminck, in southern Ghana: A baseline survey of management practices. *Tropical Animal Health and Production, 31*(4), 223–332. doi:10.1023/A:1005267110830 PMID:10504102

Africa Progress Report. (2014). *Grain, fish, money, financing Africa's green and blue revolutions*. Author.

AGRA. (2013). *Establishing the Status of Post-harvest Losses and Storage for Major Staple Crops in Eleven African Countries (phase I)*. Retrieved July 23, 2015, from file:///C:/Users/Bond/Downloads/establishing-the-status-of-postharvest-losses-and-storage-for-major-staple-crops%20(4).pdf

Agrictrade. (2011). *Informed analysis, expert opinions- executive brief: Pacific, agric- trade policy debates and developments.* Author.

Agwu, A. E., & Chukwuone, N. A. (2005). Funding of Agricultural Extension in a Democratic and Deregulated Economy: The Cost Sharing Approach. *Journal of Agricultural Extension*, *8*, 90–98.

Alimuga, L., Magala, D., & Lwasa, S. (2007). Regoverning Markets: Small-scale producers in modern agrifood markets, Uganda--The case of the Nyabyumba United Farmers Group in Kabale district. Uganda Agricultural Economics Association.

Allaire, F. R., Sterwerf, H. E., & Ludwick, T. M. (1977). Variations in removal reasons and culling rates with age for dairy females. *Dairy Science*, *60*(2), 254–267. doi:10.3168/jds.S0022-0302(77)83862-9

American Marketing Association. (2005). *Definition of Marketing.* American Marketing Association. Retrieved from https://www.ama.org/AboutAMA/Pages/Definition-of-Marketing.aspx

Amin, S. (1977). Self-reliance and the new international economic order. *Monthly Review (New York, N.Y.)*, *3*(29), 1–21. doi:10.14452/MR-029-03-1977-07_1

Amir, P. (2014). GM crops-no gain for small farmers. *SciDevNet.* Retrieved 30 May 2014 from, http://m.scidev.net/south-asia/agriculture/opinion/gm-crops-no-gain-for-small-farmers.html

Anami, S. E., Zhang, L. M., Xia, Y., Zhang, Y. M., Liu, Z. Q., & Jing, H. C. (2015). Sweet sorghum ideotypes: Genetic improvement of stress tolerance. *Food and Energy Security*, *4*(1), 3–24. doi:10.1002/fes3.54

Anandajayasekeram, P., Puskur, R., Workneh, S., & Hoekstra, D. (2008). *Concept sand practices in agricultural extension in developing countries: A source book.* Nairobi, Kenya: International Livestock Research Institute.

Anandajayasekeram, P., Torkelsson, A., & Dixon, J. (2002). *Emerging participatory approaches to technology development and transfer. What is new? Occasional paper.* Harare, Zimbabwe: FARMESA.

Anand, R., Mishra, S., & Peiris, S. J. (2013). *Inclusive Growth: Measurement and Determinants.* Washington, DC: International Monetary Fund.

Andersen, L., Breisinger, C., Mason-D'Croz, D., Jemio, L. C., Ringler, C., Robertson, R. E., . . . Wiebelt, M. (2014). *Agriculture, incomes, and gender in Latin America by 2050: An assessment of climate change impacts and household resilience for Brazil, Mexico, and Peru.* IFPRI Discussion Paper 1390. Retrieved June 15, 2015, from http://ssrn.com/abstract=2539588

Anderson, K. M., & McCall, M. (2005). *Farm Crime in Australia.* Canberra, Australia: Australian Institute of Criminology. Retrieved on October 5th, 2015 from www.aic.gov.au/publications/current%20series/cfi-120/cfi119.aspx

Anderson, S. (1997). Crime and social change in Scotland. In G. Dingwall & S. Moody (Eds.), *Crime and conflict in the countryside* (pp. 24–41). Cardiff, UK: University of Wales Press.

Andrews, W. L., & Adams, E. M. (2016). *North American Slave Narratives.* Retrieved from Documenting the American South: http://docsouth.unc.edu/neh/intro.html

Anim-Kwapong, G. J., & Frimpong, E. B. (2005). *Vulnerability of agriculture to climate change-impact of climate change on cocoa production.* Final Report submitted to the Netherlands climate change studies assistance Programme: NCAP, 2. Retrieved June 12, 2015, from http://www.nlcap.net/fileadmin/NCAP/Countries/Ghana/COCOA_DRAFT_FINAL_REPORT.pdf

Antwi, A. (1994). *The effects of water deficit on growth and development in young cocoa plants (Theobroma cacao L).* (Unpublished MPhil thesis). University of the West Indies, St. Augustine, Trinidad and Tobago.

ArcNews. (2014). *GIS Supports World Food Programme's Food Security Program.* Retrieved from http://www.esri.com/esri-news/arcnews/spring14articles/gis-supports-world-food-programmes-food-security-program

ArcUser. (2010). *The Next Generation of GIS. Dealing with design and space-time.* The Magazine for ESRI Software Users.

Areavibes. (2016). *South Bend IN, Livability.* Retrieved from http://www.areavibes.com/south+bend-in/livability/

Arias, O. (1999). *Addressing market needs through biotechnology: The example of Agribiotecnolog'ıa de Costa Rica.* Paper presented at the Agricultural Biotechnology for Latin America Part 4. Turning Priorities into Feasible Programs.

Asian Development Bank (ADB). (2013). Gender equality and food security: Women's Empowerment as a Tool against Hunger. Manila, Philippines: Author.

Asian Development Bank. (2013). *Gender equality and food security: Women's empowerment as a tool against hunger.* Author.

Badgley, C., & Perfecto, I. (2007). Can organic agriculture feed the world? *Renewable Agriculture and Food Systems, 22*(2), 80–85. doi:10.1017/S1742170507001986

Baenziger, P. S., Russell, W. K., Graef, G. L., & Campbell, B. T. (2006). Improving lives: 50 years of crop breeding, genetics, and cytology (C-1). *Crop Science, 46*(5), 2230–2244. doi:10.2135/cropsci2005.11.0404gas

Bailey, R., Froggatt, A., & Wellesley, L. (2014). Livestock – Climate change's forgotten sector. *Global Public Opinion on Meat and Dairy Consumption.* Retrieved June 11, 2015, from http://www.gastronomiaycia.com/wp-content/uploads/2014/12/estudio_consumo_carne.pdf

Bakırcı, M. (2005). The future of organic farming in Turkey and its effects to the agricultural negotiations between Turkey and European Union. *Istanbul University Geographical Review, 13*, 67–83.

Bakırcı, M. (2007). *Rural development in Turkey: Concepts, policies, practices.* Ankara, Turkey: Nobel Press.

Balakrishnan, R., & Fairbairn-Dunlap, P. (2005). Rural women and food security in Asia and the Pacific: Prospects and paradoxes. UN-FAO. RAP, 2005/30.

Balakrishnan, N. P., & Ellis, J. L. (1996). Andaman and Nicobar Islands. In *Flora of India, Part 1.* Calcutta, India: Botanical Survey of India.

Balogun, E. D. (2003). Foreign Private Investment in Nigeria. *Central Bank of Nigeria Economic and Financial Review, 37*(2), 31-81. Retrieved from http://www.cenbank.org/OUT/PUBLICATIONS/EFR/RD/1999/EFRVOL37-2-3.PDF

Bansal, K. C., Lenka, S. K., & Mondal, T. K. (2014). Genomic resources for breeding crops with enhanced abiotic stress tolerance. *Plant Breeding, 133*(1), 1–11. doi:10.1111/pbr.12117

Barclay, L. (2001). *A review of the literature on agricultural crime: report to the criminology research council.* Institute for Rural Futures, University of New England. Retrieved on December 19th, 2015 from: www.criminologyresearchcouncil.gov.au/reports/barclay.pdf

Barclay, E. M., & Donnermeyer, J. F. (2002). Property Crime and Crime Prevention on Farms in Australia. *Crime Prevention and Community Safety: An International Journal., 4*(4), 47–61. doi:10.1057/palgrave.cpcs.8140169

Barclay, E. M., & Donnermeyer, J. F. (2011). Crime and Security on Agricultural Operations. *Security Journal, 24*(1), 1–18. doi:10.1057/sj.2008.23

Barclay, E. M., Donnermeyer, J. F., Doyle, B. P., & Talary, D. (2001). *Property Crime Victimization and Crime Prevention on Farms. Report to the New South Wales Attorney General's Crime Prevention Division*. Armidale, Australia: The Institute for Rural Futures, Incorporating the Former Rural Development Centre. University of New England.

Barclay, E. M., Donnermeyer, J. F., & Jobes, P. C. (2004). The dark side of gemeinschaft: Criminality within rural communities. *Crime Prevention and Community Safety: An International Journal*, *6*(3), 7–22. doi:10.1057/palgrave.cpcs.8140191

Barone, A., Ercolano, M. R., & Langella, R., Monti, & Frusciante, L. (2005). Molecular marker-assisted selection for pyramiding resistance genes in tomato. *Advanced Horticultural Science*, *363*(19), 147–152.

Barro, R. (1998). Notes on growth accounting. *Journal of Economic Growth*, *4*(2), 119–137. doi:10.1023/A:1009828704275

Basic Statistics. (2011). Andaman and Nicobar Islands, India.

Bauman, D. E. (1989). *Biology of Bovine Somatotropin. Advanced Technologies Facing the Dairy Industry: bST*. Cornell University.

Beckford, C. L., & Campbell, D. R. (2012). *Domestic Food Production and Food Security in the Caribbean: Building Capacity and Strengthening Local Food Production Systems*. Palgrave Macmillan.

Beckford, G. (1972). *Persistent poverty: underdevelopment in plantation economies of the Third World*. Oxford, UK: Oxford University Press.

Beddington, J., Asaduzzaman, M., Fernandez, A., Clark, M., Guillou, M., Jahn, M., . . . Wakhungu, J. (2012). *Achieving food security in the face of climate change: Final report from the Commission on Sustainable Agriculture and Climate Change*. Copenhagen, Denmark: CGIAR Research Program on Climate Change, Agriculture and Food Security (CCAFS). Retrieved June 2, 2015, from http://www.ccafs.cgiar/commission

Beebe, S., Ramirez, J., Jarvis, A., Rao, I. M., Mosquera, G., Bueno, J. M., & Blair, M. (2011). Genetic improvement of common beans and the challenges of climate change. In S. S. Yadav, R. J. Redden, J. L. Hatfield, H. Lotze-Campen, & A. E. Hall (Eds.), *Crop adaptation to climate change* (pp. 356–369). Richmond, Australia: John Wiley & Sons, Ltd., Published by Blackwell Publishing Ltd. doi:10.1002/9780470960929.ch25

Beegle, C. C., & Yamamoto, T. (1992). History of *Bacillus thuringiensis* Berliner research and development. *Canadian Entomologist*, *124*(4), 587–616. doi:10.4039/Ent124587-4

Bekele, I., & Ganpat, W. (2014). Education, extension, and training for climate change. In W. Ganpat & W.P. Isaac (Eds.), Impacts of Climate Change on Food Security in Small Island Developing States (pp. 361-388). Hershey, PA: IGI Global.

Bekele, I., Singh, R., Mustapha, N., & Ramsaroop, M. (1998). Incidence of Praedial Larceny in Food Crop Production in Trinidad: Major Highlights. *Proceedings of UWI, Ag. 50.*

Belay, M., & Guri, B. (2013). Why African farmers do not want GMOs. *Truthout*. Retrieved 29 May 2014 from, http://webcache.googleusercontent.com/search?q=cache:http://www.truth-out.org/news/item/20058-why-african-farmers-do-not-want-gmos

Bennett, R., Ismael, Y., Kambhampati, U., & Morse, S. (2004). Economic impact of genetically modified cotton in India, *Agbioforum*, *7*(3), Article 1. Retrieved 2 June 2014 from, http://www.agbioforum.org/v7n3/v7n3a01-morse.htm

Berg, A. G., & Ostry, J. D. (2011). *Inequality and unsustainable growth: two sides of the same coin?* IMF Staff Discussion Note 11/08. Washington, DC: International Monetary Fund. Retrieved 15 July 2014 from, https://www.imf.org/external/pubs/ft/sdn/2011/sdn1108.pdf

Berlan, A. (2009). Child labour and cocoa: Whose voices prevail? *The International Journal of Sociology and Social Policy*, *29*(3/4), 141–151. doi:10.1108/01443330910947516

Berman, E. (2011). *Local Food Systems: Selected Resources*. Retrieved from Science and Technology Resources on the Internet: http://www.istl.org/11-summer/internet.html

Bertow, K., & Schultheis, A. (2007). *Impact of EU's agricultural trade policy on smallholders in Africa*. Germanwatch.

Bevan, M. W., Flavell, R. B., & Chilton, M. D. (1983). A chimaeric antibiotic resistance gene as a selectable marker for plant cell transformation. *Nature*, *304*(5922), 184–187. doi:10.1038/304184a0

Bharucha, Z., & Pretty, J. (2010). The roles and values of wild foods in agricultural systems. *Philosophical Transactions of the Royal Society of London. Series B, Biological Sciences*, *365*(1554), 2913–2926. doi:10.1098/rstb.2010.0123 PMID:20713393

Bilal, S., Dalleau, M., & Lui, D. (2012). *Trade liberalization and fiscal adjustments: The case of EPAs in Africa*. European Center for Development Policy Management.

Binswanger, H. P., & Deininger, K. (1997). Explaining Agricultural and Agrarian Policies in Developing Countries. *Journal of Economic Literature*, *35*(4), 1958–2005. Retrieved from http://www-wds.worldbank.org/servlet/WDSContentServer/WDSP/IB/1997/05/01/000009265_3971126124251/Rendered/PDF/multi_page.pdf

Bioversity International. (2013). *Seeds for needs project*. Retrieved June 7, 2015, from http://www.bioversityinternational.org/research/sustainable_agriculture/seeds_for_needs.html)

Black, R., Seal, S., Abubakar, Z., Nono-Womdim, R., & Swai, I. (2001). Bacterial spot (*Xanthomonas campestris* pv. *vesicatoria*) of tomato and sweet pepper in Tanzania. *Plant Pathology*, *50*(6), 20. doi:10.1046/j.1365-3059.2001.00633.x

Boland, M. J., Rae, A. N., Vereijken, J. M., Meuwissen, M. P., Fischer, A. R., van Boekel, M. A., & Hendriks, W. H. et al. (2013). The future supply of animal-derived protein for human consumption. *Trends in Food Science & Technology*, *29*(1), 62–73. doi:10.1016/j.tifs.2012.07.002

Borgen Project Report. (2015). Retrieved from http://borgenproject.org/9-facts-poverty-south-pacific/

Borlaug, N. E. (1997). Feeding a world of 10 billion people: The miracle ahead. *Biotechnology, Biotechnological Equipment*, *11*(3-4), 3–13. doi:10.1080/13102818.1997.10818934

Botella-Rodriguez, E. (2011). Cuba's inward-looking development policies: Towards sustainable agriculture (1990-2008). *Historia Agraria*, *55*(5), 135–171.

Bouffard, A. L., & Muftíc, R. L. (2006). The "Rural Mystique": Social disorganization and Violence beyond Urban Communities. *Western Criminology Review*, *7*, 56–66.

Bourne, C. (2008). *Perspectives on Enhancing Sustainable Growth and Development of Caribbean Agriculture*. Keynote Address, 44th Meeting Caribbean Food Crops Society.

Branca, G., McCarthy, N., Lipper, L., & Jolejole, M. C. (2011). *Climate-smart agriculture: a synthesis of empirical evidence of food security and mitigation benefits from improved cropland management*. Retrieved June 13, 2015, from http://www.fao.org/climatechange/29764-0aa5796a4fb093b6cfdf05558c6dd20bb.pdf

Brantingham, P. & Brantingham, P. (1995). Criminality of Place: Crime Generators and Crime Attractors. *European Journal on Criminal Policy and Research*, *3*(3).

Brantingham, P., & Brantingham, P. (1981). *Environmental Criminology*. Prospect Heights: Waveland Press.

Brenton, P., Portugal-Perez, A., & Regolo, J. (2013). *Food prices, road infrastructure, and border effects in central and eastern Africa.* Retrieved 8 July 2014 from, https://editorialexpress.com/cgi-in/conference/download.cgi?db_name=CSAE2013&paper_id=919

Brereton, B. (2010). The Historical Background to the Culture of Violence in Trinidad and Tobago. *Caribbean Review of Gender Studies, A Journal of Caribbean Perspectives on Gender and Feminism, 4*, 1-16.

Brereton, B. (1979). *Race Relations in Colonial Trinidad 1870-1900.* Cambridge, UK: Cambridge University Press.

Brethour, C., & Weersink, A. (2001). An economic evaluation of the environmental benefits from pesticide reduction. *Agricultural Economics, 25*(2-3), 219–226. doi:10.1111/j.1574-0862.2001.tb00202.x

Brookes, G., & Barfoot, P. (2013). *GM crops: global socio-economic and environmental impacts 1996-2011.* Dorchester, UK: PG Economics Ltd. Retrieved 10 July 2014 from, http://www.europabio.org/sites/default/files/report/2013globalimpactstudyfinalreport.pdf

Brown-Uddenberg, R. C., Garcia, G. W., Baptiste, Q. S., Counand, T., Adogwa, T. O., & Sampson, T. (2004). *The Agouti (Dasyprocta leporina, D. aguti) Booklet and Producers' Manual.* Trinidad and Tobago: Champs Fleurs.

Bryan, P. (2000). *The Jamaican People 1880-1902, Race, Class and Social Control.* The University Press.

Bulletin on Asia-Pacific Perspectives. (2003/04). *Beyond "subsistence affluence".* Poverty in Pacific Island Countries.

Bunei, E. K., Rono, J. K., & Chessa, S. R. (2013). Factors Influencing Farm Crime in Kenya: Opinions and Experiences of Farmers. *International Journal of Rural Criminology, 2*(1), 75–100.

Bunge, J. (2015, November 19). Food and Drug Administration Approves Genetically Modified Salmon. *The Wall Street Journal.*

Bunting, M. (2011). Small Is Beautiful—An Economic Idea that Has Sadly Been Forgotten. *Guardian UK.* Retrieved from http://www.theguardian.com/commentisfree/2011/nov/10/small-is-beautiful-economic-idea

Burlingame, B. (2000). Wild nutrition. *Journal of Food Composition and Analysis, 13*(2), 99–100. doi:10.1006/jfca.2000.0897

Bursik, R. J., & Grasmik, H. G. (1993). *Neighborhoods and Crime: Dimensions of Effective Community Control.* New York: Lexington Books.

Buttigieg, P. (2015). *South Bend's New Economic Direction.* Retrieved from Pete Buttigieg, Mayor: http://www.pete-forsouthbend.com/economicplan

CAFOD. (2014). *What is "inclusive growth"?* CAFOD Discussion Paper.

Campbell, D. A., & Barker, S. C. (1998). Selecting appropriate content method in program delivery. In Improving agricultural extension: A reference manual. FAO.

Campton, R. (2008). *The Economic Value of Wildlife in Eastern and Central Africa.* Research Paper No: 23. Institute of Development Studies (IDS), University of Dar es Salaam, Tanzania.

Cardeña, R., Oropeza, C., & Zizumbo, D. (1998). Leaf proteins as markers useful in the genetic improvement of coconut palms. *Euphytica, 102*(1), 81–86. doi:10.1023/A:1018392908569

Caribbean Food and Nutrition Institute. (2008). *Annual Report.* Retrieved from ris.paho.org/xmlui/bitstream/handle/.../AnnualReport08%20CFNI.pdf

CARIBSAVE. (2011). Caribbean Climate Change, Tourism & Livelihoods: A sectoral approach to vulnerability and resilience. The CARIBSAVE Climate Change Risk Atlas – St. Kitts Final Draft Country Risk Profile. The CARIBSAVE Partnership.

Caricom. (2011). Food Security in Caricom. *Caricom View*. Retrieved from http://www.caricom.org/jsp/communications/caricom_online_pubs/caricom_view_jul_2011.pdf

Carlton, B. C., & Gonzalez, J. M. (1985). Plasmids and δ- endotoxin production in different subspecies of *Bacillus thuringiensis*. In J. A. Hoch & P. Setlow (Eds.), *Molecular Biology of Microbial Differentiation* (pp. 246–252). Washington, DC: American Society for Microbiology.

Carozzi, N. B., Kramer, V. C., Warren, G. W., Evola, S., & Koziel, M. G. (1991). Prediction of insecticidal activity of *Bacillus thuringiensis* strains by polymerase chain reaction product profiles. *Applied and Environmental Microbiology*, *57*(11), 3057–3061. PMID:1781673

Catanzarite, M. (2016, February 24). *Mayor Buttigieg delivers "State of the City" address*. Retrieved from http://www.wndu.com/content/news/Mayor-Buttigieg-delivers-State-of-the-City-addresss-369918351.html

Causse, M., Saliba-Colombani, V., Lecomte, L., Duffé, P., Rousselle, P., & Buret, M. (2002). QTL analysis of fruit quality in fresh market tomato: A few chromosome regions control the variation of sensory and instrumental traits. *Journal of Experimental Botany*, *53*(377), 2089–2098. doi:10.1093/jxb/erf058 PMID:12324532

CDW Services. (2015). *City-Data.com*. Retrieved from South Bend, Indiana (IN) Poverty Rate: http://city-data.com/poverty/poverty-South-Bend-Indiana.html

Census. (2011). *General Census*. Office of the Register General & Census Commissioner, Ministry of Home Affairs, Government of India.

Center for the Homeless. (1988). Retrieved from Center for the Homeless: http://cfh.net/

CGIAR (Consultative Group for International Agricultural Research). (2015). *CGIAR's strategy and results framework 2016-2030*. Retrieved June 21, 2015, from http://www.cgiar.org/consortium-news/cgiars-new-strategy-and-research-programs-answering-to-poverty-health-and-climate-change/#

Chand, S., Singh, D. R., Singh, S., Srivastava, R. C., Birah, A., & Banu, S. (2012). *Vegetables production and role of farmwomen in Andaman and Nicobar Islands, India*. Paper presented at the Global Conference on women in Agriculture at New Delhi.

Chand, S., Srivastava, R. C., Krishnan, P., Jaikumar, S., Birah, A., Sethi, S. N., & Ram, N. (2011). *Impact of Tsunami Rehabilitation Programme for Reviving Livelihood of Affected People of Bay Islands, India*. 6th National Extension Education Congress, Old Goa, India.

Chandler, D., Bailey, A. S., Tatchell, G. M., Davidson, G., Greaves, J., & Grant, W. P. (2011). The development, regulation and use of biopesticides for integrated pest management. *Philosophical Transactions of the Royal Society of London. Series B, Biological Sciences*, *366*(1573), 1987–1998. doi:10.1098/rstb.2010.0390 PMID:21624919

Chandrashekara, K. N., Raju, S. R., & Chandrashekara, C. (2012). Molecular and Biotechnological Approaches in Plant Disease Management. In Ecofriendly Innovative Approaches in Plant Disease Management. International Book Distributors.

Chand, S., Krishnan, P., & Sethi, S. N. (2013). Economic impact assessment of rehabilitation scheme of Tsunami affected farmers of Bay Islands in India. *Indian Research Journal of Extension Education.*, *13*(1), 56–62.

Chand, S., Singh, S., Parappurathu, S., Roy, S. D., & Kumar, A. (2015). Explaining the status and scope of ecotourism development for livelihood security: Andaman and Nicobar Islands, India. *International Journal of Sustainable Development and World Ecology*, *22*(4), 335–345. doi:10.1080/13504509.2015.1050478

Chappell, M. J., Wittman, H., Bacon, C. M., Ferguson, B. G., Barrios, L. G., Barrios, R. G., & Perfecto, I. (2013). Food sovereignty: An alternative paradigm for poverty reduction and biodiversity conservation in Latin America. *F1000 Research*, *2*, 235. doi:10.12688/f1000research.2-235.v1 PMID:24555109

Charlton, A. J., Dickinson, M., Wakefield, M. E., Fitches, E., Kenis, M., & Han, R. et al. (2015). Exploring the chemical safety of fly larvae as a source of protein for animal feed. *Journal of Insects as Food and Feed*, *1*(1), 7–16. doi:10.3920/JIFF2014.0020

Chatenet, M., Delage, C., Ripolles, M., Irey, M., Lockhart, B. L. E., & Rott, P. (2001). Detection of sugarcane yellow leaf virus in quarantine and production of virus-free sugarcane by apical meristem culture. *Plant Disease*, *85*(11), 1177–1180. doi:10.1094/PDIS.2001.85.11.1177

Chaudhry, K. M., Muhammad, S., & Ashraf, I. (2006). Alternative Extension Approaches to Technology Dissemination for Sustainable Agriculture in The Punjab, Pakistan. *International Journal of Agriculture and Biology*, *8*(6), 832–835.

Chaudhury, M., Vervoort, J., Kristjanson, P., Ericksen, P., & Ainslie, A. (2013). Participatory scenarios as a tool to link science and policy on food security under climate change in East Africa. *Regional Environmental Change*, *13*(2), 389–398. doi:10.1007/s10113-012-0350-1

Cheng, A., Mayes, S., Dalle, G., Demissew, S., & Massawe, F. (2015). Diversifying crops for food and nutrition security – a case of teff. *Biological Reviews of the Cambridge Philosophical Society*, n/a. doi:10.1111/brv.12225 PMID:26456883

Choi, J. Y., Roh, J. Y., Kang, J. N., Shim, H. J., Woo, S. D., Jin, B. R., Je, Y. H. (2005). Genomic segments cloning and analysis of *Cotesia plutellae* polydnavirus using plasmid capture system. *Biochemical Biophysics Research Communications, 332*, 487-493.

Choi, J. Y., Roh, J. Y., Li, M. S., Shim, H. J., Kang, J. N., Woo, S. D., & Je, Y. H. et al. (2004). Cloning of small plasmids from *Bacillus thuringiensis* subsp. israelensis using plasmid capture system. *International Journal of Industrial Entomology*, *9*(2), 183–186.

Choi, Y. W., Cho, J. L., & Kang, S. M. (1994). Studies on rapid multiplication of microtubers from potatoes (*Solanum tuberosum* L.) *in vitro* and their practical use. IV. Effect of coating and foliar treatments on field performance. *Korean Journal of Horticultural Science*, *35*(2), 323–329.

Cho, K. M., & Boland, H. (2004). *Education and Extension for Multi-Functional Agriculture: Extension Concepts for Sustainable Agricultural Development in Myanmar*. Dublin, Ireland: AIAEE.

Choma, C. T., & Kaplan, H. (1992). *Bacillus thuringiensis* crystal protein: Effect of chemical modification of the cysteine and lysine residues. *Journal of Invertebrate Pathology*, *59*(1), 75–80. doi:10.1016/0022-2011(92)90114-J PMID:1541850

Christensen, C. (1997). *The Innovator's Dilemma*. Harvard Business Review Press.

Christou, P. T., Capell, A., Kohli, A., Gatehouse, J. A., & Gatehouse, A. M. (2006). Recent developments and future prospects in insect pest control in transgenic crops. *Trends in Plant Science*, *11*(6), 302–308. doi:10.1016/j.tplants.2006.04.001 PMID:16690346

Chung, T. Y. (1989). *Biotechnology in the Republic of Korea*. Paper presented at the FAO Regional Expert Consultation on Biotechnology, Bangkok, Thailand.

Citizen's Fair Housing Committee of South Bend-Mishawaka. (1963). *Transcript of public hearing concerning discrimination in the sale, rental and financing of private housing in South Bend and Mishawaka: held March 19, 1963 in Notre Dame Law School Auditorium. Discrimination in housing--Indiana--South Bend.* South Bend, IN: Citizen's Fair Housing Committee of South Bend-Mishawaka.

City of South Bend Common Council. (1965). *South Bend Fair Employment Practice Ordinance. City Ordinance No. 4091 prohibiting discrimination in employment, providing for the creation of the South Bend Fair Employment Practice Commission.* South Bend, IN: City of South Bend Common Council.

City of South Bend Indiana. (2016). *Interactive Mapping-GIS.* Retrieved January 2016, from City of South Bend: http://southbendin.gov/government/content/interactive-mapping-gis-0

Civil Rights Heritage Center. (2015). *Civil Rights Heritage Center.* Retrieved from https://www.iusb.edu/civil-rights/about-us/index.php

Clarke, J. (2013). *Sugar, Land and Sustainable Livelihoods: Global Change and Local Vulnerability in St. Kitts.* (Ph.D. Thesis). University of the West Indies, Mona Campus.

Clarke, R.V. (1995). Situational Crime Prevention. *Crime and Justice, 19*, 91-150.

Clarke, R. V., & Felson, M. (Eds.). (1993). *Routine activity and rational choice.* New Brunswick, NJ: Transaction Publishers, Inc.

Clunis, A., & Sandy, K. (2008). *A review of the reproductive performance of the dairy herd at the Eastern Caribbean Institute of Agriculture and Forestry (ECIAF).* Project submitted in partial fulfillment of the Diploma in Agriculture.

Cockcroft, C. E., Herrera-Estrella, L., & Borroto Nordelo, C. G. (2004). *Agricultural Biotechnology in Latin America and the Caribbean. Handbook of Plant Biotechnology* (Vol. 11, pp. 1–10). John Wiley & Sons.

Cohen, L. E., & Felson, M. (1979). Social change and crime rates trends: A routine activities approach. *American Sociological Review, 44*(4), 588–608. doi:10.2307/2094589

Commodities and Trade Technical Paper #8. (2006). *The agricultural dimension of the ACP-EU Economic Partnership Agreements.* United Nations, Food and Agriculture Organization.

Connell, J., & Soutar, L. (2007). Free trade or free fall? Trade liberalization and development in the Pacific and Caribbean. *Social and Economic Studies, 56*(1/2), 41-66.

Consolandi, C., Castiglioni, B., Bordoni, R., Busti, E., Battaglia, C., & De Bellis, G. (2001). Development of oligonucleotide arrays to detect mutations and polymorphisms. *Minerva Biotechnology, 13*(3), 261–268.

Cooke, J. G., & Downie, R. (2010). *African perspectives on genetically modified crops: assessing the debate in Zambia, Kenya and South Africa.* A report of the CSIS global food security project. Washington, DC: Centre for Strategic and International Studies. Retrieved 8 June 2014 from, http://csis.org/files/publication/100701_Cooke_AfricaGMOs_WEB.pdf

Cooper, H. D., Spillane, C., Hodgkin, T., & Cooper, H. (2001). *Broadening the genetic base of crops: an overview. Broadening the genetic base of crop production.* Wallingford, UK: CABI Publishing. doi:10.1079/9780851994116.0001

Cox, D. N., & Evans, G. (2008). Construction and validation of a psychometric scale to measure consumers' fears of novel food technologies: The food technology neophobia scale. *Food Quality and Preference, 19*(8), 704–710. doi:10.1016/j.foodqual.2008.04.005

Craig, P. H. (2011). *Protecting you Farm from Theft. PennState Extension, Cumberland County.* Retrieved on July 7th, 2011 from: http://extension.psu.edu/cumberland/news/2011/protecting-your-farm-from-theft

Cronauer, S. S., & Krikorian, A. D. (1984). Rapid multiplication of bananas and plantains by *in vitro* shoot tip culture. *HortScience*, *19*(2), 353–380.

Datta, K. S. (2004). Rice Biotechnologies: A need for developing countries. *AgBioForum*, *7*(1), 31–35.

Dawar, N. & Chattopadhyay, A. (2002). Rethinking Marketing Programs for Emerging Markets. *Long Range Planning, 35*(5), 457–474. Retrieved from http://www.sciencedirect.com/science/journal/00246301

Dawson, I. K., Vinceti, B., Weber, J. C., Neufeldt, H., Russell, J., Lengkeek, A. G., & Jamnadass, R. et al. (2011). Climate change and tree genetic resource management: Maintaining and enhancing the productivity and value of smallholder tropical agroforestry landscapes. A review. *Agroforestry Systems*, *81*(1), 67–78. doi:10.1007/s10457-010-9302-2

de Merode, E., Homewood, K., & Cowlishaw, G. (2004). The value of bushmeat and other wild foods to rural households living in extreme poverty in Democratic Republic of Congo. *Biological Conservation*, *118*(5), 573–581. doi:10.1016/j.biocon.2003.10.005

Debergh, P. C., & Read, P. E. (1991). *Micropropagation*. Springer Netherlands. doi:10.1007/978-94-009-2075-0_1

Dekkers, J. C. M., & Hospital, F. (2002). The use of molecular genetic in the improvement of agricultural populations. *Nature Reviews. Genetics*, *3*(1), 22–32. doi:10.1038/nrg701 PMID:11823788

Demiryürek, K. (2011). The concept of organic farming and the situation of organic farming in the world and Turkey. *Journal of Agricultural Faculty of Gaziosmanpasa University*, *28*(1), 27–36.

Demiryürek, K., Stopes, C., & Güzel, C. (2008). Organic agriculture: The case of Turkey. *Outlook on Agriculture*, *37*(4), 261–267. doi:10.5367/000000008787167754

Dempewolf, H., Eastwood, R. J., Guarino, L., Khoury, C. K., Müller, J. V., & Toll, J. (2014). Adapting agriculture to climate change: A global initiative to collect, conserve, and use crop wild relatives. *Agroecology and Sustainable Food Systems*, *38*(4), 369–377. doi:10.1080/21683565.2013.870629

Department of Community & Economic Development. (2011). *2010 Census Report: South Bend-St. Joseph County Community Overview*. South Bend, IN: City of South Bend Department of Community & Economic Development Division of Community Development. Retrieved from https://www.southbendin.gov/sites/default/files/files/CI_Census2010Report.pdf

DESA. (2013). World Economic and Social Survey 2013: Sustainable Development Challenges. New York: United Nations Publication.

Development Report. (2008). *Development Report*. Andaman and Nicobar Islands: Planning Commission, Government of India.

DFID. (2012). *Growth Building Jobs and Prosperity in Developing Countries*. Retrieved July 4, 2015, from http://www.oecd.org/derec/unitedkingdom/40700982.pdf

Dharmasiri, L. M. (2009). Measuring agricultural productivity using average productivity index. *Sri Lanka Journal of Advanced Social Studies*, *2*(1), 25–44.

Dhingra, K. (2005). *The Andaman and Nicobar Islands in the 20th Century a Gazetteer*. New Delhi: Oxford University Press.

Dickson, D. (2013). Praedial Larceny - A Serious Crime. Trinidad and Tobago Chamber of Industry and Commerce. *Feature*, *13*(1), 26–27.

Directorate of Economics and Statistics Census. (2011). *Andaman and Nicobar Islands*. Port Blair, India.

Dixie, G. (2005). Horticultural Marketing. Rome: Food and Agriculture Organization of the United Nations. Retrieved from http://www.fao.org/docrep/008/a0185e/a0185e05.htm#bm05

Dixon, J., Taniguchi, K., Wattenbach, H., & Tanyeri-Arbur, A. (2004). Smallholders, Globalization and Policy Analysis. Rome: Food and Agriculture Organization of the United Nations. Retrieved from http://www.fao.org/docrep/007/y5784e/y5784e02.htm

Dodds, J. H. (1988). Tissue culture technology: Practical application of sophisticated methods. *American Potato Journal*, *65*(4), 167–180. doi:10.1007/BF02854450

Doğanay, H. (2007). *Agricultural geography*. Erzurum, Turkey: Aktif Press.

Dollar, D., & Kraay, A. (2004). Trade growth and poverty. *The Economic Journal*, *114*(493), F22–F49.

Donnelly, D. J., Coleman, W. K., & Coleman, S. E. (2003). Potato microtuber production and performance: A review. *American Journal of Potato Research*, *80*(2), 103–115. doi:10.1007/BF02870209

Donnermeyer, J., & Scott, J. (2013). How Rural Criminology Informs Critical Thinking. *International Journal for Crime, Justice and Social Democracy*. Retrieved from http://eprints.qut.edu.au/68038/2/C1_Donnermeyer_Rural_Criminology_Pub_Paper.pdf

Donnermeyer, J. F., & Barclay, E. M. (2005). The policing of farm crime. *Police Practice and Justice Research*, *6*(1), 3–17. doi:10.1080/15614260500046913

Donovan, K. (2012). *Agricultural risk, intermediate inputs and cross-country Productivity Differences*. Job Market Paper.

dos Santos, I. C., de Almeida, A. A. F., Anhert, D., da Conceição, A. S., Pirovani, C. P., Pires, J. L., & Baligar, V. C. et al. (2014). Molecular, physiological and biochemical responses of *Theobroma cacao* L. genotypes to soil water deficit. *PLoS ONE*, *9*(12), e115746. doi:10.1371/journal.pone.0115746 PMID:25541723

DUFI. (2015, August 15). Retrieved from Distributed Urban Farm Initiative: A community and economic growth project by Advent GX: http://dufi.adventgx.com/

Durst, P. B., & Shono, K. (2010). Edible Forest Insects: exploring new Horizons and traditional practices. In P. B. Durst, D. V. Johnson, R. N. Leslie, & K. Shono. (Eds.), *Forest insects as food: humans bite back. Proceedings of a workshop on Asia -Pacific resources and their potential for food development.*

Dutta, S. (2015). Biopesticides: An ecofriendly approach for pest control. *World Journal of Pharmacy and Pharmaceutical Sciences*, *4*(6), 250–265.

Ebai, S. (2004). Global coordination towards a sustainable cocoa economy: an update. In *Proceedings of the APASTT Seminar – Exhibition entitled Re-vitalisation of the Trinidad & Tobago Cocoa Industry*, (pp. 17-21). St. Augustine, Trinidad: APASTT.

Edelman, M. (2014). Food Sovereignty: Forgotten Genealogies and Future Regulatory Challenges. *The Journal of Peasant Studies*, *41*(6), 959–978. doi:10.1080/03066150.2013.876998

Education, G. M. (2014). *A brief history of genetic modification*. Retrieved 19 July 2014 from, http://www.gmeducation.org/faqs/p149248-20brief%20history%20of%20genetic%20modification.html

Edwards, D., & Batley, J. (2010). Plant genome sequencing: Applications for crop improvement. *Plant Biotechnology Journal*, *8*(1), 2–9. doi:10.1111/j.1467-7652.2009.00459.x PMID:19906089

Edwards, M. D., Stuber, C. W., & Wendel, J. F. (1987). Molecular marker-facilitated investigations of quantitative-trait loci in maize. I. Numbers, genomic distribution and types of gene action. *Genetics*, *116*(1), 113–125. PMID:3596228

Eisner, G. (1961). *Jamaica 1830 - 1930, A Study in Economic Growth*. Manchester, UK: Manchester University Press.

Ejeta, G. (2009). Revitalizing agricultural research for global food security. *Food Security*, *1*(4), 391–401. doi:10.1007/s12571-009-0045-8

Engel, J., & Jouanjean, M.-A. (2013). *The history, impact and political economy of barriers to food trade in sub-Saharan Africa: an analytical review*. London: ODI. Retrieved 14 June 2014 from, http://www.odi.org/sites/odi.org.uk/files/odi-assets/publications-opinion-files/8803.pdf

Engledow, F. L. (1945). *West India Royal Commission: Report on Agriculture, Fisheries, Forestry and Veterinary Matters. Cmd. 6608*. London: His Majesty's Stationery Office.

Eskes, A. B., Engels, J. M. M., & Lass, R. A. (Eds.). (1998). Working procedures for cocoa germplasm evaluation and selection. In *Proceedings of the CFC/ICCO/IPGRI Project workshop* (*vol. 1*, no. 6). Rome, Italy: IPGRI.

Eskes, A. B. (2011). *Collaborative and participatory approaches to cocoa variety improvement. Final report of the CFC/ICCO/Bioversity project on cocoa productivity and quality improvement: a participatory approach (2004–2010)*. London, UK: Bioversity International.

Eskes, B., & Efron, Y. (Eds.). (2006). *Global Approaches to Cocoa Germplasm Utilization and Conservation. Final Report of CFC/ICCO/IPGRI Project on Cocoa Germplasm Utilization and Conservation (1998-2004)*. Rome, Italy: Bioversity International.

Esri. (2008). *GIS for sustainable agriculture* [PDF document]. Retrieved from http://www.esri.com/library/bestpractices/sustainable-agriculture.pdf

ESRI. (2011, Spring). *South Bend, Indiana, Uses GIS for Brownfields Inventory*. Retrieved from ArcNews: http://www.esri.com/news/arcnews/spring11articles/south-bend-indiana-uses-gis-for-brownfields-inventory.html

Ewah, S. O. E., & Ekeng, A. B. (2009). Problems and Prospects of Marketing in Developing Economies: The Nigerian Experience. *International Journal of Business and Management*, *4*(9). doi:10.5539/ijbm.v4n9p187

Fagwalawa, L. D., Kutama, A. S., & Yakasai, M. T. (2013). Current Issues in Plant Disease Control: Biotechnology and Plant Disease. *Bayero Journal of Pure and Applied Sciences*, *6*(2), 121–126. doi:10.4314/bajopas.v6i2.26

Fahlgren, N., Feldman, M., Gehan, M. A., Wilson, M. S., Shyu, C., Bryant, D. W., . . . Baxter, I. (2015). A versatile phenotyping system and analytics platform reveals diverse temporal responses to water availability in Setaria. *Molecular plant*. Retrieved June 2, 2015, from http://www.sciencedirect.com/science/article/pii/S1674205215002683

Falvery, R., Foster, N., & Greenaway, D. (2010). *Trade liberalization, economic crises, and growth*. Bond University.

Falvey, R., Foster, N., & Greenaway, D. (2012). Trade Liberalization, Economic Crises, and Growth. *World Development*, *40*(11), 2177–2193. doi:10.1016/j.worlddev.2012.03.020

Famuyiwa, B. S. (2013). *Assessment of Environmental hazards associated with cocoa farming in Nigeria*. (Unpublished Ph.D. thesis). Department of Agric., Extension and Rural Development, O.A.U. Ife.

Famuyiwa, B.S., Torimiro, D.O., & Adesoji, S.A. (2013). Determination of Cocoa Farmers' Knowledge on Environmental Hazards Associated with Cocoa Farming Operations in Nigeria. *International Journal of Knowledge, Innovation and Entrepreneurship*, *1*(1-2), 58-70.

FAO (Food and Agriculture Organization). (2004). *Carbon sequestration in dryland soils*. Retrieved August 30, 2015 from http://www.fao.org/fileadmin/templates/nr/images/resources/pdf_documents/wsrr102.pdf

FAO. (2005*).Food and Agriculture Organization of the United Nations.Environment and Natural Resources in Small Island Developing States*. Overview prepared for a Special Ministerial Conference on Agriculture in Small Island Developing States, Rome, Italy.

FAO. (2007). Adaptation to climate change in agriculture, forestry and fisheries: Perspective, framework and priorities. Rome: Interdepartmental Working Group on Climate Change, Food and Agriculture Organization of the United Nations (FAO).

FAO. (2008). *Climate change adaptation and mitigation in the food and agriculture sector. High-level conference on food security – the challenges of climate change and bioenergy*. Retrieved June 17, 2015, from ftp://ftp.fao.org/docrep/fao/meeting/013/ai782e.pdf

FAO. (2009). *High level expert forum, How to feed the world in 2050: The special challenge for sub Saharan Africa*. Retrieved 12 July 2014 from, http://www.fao.org/fileadmin/templates/wsfs/docs/Issues_papers/HLEF2050_Africa.pdf

FAO. (2009a). *Increasing crop production sustainability, the perspective of biological processes*. Retrieved June 18, 2015, from http://www.fao.org/docrep/012/i1235e/i1235e00.htm

FAO. (2009b). *How to feed the world in 2050*. Rome. Retrieved June 18, 2015, from www.fao.org/fileadmin/templates/wsfs/docs/expert_paper/How_to_Feed_the_World_in_2050.pdf

FAO. (2009c). *Climate change and biodiversity for food and agriculture: taking systemic and second order effects into account*. Rome: Commission on Genetic Resources for Food and Agriculture. Retrieved from June 18, 2015, from ftp://ftp.fao.org/docrep/fao/meeting/017/ak223e.pdf

FAO. (2010a). An international consultation on integrated crop-livestock systems for development, the way forward for sustainable production intensification. *Integrated Crop Management*, *13*, 208.

FAO. (2010b). *Gardens of biodiversity – conservation of genetic resources and their use in traditional food production systems by small farmers of the Southern Caucasus*. Rome: Food and Agriculture Organization of the United Nations.

FAO. (2011). *Save and grow: a policymaker's guide to the sustainable intensification of smallholder crop production*. Retrieved June 17, 2015, from http://www.fao.org/ag/save-and-grow/

FAO. (2012). *Statistical Yearbook, Part 4, Genetically Modified crops*. Rome. Retrieved 29 May 2014 from, http://www.fao.org/docrep/015/i2490e/i2490e04d.pdf

FAO. (2012). *Trends and Impacts of Foreign Investment in Developing Country Agriculture*. Rome: FAO. Retrieved from http://www.fao.org/fileadmin/templates/est/INTERNATIONAL-TRADE/FDIs/Trends_publication_12_November_2012.pdf

FAO. (2012). *What is conservation agriculture?* Retrieved June 18, 2015, from http://www.fao.org/ag/ca/1a.html

FAO. (2013). *A Tool for Gender-sensitive Agriculture and Rural Development Policy and Programme Formulation: Guidelines for Ministries of Agriculture and FAO*. Rome: FAO.

FAO. (2013). *Climate-smart agriculture sourcebook*. Rome: Food and Agriculture Organization of the United Nations.

FAO. (2013). *FAO Policy on Gender Equality: Attaining Food Security Goals in Agriculture and Rural Development*. Rome: FAO.

FAO. (2014). *The state of food insecurity in the world*. Rome. Retrieved June 17, 2015, from http://www.fao.org/3/a-i4030e.pdf

FAO. (2015). *Five agents of change for a sustainable world.* Accessed on 24 September 2015 from, http://www.fao.org/post-2015-mdg/news/detail-news/en/c/281554/

FAO. (2015). *Saint Kitts and Nevis and FAO.* Food and Agriculture Organization of the United Nations. Retrieved from http://www.fao.org/3/a av273e.pdf

FAO. (2015). The State of Food Insecurity in the World 2015. *Meeting the 2015 International Hunger Targets: taking stock of uneven progress.* Rome: FAO. Retrieved 26 January 2016, from http://www.fao.org/3/a-i4646e.pdf

FAO/WUR. (2014). Insects to feed the world: summary report. In P. Vantomme, C. Münke, A. van Huis, J. van Itterbeeck, & A. Hakman (Eds.), *Insects to Feed the World.* Ede, Netherlands: FAO/WUR.

FAO-PAR (Platform for Agrobiodiversity). (2011). *Biodiversity for food and agriculture: contributing to food security and sustainability in a changing world.* Retrieved June 16, 2015, from http://agrobiodiversityplatform.org/files/2011/04/PAR-FAO-book_lr.pdf)

Farming Matters. (2013). *Family farming – a way of life.* ILEIA, Center for Learning and Sustainable Agriculture. Retrieved from www.farmingmatters.org

Felipe, J., & Hasan, R. (2006). *The Challenge of job creation in Asia.* Asian Development Bank: ERD Policy Brief No. 44.

Felson, M. (2002). *Crime and everyday life* (3rd ed.). London: Sage.

Fereres, E., Orgaz, F., & Gonzalez-Dugo, V. (2011). Reflections on food security under water scarcity. *Journal of Experimental Botany*, *62*(12), 4079–4086. doi:10.1093/jxb/err165 PMID:21624976

Fernandes, J. M., Cunha, G. R., Del Ponte, E., Pavan, W., Pires, J. L., Baethgen, W., ... Travasso, M. I. (2004). Modelling Fusarium head blight in wheat under climate change using linked process-based models. In *Proceedings of the 2nd International Symposium on Fusarium Head Blight* (pp. 441-444). East Lansing, MI: Michigan State University.

Finger, R., El Benni, N., Kaphengst, T., Evans, C., Herbert, S., Lehmann, B., . . . Stupak, N. (2011). A meta analysis on farm-level costs and benefits of GM crops. *Sustainability, 3*, 743–62. Retrieved 6 July 2014 from, http://www.mdpi.com/2071-1050/3/5/743

Flachsbarth, I., Willaarts, B., Xie, H., Pitois, G., Mueller, N. D., Ringler, C., & Garrido, A. (2015). The role of Latin America's land and water resources for global food security: Environmental Trade-Offs of Future Food Production Pathways. *PLoS ONE, 10*(1), e0116733. doi:10.1371/journal.pone.0116733 PMID:25617621

Food & Agriculture Organization (2013). *Praedial Larceny in the Caribbean.* Issue Brief #3.

Food and Agriculture Organization (FAO). (1996). *Global Plan of Action for the Conservation and Sustainable Utilization of Plant Genetic Resources for Food and Agriculture.International Technical Conference on Plant Genetic Resources Leipzig.*

Food and Agriculture Organization (FAO). (2005). *Impact of Climate Change, Pests and Diseases on Food Security and Poverty Reduction.* Special Event Background Document for the 31st Session of the Committee on World Food Security.

Food and Agriculture Organization (FAO). (2006). *World agriculture: towards 2030/2050. Food and Agriculture Organization of the United Nations.* Rome: FAO.

Food and Agriculture Organization (FAO). (2009a). How to Feed the World in 2050. Rome, Italy, Food and Agriculture Organization.

Food and Agriculture Organization (FAO). (2009b). The State of Food Insecurity in the World. Rome, Italy: Food and Agriculture Organization.

Food and Agriculture Organization (FAO). (2013). *Edible insects: Future prospects for food and feed security*. Forestry paper, 171. FAO.

Food and Agriculture Organization of the United Nations (FAO). (2014). *Statistical yearbook 2014: Europe and Central Asia food and agriculture*. Retrieved from http://www.fao.org/3/a-i3621e.pdf

Food and Agriculture Organization. (2003). Trade reforms and food security: Conceptualizing the linkages. United Nations, FAO.

Food and Agriculture Organization. (2006). *Agriculture, trade negotiations and gender*. Rome: United Nations FAO Publications.

Food Bank of Northern Indiana. (2016). *Foodbank of Northern Indiana*. Retrieved from http://www.feedindiana.org/mission

Foodbank of Northern Indiana. (2015). *About the Food Bank*. Retrieved from Foodbank of Northern Indiana: http://feedindiana.org/

FoodTank.com. (2013, October 12). Retrieved from Five Examples of Successful Urban Agriculture Done Differently Around the World: http://foodtank.com/news/2013/10/five-different-examples-of-urban-agriculture-from-around-the-world

Forsberg, C. W., Crosby, B., & Thomas, D. Y. (1986). Potential for manipulation of the rumen fermentation through the use of recombinant DNA techniques. *Journal of Animal Science*, *63*, 310–325. PMID:3525493

Fraley, R. T., Rogers, S. G., Horsch, R. B., Sanders, P. R., Flick, J. S., Adams, S. P., & Fry, J. S. (1983). Expression of bacterial genes in plant cells. *Proceedings of the National Academy of Sciences of the United States of America*, *80*(15), 4803–4807. doi:10.1073/pnas.80.15.4803 PMID:6308651

Fujardo, J. (2010). Quality Declared Planting Material: Protocols and Standards for Vegetatively Propagated Crops.[). Rome]. *FAO Plant Production and Protection Paper*, *195*, 156.

Fulginiti, L., & Perrin, R. (1998). Agricultural productivity in developing countries. *Agricultural Economics*, *19*(1-2), 45–51. doi:10.1016/S0169-5150(98)00045-0

Fussel, H. M., & Klein, R. J. T. (2006). Climate change vulnerability assessments: An evolution of conceptual thinking. *Climatic Change*, *75*(3), 301–329. doi:10.1007/s10584-006-0329-3

Galyuon, I. K. A., McDavid, C. R., Lopez, F. B., & Spence, J. A. (1996a). The effect of irradiance level on cocoa (Theobroma cacao L.): I. Growth and leaf adaptations. *Tropical Agriculture, 73*(1), 23-28.

Galyuon, I. K. A., McDavid, C. R., Lopez, F. B., & Spence, J. A. (1996b). The effect of irradiance level on cocoa (Theobroma cacao L.): II. Gas exchange and chlorophyll fluorescence. *Tropical Agriculture, 73*(1), 29-33.

Ganpat, W. G., & Isaac, W. P. (Eds.). (2014). *Impacts of climate change on food security in Small Island Developing States*. Hershey, PA: IGI Global.

Garcia, G. W., Young, G. M., Amour, K. M., James, D., Lallo, C. H. O., & Mollineau, W. et al. (2005). *The Collared Peccary/Javelina/Sajina/Poco do Monte/Wild Hog/Pakira/ Patira/Taitetu/Catete/Catto/Quenk (Tayassu tajacu, Pecari tajacu) Booklet and Producers' Manual. GWG Publications.*

Garcia-Gonzales, R., Quiroz, K., Carrasco, B., & Caligari, P. (2010). Plant tissue culture: Current status, opportunities and challenges. *Cien. Inv. Agr.*, *37*(3), 5–30. doi:10.4067/S0718-16202010000300001

Garriott, W. (2011). *Policing methamphetamine: Narcopolitics in rural America*. New York University Press.

Gennaro, A., Forte, P., Carozza, R., Savo Sardaro, M. L., Ferri, D., Bitti, A., & Ceoloni, C. et al. (2007). Pyramiding different alien chromosome segments in durum wheat: Feasibility and breeding potential. *Israel Journal of Plant Sciences*, *55*(3-4), 267–276. doi:10.1560/IJPS.55.3-4.267

Ghanem, M. E., Marrou, H., & Sinclair, T. R. (2015). Physiological phenotyping of plants for crop improvement. *Trends in Plant Science*, *20*(3), 139–144. doi:10.1016/j.tplants.2014.11.006 PMID:25524213

Girvan, N. (1973). The development of dependency economics in the Caribbean and Latin America: Review and comparison. *Social and Economic Studies*, *1*(22), 1–33.

Global Hunger Index (GHI). (2014). *The Challenge of Hidden Hunger*. Welthungerhilfe, IFPRI, Concern Worldwide.

Glover, D., & Sexton, A. (2015). *Edible Insects and the Future of Food: A Foresight Scenario Exercise on Entomophagy and Global Food Security. IDS Evidence Report 149*. Brighton, UK: IDS.

GMDC. (2015, September 10). *Greenpoint Manufacturing and Design Center*. Retrieved from http://www.gmdconline.org/buildings/810-humboldt-street

Gök, S. A. (2008). *An assessment of trade in Turkish organic products in the expanding market of European Union*. (Expertise thesis). Retrieved from Republic of Turkey Ministry of Food, Agriculture and Livestock expertise thesis database. (50)

Gómez Kosky, R., Manuel de Feria, S., Posada Pérez, L., Gilliard, T., Martínez, F. B., Maritza Reyes, V., & Mendoza, E. Q. (2002). Somatic embryogenesis of the banana hybrid cultivar FHIA-18 (AAAB) in liquid medium and scaled-up in a bioreactor. *Plant Cell, Tissue and Organ Culture*, *68*(1), 21–26. doi:10.1023/A:1012905825307

Göney, S. (1979). *Geographical essentials of Turkey's agriculture*. Istanbul, Turkey: Istanbul University Press.

Gonzalez, J. M. J., Brown, B. J., & Carlton, B. C. (1982). Transfer of *Bacillus thuringiensis* plasmids coding for δ- endotoxin among strains of *B. thuringiensis* and *B. cereus*. *Proceedings of the National Academy of Sciences of the United States of America*, *79*(22), 6951–6955. doi:10.1073/pnas.79.22.6951 PMID:6294667

Gonzalez, J. M. J. Jr, & Carlton, B. C. (1980). Patterns of plasmid DNA in crystalliferous and acrystalliferous strains of *Bacillus thuringiensis*. *Plasmid*, *3*(1), 92–98. doi:10.1016/S0147-619X(80)90038-4 PMID:7335824

Goodison, P. (2007). EU trade policy and the future of Africa's trade relationship with the EU. *Review of African Political Economy*, *34*(112), 247–266. doi:10.1080/03056240701449646

Government Technology. (2016). *Disruptive Technology that Could Transform Government-Citizen Relationships*. Retrieved from Government Technology: http://www.govtech.com/budget-finance/The-Disruption-Pipeline-Government-Citizen-Relationships.html

Grassman, J. (2011). *War of Words, War of Stories: Rural Thought and Violence in Colonial Zanzibar*. Indiana University Press.

Green, C. (1992). An overview of production and supply trends in the U.S. specialty vegetable market. *Acta Horticulturae*, *318*(318), 41–45. doi:10.17660/ActaHortic.1992.318.5

Greensgrow. (2016). *What is Urban Farming*. Retrieved from http://www.greensgrow.org/: http://www.greensgrow.org/urban-farm/what-is-urban-farming/

Grigg, D. (1995). *An introduction to agricultural geography*. London: Routledge.

Grossman-Greene, S., & Bayer, C. (2009). *A History of child labor, child rights, & the Harkin-Engel Protocol*. New Orleans, LA: Tulane University.

Grossman, L. S. (1997). *The Political Ecology of Bananas: Contract Farming, Peasants and Agrarian Change in the Eastern Caribbean*. Chapel Hill, NC: The University of North Carolina Press.

Guarino, L., & Lobell, D. B. (2011). A walk on the wild side. *Nature Climate Change*, *1*(8), 374–375. doi:10.1038/nclimate1272

Gunatilke, W. D. (2003). Agricultural marketing system in Sri Lanka. *Agricultural Marketing*, *45*(4), 55.

Gundersen, C., Kreider, B., & Pepper, J. (2011). The Economics of Food Insecurity in the United States. *Applied Economic Perspectives and Policy*, *33*(3), 281–303. doi:10.1093/aepp/ppr022

Gupta, S., & Dikshit, A. K. (2010). Biopesticides: An ecofriendly approach for pest control. *Journal of Biopesticides*, *4*(6), 186–188.

Gur, A., & Zamir, D. (2004). Unused natural variation can lift yield barriers in plant breeding. *PLoS Biology*, *2*(10), 1610–1615. doi:10.1371/journal.pbio.0020245 PMID:15328532

Gustavsson, J., Cederberg, C., Sonesson, U., van Otterdijk, R., & Meybeck, A. (2011). Global Food Losses and Food Waste: Extent Causes and Prevention. Rome: Food and Agriculture Organization (FAO) of the United Nations.

Haddad, W. (2007). *ICTs for Education: A Reference Handbook*. Accessed on October 12, 2008 from, http://www.ictinedtoolkit.org/usere/p_page.php?section_number=0

Hafez, J. (1987). *Reproduction in Farm Animals* (5th ed.). Lea and Febiger.

Han, S. Y., & Kim, T. J. (1989). Can expert systems help with planning? *Journal of the American Planning Association*, *55*(3), 296–308. doi:10.1080/01944368908975417

Harvey, C. A., Chacon, M., Donatti, C. I., Garen, E., Hannah, L., Andrade, A., & Wollenberg, E. et al. (2014). Climate-Smart Landscapes: Opportunities and Challenges for Integrating Adaptation and Mitigation in Tropical Agriculture. *Conservation Letters*, *7*(2), 77–90. doi:10.1111/conl.12066

Hausmann, R., Rodrik, D., & Velasco, A. (2005). *Growth Diagnostics.* Inter-American Development Bank. Retrieved October 9, 2013, from http://www.hks.harvard.edu/fs/rhausma/new/growthdiag.pdf

Havnevik, K., Bryceson, D., & Birgegard, L.-E. (2007). *African agriculture and the World Bank: development or improvement?* Retrieved from www.eldis.org/go/topics&id=34737&ttpe=Document#.Vws_DjY1gwg

Hawley, A. (1950). *Human Ecology: A Theory of Community Structure*. New York: Ronald Press.

Heaven, D. (2010). *Local sugar facing an uncertain future-divestment not as sweet as was projected.* Retrieved from http://jamaica-gleaner.com/article/news/20151002/local-sugar-facing-uncertain-future-divestment-not-sweet-was-projected

Heisey, P. W., & Rubenstein, K. D. (2015). *Using Crop Genetic Resources To Help Agriculture Adapt to Climate Change: Economics and Policy (No. 202351).* Washington, DC: United States Department of Agriculture, Economic Research Service.

Henderson, E., & Van En, R. (2007). *Sharing the harvest: a citizen's guide to Community Supported Agriculture*. Chelsea Green Publishing.

Hittalmani, S., Parco, A., Mew, T. V., Zeigler, R. S., & Huang, N. (2000). Fine mapping and DNA marker-assisted pyramiding of the three major genes for blast resistance in rice. *Theoretical and Applied Genetics*, *100*(7), 1121–1128. doi:10.1007/s001220051395

Hodgson, K., Campbell, M. C., & Bailkey, M. (2011). *Urban Agriculture: Growing Healthy, Sustainable Places*. APA Planning Advisory Service.

Hosein, A., Paul, C., Roach-Benn, C., Borely, J., Blair, T. M., & Fearon, A. (2013). The Small Ruminant Industry in CARICOM countries with particular reference to Jamaica and Trinidad & Tobago. St Augustine, Trinidad and Tobago: Caribbean Agricultural Research and Development Institute.

Hu, A., Acosta, A., McDaniel, A., & Gittelsohn, J. (2013, January). Community Perspectives on Barriers and Strategies for Promoting Locally Grown Produce from an Urban Agriculture Farm. *Health Promotion Practice*, *14*(1), 69–74. doi:10.1177/1524839911405849 PMID:22102602

Huang, J., Pray, C., & Rozelle, S. (2002). *Enhancing the crops to feed the poor. Nature, 418,* 678–84. Retrieved 12 July 2014 from, http://www.nature.com/nature/journal/v418/n6898/pdf/nature01015.pdf

Huang, J., Hu, R., Rozelle, S., & Pray, C. (2005). Insect-resistant GM rice in farmers' fields: Assessing productivity and health effects in China. *Science*, *308*(5722), 688–690. doi:10.1126/science.1108972 PMID:15860626

Huang, N., Angeles, E. R., Domingo, J., Magpantay, G., Singh, S., Zhang, G., & Khush, G. S. (1997). Pyramiding of bacterial blight resistance genes in rice: Marker-assisted selection using RFLP and PCR. *Theoretical and Applied Genetics*, *95*(3), 313–320. doi:10.1007/s001220050565

Hubert, B., Rosegrant, M., Van Boekel, M. A., & Ortiz, R. (2010). *The future of food: scenarios for 2050*. Retrieved June 11, 2015, from https://dl.sciencesocieties.org/publications/cs/abstracts/50/Supplement_2/S-33

Human Relations Commission Praises County Housing Position. (1968, March 10). *The Reformer Newspaper*.

Hunt, D., & Lipton, M. (2011). *Green Revolutions for Sub-Saharan Africa?* Briefing Paper AFP BP 2011/01. Chatham House, Africa Programme. Retrieved 17 July 2014 from, www.chathamhouse.org/sites/default/files/public/Research/Africa/bp0111greenrevolution.pdf

Hurungo, J. (2006). An outline and analysis of EU export subsidies on production of export interest to ESA countries. Academic Press.

Hutchinson, S., Gomes, C., Alleyne, D., & Phillips, W. (2013). *An assessment of the economic and social impacts of climate change on the agriculture sector in the Caribbean*. Port of Spain: UNECLAC. Retrieved September 27, 2015, from http://200.9.3.103/portofspain/noticias/documentosdetrabajo/5/49705/Agriculture.pdf

Ianchovichina, E., & Lundstrom, S. G. (2009). *Inclusive Growth analytics framework and application*. Policy Research Working Paper 4851. World Bank.

ICCO. (2008). *Manual of best known practices in cocoa production. Version 1*. Consultative Board on the world cocoa economy sixteenth meeting Berlin, Germany.

IFAD. (2013). *Smallholder farmers key to lifting over one billion people out of poverty*. International Fund for Agricultural Development. Retrieved from http://www.ifad.org/media/press/2013/27.htm

IFPRI (International Food Policy Research Institute). (2009). *Climate change: impact on agriculture and costs of adaptation*. International Food Policy Research Institute. Retrieved July 1, 2015, from http://www.ifpri.org/publication/climate-change-impact-agriculture-and-costsadaptation

IFPRI. (2008). *Land Tenure & Agricultural Productivity in Uganda*. Kampala: IFPRI.

IFPRI. (2011). *Priority setting Research & extension adoption evaluation; Engendering agricultural research development, and extension*. 10.2499/9780896291904as

IFPRI. (2012). *Rural-Urban Transformation in Uganda.* Kampala: IFPRI. Retrieved December 10, 2014, from http://www.ifpri.org/sites/default/files/publications/usspwp10.pdf

IFPRI. (2013). *Genetically Modified Crops in Africa: Economic and Policy Lessons from Countries South of the Sahara.* International Food Policy Institute Issue Brief 80. Retrieved 12 August 2014 from, http://www.ifpri.org/sites/default/files/publications/ib80.pdf

IFPRI. (2013). *Total and Partial Factor Productivity.* Retrieved March 4, 2015, from http://www.ifpri.org/sites/default/files/gfpr2012_indicators_tfp.pdf

Indiana Demographics. (2016). *Indiana Cities by Population.* Retrieved from http://www.indiana-demographics.com/cities_by_population

Institute, I. F. P. R. (2000). 2000 IFPRI Annual Report *Annual report.*Izquierdo, J., & de la Riva, G. A. (2000). Plant biotechnology and food security in Latin America and the Caribbean. *Electronic Journal of Biotechnology, 3*(1), 1–8.

International Food Policy Research Institute (IFAD). (2007). Impact of trade liberalization on agriculture in the Near East and North Africa. Rome, Italy: Author.

International Labor Organization (ILO). (2014). *Global Employment Trends 2014: The risk of a Jobless Recovery.* Accessed March 30, 2015, from http://www.ilo.org/global/research/global-reports/global-employment-trends/2014/WCMS_234107/lang--en/index.htm

International Labor Organization (ILO). (2015). *Agriculture; Plantations; other rural sectors.* Accessed March 10, 2015, from http://ilo.org/global/industries-and-sectors/agriculture-plantations-other-rural-sectors/lang--en/index.htm

International Service for the Acquisition of Agri-Biotech Applications (ISAAA). (2010). *Are GM crops Africa's path to food security?* Retrieved 6 June 2014 from, http://www.voanews.com/content/are-gm-crops-africas-path-to-food-security-64567512/157007.html

International Service for the Acquisition of Agri-Biotech Applications (ISAAA). (2012). *S. Africa GM crop area hit record in 2011: Report.* Reuters. Retrieved 12 August 2014 from, http://af.reuters.com/article/investingNews/idAFJOE82705320120308

IPCC (Intergovernmental Panel on Climate Change). (2007). Technical Summary. In *Climate change (2007): mitigation. Contribution to Working Group III to the Fourth Assessment Report of IPCC.* Cambridge, UK: Cambridge University Press.

IPCC. (2001). *Climate Change (2001): The Scientific Basis.* Cambridge, UK: Cambridge University Press.

IPCC. (2013). *Intergovernmental Panel on Climate Change.* Retrieved June 2, 2015, from http://www.ipcc.ch/

Isaac, W. A. P., Brathwaite, R. A., Cohen, J. E., & Bekele, I. (2007). Effects of alternative weed management strategies on *Commelina diffusa* Burm. infestations in Fairtrade banana (*Musa* spp.) in St. Vincent and the Grenadines. *Crop Protection (Guildford, Surrey), 26*(8), 1219–1225. doi:10.1016/j.cropro.2006.10.019

Isaac, W., Joseph, M., Ganpat, W., Wilson, M., & Brathwaite, R. (2012). The Caribbean's Windward Islands banana industry: A heritage of dependency. *The Journal of Rural and Community Development, 7*(2), 98–117.

Istanbul Chamber Commerce. (2006). *Analysis of organic agriculture and EU market opportunity.* Istanbul, Turkey: Birlik Press.

ITU. (2015). *ICT facts and figures 2015.* Retrieved from http://www.itu.int/ict

Iwaro, A. D., Bekele, F. L., Butler, D. R., Singh, V., Holder-John, A., Bharath, S., & Bidaisee, G. G. et al. (2010). Recent progress in breeding for specific traits in cocoa to meet challenges to production. In *Proceedings of the International Congress on Tropical Agriculture: Overcoming Challenges to Developing Sustainable Agri-Food Systems in the Tropics* (pp. 43-52). Port of Spain, Trinidad: The University of the West Indies.

Jaff'e, W., & Rojas, M. (1994). Abiotic stress and biotechnology in Latin America. *Biotechnological Development Monitor, 18*(1), 6–7.

James, C. (2008). *Global status of commercialized biotech/GM crops: 2008* (ISAAA Brief No. 39). Ithaca, NY: International Service for the Acquisition of Agri-biotech Applications. Retrieved 9 July 2014 from, https://isaaa.org/resources/publications/briefs/39/download/isaaa-brief-39-2008.pdf

Jansson, A., & Berggren, A. (2015). Insects as Food – Something for the Future? A report from Future Agriculture. Uppsala, Sweden: Swedish University of Agricultural Sciences (SLU).

Jarvis, A., Upadhyaya, H. D., Gowda, C. L. L., Agrawal, P. K., Fujisaka, S., & Anderson, B. (2008). *Climate change and its effect on conservation and use of plant genetic resources for food and agriculture and associated biodiversity for food security*. Retrieved June 4, 2015, from http://oar.icrisat.org/5810/

Jarvis, A., Ramirez-Villegas, J., Campo, B. V. H., & Navarro-Racines, C. (2012). Is cassava the answer to African climate change adaptation? *Tropical Plant Biology*, *5*(1), 9–29. doi:10.1007/s12042-012-9096-7

Jauhar, P. (2006). Modern biotechnology as an integral supplement to conventional plant breeding: The prospects and challenges. *Crop Science*, *46*(4), 1841–1859. doi:10.2135/cropsci2005.07-0223

Jha, U. C., Bohra, A., & Singh, N. P. (2014). Heat stress in crop plants: Its nature, impacts and integrated breeding strategies to improve heat tolerance. *Plant Breeding*, *133*(6), 679–701. doi:10.1111/pbr.12217

Jiang, G. H., Xu, C. G., Tu, J. M., Li, X. H., He, Y. Q., & Zhang, Q. F. (2004). Pyramiding of insect- and disease-resistance genes into an elite indica, cytoplasm male sterile restorer line of rice, '*Minghui 63*'. *Plant Breeding*, *123*(2), 112–116. doi:10.1046/j.1439-0523.2003.00917.x

Johnson, D. V. (2010). The contribution of edible insects to human nutrition and to forest management: Current status and future potential. In P.B. Durst, D.V. Johnson, R.N. Leslie, & K. Shono (Eds.), *Forest insects as food: Humans bite back. Proceedings of a workshop on Asia -Pacific resources and their potential for development*. Chiang Mai, Thailand. FAO. doi:10.1146/annurev-ento-120811-153704

Jones, J. (2008). *Farm crime on Anglesey: Local partner's and organisations*. Views on the Issue, second report, January 2008. Retrieved on December 12th, 2015 from http://www.aber.ac.uk/en/media/jane-jones---second-report.pdf

Jones, R. A., & Scott, S. J. (1983). Improvement of tomato flavor by genetically increasing sugar and acid contents. *Euphytica*, *32*(3), 845–855. doi:10.1007/BF00042166

Juana, J. S., Kahaka, Z., & Okurut, F. N. (2013). Farmers' Perceptions and Adaptations to Climate Change inSub-Sahara Africa: A Synthesis of Empirical Studies and Implications for Public Policy in African Agriculture. *The Journal of Agricultural Science*, *5*(4), 121–135.

Juma, C. (2014). Feeding Africa: why biotechnology sceptics are wrong to dismiss GM. *Guardian Professional*. Retrieved 12 June 2014 from, http://www.theguardian.com/global-development-professionals-network/2014/may/27/gm-crops-food-security-calestous-juma-africa

Juma, C., & Gordon, K. (2014). *Leapfrogging African agriculture*. Retrieved 15 June 2014 from, http://www.brookings.edu/~/media/Research/Files/Reports/2014/foresight%20africa%202014/06%20foresight%20african%20agriculture%20juma%20gordon.pdf

Karingi, S. (2008). *Webpate*. Retrieved from http://www.oecd.org/tad/agricultural-policies/39596692.pdf.karingi

Kelly, C., & Schulschenk, J. (2011). Assessing the vulnerability of Stellenbosch's food system and possibilities for a local food economy. *Development Southern Africa*, *28*(4), 563–578. doi:10.1080/0376835X.2011.605575

Kemp, S. (2015). *Digital, social and mobile worldwide in 2015*. Accessed on August 12, 2015 from, http://wearesocial.net/blog/2015/01/digital-social-mobile-worldwide-2015/

Khusro, M., Andrew, N. R., & Nicholas, A. (2012). Insects as poultry feed: A scoping study for poultry production systems in Australia. *World's Poultry Science Journal*, *68*(03), 435–446. doi:10.1017/S0043933912000554

Knapp, S. (2005). *Rich chocolate and poor farmers*. Wellington Aotearoa, New Zealand: Dev-Zone.

Knowles, R., & Wareing, J. (1983). *Economic and social geography*. London: Heinemann.

Knox, J., Hess, T., Daccache, A., & Wheeler, T. (2012). Climate change impacts on crop productivity in Africa and South Asia. *Environmental Research Letters*, *7*(3), 034032. doi:10.1088/1748-9326/7/3/034032

Kodym, A., & Zapata-Arias, F. A. (1999). Natural light as an alternative light source for the *in vitro* culture of banana (*Musa acuminata* c.v. 'Grande Naine'). *Plant Cell, Tissue and Organ Culture*, *55*(2), 141–145. doi:10.1023/A:1006119114107

Koroma, S., & Deep Ford, J. R. (2006). *The Agricultural dimension of the ACP-EU Economic Partnership Agreements*. Rome, Italy: United Nations, Food and Agriculture Organization.

Kumar, A. (2015). *Lecture 2 Major characteristics of developing countries*. Retrieved from http://web.uvic.ca/~kumara/econ420/characteristics-dev.pdf

Kumar, R. (2015, October 5). *How Urban Agriculture Is Revitalizing Local Economies*. Retrieved from The Blog: http://www.growingpower.org/

Kumar, S. (2014). Plant Disease Management in India: Advances and Challenges. *African Journal of Agricultural Research*, *9*(15), 1207–1217. doi:10.5897/AJAR2014.7311

Kundu, M. S., De, A. K., & Srivastava, R. C. (2010). *Livestock and poultry production policy for Andaman and Nicobar Islands: A scientific perspective*. Port Blair: CARI.

Kutama, A. S., Emechebe, A. M., & Aliyu, B. S. (2011a). Evaluating the efficacy of seed treatment fungicides in the control of sorghum head smut caused by Sporisorium reilianum, in the Sudan savanna region of Nigeria. *Journal of Phytopathology and Plant Health*, *6*(2), 93–98.

Kuyek, D. (2002). The past predicts the future: GM crops and Africa's farmers. *Seedling*. Retrieved 15 July 2014 from, http://www.grain.org/article/entries/344-the-past-predicts-the-future-gm-crops-and-africa-s-farmers

Läderach, P., Martinez-Valle, A., Schroth, G., & Castro, N. (2013). Predicting the future climatic suitability for cocoa farming of the world's leading producer countries, Ghana and Côte d'Ivoire. *Climatic Change*, *119*(3-4), 841–854. doi:10.1007/s10584-013-0774-8

Lambert, B., & Peferoen, M. (1992). Insecticidal promise of Bacillus thuringiensis. Facts and mysteries about a successful biopesticide. *Bioscience*, *42*(2), 112–122. doi:10.2307/1311652

Lasky, J. R., Upadhyaya, H. D., Ramu, P., Deshpande, S., Hash, C. T., Bonnette, J., . . . Morris, G. P. (2015). Genome-environment associations in sorghum landraces predict adaptive traits. *Science Advances, 1*(6), e1400218. Retrieved August 5, 2015, from http://advances.sciencemag.org/content/1/6/e1400218.full-text.pdf+html

Lee, S. (2013, June 19). *The Chicago Council on Global Affairs*. Retrieved June 25th, 2015, from Global Food For Thought: http://globalfoodforthought.typepad.com/global-food-for-thought/2013/06/tackling-urban-food-security.html

Leishman, M. R., & Gallagher, R. V. (2015). Will there be a shift to alien-dominated vegetation assemblages under climate change? *Diversity and Distributions*. Retrieved August 12, 2015, from http://onlinelibrary.wiley.com/doi/10.1111/ddi.12338/full

Lewis, A. (1978a). Growth and Fluctuations 1870-1913. London: Allen & Unwin.

Lewis, A. (1978b). The Evolution of the International Economic Order. Princeton University Press.

Lewis, W. A. (1954). Economic development with unlimited supplies of labour. *The Manchester School of Economic and Social Studies, 22*(2), 139–191. doi:10.1111/j.1467-9957.1954.tb00021.x

Lin, B. B. (2011). Resilience in agriculture through crop diversification: Adaptive management for environmental change. *Bioscience, 61*(3), 183–193. doi:10.1525/bio.2011.61.3.4

Lindiwe, M. S. (2011). *FANRPAN0 2011*. Academic Press.

Links Incorporated. (2015). *The South Bend Area (IN) Chapter of The Links, Incorporated*. Retrieved from The Links Incorporated: http://southbendarealinksinc.org/About_Us.php

Little, D. (2011). *Praedial Larceny: Its Consequences For Caribbean Agriculture*. Caricom View. Retrieved on June 3rd, 2015 from http://www.caricom.org/jsp/communications/caricom_online_pubs/caricom_view_jul_2011.pdf

Liu, P., Byers, A., & Giovannucci, D. (2008). *Value-adding standards in the North American food market-trade opportunities in certified products for developing countries*. Rome, Italy: FAO Trade and Markets Division.

Livak, K. J. (1999). Allelic discrimination using fluorogenic probes and the 52 nuclease assay. *Genetic Analysis, 14*(5), 143–149. doi:10.1016/S1050-3862(98)00019-9 PMID:10084106

Liverpool, S. L. O., & Winter-Nelson, A. (2010). *Poverty status and the impact of social networks on smallholder technology adoption in rural Ethiopia*. IFPRI Discussion Paper 970. Washington, DC: IFPRI.

Lobell, D. B., Burke, M. B., Tebaldi, C., Mastrandrea, M. D., Falcon, W. P., & Naylor, R. L. (2008). Prioritizing climate change adaptation needs for food security in 2030. *Science, 319*(5863), 607–610. doi:10.1126/science.1152339 PMID:18239122

Lorenzo, J. C., Blanco, M., Pel'aez, O., Gonz'alez, A., & Cid, M. (2001). Sugarcane micropropagation and phenolic excretion. *Plant Cell, Tissue and Organ Culture, 65*(1), 1–8. doi:10.1023/A:1010666115337

Lotter, D. W. (2003). Organic Agriculture. *Journal of Sustainable Agriculture, 21*(4), 59–128. doi:10.1300/J064v21n04_06

Ludena, C. E. (2010). *Agricultural productivity growth, efficiency change and technical progress in Latin America and the Caribbean. Inter-American Development Bank*. IDB Working paper 61, Series No. IDB-WP-186. Washington, DC: Inter-American Development Bank.

Lukefahr, S. D. (2005). Sustainable and alternative systems of rabbit production. In *Proceedings of the 8th World Rabbit Congress* (pp. 1452-1464). World Rabbit Science Association (WRSA).

Madhulatha, P., Anbalagan, M., Jayachandaran, S., & Sakthivel, N. (2004). Influence of liquid pulse treatment with growth regulators on *in vitro* propagation of banana (*Musa* sp. AAA). *Plant Cell, Tissue and Organ Culture*, *76*(2), 189–192. doi:10.1023/B:TICU.0000007291.31439.6c

Madukwe, M.C. (2008). *Practice without policy: The Nigeria agricultural extension services.* 30th Inaugural Lecture of the University of Nigeria.

Mafuru, J. M., Babu, A. K., & Matutu, T. F. (2007). Tanzania: The impact of market links on horticultural production in the Mara region. *Regoverning Markets.* Lake Zone Agricultural Research and Development Institute. Retrieved from <http://pubs.iied.org/pdfs/G03265.pdf>

Maggio, A., Van Criekinge, T., & Malingreau, J. P. (2015). *Global Food Security 2030. Assessing trends with a view to guiding future EU policies.* Foresight Series. European Commission DG Joint Research Centre. Retrieved on August 18, 2015, from http://publications.jrc.ec.europa.eu/repository/bitstream/JRC94867/lbna27252enn.pdf

Maharaj, K., Maharaj, P., Bekele, F. L., Ramnath, D., Bidaisee, G. G., Bekele, I., & Sankar, R. et al. (2011). Trinidad selected hybrids: An investigation of the phenotypic and agro-economic traits of 20 selected cacao cultivars. *Trop. Agric*, *88*(4), 175–185.

Mainka, S. A., & Trivedi, M. (2002). *Links between Biodiversity Conservation, Livelihoods and Food Security: The Sustainable Use of Wild Species for Meat.* Cambridge, UK: IUCN.

Makhzoumi, J., & Pungetti, G. (1999). *Ecological landscape design and planning: The Mediterranean context.* London: Spon Press. doi:10.4324/9780203223253

Malua, M. B. (2003). Case Study: The Pacific Island in *Turning Losses into Gains.* SIDS and Multilateral Trade Liberalization in Agriculture. *UN Conference on Trade Development*, (pp. 171-204).

Marshall, B., & Shane, J. (2005). *Crime in rural areas: A review of the literature for the Rural and Evidence Research Centre.* London: University College London.

Marten, G. G. (2009). *Human ecology.* Earthscan Publications.

Martinelli, L. A., Naylor, R., Vitousek, P. M., & Moutinho, P. (2010). Agriculture in Brazil: Impacts, costs, and opportunities for a sustainable future. *Current Opinion in Environmental Sustainability*, *2*(5), 431–438. doi:10.1016/j.cosust.2010.09.008

Matthews, C. (2014, October 31). *Wealth inequality in America: It's worse than you think.* Retrieved from Fortune: http://fortune.com/2014/10/31/inequality-wealth-income-us/

Matuschke, I. (2009). *Rapid urbanization and food security: using food density maps to identify future food security hotspots.* Paper prepared for presentation at the International Association of Agricultural Economists Conference, Beijing, China.

Maunder, A. H. (Ed.). (1973). *Agricultural extension: A reference manual.* Rome: Food and Agriculture Organisation of the United Nations.

Mawby, R. I. (2004). Myth and reality in rural policing: Perceptions of the police in a rural country of England. *Policing: An International Journal of Police Strategies & Management*, *27*(3), 431–446. doi:10.1108/13639510410553158

Maximay, S. (2014). The Caribbean's Response to Climate Change Impacts. In W. Ganpat & W. P. Isaac (Eds.), *Impacts of Climate Change on Food Security in Small Island Developing States.* Hershey, PA: IGI Global.

Mayor's Biracial Committee Progress Report. (1963). *Progress report, July-October 1963, Mayor's Biracial Committee*. Retrieved from Michiana Memory: http://michianamemory.sjcpl.org/cdm/compoundobject/collection/p16827coll4/id/2286/rec/9

Mba, C., Guimaraes, E. P., & Ghosh, K. (2012). Re-orienting crop improvement for the changing climatic conditions of the 21st century. *Agriculture and Food Security, 1*(7).

McCall, M. (2003). *Results from the 2001-02 National Farm Crime Survey*. Australian Institute of Criminology, Trends and Issues in Crime and Criminal Justice, No.266.

McCouch, S., Baute, G. J., Bradeen, J., Bramel, P., Bretting, P. K., Buckler, E., & Zamir, D. et al. (2013). Agriculture: Feeding the future. *Nature, 499*(7456), 23–24. doi:10.1038/499023a PMID:23823779

McElwee, G. (2006). Farmer's as entrepreneurs: Developing competitive skills. *Journal of Developmental Entrepreneurship, 11*(3), 187–206. doi:10.1142/S1084946706000398

McElwee, G., & Vik, J. (2011). Diversification and entrepreneurial motivations of farmers in Norway. *Journal of Small Business, 49*(3), 390–410. doi:10.1111/j.1540-627X.2011.00327.x

McKenzie & Woodruff. (2015). Business Practices in Small Firms in Developing Countries. *International Labour Organization*. Retrieved from http://www.ilo.org/wcmsp5/groups/public/---ed_emp/documents/publication/wcms_360368.pdf

Mears, D. P., Scott, M. L., & Bhati, A. S. (2007a). Opportunity theory and agricultural crime victimization. *Rural Sociology, 72*(2), 151–184. doi:10.1526/003601107781170044

Mears, D. P., Scott, M. L., & Bhati, A. S. (2007b). *Policy, theory and research from an evaluation of an agricultural crime prevention program*. Washington, DC: The Urban Institute.

Mendelsohn, R., & Dinar, A. (1999). Climate change, agriculture, and developing countries: Does adaptation matter? *The World Bank Research Observer, 14*(2), 277–293. doi:10.1093/wbro/14.2.277

Mendoza, A., & Machado, R. (2009). *The Escalation in World Food Prices and Its Implications for the Caribbean* (vol. 2). Economic Commission for Latin America and the Caribbean, Caribbean Development Report.

Meyer, R.S., DuVal, A.E., & Jensen, H.R. (2012). Patterns and processes in crop domestication: an historical review and quantitative analysis of 203 global food crops. *New Phytologist, 196*, 29–48. doi:.2012.04253. x10.1111/j.1469-8137

MFPED. (2015). *Maintaining infrastructure investment and promoting excellence in public service delivery. Background to the Budget, 2015/16*. Kampala: MFPED.

Midega, C. A., Bruce, T. J., Pickett, J. A., Pittchar, J. O., Murage, A., & Khan, Z. R. (2015). Climate-adapted companion cropping increases agricultural productivity in East Africa. *Field Crops Research, 180*, 118–125. doi:10.1016/j.fcr.2015.05.022

Millard, E. (2011). Incorporating agroforestry approaches into commodity value chains. *Environmental Management, 48*(2), 365–377. doi:10.1007/s00267-011-9685-5 PMID:21556936

Millennium Development Goals (MDG) Report. (2013). *Section III-food security in Africa-issues, challenges, lessons, Assessing progress in Africa towards the millennium development*. Retrieved 6 June 2014 from, goals.http://www.afdb.org/fileadmin/uploads/afdb/Documents/Publications/Millennium%20Development%20Goals%20%28MDGs%29%20Report%202013%20-%20Section%20III%20-%20Food%20Security%20in%20Africa%20-%20Issues-Challenges-Lessons.pdf

Millennium Project. (2006). *Millennium Development Goals – What they are.* Accessed on September 24 2015 from, http://www.unmillenniumproject.org/goals/

Miller, M. (2007). *Genetically modified crops in Africa.* UNEP. Retrieved 5 May 2014 from http://www.eoearth.org/view/article/152943

Miller, S. A., & Martin, R. R. (1988). Molecular Diagnosis of Plant-disease. *Annual Review of Phytopathology, 26*(1), 409–432. doi:10.1146/annurev.py.26.090188.002205

Milner, C., & Morrissey, O. (1999). Measuring trade liberalization in Africa. doi:10.1007/978-1-349-14307-8_3

Minister of Development. (2014a). *Tenth development plan (2014-2018): Specialization commission report about agricultural structure and food security.* Retrieved from http://www.kalkinma.gov.tr

Minister of Development. (2014b). *Tenth development plan (2014-2018): Working group report about sustainable agricultural land use.* Retrieved from http://www.kalkinma.gov.tr

Ministry of Food, Agriculture, and Livestock. (2012). *Strategic plan for organic farming in Turkey (2012-2016).* Retrieved from http://www.trakya2023.com/uploads/docs/2806201331nTii.pdf

Ministry of Food, Agriculture, and Livestock. (2014). *Activities of general directorate of vegetative production* [PDF document]. Retrieved from http://www.tarim.gov.tr/sgb/Belgeler/SagMenuVeriler

Mollineau, W. M., Adogwa, A. O., & Garcia, G. W. (2007a). A preliminary technique for electro-ejaculation of the agouti (Dasyprocta leprorina). *Animal Reproduction Science, 108*(1-2), 92–97. doi:10.1016/j.anireprosci.2007.07.017 PMID:17889458

Mollineau, W. M., Adogwa, A. O., & Garcia, G. W. (2007b). Spermatozoal morphologies and fructose and citric acid concentrations in agouti (*Dasyprocta leprorina*) semen. *Animal Reproduction Science, 105*(3-4), 378–383. doi:10.1016/j.anireprosci.2007.08.009 PMID:17919856

Mollineau, W. M., Garcia, G. W., Samayah, D., Kissoonsingh, W., & Procope-Garcia, A. (2000). The wildlife industry in Trinidad: a case study towards developing a sustainable industry model for a small Twin Island State (Trinidad and Tobago). In *Proceedings of the Conference on Managing Space for Sustainable Living in Small Island Development States.* St. Augustine: The University of the West Indies.

Montgomery, M. R., & Hewett, P. C. (2005). Urban poverty and health in developing countries: Household and neighborhood Effects. *Demography, 42*(3), 397–425. doi:10.1353/dem.2005.0020 PMID:16235606

Morrissey. (Ed.). (n.d.). *Evaluating economic liberalization.* London: MacMillan.

Morse, S., Bennet, R., & Ismael, Y. (2013). *Impacts of GM cotton and maize on household livelihoods in South Africa.* University of Reading. Retrieved 9 July 2014, from bch.cbd.int/database/attachment/?id=1575

Mucavele, S. (2015). The role of rural women in Agriculture. In *MuGeDe - Women, Gender and Development, Republic of Mozambique - Southern Africa.* Retrieved from http://www.wfo-oma.com/women-in-agriculture/articles/the-role-of-rural-women-in-agriculture.html

Muhammad, B.L. (2002). *Rural Crime and Rural Policing Practices (Multicultural Law Enforcement).* An applied research project submitted to the Department of Interdisciplinary Technology as part of the School of Police Staff and Command Programme.

Muhammad, A., Hussain, I., Saqlan, S. M., Hamid, R., & Hamid, N. (2004). Banana plantlet production through tissue culture. *Pakistan Journal of Botany, 36*(3), 617–620.

Mullen, J. D. (2007). Productivity growth and returns from public investments in R & D in Australian broad-care agriculture. *The Australian Journal of Agricultural and Resource Economics*, *18*(1), 18–29.

Murashige, T. (1974). Plant propagation through tissue culture. *Annual Review of Plant Physiology*, *25*(1), 135–136. doi:10.1146/annurev.pp.25.060174.001031

Murnaghan, I. (2014). *Development and history of GM foods*. Retrieved 10 August 2014, from http://www.geneticallymodifiedfoods.co.uk/development-history-gm-foods.html

Murphy, S. (2001). The Global Food Basket. *Forum for Applied Research and Public Policy*, *16*(2), 36–42.

Mustapha, K. (2009). Taste of child labor not so sweet: A Critique of Regulatory Approaches to Combating Child Labor Abuses by the US Chocolate Industry. *Wash. UL Rev.*, *87*, 1163.

Mwangi, M., & Kariuki, S. (2015). Factors determining adoption of new agricultural technology by smallholder farmers in developing countries. Journal of Economics and Sustainable Development, 6(5), 208-217.

Mwangi, J. K., Nyende, A. B., Demo, P., & Matiru, V. N. (2008). Detection of latent infection by *Ralstonia solanacearum* in potato (*Solanum tuberosum*) using stems instead of tubers. *African Journal of Agricultural Research*, *7*(11), 1644–1649.

Myers, N. (2003). Biodiversity hotspots revisited. *Bioscience*, *53*(10), 916–917. doi:10.1641/0006-3568(2003)053[0916:BHR]2.0.CO;2

Myint, T. (2003). Agricultural marketing information system in Myanmar. *Agricultural-Marketing*, *XLV*(4), 16–28.

N.H.B. (2014). *National Horticulture Database, National Horticulture Board*. Harayana, India: Guargaon.

Nabbumba, R., & Bahiigwa, G. (2003). Agricultural Productivity Constraints in Uganda: Implications for Investments. Kampala: Uganda.

Nahms, N. A. (1996). *Dairy management practices*. Washington, DC: USDA.

Naik, P. S. (2005). Achieving self sufficiency in quality seed production in roots and tuber crops with special reference to potato. In K. L. Chadha, B. S. B.S. Ahloowalia, K. V. Prasad, & S. K. Singh (Eds.), Crop improvement and production technology of horticultural crops (Vol. 1, pp. 243-255). New Delhi: The Horticultural Society of India.

Nandwani, D., Zehr, U., Zehr, B. E., & Barwale, R. B. (2000). Mass propagation and ex vitro survival of banana cv. '*Basrai*' through tissue culture. *Gartenbauwissenschaft*, *65*(6), 237–240.

Naraine, L. (2005). *Agricultural Diversification in St. Kitts: Policy and Implementation. St. Kitts-Nevis*, Department of Agriculture.

Nebel, B., & Wright, R. (1996). *Environmental Science: The Way the World Works*. Upper Saddle River, NJ: Prentice Hall.

Neilson, J., Susilo, A., Mulia, S., Guest, D., Lambert, S., Munawar, N. L., . . . Muhajir, A. S. (2014). *Improving cocoa production through farmer involvement in demonstration trials of potentially superior and pest/disease resistant genotypes and integrated management practices*. Retrieved June 29, 2015, from http://aciar.gov.au/files/smar-2005-074_final_report.pdf

NEPAD. (2013). *African agriculture: Transformation and outlook*. Johannesburg, South Africa: NEPAD.

Neudeck, L., Avelino, L., Bareetseng, P., & Ngwenya, B.N., Teketay, D., & Motsholapheko, M.R. (2012). The contribution of edible wild plants to food security, dietary, diversity and income of households in Shorobe Village, Northern Botswana. *Ethnobotany Research and Applications*, *10*, 449–462.

Neufeldt, H., Jahn, M., Campbell, B. M., Beddington, J. R., DeClerck, F., De Pinto, A., & Zougmoré, R. et al. (2013). Beyond climate-smart agriculture: Toward safe operating spaces for global food systems. *Agric Food Secur*, *2*(12), 1180–1186.

New Partnership for Africa's Development (NEPAD). (2010). *Status of crop biotechnology in Africa*. Retrieved 4 July 2014 from, http://www.nepadbiosafety.net/subjects/biotechnology/status-of-crop-biotechnology-in-africa

Newman, O. (1972). *Defensible Space*. New York: Macmillan.

Newsweek. (2011, January 21). *America's Dying Cities*. Retrieved from http://www.newsweek.com/americas-dying-cities-66873

Noble, I. R., Huq, S., Anokhin, Y. A., Carmin, J., Goudou, D., Lansigan, F. P., & Villamizar, A. et al. (2014). Adaptation needs and options. In C. B. Field, V. R. Barros, D. J. Dokken, K. J. Mach, M. D. Mastrandrea, T. E. Bilir, & L. L. White et al. (Eds.), *Climate Change 2014: Impacts, Adaptation, and Vulnerability. Part A: Global and Sectoral Aspects* (pp. 833–868). Cambridge, UK: Cambridge University Press.

Nogueira, S. S. C., & Nogueira-Filho, S. L. G. (2011). Wildlife farming: An alternative to unsustainable hunting and deforestation in Neotropical forests? *Biodiversity and Conservation*, *20*(7), 1385–1397. doi:10.1007/s10531-011-0047-7

North, D. C. (1990). *Institutions, Institutional Change and Economic Performance*. Cambridge University Press. doi:10.1017/CBO9780511808678

Nwanze, K. F., da Silva, J. G., Cousin, E., & Frison, E. (2012). *Opinion: Food Security Must Be On The Table At Rio+20*. Retrieved February 13, from http://www.wfp.org/news/news-release/food-security-must-be-table-rio20

Oakley, P., & Garforth, C. (1997). Extension methods. In *Guide to Extension Training*. FAO. Retrieved from www.fao.org/docrep/t0060e/T0060E07.htm#5

OECD. (2012). *Promoting Inclusive Growth: Challenges and Policies*. Paris: OECD.

Official website of ICAR-Central Island Agricultural Research Institute, Port Blair, Andaman and Nicobar Islands, India. (n.d.). Retrieved from http://icar-ciari.res.in

Ofori, A., Padi, F. K., Acheampong, K., & Lowor, S. (2015). Genetic variation and relationship of traits related to drought tolerance in cocoa (Theobroma cacao L.) under shade and no-shade conditions in Ghana. *Euphytica*, *201*(3), 411–421. doi:10.1007/s10681-014-1228-8

Ogero, K. O., Mburugu, G. N., Mwangi, M., Ngugi, M. M., & Ombor, O. (2012). Low cost tissue culture technology in the regeneration of sweet potato (*Ipomoea batatas* (L) Lam). *Research Journal of Biology*, *2*(2), 51–58.

Ogunjimi, S. I., & Farinde, A. J. (2012). Farmers' Knowledge Level of Precautionary Measures in Agro-Chemicals Usage on Cocoa Production in Osun and Edo States, Nigeria. International. *Journal of Agriculture and Forestry.*, *2*(4), 186–194. doi:10.5923/j.ijaf.20120204.10

Olaniyi, O. A. (2010). *Assessment of Utilization of Agricultural Information on Selected Arable crops among Rural youth in Oyo and Osun states, Nigeria*. (Unpublished Ph.D Thesis). Department of Agricultural Extension and Rural Development, Ladoke Akintola University of Technology, Ogbomoso, Nigeria.

Olaniyi, O. A., Ogunleye, Y. K., & Eniola, P. O. (2006). Extension Delivery Methods Preferred by Women Farmers in Oyo State, Nigeria. *Journal of Agricultural Extension Management*, *7*(2), 1–6.

Olayide, S. O., & Heady, E. (1982). *Introduction to Agricultural Production Economics* (1st ed.). Ibadan: Ibadan University Press.

Omiti, J. (2007). *Participatory prioritization of issues in smallholder agricultural commercialization in Kenya.* Nairobi: KIPPRA.

Organisation for Economic Co-operation and Development (OECD). (2011). *Evaluation of agricultural policy reforms in Turkey*. .10.1787/9789264113220-cn

Overman, H. G., & Venables, A. J. (2005). *Cities in the Developing World*. CEP Discussion Paper No 695. Available from: http://cep.lse.ac.uk/pubs/ download/dp0695.pdf

Oxfam. (2014). *Oxfam: 85 richest people as wealthy as poorest half of the world.* World Economic Analysis. Accessed July 4, 2015, from http://www.theguardian.com/business/2014/jan/20/oxfam-85-richest-people-half-of-the-world

Oyekale, A. S., Bolaji, M. B., & Olowa, O. W. (2009). The effects of climate change on cocoa production and vulnerability assessment in Nigeria. *Agricultural Journal*, *4*(2), 77–85.

Ozkan, M. (2003). Organic agriculture and national legislation in Turkey. In Organic agriculture: Sustainability, markets and policies (pp. 289-294). Wallingford, UK: CABI Publishing.

Paarlberg, R. (2010, November30). GMO foods and crops: Africa's choice. *New Biotechnology*, *27*(5), 609–613. doi:10.1016/j.nbt.2010.07.005 PMID:20637906

Palombi, L., & Sessa, R. (2013). Climate-smart crop production system Module 7. In *Climate-smart agriculture: Sourcebook*. Retrieved June 23, 2015, from http://www.fao.org/3/a-i3325e/i3325e07.pdf

Pal, P., & Roy, S. (2014). Edible Insects: Future of Human Food - A Review. *International Letters of Natural Sciences*, *26*, 1–11. doi:10.18052/www.scipress.com/ILNS.26.1

Pan-American Health Organization (PAHO). (2014). *Countries pledge action to reduce child obesity in the Americas*. Author.

Pandey, C. B., & Singh, R. K. (2010). Organic black pepper (*Piper nigrum* L.) cultivation on glyricidia standards in coconut (*Cocos nucifera* L.) plantations in South Andaman: Organic matter production and recycling of nutrients. *Indian Journal of Agricultural Science*, *80*(11), 975–980.

Pandey, C. B., Venkatesh, A., & Medhi, R. P. (2006). Diversity and species structure of home gardens in South Andaman. *Tropical Ecology*, *47*(2), 251–258.

Papp, A. C., Pinsonneault, J. K., Cooke, G., & Sadee, W. (2003). Single nucleotide polymorphism genotyping using allele-specific PCR and fluorescence melting curves. *BioTechniques*, *34*(5), 1068–1072. PMID:12765033

Park, J., & Seaton, R. A. F. (1996). Integrative research and sustainable agriculture. *Agricultural Systems*, *50*(1), 81–100. doi:10.1016/0308-521X(94)00050-2

Pay, E. (2009). *The market for organic and fair-trade cocoa.* Rome, Italy: FAO. Retrieved August 5, 2015, from: http://www.fao.org/fileadmin/templates/organicexports/docs/Market_Organic_FT_Cocoa.pdf

Peel, C. J., Hurd, D. L., Madsen, K. S., & de Kerchove, G. (1989). Monsanto Agricultural Company. In *Proceedings, Monsanto Technical Symposium*. The Monsanto Company.

Perez, C., Jones, E. M., Kristjanson, P., Cramer, L., Thornton, P. K., Förch, W., & Barahona, C. (2015). How resilient are farming households and communities to a changing climate in Africa? A gender-based perspective. *Global Environmental Change*, *34*, 95–107. doi:10.1016/j.gloenvcha.2015.06.003

Peterson, S. (2011). International technology transfer-not quite the silver bullet in international climate policy. *Climate Change: Current Issues*, 20-21. Retrieved August 13, 2015, from https://www.ifw-kiel.de/pub/e-books/climate_change.pdf

Point2Homes. (2016). *South Bend.* Retrieved from pointhomes.com: http://www.point2homes.com/US/Neighborhood/IN/South-Bend-Demographics.html

Pollack, A. (2015, November 20). Genetically Engineered Salmon Approved for Consumption. *Associated Press.*

Pretty, J. (2003). *Social capital and connectedness: Issues and implications for agriculture, rural development and natural resource management in ACP countries.* Review paper for CTA. CTA Working Document Number 8032.

Pretty, J. (2008). Agricultural sustainability: Concepts, principles and evidence. *Philosophical Transactions of the Royal Society of London. Series B, Biological Sciences, 363*(1491), 447–465. doi:10.1098/rstb.2007.2163 PMID:17652074

Pretty, J. N. (1995). Participatory learning for sustainable agriculture. *World Development, 23*(8), 1247–1263. doi:10.1016/0305-750X(95)00046-F

Priya, V. P., Celine, V. A., Gokulapalan, C., & Rajamony, L. (2007). Screening amaranth genotypes (Amaranthus spp.) for yield and resistance to leaf blight caused by *Rhizoctonia solani* Kuhn. *Plant Genetic Resources Newsletter (Rome, Italy), 147*, 1–4.

Pschorn-Strauss, E. (2005). Bt Cotton in South Africa: the case of the Makhatini farmers. *Seedling*, 13–24. Retrieved 10 July 2014 from, www.grain.org/article/entries/492-bt-cotton-in-south

Pugliese, P., Bteich, M. R., & Al-Bitar, L. (2014). *Mediterranean organic agriculture: Key features, recent facts, latest figures Report 2014.* Retrieved from http://moan.iamb.it/

Purcell, D. L., & Anderson, J. R. (1997). *Agricultural Research and extension: Achievementand problems in national systems. World Bank Operations Evaluation Study.* Washington, DC: World Bank. doi:10.1596/0-8213-3878-1

Radican, E. A. (2010). *Growing Food Where it is Needed.* Indiana University, South Bend. South Bend Indiana: Center for a Sustainable Future (CSF). Retrieved November 10, 2015, from https://www.iusb.edu/csfuture/docs/WHITE%20PAPER%20NO.%202.pdf

Rajack-Talley, T. A. (2014). *Poverty is an economic crime not a moral sin: A history of Globalized food production, foreign dominated economics, trade liberalization and dependency in the Caribbean.* Paper presented at SALISES 15th Annual Conference, St. Augustine, Trinidad.

Rajack-Talley, T. A. (2015). The role of the small farmer, family farms and women in food production, food security, and sustainable development in the Caribbean. In Sustainable Food Production Practices in the Caribbean (vol. 2). Ian Randle Publishers.

Rajack-Talley, T.A., & Talley, C. (2000). The role of farming women in Caribbean agriculture: A historical and contemporary perspective. *Caribbean Journal for Agriculture and Natural Resources, 2*(2), 4-13.

Ramdwar, M. N. A., Maharaj, R., & Siew, N. (2015). Sustaining The Environment: Farm and Beyond the Farm. Chapter 7 Pp 213-246. In W.G. Ganpat & W.P. Isaac (Eds.), Sustainable Food Production Practices in the Caribbean (vol. 2). Randle.

Ramgattie, R., Siew, N., Diptee, M., Stoute, V., & Knights. (2014). Effect of mammary stimulation on dairy cows and heifers exposed to a lactation induction protocol. *Open Journal of Animal Sciences, 4*(1). DOI:10.4236/ojas.2014.41001

Ramirez-Villegas, J., & Thornton, P. K. (2015). *Climate change impacts on African crop production.* CCAFS Working Paper no. 119. Copenhagen, Denmark: CGIAR Research Program on Climate Change, Agriculture and Food Security (CCAFS). Retrieved June 22, 2015, from https://cgspace.cgiar.org/bitstream/handle/10568/66560/WP119_FINAL.pdf?sequence=1

Rasul, G., & Sharma, B. (2015). The nexus approach to water–energy–food security: an option for adaptation to climate change. *Climate Policy*, 1-21. Retrieved August 11, 2015, from http://www.tandfonline.com/loi/tcpo20

Reardon, T., & Berdeagué, J. A. (2002). The Rapid Rise of Supermarkets in Latin America: Challenges and Opportunities for Development. *Overseas Development Institute: Development Policy Review*, *20*(4), 371–388.

Reddy, S. (2007). Mega tourism in Andaman and Nicobar Islands: Some concerns. *Journal of Human Ecology*, *21*(3), 231–239.

Rehber, E., & Turhan, Ş. (2002). Prospects and challenges for developing countries in trade and production of organic food and fibers: The case of Turkey. *British Food Journal*, *104*(3/4/5), 371-390. doi:10.1108/00070700210435380

Reichert, J. M., Rodrigues, M. F., Awe, G. O., Riquelme, U. F. B., Kaiser, D. R., & Reinert, D. J. (2015). Common bean in highly variable weather conditions, on sandy soils, and food security in a subtropical environment. In *Food and Energy Security*. John Wiley & Sons Ltd. and the Association of Applied Biologists. Retrieved August 7, 2015, from http://onlinelibrary.wiley.com/doi/10.1002/fes3.65/epdf

Reis, V. M., Baldani, J. I., Baldani, V. L. D., & Dobereinera, J. (2000). Biological dinitrogen fixation in *gramineae* and palm trees. *Critical Reviews in Plant Sciences*, *19*(3), 227–247. doi:10.1016/S0735-2689(00)80003-9

Rellstab, C., Gugerli, F., Eckert, A. J., Hancock, A. M., & Holderegger, R. (2015). A practical guide to environmental association analysis in landscape genomics. *Molecular ecology*. Retrieved August 9, 2015, from http://onlinelibrary.wiley.com/doi/10.1111/mec.13322/epdf

Rengarajan, S., Veeraragavan, S., Thangaraj, M., & Ramachandran, R. (2014). Geographical analysis of tourism sites in Andaman Archipelago (India) and ecotourism development for Smith Island of North Andaman. *International Journal of Sustainable Development and World Ecology*, *21*(5), 449–455. doi:10.1080/13504509.2014.958118

Ribaut, J. M., de Vicente, M. C., & Delannay, X. (2010). Molecular breeding in developing countries: Challenges and perspectives. *Current Opinion in Plant Biology*, *13*(2), 213–218. doi:10.1016/j.pbi.2009.12.011 PMID:20106715

Ribaut, J. M., & Ragot, M. (2007). Marker-assisted selection to improve drought adaptation in maize: The backcross approach, perspectives, limitations, and alternative. *Journal of Experimental Botany*, *58*(2), 351–360. doi:10.1093/jxb/erl214 PMID:17158111

Rijsberman, F. (2012). *CGIAR: A global research partnership for a food secure future*. Retrieved 25th December 2015, from http://www.cgiar.org/consortium-news/cgiar-global-research-partnership-for-a-food-secure-future/

Rijsberman, F. (2015). *CGIAR's new strategy and research programs: Answering to poverty, health and climate change*. News from the CGIAR Consortium. Retrieved August 3, 2015, from http://www.cgiar.org/consortium-news/cgiars-new-strategy-and-research-programs-answering-to-poverty-health-and-climate-change/

Rivest, S. (2015, November 5). *ABC57 News*. Retrieved from www.abc57.com: http://www.abc57.com/story/30451531/local-businesses-work-to-combat-food-insecurity

Roberts, P. K. C., Shyam, & Rastogi, C. (2006). Rural access index: a key development indicator. *Transport Papers 10*. Washington, DC: World Bank. Retrieved 5 July 2014 from, www.worldbank.org/transport/transportresults/headline/rural-access/tp-10-final.pdf

Roberts-Nkrumah, L. B., & Legall, G. (2013). Breadfruit (*Artocarpus altilis*, Moraceae) and Chataigne (*A. camansi*) for Food Security and Income Generation: The Case of Trinidad and Tobago. *Economic Botany*, *67*(4), 324–334. doi:10.1007/s12231-013-9250-7

Roberts, T. G., Ganpat, W. G., Narine, L., Heinert, S. B., & Rodriguez, M. T. (2015). Trinidad Food Producers' Training Needs for Food Security and Implications for Extension. *Journal of International Agricultural and Extension Education, 22*(1), 7–20. doi:10.5191/jiaee.2015.22101

Roberts, T. G., & Rodriguez, M. T. (2014). An Overview of climate change and impacts on food security in small island developing states. In W. Ganpat & W. P. Isaac (Eds.), *Impacts of Climate Change on Food Security in Small Island Developing States* (pp. 1–30). Hershey, PA: IGI Global.

Robinson, G. (2015). *Better Homes of South Bend: An American Story of Courage*. Charleston, SC: The History Press.

Rodney, W. (1974). *How Europe underdeveloped Africa*. Washington, DC: Howard University Press.

Rodriguez, F., & Rodrik, D. (2000). *Trade policy and economic growth: A skeptic's guide to the cross-national evidence*. Academic Press.

Roe, D. (2008). *Trading Nature. A report, with case studies, on the contribution of wildlife trade management to sustainable livelihoods and the Millennium Development Goals*. Switzerland: TRAFFIC International and WWF International, Cambridge, UK and Gland.

Roling, N. (1997). *Extension Role in Sustainable Agricultural Development in Improving Agricultural Extension a Reference Manual*. FAO Rome.

Rosegrant, M. W. (2011). Impacts of climate change on food security and livelihoods. In *Food security and climate change in dry areas* (pp. 24-26). Aleppo: International Center for Agricultural Research in the Dry Areas (ICARDA).

Rubenstein, H. (1975). The Utilization of Arable Land in an Eastern Caribbean Valley. *Canadian Journal of Sociology, 1*(2), 157–167. doi:10.2307/3339805

Rubenstein, H. (1987). Remittances and Rural Underdevelopment in the English Speaking Caribbean. *Human Organization, 42*(4), 295–306. doi:10.17730/humo.42.4.0709108414538471

Rumpold, B. A., & Schlüter, O. K. (2013). Potential and challenges of insects as an innovative source for food and feed production. *Innovative Food Science & Emerging Technologies, 17*, 1–11. doi:10.1016/j.ifset.2012.11.005

Sachs, J. D., & Warner, A. (1995). Economic reform and the process of integration. Brookings Papers on Economic Activity.

Sadowsky, G. (1996). *The Internet Society and Developing Countries*. Accessed on October 25, 2015 from, http://www.isoc.org/oti/printversions/1196prsadowsky.html

Saibo, N. J., Lourenço, T., & Oliveira, M. M. (2009). Transcription factors and regulation of photosynthetic and related metabolism under environmental stresses. *Annals of Botany, 103*(4), 609–623. doi:10.1093/aob/mcn227 PMID:19010801

Salam, M. A. (2010). Horticultural sector in Andaman & Nicobar. In *Proceedings of National Workshop cum Seminar on Status and Future strategies for Horticulture Development in A& N Islands*. CARI.

Salem, M. A., Kakani, V. G., Koti, S., & Reddy, K. R. (2007). Pollen-based screening of soybean genotypes for high temperatures. *Crop Science, 47*(1), 219–231. doi:10.2135/cropsci2006.07.0443

Sanchez, A. C., Brar, D. S., Huang, N., Li, Z., & Khush, G. S. (2000). Sequence tagged site marker-assisted selection for three bacterial blight resistance genes in rice. *Crop Science, 40*(3), 792–797. doi:10.2135/cropsci2000.403792x

Sandler, T. (1992). *Collective Action: Theory and Applications. University of Michigan Press.*

Sanginga, P.C., Lilja, N. &Tumwine, J. (2001). *Assessing the quality of participation in farmers' research groups in the Highlands of Kabale, Uganda*. PRGA; CGIAR, Future Harvest. Working document/PRGA 19.

Saravanan, R. (2011a). Tribal Farm Youth for Facilitating Agricultural Advisory Services by ICTs: A Success Story from North-East India. *Abstract Volume of the National Seminar on Attracting Farm Youth to Sustainable Agriculture.*

Saravanan, R. (2011b). *e-Arik: Using ICTs to Facilitate "Climate-Smart Agriculture" among Tribal Farmers of North-East India.* ICTs and Agricultural Adaptation to Climate Change Case Study. Centre for Development Informatics, University of Manchester. Accessed on April 5, 2012 from www.niccd.org/NICCD_AgricAdapt_Case_Study_eArik.pdf

Saravanan, R. (2012). e-Initiative for Agricultural Extension: Browsing for Logout? *Proceedings of the AFITA/WCCA 20012-8th Asian Conference for Information Technology in Agriculture (AFITA) and World Conference on Computer in Agriculture (WCCA).* Retrieved from http://www.afita.org/graph/web_structure//files/Seminar%20%2807%29-01%281%29.pdf

Saravanan, R., & Suchiradipta, B. (2015a). Role of ICTs in family farming: Experiences and way forward. In Family Farming and Rural Economic Development. New India Publishing Agency.

Saravanan, R., & Suchiradipta, B. (2015b). Invest in IT to boost agricultural services. *Smart Agri Post: Empowering Agripreneurs, 1,* 22-24. Retrieved from http://www.smartagripost.com/wp-content/uploads/2015/07/Smart-Agripost-July-20153.pdf

Saravanan, R., & Suchiradipta, B. (2015c). *mExtension – Mobile Phones for Agricultural Advisory Services.* Note 17. GFRAS Good Practice Notes for Extension and Advisory Services. GFRAS: Lindau, Switzerland. Retrieved from www.g-fras.org/en/download.html?download=349:ggp-note-17-mextension-mobile-phones-for-agricultural-advisory-services

Saravanan, R., Suchiradipta, B., Chowdhury, A., Hambly Odame, H., & Hall, K. (2015c). *Social Media for Rural Advisory Services.* Note 15. GFRAS Good Practice Notes for Extension and Advisory Services. GFRAS. Retrieved from www.g-fras.org/en/download.html?download=355:ggp-note-15-social-media-for-rural-advisory-services

Saravanan, R., Suchiradipta, B., Meera, S. N., Kathiresan, C., & Anandaraja, N. (2015b). *Web Portals for Agricultural Extension and Advisory Services.* Note 16. GFRAS Good Practice Notes for Extension and Advisory Services. GFRAS. Retrieved from www.g-fras.org/en/download.html?download=356:gfras-ggp-note-16-web-portals-for-agricultural-extension-and-advisory-services

Saravanan, R., Sulaiman, R. V., Davis, K., & Suchiradipta, B. (2015a). *Navigating ICTs for Extension and Advisory Services.* Note 11. GFRAS Good Practice Notes for Extension and Advisory Services. GFRAS. Retrieved from www.g-fras.org/en/download.html?download=351:ggp-note-11-navigating-icts-for-extension-and-advisory-services

Saravanan, R. (2013). e-Agriculture prototype for knowledge facilitation among tribal farmers of North-East India: Innovations, impact and lessons. *Journal of Agricultural Education and Extension, 19*(2), 113–131. doi:10.1080/1389224X.2012.718247

Saravanan, R., & Suchiradipta, B. (2015d). *Social media: Shaping the future of agricultural extension and advisory services. GFRAS interest group on ICT4RAS discussion paper.* Lindau, Switzerland: GFRAS.

Sasson, A. (1993). Biotechnologies in developing countries: present and future. 1. Regional and national survey. *Biotechnologies in developing countries: present and future. 1. Regional and national survey.*

Sasson, A. (2012). Food security for Africa: an urgent global challenge. *Agriculture & Food Security, 1*(2), 1-16. Retrieved June 12, 2015, from http://www.biomedcentral.com/content/pdf/2048-7010-1-2.pdf

Saugeres, L. (2002). Of Tractors and Men: Masculinity, Technology and Power in a French Farming Community. *Sociologia Ruralis, 42*(2), 143–159. doi:10.1111/1467-9523.00207

Scherr, S. J., Shames, S., & Friedman, R. (2012). From climate-smart agriculture to climate-smart landscapes. *Agriculture and Food Security, 1*, 12. Retrieved June 17, 2015, from http://www.biomedcentral.com/content/pdf/2048-7010-1-12.pdf

Schnell, S. M. (2007). *Food with a Farmer's Face: Community Supported Agriculture in the United States*. Retrieved from Researchgate.net: https://www.researchgate.net/publication/227701611_Food_with_a_Farmer's_Face_Community_Supported_Agriculture_in_the_United_States

Schrage, E. J., & Ewing, A. P. (2005). The cocoa industry and child labour. *Journal of Corporate Citizenship, 2005*(18), 99–112. doi:10.9774/GLEAF.4700.2005.su.00013

Scollay, R. (2010). *South-South and North-South trade agreements: The Pacific Islands Case*. United Nations University.

Secretariat of the Convention on Biological Diversity (SCBD). (1992). *The Convention on Biological Diversity*. Montreal, Canada: Text and Annexes.

Sergi, A., & Lavorgna, A. (2012, September). Trade Secrets: Italian Mafia expands its illicit business. *Janes Intelligence Review,* 44-47.

Shankar, S. C., Balaji, P., & Sathish, S. D. (2014). Mass Propagation of Banana (*Musa sp.*) cv. Grand Naine through direct organogenesis by benzyl adenine purine and kinetin. *Journal of Academia and Industrial Research, 3*(2), 92–97.

Shanley, P., & Luz, L. (2003). The impacts of forest degradation on medicinal plant use and implication for health care in Eastern Amazonia. *Bioscience, 53*(6), 573–584. doi:10.1641/0006-3568(2003)053[0573:TIOFDO]2.0.CO;2

Sharma, P. N., Torii, A., Takumi, S., Mori, N., & Nakamura, C. (2004). Marker-assisted pyramiding of brown planthopper (*Nilaparvata lugens Stål*) resistance genes *Bph1* and *Bph2* on rice chromosome 12. *Hereditas, 140*(1), 61–69. doi:10.1111/j.1601-5223.2004.01726.x PMID:15032948

Shava, S. (2005). Research on indigenous knowledge and its application: A case of wild food plants of Zimbabwe. *Southern African Journal of Environment Education, 22*, 73–86.

Shephard, C. Y. (1945). *Peasant Agriculture in the Leeward and Windward Islands*. Trinidad: Imperial College of Tropical Agriculture.

Shepherd, A. W. (2005). *Associations of market traders: Their roles and potential for further development*. AGSF Occasional Paper 7. FAO.

Sherman, L. W., Gartin, P. R., & Buerger, M. E. (1989). Hot spots of predatory crime: Routine activities and the criminology of place. *Criminology, 27*(1), 27–56. doi:10.1111/j.1745-9125.1989.tb00862.x

Simoes, A. (2010). Enhancing adaptive capacity to climate change: The case of smallholder farmers in the Brazilian semi-arid region. *Environmental Science and Policy,* (13), 801-808. Retrieved from www.sciencedirect.com

Singh, D. R., & Singh, S. (2014). Horticulture as an indispensable livelihood option in tropical island region. In *Souvenir: Indian Horticulture Congress*. HIS Westville Publishing House.

Singh, S., Sankaran, M., Singh, D.R., & Roy, S.D. (2014, June 22). CIARI Promoting Island horticulture to eradicate micronutrient deficiency. *The Daily Telegrams.*

Singh, D. B., Attri, B. L., & Singh, C. (2012b). Underutilized fruit crops of Andaman: Its biodiversity and role in environmental conservation and nutritional security. *Indian Forester, 138*(1), 27–30.

Singh, D. M., Adogwa, O. A., Mollineau, W., & Garcia, G. (2014). Gross and microscopic anatomy of the reproductive tract of the female agouti (*Dasyprocta leporina*): A neotropical rodent with potential for food production. *Tropical Agriculture (Trinidad), 9*(1), 38–46.

Singh, D. R., Singh, S., & Roy, S. D. (2012a). *Noni (Morinda citrifolia L.): wonderful medicinal plant*. Port Blair: Published by Director CARI.

Singh, D. R., Singh, S., Salim, M. K., & Srivastava, R. C. (2011b). Estimation of phytochemicals and antioxidant activity of underutilized fruits of Andaman & Nicobar Islands (India). *International Journal of Food Sciences and Nutrition*, *63*(4), 446–452. doi:10.3109/09637486.2011.634788 PMID:22080844

Singh, H. P. (2011). Horticulture—a best livelihood option in tropical islands. In R.K. Gautam, M. Sankaran, S. Jaikumar, D. Kumar, S.K. Verma, P. Krishanan, G. George, & S.K. Ahmed (Eds.), *International Conference on Tropical Island Ecosystem—Issues Related to Livelihood, Sustainable Development and Climate Change*. CARI.

Singh, J., & Dhillion, S. S. (2000). *Agricultural Geography* (2nd ed.). New Delhi: McGraw Hill.

Singh, S., & Bainsla, N. K. (2015a). Analysis of climate change impacts and their mitigation strategies on vegetable sector in tropical islands of Andaman and Nicobar Islands, India. *Journal of Horticulture*, *2*(1). doi:10.4172/2376-0354.1000126

Singh, S., Sidhu, J. S., Huang, N., Vikal, Y., Brar, Z. L. D. S., Dhaliwal, H. S., & Khush, G. S. (2001). Pyramiding three bacterial blight resistance genes (*xa5, xa13* and *Xa21*) using marker-assisted selection into indica rice cultivar PR106. *Theoretical and Applied Genetics*, *102*(6-7), 1011–1015. doi:10.1007/s001220000495

Singh, S., Singh, D. R., Chand, S., Birah, A., & Roy, S. D. (2015a). Analysis of perspectives of self-sufficiency in vegetable production under tropical conditions. *International Journal of Vegetable Science*, *21*(1), 53–68. doi:10.1080/19315260.2013.823585

Singh, S., Singh, D. R., Salim, M. K., Singh, L. B., Srivastava, A., & Srivastava, R. C. (2011a). Estimation of proximate composition, micronutrients and phytochemical compounds in traditional vegetables from Andaman & Nicobar Islands. *International Journal of Food Sciences and Nutrition*, *62*(7), 765–773. doi:10.3109/09637486.2011.585961 PMID:21615278

Singh, S., Singh, D. R., Singh, L. B., Chand, S., & Roy, S. D. (2013). Indigenous vegetables for food and nutritional security in Andaman and Nicobar Islands, India. *International Journal of Agriculture and Food Science Technology.*, *4*(5), 503–512.

Singh, S., Swain, S., Singh, D. R., Salim, K. M., Nayak, D., & Roy, S. D. (2015b). Changes in phytochemicals, anti-nutrients and antioxidant activity in leafy vegetables by microwave boiling with normal and 5% NaCl solution. *Food Chemistry*, *176*, 244–253. doi:10.1016/j.foodchem.2014.12.068 PMID:25624230

Smart Growth America. (2016). *The Best Complete Streets Policies of 2015*. Retrieved from smartgrowthamerica.org: http://www.smartgrowthamerica.org/best-complete-streets-policies-of-2015

Smith, R. (2010). Policing the Changing Landscape of Rural Crime: A Case Study from Scotland. *International Journal of Police Science & Management*, *12*(3), 18–30. doi:10.1350/ijps.2010.12.3.171

Smith, R. S. (1992). Legume inoculant formulation and application. *Canadian Journal of Microbiology*, *38*(6), 485–492. doi:10.1139/m92-080

Smith, R., Laing, A., & Mcelwee, G. (2013). The Rise of Illicit Rural Enterprise within the Farming Industry. *International Journal of Agricultural Management*, *2*(4), 185–188. doi:10.5836/ijam/2013-04-01

Smith, R., & McElwee, G. (2013). Confronting Social Constructions of Rural Criminality: A Case Story on "Illegal Pluriactivity" in the Farming Community. *Sociologia Ruralis*, *53*(1), 112–134. doi:10.1111/j.1467-9523.2012.00580.x

Sonnino, R., Moragues Faus, A., & Maggio, A. (2014). Sustainable food security: An emerging research and policy agenda. *International Journal of Sociology of Agriculture and Food*, *21*(1), 173–188.

Sosan, M. B., & Akingbohungbe, A. E. (2009). Occupational Insecticide Exposure and Perception of Safety Measures among Cacao Farmers in Southwestern Nigeria. *Archives of Environmental & Occupational Health*, *64*(3), 185–193. doi:10.1080/19338240903241077 PMID:19864221

South Bend Region Economic Development. (2016). *Economic Profile*. Retrieved from southbendregion.com: http://southbendregion.com/index.php?src=gendocs&ref=EconomicProfile&category=Main

South Bend Tribune. (2014). *Can We Dream Big?* Retrieved from http://www.southbendtribune.com/app/economy/

South Bend Tribune. (2015, May 17). *South Bend's History Seen in Five Periods of Settlement*. Retrieved from http://www.southbendtribune.com/news/local/history/sb150/south-bend-s-history-seen-in-five-periods-of-settlement/article_79cb9dc6-f99e-11e4-b2c4-cbe41056da99.html

South Bend Tribune. (2016). *Did South Bend eserve the Harvard Ranking?* Retrieved from http://www.southbendtribune.com/news/local/did-south-bend-deserve-the-harvard-ranking/article_2000116e-5218-5175-8299-1d3a732165b2.html

Spore Magazine. (2009). *Theft: Tactics for battling crime*. Technical Centre for Agricultural Land Rural Cooperation (CTA). Retrieved on March 21st, 2016 from: http://spore.cta.int/images/stories/pdf/SE139-web.pdf

Srivastava, R. C., Singh, D. R., & Singh, S. (2010). Horticultural scenario in Andaman & Nicobar. In *Proceedings of National Workshop cum Seminar on Status and Future strategies for Horticulture Development in A& N Islands*. CARI.

Srivastava, R. C., Singh, D. R., Singh, S., & Srivastava, A. K. (2009). Andaman & Nicobar Islands: Potential Hot Spot for Genetic Resources of Horticultural Crops. Presented and published in 9th Agricultural Science Congress.

Srivastava, R. C., & Ambast, S. K. (2011). *Water policy of Andaman and Nicobar Islands: a scientific perspective*. Port Blair, India: CARI.

Srivastava, S. K., Babu, N., & Pandey, H. (2009). Traditional insect bioprospecting - As human food and medicine. *Indian Journal of Traditional Knowledge*, *8*(4), 485–494.

STATSINDIANA. (2015, November). *STATSINDIANA Indiana's Public Data Utility*. Retrieved from Indiana Indepth Profile: http://www.stats.indiana.edu/profiles/profiles.asp?scope_choice=a&county_changer=18141

Steinfeld, H., Gerber, P., Wassenaar, T., Castel, V., Rosales, M., & De Haan, C. (Eds.). (2006). *Livestock's long shadow. Environmental issues and options*. Rome, Italy: Food and Agriculture Organization of the United Nations.

Stigter, K. (2008). Cocoa and climate change. *Can the lame help the blind*. Retrieved June 19, 2015, from http://sitp.rpn.co.id/uploads/riset/kakao/Prosiding%20Kakao%202008%20Cocoa%20And%20Climate%20Change.pdf

Stockbridge, M., Dorward, A., & Kydd, J. (2003). *Farmer organizations for market access: A briefing paper*. London: Imperial College.

Suburban Stats. (2016). *Current South Bend, Indiana Population, Demographics and stats in 2016, 2015*. Retrieved from suburbanstats.org: https://suburbanstats.org/population/indiana/how-many-people-live-in-south-bend

Suchiradipta, B., & Saravanan, R. (2016). *Social media: Shaping the future of agricultural extension and advisory services*. GFRAS interest group on ICT4RAS discussion paper. GFRAS. Available at www.g-fras.org/en/knowledge/gfras-publications.html?download=414:social-media-shaping-the-future-of-agricultural-extension-and-advisory-services

Suchiradipta, B. (2012). *Mobiles for mobilizing Agricultural Extension in India*. Credit Seminar presented in School of Social Sciences. *College of Post Graduate Studies, Central Agricultural University, Umiam, Meghalaya*, *793*, 103.

Suchiradipta, B., & Saravanan, R. (2013). Youth and ICTs for Agricultural Development. In K. Narayana Gowda, M. S. Nataraju, & V. Veerabhdraiah (Eds.), *Youth in Agriculture and Rural Development*. New Delhi: New India Publishing Agency.

Sugden, G. (1999). Farm crime: out of sight, out of mind: a study of crime on farms in the county of Rutland, England. *Crime Prevention and Community Safety: An International Journal*, *1*(3), 29–36. doi:10.1057/palgrave.cpcs.8140023

Summers, R., & Heston, A. (1991). *The Penn World Table (Mark5): An Expanded Set of International Comparisons, 1950-1987*. NBER Working Paper No. R1562.

Swaminathan, M. (1982). Biotechnology research and third world agriculture. *Science*, *218*(4576), 967–972. doi:10.1126/science.218.4576.967 PMID:17790573

Swaminathan, M. S. (2012). Role of genetic modification in developing climate smart agriculture to ensure sustained food security. *Agricultural Research*, *1*(4), 295–298. doi:10.1007/s40003-012-0041-7

Swanson, B., Bentz, R., & Sofranko, A. (1997). *Improving agricultural extension: A reference manual*. Rome, Italy: Food and Agriculture Organization of the United Nations.

Swanson, C. R., Chamelin, N. C., & Territo, L. (2000). *Criminal investigation*. Boston: McGraw Hill.

Swanson, C. R., & Territo, L. (1980). Agricultural Crime: It's Extent, Prevention and Control. *FBI Law Enforcement Bulletin*, *49*(5), 8–12.

Technical Center for Agriculture and Rural Development (CTA). (2011). *Policy brief 3, improving nutrition in AC countries and the role of agriculture*. Retrieved from http://www.cta.int/images/docman-files/policypapers/3_Improving_nutrition_in_ACP_countries.pdf

Tester, M., & Langridge, P. (2010). Breeding technologies to increase crop production in a changing world. *Science*, *327*(5967), 818–822. doi:10.1126/science.1183700 PMID:20150489

The Chicago Council on Global Affairs. (2013). *Feeding an Urban World: A Call to Action*. Chicago: The Chicago Council on Global Affairs.

The History Museum. (2016). *Local African-American History*. Retrieved from historymuseusb.org: http://historymuseumsb.org/local-african-american-history/#

The Union of Chambers and Commodity Exchanges of Turkey. (2013). *Agricultural sector report of Turkey*. Ankara, Turkey: Afşaroğlu Publishing.

Thomas, C. Y. (1982). From colony to state capitalism: Alternative paths of development in the Caribbean. *Transition*, *5*, 1–20.

Thornton, P. K., & Herrero, M. (2015). Adapting to climate change in the mixed crop and livestock farming systems in sub-Saharan Africa. *Nature Climate Change*, *5*(9), 830–836. doi:10.1038/nclimate2754

Thorpe, T. (2007). History of plant tissue culture. *Journal of Molecular Biotechnology*, *37*(1), 169–180. doi:10.1007/s12033-007-0031-3 PMID:17914178

Tilman, D. (1999). Global environmental impacts of agricultural expansion: The need for sustainable and efficient practices. *Proceedings of the National Academy of Sciences of the United States of America*, *96*(11), 5995–6000. doi:10.1073/pnas.96.11.5995 PMID:10339530

Timmer, C. P., Falcon, W. P., & Pearson, S. R. (1983). *Food Policy Analysis*. The John's Hopkins University Press. Retrieved from http://web.stanford.edu/group/FRI/indonesia/documents/foodpolicy/chapt4.fm.html

Tol, R. S. J. (2002). Estimates of the damage costs of climate change, Part I: Benchmark Estimates. *Environmental and Resource Economics*, *21*(1), 47–73. doi:10.1023/A:1014500930521

Topp, C. F. E., Stockdale, E. A., Watson, C. A., & Rees, R. M. (2007). Estimating resource use efficiencies in organic agriculture: A review of budgeting approaches used. *Journal of the Science of Food and Agriculture*, *87*(15), 2782–2790. doi:10.1002/jsfa.3059

Toulmin, C. (2009). *Climate change in Africa*. London: Zed Books.

Trauger, A. (2009). Social agency and networked spatial relations in sustainable agriculture. *Area*, *41*(2), 117–128. doi:10.1111/j.1475-4762.2008.00866.x

Trueblood, M., & Shapouri, S. (2001). Implications of Trade Liberalizationon Food Security of Low-income Countries. United States Department of Agriculture. Agriculture Information Bulletin No. 765-5.

Turhan, Ş. (2005). Sustainability in agriculture and organic farming. *Turkish Journal of Agricultural Economics*, *11*(1), 13–24.

Turkish Statistical Institute (TUIK) Databases. (n.d.). Retrieved from http://www.turkstat.gov.tr

Turkish Statistical Institute. (2012). *The summary of agricultural statistics 2011*. Ankara, Turkey: Turkish Statistical Institute Press.

Turkish Statistical Institute. (2014). *Turkey in statistics 2013*. Ankara, Turkey: Turkish Statistical Institute Press.

U.S. Bureau of Census. (1998, June 15). *Population of the 100 Largest Urban Places: 1920*. Retrieved from https://www.census.gov: https://www.census.gov/population/www/documentation/twps0027/tab15.txt

UBC News. (2014). *Study Reveals Unhappiest Cities in the US*. Retrieved from http://news.ubc.ca/2014/07/21/unhappy-cities/

UBOS. (2014). *2014 Statistical Abstract*. Kampala: Uganda Bureau of Statistics.

UBOS. (2015). *2015 Statistical Abstract*. Kampala: Uganda Bureau of Statistics.

UNCTAD. (2013). *Global supply chains: trade and economic policies for developing countries*. Geneva: United Nations.

UNDP. (2012). *Africa human development report: towards a food secure future*. New York: UNDP Regional Bureau for Africa. Retrieved 4 July 2014 from, http://www.undp.org/content/dam/undp/library/corporate/HDR/Africa%20HDR/UNDP-Africa%20HDR-2012-EN.pdf

UNDP. (2013a). *Assessing Progress in Africa toward the Millennium Development Goals. MDG Report 2013*. New York: UNDP.

UNDP. (2013b). *Human Development Report 2013: The Rise of the South*. New York: UNDP.

UNFCCC (United Nations Framework Convention on Climate Change). (2007). *Climate change: impacts, vulnerabilities and adaptation in developing countries*. Bonn, Germany: UNFCCC Secretariat. Retrieved June 4, 2015, from http://unfcc.int/resource/docs/publications/impacts.pdf

United Nation. (2015). *Sustainable Development Goal*. Retrieved from http://www.un.org/sustainabledevelopmentas

United Nations, Department of Economic and Social Affairs, Population Division. (2015). *World Population Prospects: The 2015 Revision, Key Findings and Advance Tables*. Working Paper No. ESA/P/WP.241.

Unity Gardens. (2009). *Unity Mission/Story*. Retrieved from Unity Gardens: http://theunitygardens.blogspot.com/p/unity-mission-story.html

USA.com. (2016). *South Bend, IN Income and Careers*. Retrieved from USA.com: http://www.usa.com/south-bend-in-income-and-careers.htm#Poverty Level

USDA. (2015, September 8). *USDA ERS-Food Security in the US*. Retrieved from United States Department of Agriculture Economic Research Service: http://www.ers.usda.gov/topics/food-nutrition-assistance/food-security-in-the-us.aspx

Van den Ban, A. W., & Hawkins, H. S. (2002). Agricultural Extension (2nded). Blackwell Science publishers.

van Huis, A., van Itterbeeck, J., Klunder, H., Mertens, E., Halloran, A., Muir, G., & Vantomme, P. (2013). *Edible Insects: Future Prospects for Food and Feed Security. FAO Forestry Paper 171*. Rome: Food and Agriculture Organization of the United Nations.

van Leeuwen, M. A. E., van der Voort, M. P. J., Sukkel, W., & Balci, S. (2008). *Organic agriculture in Turkey: trade opportunities for organic fruit and vegetables*. Retrieved from http://edepot.wur.nl/29925

Varalakshmi, B. (2004). Characterization and preliminary evaluation of vegetable amaranth (Amaranthus spp.) germplasm. *Plant Genetic Resources Newsletter (Rome, Italy)*, *137*, 55–57.

Varshney, R. V., Hoisington, D. A., & Tyagi, A. K. (2006). Advances in cereal genomics and applications in crop breeding. *Trends in Biotechnology*, *24*(11), 490–499. doi:10.1016/j.tibtech.2006.08.006 PMID:16956681

Vervisch, T. G. A., Vlassenroot, K., & Braeckman, J. (2013). Livelihoods, power, and food insecurity: Adaptation of social capital portfolios in protracted crises—case study Burundi. *Disasters*, *37*(2), 267–292. doi:10.1111/j.1467-7717.2012.01301.x PMID:23278356

Vervoort, J. M., Thornton, P. K., Kristjanson, P., Förch, W., Ericksen, P. J., Kok, K., & Jost, C. et al. (2014). Challenges to scenario-guided adaptive action on food security under climate change. *Global Environmental Change*, *28*, 383–394. doi:10.1016/j.gloenvcha.2014.03.001

Vitale, J. D., Vognan, G., Ouattarra, M., & Traore, O. (2010). The Commercial Application of GMO Crops in Africa: Burkina Faso's Decade of Experience with Bt Cotton. *AgBioForum*, *13*(4), 320-332. Retrieved 10 August 2014 from, http://soybeaninnovationlab.illinois.edu/sites/soybeaninnovationlab.illinois.edu/files/cotton%20prfotiabilityv13n4a05-vitale.pdf

Vitale, J., Glick, H., Greenplate, J., & Traore, O. (2008). The Economic Impact of Second Generation Bt Cotton in West Africa: Emperical Evidence from Burkina Faso. *International Journal of Biotechnology*, *10*(2-3), 167–183. doi:10.1504/IJBT.2008.018352

Waldron, A., Justicia, R., Smith, L., & Sanchez, M. (2012). Conservation through chocolate: A win-win for biodiversity and farmers in Ecuador's lowland tropics. *Conservation Letters*, *5*(3), 213–221. doi:10.1111/j.1755-263X.2012.00230.x

Walters, L. M., & Keithly, J. G. (2012). *Caribbean Food Import Demand: Influence of the Changing Dynamics of the Caribbean Economy*. Presentation at the Southern Agricultural Economics Association Annual Meetings, Birmingham, AL.

Wang, P. J., & Hu, C. Y. (1982). *In vitro* mass tuberization and virus-free seed potato production in Taiwan. *American Potato Journal*, *59*(1), 33–39. doi:10.1007/BF02854881

Waugh, D. (2009). *Geography: An Integrated Approach*. Cheltenham, UK: Nelson Thornes.

Weisheit, R. A., Falcone, D. N., & Wells, L. E. (2006). Crime and policing in rural and small-town America (3rd ed.). Waveland Press.

West African Rice Development Association. (2001). NERICA: Rice for life. *WARDA/ADRAO, Bouaké*. Retrieved June 18, 2015, from http://www.warda.org/publications/NERICA.pdf

Wheeler, T., & von Braun, J. (2013). Climate change impacts on global food security. *Science*, *341*(6145), 508–513. doi:10.1126/science.1239402 PMID:23908229

Wickham, L. D. (1992). The potential for increasing the productivity and quality of vegetables produced in the Caribbean region: UWI research and perspectives. In G. Villanueva, J. Corven & A. Campos (Eds.), Proceedings of Taller regional centroamericano y consulta sobre planificacion de investigacion horticola (pp. 245-251). San José, Costa Rica: IICA.

Wikipedia. (2016, March 19). *Michiana*. Retrieved from https://en.wikipedia.org/wiki/Michiana

Wikiversity. (2015). *Characteristics of developing countries*. Retrieved from https://en.wikiversity.org/wiki/Characteristics_of_developing_economies

Wilcox, P., Land, K. C., & Hunt, S. (2003). *Criminal Circumstance: A Dynamic Multicontextual Criminal Opportunity Theory*. New York: Walter de Gruyter.

Wilkie, D., & Carpenter, J. F. (1999). Bushmeat hunting in the Congo Basin: As assessment of impacts and options for mitigation. *Biodiversity and Conservation*, *8*(7), 927–955. doi:10.1023/A:1008877309871

Wilkinson, M., Craig, G., & Gaus, A. (2010). Forced Labour in the UK and Gangmaster Licensing Authority. Report for the Contemporary Research Centre, University of Hull.

Wilkinson, R., & Pickett, K. (2009). *The Spirit Level: Why More Equal Societies Almost Always Do Better*. Allen Lane.

Willer, H., & Lernoud, J. (Eds.). (2014). *The world of organic agriculture: Statistics and emerging trends 2014*. Retrieved from https://www.fibl.org/fileadmin/documents/shop/1636-organic-world-2014.pdf

Williams, S.K., Fenley, T.M., & Williams, C.E. (1981). *A Manual for Agricultural Extension Workers in Nigeria*. Academic Press.

Wilson, W. J. (1987). *The Truly Disadvantaged*. Chicago, IL: Chicago Press.

Windapo, O., & Afolayan, S. O. (2005). Group Dynamics and Leadership in Agriculture, Extension. In S. F. Adedoyin (Ed.), *Agricultural Extension in Nigeria* (pp. 134–138). Ilorin Agricultural Extension Society of Nigeria.

Winer, L. (2009). *Dictionary of the English / Creole of Trinidad and Tobago*. McGill-Queen's University Press.

Witt, H., Patel, R., & Schnurr, M. (2006). Can the poor help GM crops, Technology, representation & cotton in the Makhathini Flats, South Africa. *Review of African Political Economy*, *109*, 497–513. Retrieved 12 June 2014 from, http://rajpatel.org/wp-content/uploads/2009/11/Can-the-Poor-help-GM-Crops_final-printed-version.pdf

WNDU. (2010, February 1). *Delve into South Bend's past on this Black History Month*. Retrieved from WNDU 16 South Bend Indiana: http://www.wndu.com/home/headlines/83231437.html

Wong, W. C. (1986). *In vitro* propagation of banana (Musa sp.): Initiation, proliferation and development of shoot-tip cultures of defined media. *Plant Cell, Tissue and Organ Culture*, *6*(2), 159–166. doi:10.1007/BF00180799

World Agriculture Towards 2030/2050. (n.d.). *The2012revision ESA E Working Paper No. 12-03*. Retrieved from http://www.fao.org/economic/esa/esag/en/

World Bank. (2008). *Agriculture for development. World Development Report 2008*. Washington, DC: World Bank.

World Bank. (2010). *Enterprise Survey: Saint Kitts and Nevis*. The World Bank. Retrieved from http://www.caribank.org/uploads/2014/11/2014-St.-Kitts-and-Nevis-PSAR.pdf

World Bank. (2011). *ICT in agriculture: Connecting small holders to knowledge, networks, and institutions*. e-Source Book. Report no. 64605. Washington, DC: The World Bank. Available at: http://www.ictinagriculture.org/content/ict-agriculture-sourcebook

World Bank. (2011). *Uganda: Growing Out of Poverty. Poverty Reduction and equity*. Accessed July 20, 2015, from http://web.worldbank.org/WBSITE/EXTERNAL/TOPICS/EXTPOVERTY/EXTPA/0,contentMDK:20204772~menuPK:435735~pagePK:148956~piPK:216618~theSitePK:430367,00.html

World Bank. (2012). *Little Data book*. Author.

World Bank. (2012). *Press Release*. Retrieved from http://www.worldbank.org/en/news/press-release/2012/10/24/africa-can-feed-itself-earn-billions-avoid-food-crises-unblocking-regional-food-trade

World Bank. (2013). *End Extreme Poverty 2030 Promote Shared PROSPERITY. Annual Poverty Report 2013*. Washington, DC: World Bank.

World Bank. (2014). *World Bank and Agriculture in Africa*. Accessed June 20, 2014, from http://web.worldbank.org/WBSITE/EXTERNAL/COUNTRIES/AFRICAEXT/0,contentMDK:21935583~pagePK:146736~piPK:146830~theSitePK:258644,00.html

World Bank. (2015). *Data: Agriculture and Rural Development*. The World Bank. Retrieved from http://data.worldbank.org/topic/agriculture-and-rural-development

World Bank. (2015). Retrieved from http://www.worldbank.org/en/topic/poverty/overview

World Bank. (2015a). Agriculture and rural development. *World Development Indicators*. Accessed July 1, 2015, from http://data.worldbank.org/topic/agriculture-and-rural-development

World Bank. (2015b). *Can Uganda's land support its prosperity drive?* World Bank Report No 99060. Washington, DC: World Bank.

World Cocoa Foundation (WCF). (2015). *President's message*. Retrieved June 27, 2015, from: http://worldcocoafoundation.org/wcf-may-june-2015-2/#2&utm_source=WCF+May+%26+June+2015+Newsletter&utm_campaign=WCF+Newsletter&utm_medium=email

World Commission on Environment and Development. (1987). *The Bruntland Report: Our Common Future*. Oxford, UK: Oxford University Press.

World Economic Forum. (WEF). (2015). *The Inclusive Growth and Development Report 2015*. Geneva: World Economic Forum.

World Health Organization (WHO). (2014). *Frequently asked questions on genetically modified foods*. Available from http://www.who.int/foodsafety/areas_work/food-technology/Frequently_asked_questions_on_gm_foods.pdf

World Health Organization. (2015). *Trade, foreign policy, diplomacy and health*. Retrieved from WHO Food Security: who.int/trade/glossary/story028/en

Worldatlas. (2015). *Uganda*. Retrieved August 4, 2015, from http://www.worldatlas.com/webimage/countrys/africa/ug.htm

Wu, H., Sun, M., Yue, S., Sun, H., Cai, Y., & Huang, R. et al. (2000). *Field evaluation of an Amaranthus genetic resource collection in China*. Genetic Resources.

XiangYan, Z., ZengYan, Z., LiPu, D., ZhiYong, X., & Xiao, C. (2005). Development of wheat germplasms with multi-resistance to powdery mildew, stripe rust and yellow dwarf virus by molecular markerassisted selection. *Science and sustainable Agriculture, 38*(12), 2380-2386.

Yang, W. C., & Francis, D. M. (2005). Marker-assisted selection for combining resistance to bacterial spot and bacterial speck in tomato. *Journal of the American Society for Horticultural Science*, *130*(5), 716–721.

Yarwood, R., & Gardener, G. (2005). Fear of Crime, Cultural Threat and the Countryside. *Area*, *32*(4), 403–411. doi:10.1111/j.1475-4762.2000.tb00156.x

Yorobe, J. M., Jr., & Quicoy, C. B. (2006). Economic Impact of Bt Corn in the Philippines. *The Philippine Agricultural Scientist*, *89*(3), 258–67. Retrieved 5 July 2014 from, https://www.isaaa.org/programs/impact_assessment_of_crop_biotechnology/download/Economic%20impact%20of%20Bt%20corn%20in%20the%20Philippines.pdf

Zepeda, L. (2001). *Agricultural Investment, Production Capacity and Productivity*. Geneva: FAO.

zipcode.org. (2016). *City of South Bend, IN Zip Codes*. Retrieved from zipcode.org: zipcode.org/city/IN/SOUTHBEND

About the Contributors

Wayne Ganpat (PhD Agricultural Extension) is a Caribbean extension professional. Presently, he is a Senior Lecturer in Agricultural Extension and Communications at the University of the West Indies (UWI), St Augustine, Trinidad and Tobago and teaches courses at both the undergraduate and graduate levels. Prior to joining the UWI, he worked as an Extension Officer in the government extension service for over 20 years and left as the Deputy Director of Extension Services. He has worked as a consultant for governments in most Caribbean countries over the last 20 years doing work both in Extension and Communications.

Wendy-Ann Isaac is a Lecturer in Crop Production in the Department of Food Production, Faculty of Food and Agriculture at the University of the West Indies (UWI), St. Augustine Campus, Trinidad and Tobago. Dr. Isaac received her BSc. MSc., and Ph.D. degrees from the UWI, St. Augustine and M.Appl.Sci. in Agronomy from Lincoln University, New Zealand. Her research interests include Sustainable crop production systems and Participatory action research techniques with small farmers. She has worked with Fairtrade banana small farmers in St. Vincent and The Grenadines using the Participatory Research methods as part of her PhD. studies. She has authored and co-authored papers on these topics, which have been published in both regional and international peer-reviewed journals and is the co-editor on three books including *Sustainable Food Production Practices in the Caribbean* Volumes 1 and 2 *and The Impact of Climate Change on Food Security in Small Island Developing States.*

***

Solomon Adedapo Adesoji hails from southwestern part of Nigeria. He is an experienced agriculturist and lecturer. He obtained his Bachelor's degree in Agricultural Extension Services from the University of Ibadan, Nigeria. He also obtained Masters of Science and Doctor of Philosophy in Agricultural Extension and Rural Sociology from the Obafemi Awolowo University, Ile – Ife, Nigeria. He worked with the State's Agricultural Development Programme as an Agricultural Extension Officer and later as a training officer in the Ministry of Agriculture before he joined the academia. Dr. Adesoji is currently an Assistant Professor in the Department of Agricultural Extension and Rural Development of the Obafemi Awolowo University, Ile – Ife, Nigeria. He has widely published in reputable local and international high impact journals. His research interest span through evaluation of development programmes, impacts on livelihood of rural dwellers and policy. He has worked in many international sponsored researches. He is a fellow of the Institute of Climate Change and its Impacts of the Brown International Advanced Research Institute (BIARI) of the United State of America. He is a member of Agricultural Extension

Society of Nigeria (AESON), Rural Sociological Association of Nigeria (RuSAN), member of National Research network of Children and Youth in Agriculture Programme (CYIAP) and Global Development Network. Dr. Adesoji has attended many conferences at both local and international levels. Dr. Adesoji is also a non – stipendiary priest of the Church of Nigeria (Anglican Communion).

William Amone has been working at Gulu University in Uganda as a Lecturer of Economics since the year 2006. He also serves as a Consultant in Business Management at Uganda Management Institute. William has been involved in many researches and trainings in Uganda, Kenya, Tanzania and Canada. He is currently a PhD candidate at Mbarara University of Science and Technology in Uganda and a researcher in Development Economics. William has published 4 peer-reviewed papers, including 3 chapters in Business Reference Books with IGI Global, and 1 article in the Commonwealth Youth and Development journal.

Frances Bekele is a Research Fellow (Germplasm Specialist) at the Cocoa Research Centre, The University of the West Indies, St. Augustine, Trinidad & Tobago where she co-ordinates botanical characterization and evaluation of cacao germplasm. She has also been affiliated with studies on heavy metal and Ochratoxin A contamination and flavour chemicals and volatiles in cocoa, farmer field school training and several collaborative research projects. Frances serves as a reviewer of scientific publications, has published several scientific manuscripts and was editor for the International Group for the Genetic Improvement of Cocoa from 1994 – 2012.

Isaac Bekele is a Senior Lecturer in the Faculty of the Food and Agriculture, The University of the West Indies, St. Augustine, Trinidad and Tobago. He is a Biometrician with several publications relevant to the topic covered. Isaac served as Dean of the Faculty of Food and Agriculture at The University of the West Indies between 2013 and 2016, and has lectured on Statistics and Mathematics for more than 30 years. He has lent support in statistical analysis and experimental design to many staff members and students at the St Augustine Campus of The University of the West Indies.

Busayo Solomon Famuyiwa, (Ph.D.), was born Nov. 29th 1963 at Markurdi, Nigeria. He attended University of Ibadan and graduated with Bsc. Agriculture (Special honour in Extension Services) in 1986. He had his Master's degree in Business Administration (Management and Accounting) in 1997, Msc. Agricultural Extension and Rural Sociology in 2008 and Ph.D. Agricultural Extension and Rural Sociology in 2013 from Obafemi Awolowo University, Ife. He joined CRIN in 2010 as Research Officer 1 in Extension and has 23 journal papers, 7 others in proceedings and books. Professionally, he has worked assiduously on agricultural health study, Agricultural knowledge management platforms and extension methodologies. He has also served as Reviewer in some scientific journals. He is the CRIN Environmental Safeguard officer on West African Agricultural Productivity Programme (WAAPP). He is the present Alternate Program Leader of coffee and has facilitated survey, seminars and training of farmers in cocoa and coffee. He has attended Conferences both locally and international; he is a member of Agricultural Extension Society of Nigeria (AESON), Rural Sociology Association Nigeria (RuSAN) and Children – Youth in Agriculture Programe CYAP. He is a Senior Research Officer in CRIN.

Vidya de Gannes is affiliated with the Department of Food Production, Faculty of Food and Agriculture, The University of the West Indies, St. Augustine, Trinidad and Tobago. She holds a Ph.D. (high commendation) in Soil Science. Her research areas have focused on Biotechnology and Environmental Microbiology where cutting edge technologies such as 454-pyrosequencing and Illumina sequencing have been applied to unravel microbial communities to unprecedented depth in tropical composts and soils. She has published in high impact factor journals spanning areas such as compost health and safety to microbial ecology. Dr. de Gannes have recently received rigorous training in the areas of Molecular Biology, Bioinformatics and Plant Tissue Culture (University of Wisconsin, Madison); Molecular Diagnosis of Plant Diseases (Centro de Investigación Científica de Yucatán, Mexico) and Diagnosis of Plant Diseases through Serological Tests (University of Florida). Her most recent accolades include: Most Outstanding Graduate Researcher (Sciences), best PhD thesis for the academic year 2014 and 1st place in the UWI Employee Excellence award, Technical category.

Edwin Joseph is the Director of the Sustainability Program at Indiana State University in South Bend (IUSB). He was a former professor at Grand Valley State University, Allendale Michigan, and Eastern Michigan University in Ypsilanti Michigan. He is a Geographic Information Systems (GIS) Specialist and a former Environmental Systems Research Institute (ESRI) authorized GIS instructor (10 years). A graduate from the Eastern Caribbean Institute of Agriculture, University of the West Indies, and the University of Wisconsin-Madison, Edwin pursued agriculture (Diploma), agricultural extension (Diploma), rural sociology (BS), adult education (MS), and development (PhD). He currently teaches courses in sustainability and conducts research in multimedia mapping using mapping drones.

Dick Nuwamanya Kamuganga is a teacher. He teaches Economics both at Undergraduate and graduate level at Uganda Technology & Management University (UTAMU & in the Department of Development Studies, School of Social Sciences and Performing Arts, College of Humanities and Social Sciences, Makerere University Kampala, Uganda. He is also the Dean of the School of Business and Management, at UTAMU). He obtained his PhD in International Economics from the Graduate Institute of International and Development Studies, Geneva Switzerland, Masters & Bachelors of Science in Agriculture Economics from Makerere University, Kampala Uganda. His fields of expertise are: international trade economics; microeconomics of development; Agricultural development in Developing Countries; and development finance. His academic research is focused on Africa and global trade governance in the 21st century, Gender and finance for development in Africa, Regional trade & Economic Integration in Africa, Financial Integration. For the wider public he writes widely on Africa's inclusive sustainable development. Media publications of opinion pieces have appeared in African Leading daily newspapers including the East African. He has advised the UN-Food Agriculture Organization, The World Trade Organization, and the Commonwealth Secretariat, Southern African Development Community (SADC) and the East African Community. He is an active member of the African Economic Research Consortium (AERC), Nairobi Kenya.

Aylin Yaman Kocadağlı received her B.Sc. in Geography from Istanbul University, M.Sc. and Ph.D. in Human Geography from the Institute of Social Sciences at Istanbul University. She worked as a research assistant in the Department of Geography at Istanbul University from 2004-2013. During the 2008 - 2009 academic year, she worked as a research fellow in the Department of Geography at Passau University in Passau-Germany. During the 2009 - 2010 academic year, she worked as a research fellow in the Faculty of Earth Science and Environmental Management at Wroclaw University in Wroclaw-Poland. She works as an assistant professor in the Department of Geography at Istanbul University and is currently working as a visiting scholar in the Department of Geography at Texas A&M University.

Godswill Makombe has more than 20 years of research on development issues in Africa, mainly from the Sub-Saharan African countries. He has worked with academics, international and national institutions in Sub-Saharan Africa, namely University of Zimbabwe, Government of Zimbabwe, The International Water Management Institute, The European Union, The Africa Rice Centre and the University of Limpopo. He has field experience in design, implementation and analysis of large data sets. He still enjoys doing field work. He has strong quantitative analysis skills and he has both the training and experience of working in multidisciplinary teams. He has taught economics, research and development classes at different levels. Godswill has been involved with project impact assessment and monitoring and evaluation. He has always worked in interdisciplinary teams.

Kevin Meehan is Professor of English and Director of the Haitian Studies Project at the University of Central Florida in Orlando, FL. He specializes in Caribbean and multiethnic US literature, as well as climate change education for sustainable development (CCESD). His book, People Get Ready: African American and Caribbean Cultural Exchange (University Press of Mississippi, 2009) traces a long history of decolonizing cultural dialogue and activism in the Americas, and he has published scholarly articles on topics including Caribbean women writers, C.L.R. James and radical fieldwork, the globalized political economy of independent music, and climate change adaptation in the small-scale Caribbean agriculture sector. He has been a visiting professor at the University of Nouvelle Grand'Anse in Haiti and the Clarence Fitzroy Bryant College in St. Kitts, and from 2008-2014 he co-directed a study abroad program that brought students from UCF to study and work in St. Kitts-Nevis. He has received grants from the National Science Foundation, the National Endowment for the Humanities, and the Organization of American States. From 2013-2016, he was the coordinator of communication and documentation for an OAS-funded initiative called The Provisions Project, which replicated a model hydroponic shade house growing system in Nevis, Guyana, Barbados, and Trinidad. Dr. Meehan's short subject videos from the Provisions Project can be viewed on the Provisions Project YouTube channel, and on his own Sugar City Music YouTube channel.

Leighton Naraine is the Director of Employee and Program Development at the Clarence Fitzroy College (CFBC), Basseterre, St. Kitts-Nevis, where his career, since 2002, included lecturing in Geography, Environmental Science, Agricultural Studies, and Teacher Education, and where he was also the Vice Principal for several years. With his specializations in Geography, Urban and Regional Planning, and Policy and Governance, he has a preference for technical work in developing programs for competency-based education and training, including sustainable agriculture. He developed and implemented the first tertiary level program in agriculture in St. Kitts-Nevis. Dr. Naraine is actively involved in community outreach and development through project work in such areas as renewable energy, and climate change education for sustainable development under the UNESCO Climate Change Agenda for the Caribbean and Small Island Developing States (SIDS). He planned and chaired two sub-regional conference for UNESCO on "Adaptation to Climate Change: Science, Ethics, and Policies," and also on "Environmental Policy and Planning for the Caribbean." He also made expert presentations at various regional conferences in the Caribbean and Latin America on behalf of UNESCO and the Organization of American States (OAS).

He was the project team leader for a CFBC OAS-funded project in hydroponics and organoponics which he and his colleagues conceptualized as an adaptation strategy to climate change, footage of which can be found in the Provisions Project on YouTube Channel and FaceBook. Dr. Naraine is co-author on a book chapter: Agricultural Diversification and Non-Traditional Agricultural Systems for the Caribbean" in "Sustainable Food Production Practices for the Caribbean" Volume 2, Editors: Wayne Ganpat and Wendy-Ann Isaac. Dr. Naraine is a farmer who is developing a semi-commercial integrated farming model in St. Kitts-Nevis which he intends to publish to facilitate upscaling and replication.

Elizabeth O'Dea is completing her senior year at Indiana University South Bend. She is pursuing a double major in Sustainability and Anthropology with a concentration in international studies. Her interest is in progressing and promoting sustainable practices in land and water management globally. She became involved in sustainable studies in the spring semester of 2014 when she became increasingly aware of the persistent negative impacts of overconsumption and environmental degradation. Elizabeth aspires to pursue a career in international development. When she is not busy reading about global affairs, she enjoys hiking with her two daughters and training for marathons.

Olumuyiwa Akin Olaniyi is a Doctoral degree holder and currently an Associate Professor in the Department of Agricultural Extension and Rural Development., Ladoke Akintola University of Technology, Ogbomoso, Nigeria. He specializes in Extension Communication and application of Information and Communication Technologies (ICTs) to development. Dr. Olaniyi is a member of reputable national and international professional associations including the Agricultural Extension Society of Nigeria (AESON), Agricultural Society of Nigeria (ASN), and International Association of Agricultural Information Specialists (IAALD).

Theresa Ann Rajack-Talley is a distinguished academic of the Caribbean. She has worked with many Caribbean and International research and development institutions over the last 30 years. Her mainstay has been with university systems: the University of the West Indies and currently with the University of Louisville in the United States where she serves in the College of Arts & Sciences as the Associate Dean for International, Diversity and Community Engagement Programs and as an Associate Professor in the Department of Pan-African Studies. Her primary research interest and experience is focused on race/ethnicity, gender, class and social inequality with an emphasis on poverty, human and economic development. She has researched, presented and published numerous technical reports, academic papers and articles. Her most recent book is titled "Poverty is a Person: Human Agency, Women and Caribbean Households and she just completed a study on Rural Employment & Rural Development in the Caribbean published by the International Labour Organization.

Nira Ramachandran is currently an independent researcher and consultant based in New Delhi, India. As ICSSR Senior Fellow at the Institute of Economic Growth, New Delhi her research centred on the fine tuning of interventions to accelerate improvement in nutrition outcomes. Her interest areas include food security, nutrition, gender, child labour and related issues. She has provided consultancy to a number of International organizations including the FAO, UN World Food Programme, UNICEF and The World Bank. She devotes part of her time to a research NGO- the Bhoovigyan Vikas (Earth Care) Foundation. She holds a Ph.D from the Jawaharlal Nehru University, Delhi and has also been trained abroad. She has lectured extensively and conducted training programmes in the areas of food security, environmental sustainability, primary education and project management techniques. She has to her credit several articles in peer reviewed journals as well as a number of books including *Persisting Undernutrition in India; Causes, Consequences and Possible Options* (Springer 2014), *Towards a Food Secure India: Issues and Policies, Coming to Grips with Rural Child Work: A Food Security Approach, Carrying Capacity of Mussoorie,* and *Monitoring Sustainability: Indices and Techniques of Analysis.* She has held a range of teaching and research positions at well known academic institutions including the Indian Institute of Management, Bangalore, Bangalore University, Delhi University and the LBS Academy of Administration, Mussoorie, the premier training institution for the Indian Administrative Service.

Raj Saravanan is the Director (Agricultural Extension) at National Institute of Agricultural Extension Management (MANAGE), Hyderabad, India (saravananraj@hotmail.com/saravanan.raj@manage.gov.in).

Bhattacharjee Suchiradipta is working as an Independent Researcher after completing her Ph.D. in Agricultural Extension from Central Agricultural University, Imphal, India (suchiradipta@hotmail.com/suchiradipta.cau@gmail.com).

Index

A

B

C

D

E

T

U

W

Y

www.ingramcontent.com/pod-product-compliance
Lightning Source LLC
LaVergne TN
LVHW060410300926
849162LV00006B/1389